BIOCHEMISTRY RESEARCH TRENDS

# HYDROGELS

## SYNTHESIS, CHARACTERIZATION AND APPLICATIONS

## BIOCHEMISTRY RESEARCH TRENDS

Additional books in this series can be found on Nova's website under the Series tab.

Additional e-books in this series can be found on Nova's website under the e-book tab.

## PHARMACOLOGY - RESEARCH, SAFETY TESTING AND REGULATION

Additional books in this series can be found on Nova's website under the Series tab.

Additional e-books in this series can be found on Nova's website under the e-book tab.

BIOCHEMISTRY RESEARCH TRENDS

# HYDROGELS

## SYNTHESIS, CHARACTERIZATION AND APPLICATIONS

FABRICIO VITOR CÂMARA
AND
LEANDRO J. FERREIRA
EDITORS

*New York*

For permission to use material from this book please contact us:
Telephone 631-231-7269; Fax 631-231-8175
Web Site: http://www.novapublishers.com



**LIBRARY OF CONGRESS CATALOGING-IN-PUBLICATION DATA**

Library of Congress Control Number: 2011946080
ISBN: 978-1-61942-842-3

*Published by Nova Science Publishers, Inc. † New York*

# CONTENTS

# PREFACE

Hydrogels are polymeric networks with three-dimensional configurations capable of imbibing high amounts of water or biological fluids. The application of hydrogels technology is being widely used in the pharmaceutical industry as drug delivery systems. In this book, the authors present current research in the study of the synthesis, characterization and applications of hydrogels. Topics include the synthesis of protein-based hydrogels for biomedical applications; the role of hydrogels in the development of artificial bone substitutes; protein-based hydrogels from fish industrialization wastes; silk fibroin hydrogels; PVP-Based hydrogels; and the tribological properties of polyvinyl alcohol hydrogels as artifical articular cartilage.

Chapter 1- Natural polymers, such as proteins, are materials extensively investigated due to their biocompatibility, biodegradability and non-toxic and non-immunogenic characteristics. Enclosing these biomacromolecules in a complex structure, these features can be transferred to a biomaterial in order to extend the performance of the device. Thus, the covalent conjugation of a biodegradable macromolecule, as a protein, represents a versatile strategy to produce intelligent biodegradable hydrogels, suitable for pharmaceutical and biomedical applications. The cross-linking of biomacromolecules is somewhat important in these fields because it increase the mechanical properties of the hydrogels without to affect, in severe degree, the biodegradability. Basically, albumin and gelatin represents the protein mainly used in the biomedical field and their insertion in a polymeric network was performed employing different synthetic strategy involving the functional groups in the side chain of the polypeptides. Hydrogel containing proteic moieties in the polymeric network can be achieved by thermal denaturation of the protein either by direct reaction between functional groups (usually carboxyl and amino goups) in the polypeptide side chains and also by use chemical cross-linking agents as bifunctional carbonyl reagents. Finally, involving the protein in a radical process innovative synthetic procedures were proposed in order to modify polymeric network composition and to design hydrogels with appropriate and mouldable physicochemical properties.

Chapter 2 - Biomedical hydrogels for tissue engineering and more specifically for bone regeneration purposes are investigated throughout this book chapter. The most commonly utilised natural and synthetic polymer materials researched in the development of such scaffolds is initially documented. Synthesis techniques and in particular the role of photopolymerisation in the design of these devices is also explored. The importance of the physical and mechanical performance of the hydrogel scaffolds is examined, while current mechanisms of improving mechanical performance including the use of composite materials are considered. The importance of porosity, scaffold surface modification, degradability and degradation rate is also highlighted. Biological considerations in terms of preliminary toxicological analysis, cell isolation and loading, growth factor incorporation and release, as well as other advanced clinical requirements necessary for bone regeneration scaffolds are evaluated. While it is clear following critical appraisal that hydrogels alone will not suffice as load bearing bone tissue engineering devices, hydrogel composites look set to play a vital role in the development of the next generation of bone substitute materials.

Chapter 3 - As a great challenge for any country today, billions are being invested in the development of technology to obtain more efficient and competitive products that may bear fruits and boost the economy of the country. The application of technological innovation is employed in several areas aimed at obtaining products, structures, devices and / or systems that help have more sophisticated control of specific properties such as use, size and shape. The application of hydrogels technology is being widely used in pharmaceutical industry as drug delivery systems that have physical, chemical and biological characteristics which make them very interesting vehicles as carriers of drugs, medicines and cosmetics. The hydrogels can be used in all routes of administration for presenting compatibility with the tissues, low toxicity and generally for being biodegradable. They are made of polymeric materials in order to aggregate the different properties between the constituent polymers and possess the convenience of having characteristics that may be shaped according to their target application. Hydrogels synthesized through blends of synthetic and natural polymers have been an alternative of interest to merge interesting properties. The response to stimuli in polymeric hydrogels may increase or decrease according to the three-dimensional shape of the polymer chain and it also may change in environmental conditions. In order to guarantee the properties that a perfect carrier can hold, the quality control of these systems from their pre-formulation to the final product it is crucial to assure their efficiency as a controlled release system.

Chapter 4 - This chapter describes and discusses (1) the disadvantage of formulating an acidic drug with hydrogel using traditional approach, (2) preformulation of niacin including drug particle characterization using Accusizer™ 780 A, (3) high performance liquid chromatographic and UV-Vis spectrophotometric assay developments for the active pharmaceutical ingredient, and the correlation of the two assays, (4) formulation procedures, (5) proposals of two pH stage *in vitro* drug release study to simulate fasting and fed conditions, (6) application of FDA similarity factor to determine the similarity of release profiles, (7) percentage of drug remained in the accelerated temperature testing, (8) thermal analysis using a differential scanning calorimeter to examine whether there were melting point shifting in drug and three important excipients contributing the inert core in the samples underwent accelerated stability study, and (9) challenges encountered in selecting excipients for making drug dough extrudable into pellets, but facing accelerated temperature test in the later stage to determine shelf life.

Chapter 5 - Chemical modification of proteins has been largely used with the subject of expose functional groups "hidden" before into the protein chains, determinate functional groups of enzymes and also help in understanding of drug action. Animal proteins are rich in lysine, which is an essential amino acid with many functional properties, to the opposite of vegetal proteins, such as soybean. The fish proteins tend to be very big, with an inherent capacity of water uptake. For the application in hydrogel, the physical properties of the fish polymers are more superior to vegetable polymers. Hydrogel is a light and wet class material, whose properties depend on the polymer net constructed and the water content. Proteins permeability of different molecular weight in a biodegradable hydrogel has been researched. Many synthetic polymers have limited structural and functional properties, while the natural polymers are unique, with various functional and structural properties. The mechanical properties and swelling properties of hydrogel made by using natural polymers can be improved by chemical or physical modification of the functional groups. Modification does not alter the biodegradable and biocompatible characteristics of the proteins. Therefore, if suitable functionality can be imparted, the natural polymers could replace some potentially toxic synthetic polymers, which are unsuitable for many biological applications. The hydrogel with high water retention capacity has a great number of attributes that make them attractive in several applications. The basic property of uptake water has been suggesting the use of absorbent materials in many applications, such as towel paper, surgery sponge, meat tray, bath rug and medical wound. One of the greatest challenges that confront the science nowadays is the development of a new generation of biomaterials for human organism repair and also to combat environmental problems relate with pollution.

Chapter 6 - In nature polysaccharide-based 3D-hydrogel structures, such as cell walls, biofilms, and extracellular matrices, play a fundamental role in providing mechanical support and protection to biological entities, also controlling the diffusion of metabolites and potentially harmful molecules. Hence, natural saccharidic 3D-structures have always been regarded both as important models and as excellent materials for human applications (food, pharmaceutical, and biomedical).

Alginate and chitosan are two families of polysaccharides widely studied and employed as natural materials for bio-oriented applications, and they are exploited as immobilization and protective structures for sensible materials (from drug molecules and protein to cells).

Moreover, polysaccharide-based hydrogels are characterized by the possibility to engineer the polymer-based structure, thus controlling the resulting biological, mechanical, and transport properties. The encapsulation in polysaccharide matrices represents since ever the best example of protection. A variety of methods to encapsulate has been developed, by exploiting different materials and physico-chemical principles. In this paper, some relevant examples of micro- and nano-encapsulation based on alginate and chitosan in several bio-oriented technological fields are reviewed.

Chapter 7 - Silk fibroin (SF) is a protein that is spun into fibers by arthropods such as silkworms, spiders, scorpions, mites and flies. The process to obtain SF solution consists on fiber dissolution in concentrated saline solutions, followed by dialysis in water. SF aqueous solution (after dialysis) is thermodynamically metastable and may undergo sol–gel transition due to the lack of ions available to promote its solvation. The hydrogel formation occurs because SF chains tend to aggregate, passing from an amorphous conformation (random coil) to a more stable structure (β-sheet). The formation of β-sheets stabilizes the hydrogel and is irreversible under physiological conditions. SF hydrogels can be obtained by changing the protein concentration, pH or temperature, and by the addition of salts, metal ions and other components. This review describes the several methods available for SF hydrogel production and relates these methods with the final hydrogel properties and applications, focusing on biomaterials field.

Chapter 8 - Hydrogels are three dimensionally crosslinked polymeric network which can retain a huge amount of water but do not dissolve in water. Polyvinylpyrrolidone (PVP) is a synthetic polymer with good biocompatibility and transparency and with the action of different stimuli (radiation, heat, pressure, chemicals, etc.) undergoes crosslinking, and can form hydrogels. Due to their tissue compatibility and tissue like consistency, PVP hydrogels are very promising for different biomedical applications. However, because of inferior swelling capacity and poor mechanical property, the use of pure PVP hydrogels is limited. To overcome this problem, PVP is blended with different polysaccharides or other polymers according to requirement and from the standpoint of applications. This review article mainly focuses on different kinds of PVP based hydrogels; their modes of synthesis, properties and a range of applications of these blend hydrogels.

Chapter 9 - As one of hot research topics in polymer science, hydrogels attract increasing attentions. Following traditional preparation methods of hydrogels such as conventional chemical and physical crosslinks, some new strategies to fabricate hydrogels come forth recently. This review describes some novel hydrogels with excellent performances, which include slide-ring hydrogels, double-network hydrogels, nano-(micro-)composite hydrogels, ideal network hydrogels, self-oscillating hydrogels, non-covalently crosslinked hydrogels, and reversible covalently crosslinked hydrogels.

Chapter 10 - Recently, the applications of hydrogels as a carrier vehicle for the controlled release of fertilizers in the agricultural field have been extensively proposed due to the unique properties of these materials, such as their biodegradability, low-toxicity, high water uptake and reversible properties. The major benefit of these systems when applied in the agricultural setting is the gradual release of a loading, which maximizes that loading's concentration in the soil over a long time period. As a direct consequence, the number and frequency of the fertilizer applications to the soil are reduced. This diminishes the cost of agricultural production as well as the risks of human contamination and environmental pollution. This chapter is intended to describe the preparation of novel hydrogels composed of poly(acrylamide) (PAAm) / poly(methacrylic acid) (PMAA)/carboxymethyl cellulose (CMC) synthesized by a free-radical polymerization reaction catalyzed by temperature for applications in the controlled release of fertilizers. Fourier transform infrared (FTIR) spectroscopy confirmed the formation of the PAAm-PMAA-CMC hydrogels. The hydrophilic properties, determined by the degree of swelling, were dependent on the pH of the external swelling medium, indicating that this hydrogel is pH sensitive. The pKa value of the hydrogel estimated from the sigmoid-curve was approximately 4.0. It was also possible to control the amount and rate of water absorbed by controlling the monomer and polysaccharide compositions. The amount of water absorbed is directly related to the network rigidity. The Freundlich model performed better than the Langmuir model in describing the adsorption isotherm of the PAAm-PMAA-CMC/$(NH_4)_2SO_4$ fertilizer system. The $K_F$ and 1/n Freundlich constants were 3.242 L $g^{-1}$ and 0.4143, respectively. Thermodynamic investigations proved that the adsorption of the fertilizer in the hydrogels is spontaneous. An investigation of the cumulative desorption confirmed that these hydrogels have potential applications in the agricultural field.

Chapter 11 - Thermoresponsive hydrogels have been widely investigated in the past few years for various biomedical purposes including controlled delivery of drugs and vaccines. This is due to their unique characteristics of being able to exist as a solution at low temperature and form a solid gel at physiological temperature. Among thermoresponsive hydrogels, chitosan and poloxamer 407 are the two polymers that have attracted most attention for drug and vaccine delivery. Chitosan is a polysaccharide derived from crustacean shells and poloxamer 407 consists of the non-ionic triblock co-polymers, poly(ethylene oxide) (PEO) and poly(propylene oxide) (PPO). The following review focuses on the synthesis and characterisation of hydrogels based on these polymers or modified polymer/co-polymer systems. Data on the use of these systems and future prospects of hydrogels for drug and vaccine delivery will be presented.

Chapter 12 - Sol-gel encapsulation of living cells within inorganic matrices is an area of great scientific and technological interest. However, only recently a two-step procedure allowing population growth inside the hydrogel matrix has been proposed. In conventional one-pot encapsulations, space limitation suffered by confined cells makes their division completely impossible, and only intrinsic resistant microorganisms (mainly bacteria and yeasts) were encapsulated using extremely biocompatible synthesis procedures in order to obtain an acceptable viability in the resulting biomaterial.

The two-step procedure based on a pre-encapsulation of the biological guest within Ca(II)-alginate, has proved to be a versatile method for advanced biomaterials development. For instance, in applications were it is necessary to keep a large number of metabolically active cells to render them efficient, or where the encapsulation of actively growing cells may be required. Moreover, it opened up many different possibilities to biomaterials, allowing the encapsulation of higher organisms cells (more sensitive than microorganism) as carrot phloematic tissue and tobacco BY2 Line or even of multicellular organisms, such as filamentous fungi.

On the other hand, while one-pot encapsulations uses almost exclusively silica hydrogel hosts, the protection conferred to the cells by the pre-encapsulation step during the sol-gel synthesis procedure allows to extend this approach to other sol–gel materials, as has recently been addressed for Al(III) and Zr(IV) oxohydroxide hydrogels. This work constitutes the first observation of *Escherichia coli* growth within such metal oxide hydrogels, giving clear evidence that sol–gel based cell encapsulation can now be envisioned within a wide variety of metal oxide hosts through the optimization of the pre-encapsulation environment.

Chapter 13 - An exhaustive scientific investigation has been performed on polymeric hydrogels as potential materials to be used in cartilage tissue repair or replacement. Polyvinyl alcohol (PVA) possesses excellent properties as biocompatibility, low coefficient of friction, low interfacial tension and high permeability to fluids, emerging as an eligible material to be studied for biomedical applications.

The main goal of this study was the tribological characterization of PVA hydrogels in lubricated sliding with distilled water and phosphate buffered saline solution (PBS). The influence of hydrogels mechanical properties and operating conditions were considered. Tribological tests were performed on a pin-on-plate tribometer with a linear reciprocating geometry using the PVA/stainless steel 316L and PVA/bovine articular cartilage pairs. Experiments were performed at 37 °C with a constant reciprocating sliding frequency of 1 Hz and applied load of 1 MPa, 2 MPa and 3.5 MPa. For each experiment the friction coefficient was determined and the dominating wear mechanisms analysed by scanning electron microscopy and EDS. PVA showed an excellent performance as possible material for the repair of articular cartilage due to the high level of surface preservation and the very low friction coefficient presented when PBS was used as fluid lubricant. The performed tribological study revealed the high potentiality of PVA hydrogels to be used as artificial articular cartilage.

Chapter 14 - This chapter presents the development of a hydrogel matrix with 3D gradient of mechanical properties. It can have several applications to the study of phenomena related to cell mechanotaxis (cells tend to migrate from soft regions to more rigid ones when they are seeded on a substrate with a gradient of stiffness), or differentiation on different stiff substrates. The authors performed mechanical characterization of alginate hydrogels at different concentrations, cross-linked with several concentrations of calcium chloride. The samples, produced by spraying the calcium chloride solution on a layer of alginate solution, were subjected to compression tests to obtain a "map" of elastic modulus as function of alginate and calcium chloride concentration. Then the authors realized alginate matrices characterized by a gradient of mechanical properties. To do that the authors used a device purposely developed, able to impose a three-dimensional gradient of calcium chloride concentration to alginate solution, allowing a differential crosslinking in the matrix. The concentration profile was predicted before with computational fluid dynamics software. Compression tests were made on these matrices, and the the formation of an elastic modulus gradient was observed. However hydrogel skin formation effects limit the whole efficiency of the device, that worked properly only in the middle of the matrix. The important aspect of this research is that the authors can construct hydrogel matrix with a gradient of mechanical properties, in which cells can be embedded in order to analyze cell behavior as function of substrate mechanical properties.

Hydrogels: Synthesis, Characterization and Applications
Editors: F. V. Câmara and L. J. Ferreira
ISBN: 978-1-61942-842-3


*Chapter 1*

# SYNTHESIS OF PROTEIN-BASED HYDROGEL FOR PHARMACEUTICAL AND BIOMEDICAL APPLICATIONS

***U. Gianfranco Spizzirri, Giuseppe Cirillo, Ortensia I. Parisi and Francesca Iemma****
Pharmaceutical Science Department,
University of Calabria, Rende (CS), Italy

## ABSTRACT

Natural polymers, such as proteins, are materials extensively investigated due to their biocompatibility, biodegradability and non-toxic and non-immunogenic characteristics. Enclosing these biomacromolecules in a complex structure, these features can be transferred to a biomaterial in order to extend the performance of the device. Thus, the covalent conjugation of a biodegradable macromolecule, as a protein, represents a versatile strategy to produce intelligent biodegradable hydrogels, suitable for pharmaceutical and biomedical applications. The cross-linking of biomacromolecules is somewhat important in these fields because it increase the mechanical properties of the hydrogels without to affect, in severe degree, the biodegradability. Basically, albumin and gelatin represents the protein mainly used in the biomedical field and their insertion in a polymeric network was performed employing different synthetic strategy involving the functional groups in the side chain of the polypeptides. Hydrogel containing proteic moieties in the polymeric network can be achieved by thermal denaturation of the protein either by direct reaction between functional groups (usually carboxyl and amino goups) in the polypeptide side chains and also by use chemical cross-linking agents as bifunctional carbonyl reagents. Finally, involving the protein in a radical process innovative synthetic procedures were proposed in order to modify polymeric network composition and to design hydrogels with appropriate and mouldable physicochemical properties.

---

* Corresponding author: francesca.iemma@unical.it

# 1. INTRODUCTION

The integration of the structural and functional properties of peptides and proteins with the versatility of synthetic polymers has gained significant interest in materials design and application [1]. Although polymers possess a broad and versatile range of properties (including stimuli-responsiveness, biocompatibility, degradability, conductivity, and mechanical strength), many compositions (owing to the availability of many different monomers) and varied architectures (from the wide variety of topologies including linear, block, graft, and hyperbranched polymers), they are not capable of the folding and recognition of peptides and proteins. [2] Considerable interest in recent years has been shown in the use of natural polymer to produce biomedical devices. Thus, the crosslinking of a biodegradable macromolecule, as a protein, represents a versatile strategy to produce intelligent biodegradable hydrogels, suitable for pharmaceutical and biomedical applications [3]. The cross-linking of biomacromolecules is somewhat important in pharmaceutical field because it increase the mechanical properties of the hydrogels without to affect, in severe degree, the biodegradability [4, 5]. Basically, the insertion of natural polymers in a macromolecular network, can be achieved employing different synthetic approaches. The first method involves the chemical modification of the biomacromolecules either by introducing functionality able to undergo radical polymerization reactions or by involving the protein, without any functional changes, in graft radical polymerization reactions that interest the heteroatoms of the substrates. In addition, proteins can take part in crosslinking processes by direct reaction between functional groups (usually carboxyl and amino groups) in the polypeptide side chains and also by use chemical cross-linking agents as bifunctional carbonyl reagents.

The gamut of applications of protein-based hydrogels has gone beyond biomedical applications to their use for biosensors, phase-separation bio-assays, novel bioseparation protocols, bio-catalysis, artificial enzymes, biometrics, light harvesting systems, photonics, and nanoelectronic devices [6]. In particular, the combination of synthetic polymers and reconstituted proteins for scaffold design may help broaden the usefulness of natural protein-based scaffolds in tissue engineering and regeneration. The versatility and utility of protein-based hydrogels were also investigated to be used as carrier for drug release due to the insertion of a protein in a polymeric network can establish specific drug-carrier interactions. These characteristics are complemented by the fact that by altering the crosslinking density of the hydrogel carrier, the period of drug release can be regulated, enabling investigators to tailor this biomaterial with the optimal release characteristics required for each individual application.

## 2.1. Albumin

Albumin is an attractive macromolecular specie to prepare pharmaceutical carriers used for the sustained delivery of therapeutic agents since this protein is biodegradable, biocompatible, non-toxic and non-immunogenic. Chemical groups susceptible to radical polymerization were introduced onto bovine serum albumin (BSA) by acylation reaction the BSA with methacrylic anhydride (MA) in water at 0°C and neutral pH [7]. The nucleophilic chemical groups in BSA that could react with MA are the thyolic groups of cysteine, hydroxyl groups of serine and tyrosine, and amino groups in the side chain of lysine. The first are involved in disulfide bridges, except cys-34, the latter are the least nucleophilic, and do not react in mild experimental conditions. Then, the sterically accessible amino groups of lysine only react chiefly with acylating agent at controlled pH and temperature to produce water-soluble MA-BSA (Figure 1). If the reaction is carried out without pH and temperature control, denaturation of BSA was observed, and its water solubility is lost. The water-solubility of the derivatized protein was essential to provide spherical polymeric microparticles via reverse phase suspension copolymerisation with DMAA, using APS and TEMED as initiator system. The reverse-phase suspension polymerisation technique represents a simple method for obtaining spherical microparticles whose size can be conveniently varied, according to particular needs, by changing the reaction conditions (speed of stirring, shape and dimension of reactor, shape of stirring-rod).

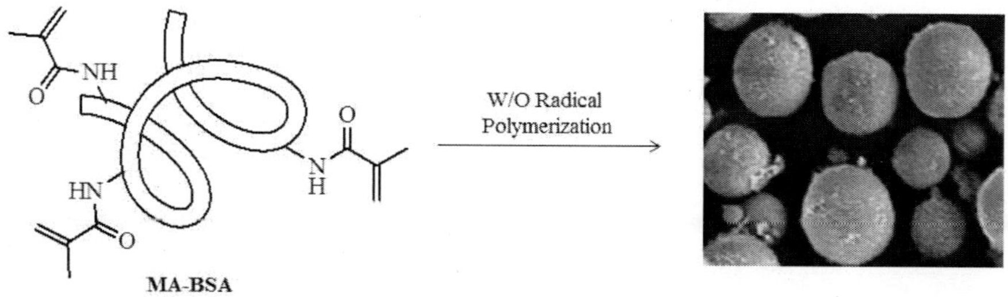


Figure 1.

The polymerization reaction, owing to steric and geometric constraints, involves only the methacrylic functions of MA-BSA which are accessible to the growing chains. The microparticle structure is characterised by a network where the BSA chains are linked by some hydrocarbon bridges. The MA-BSA samples were also polymerized in the presence of various amounts of comonomer (DMAA) in order to study the effects of a different chemical structure and degree of crosslinking on the physical properties of the microparticles. It can be supposed that in the copolymerisation reaction the chains obtained consist of DMAA units randomly interrupted by methacrylic BSA-MA functions which are sterically and geometrically attainable. The applicability of these materials as drug delivery systems has been evaluated by loading drugs with different chemical properties by a soaking procedure. In particular, for this material, the drug release features depend principally on crosslinking degree, ratio among albumin and DMAA, and interactions of loaded drug-beads [8].

Using comonomers with different chemical properties, the polymerization technique allowed to synthesize versatile materials for biomedical and pharmaceutical applications. In order to prepare pH-sensitive microspheres, suitable for oral drug administration, MA-BSA was copolymerized with methacrylic acid sodium salt (NaMA) by reverse phase suspension copolymerization [9]. Additionally, thermo-responsive hydrophilic microspheres were prepared by free radical polymerization of MA-BSA and NIPAAm, as cross-linker and functional monomer, respectively [10]. Thermal analyses of the samples showed negative thermo-responsive behaviour with pronounced water affinity of microspheres at temperature lower than LCST. Furthermore, the influence of the hydrophilic/hydrophobic balance of the monomers in the polymerization feed on the transition temperature of the macromolecular device was investigated [11]. Thermoresponsive microspheres were synthesized by copolymerization of hydroxyethylmethacrylate (HEMA) and NIPAAm, as hydrophilic and thermoresponsive monomers respectively, in presence of a proteic crosslinker, such as MA-BSA. The synthetic approach allows to modify the polymeric network composition producing hydrogels with appropriate and modulable physicochemical properties and a LCST close to the physiological temperature. Finally, dual stimuli-responsive microspheres (pH and temperature) were prepared by free radical polymerization of BSA-MA and NaMA and NIPAAm, as hydrophilic/pH-sensitive and thermo-responsive monomers, respectively [12]. In order to test the preformed materials as site-specific (oral or transdermal) drug carriers, *in vitro* experiments using diclofenac diethyl ammonium-loaded microgels were performed.

The conjugation of polyethylenoxide diacrylated ($PEGDA_1$ and $PEGDA_2$) to the modified serum albumin allowed the introduction of polymerizable groups on the proteinous backbone [13, 14]. This reaction involves the preliminary reduction of the disulphide bridges to thiolic groups able to react with PEGDAs by a Michael reaction (Figure 2). For this purpose tris(2-carboxyethyl) phosphine hydrochloride was employed as reducting agent.

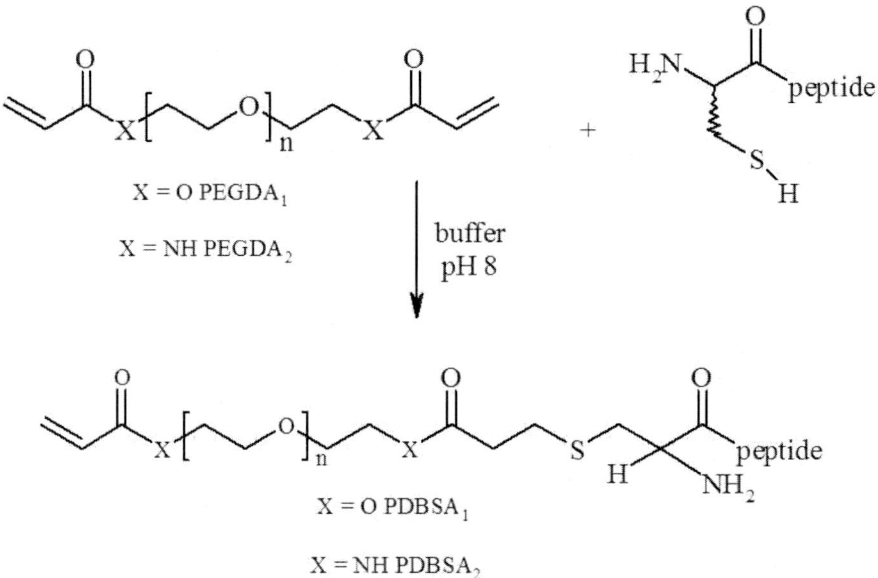


Figure 2.

The photopolymerization reaction of the PD-$BSA_1$ conjugate carried out to a protein-based biosynthetic hydrogel for the purpose of creating scaffolds that emulate the biological characteristics of the native natural extracellular matrix with added control over structural properties and biodegradation [15]. The hydrogel were characterized and tested with smooth muscle cells in 2-D and 3-D cultures for up to one week. This biomaterial maintains the cell adhesion and support proteolytic degradability based on the specific characteristics of the protein backbone. The synthetic PEG constituent provides the possibility of controlling the enzymatic biodegradation and structural properties of the materials without introducing additional cross-linking steps.

PD-$BSA_2$ was photopolymerizated to give hydrogel for drug delivery applications where controlled release is accomplished using the natural affinity of the drugs to the serum albumin

[16]. Two model drugs, Warfarin and Naproxen, were used for equilibrium dialysis and release experiments from the hydrogels, both having relatively low molecular weights and a known high affinity for albumin. Equilibrium dialysis experiments showed that multi-PEGylation of albumin significantly decreased the drug affinity to the protein compared to non-PEGylated controls, irrespective of the PEG molecular weight. Additionally, these hydrogels were used as a basis for drug carrying tissue engineering scaffold materials, based on the natural affinity of various drugs and compounds for the tethered albumin in the polymer network [17]. The results of the drug release validation experiments showed that the release kinetics of the drugs from the hydrogels based on PD-$BSA_2$ were controlled by the molecular weight of $PEGDA_2$ conjugated to the protein, the drug molecular weight and its affinity for albumin. Additionally composite hydrogels containing PD-$BSA_2$ copolymerized with PEGylated and acrylated fibrinogen were used specifically for 3-D cell culture scaffolds, with inherent bioactivity, proteolytic biodegradability and controlled drug release properties [17]. The new composite hydrogel combines the biocompatibility of fibrinogen, for tissue regeneration, and the high affinity for certain drug molecules of albumin, for controlled drug release. The main advantage of the composite biomaterial is the ability to provide sustained drug release using affinity and simultaneous application of the material as a bioactive tissue regeneration matrix.

Acrylated BSA (A-BSA) (Figure 3), as referred by Tada et al., was prepared via reaction of albumin with N-succinimidylacrylate in aqueous medium at room temperature [18]. BSA with acrylate groups was then copolymerized with acrylamide (AAm) and N,N-methylenebis (acrylamide) (MEBA), using APS/TEMED as initiator system, to produce a biodegradable hydrogel potentially useful as sustained drug release carrier for albumin binding substances.

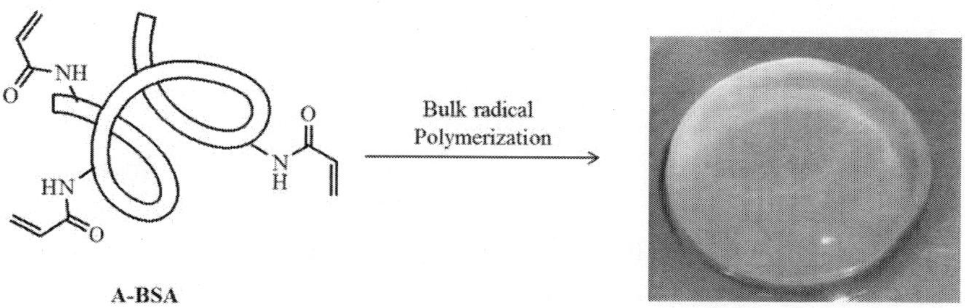


Figure 3.

However, the authors affirm that acrylamide monomer is likely to be harmful if it remained in a hydrogel and BSA possibly causes immunogenic adverse effects for human. Therefore, they replaced BSA and acryl amide with recombinant human serum albumin (rHSA) and alginic acid (ALG), respectively [19]. Hydrogels (HSA-ALG) consisting of rHSA and ALG were prepared according to the procedure shown in Figure 4. ALG was firstly converted to its activated ester by dehydrating condensation with N-hydroxysuccinimide (NHS) by the aid of water-soluble 1-ethyl-3-(3-dimethylaminopropyl)carbodiimide (EDC). Subsequently, ALG chains were crosslinked with HSA by the reaction of amino residues in HSA with activated ester groups in ALG. Thus, hydrogels consisting of recombinant HSA and ALG have been evaluated as carrier of cationic drugs as salicylic acid, o-anisic acid, benzoic acid and dibucaine hydrochloride.

Tissue adhesives consisting of human serum albumin, and citric acid derivative (CAD), as a crosslinking reagent, were proposed [20]. The crosslinking reagent CAD was prepared using citric acid as a starting material, modified with NHS in the presence of EDC. The CAD-HSA glue showed excellent biocompatibility and high ability of wound closure similar to that of cyanoacrylate-based adhesive when glues were applied to the mouse skin. The same synthetic strategy was employed to prepare an injectable in situ forming drug delivery system (DDS) using HSA and tartaric acid derivative (TAD), as showed in Figure 5 [21]. The synthesis of TAD was carried out by the standard protocol using NHS and N-N'-dicyclohexylcarbodiimide (DCC) [20].

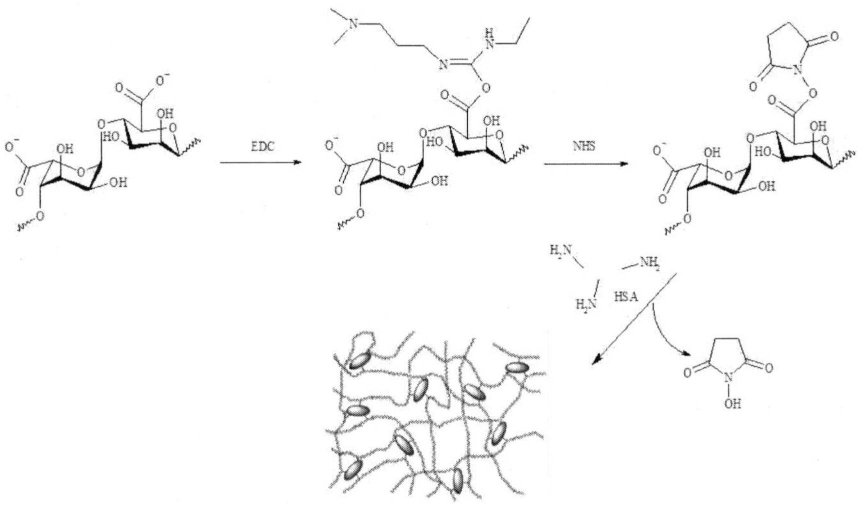


Figure 4.

The resulting tissue adhesive forms in situ as a gel and has high bonding strength for living tissue as well as it has good biocompatibility and biodegradability. TAD-HAS hydrogel was proposed for cancer chemotherapy and carcinostatic doxorubicin hydrochloride (DOX) was employed in this preparation because it is widely active for several solid tumors. The release rate of DOX from HSA–TAD gel can be controlled by the matrix composition. The release of DOX from HSA–TAD gels reached a maximum after 24 hours, after which the concentrations were maintained for a long period, approximately 100 h.

Recombinant rHSA hydrogel as a carrier for the controlled drug release was prepared by reducing HSA with dithiothreitol (DTT) [22]. The gelation takes place through partial denaturation of HSA, which is caused mainly by the intermolecular interaction of hydrophobic residues through the cleavage of disulfide linkages. However, the drug binding ability of albumin still remained after hydrogel formation. Therefore, a higher amount of warfarin, a drug having the highest affinity for albumin, was loaded on thus prepared rHSA hydrogel and its release from the hydrogel sustained longer comparing with benzoic acid which has negligible binding affinity for albumin.

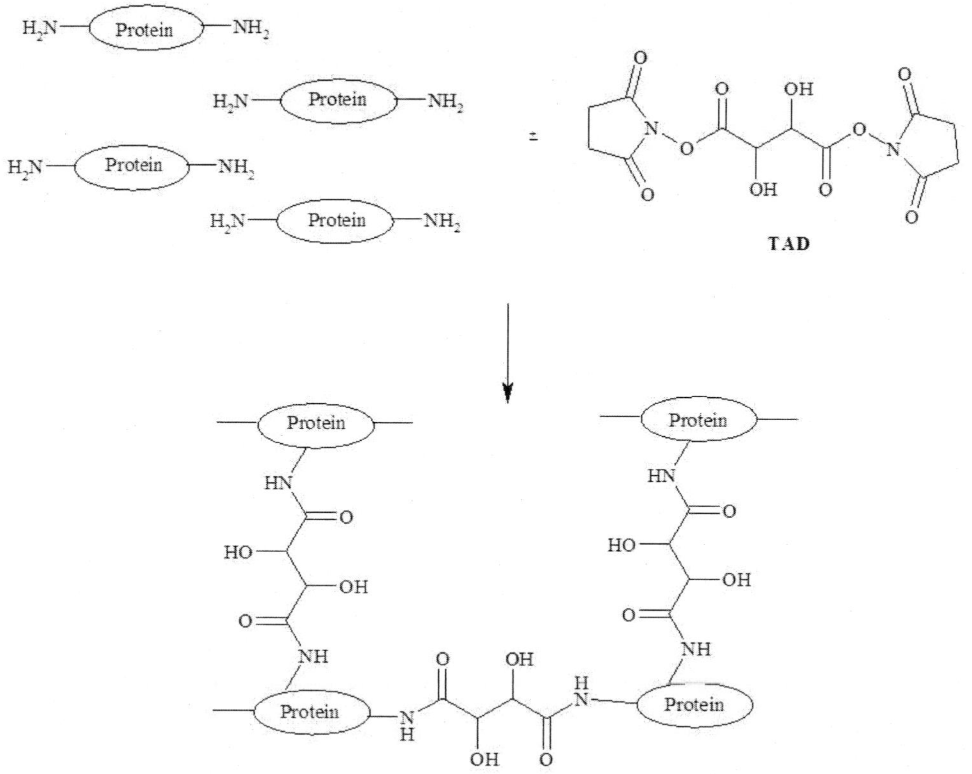


Figure 5.

## 2.2. Gelatin

Gelatin (GL) is a natural protein which is derived from collagen with many desirable properties for application as a biomaterial, including scaffolds for tissue engineering. Additionally, different kinds of GL carrier matrices can be employed as drug delivery devices, because of their ability to sorb biomolecules. As will be seen, depending on the fabrication method, variations in the electrical and physical properties of GL-based controlled-release systems can be achieved. It is this flexibility in processing that has allowed GL-based controlled-release systems to find diverse applications in fields ranging from tissue engineering, to drug delivery and gene therapy [5].

Gelatin-based hydrogels can be prepared by crosslinking reaction in order to insolubilize the carrier in water, improve the thermal and mechanical stability of the carrier under physiological conditions, as well as control the rate of degradation in vivo [23]. Chemical reactions involving radical or condensation processes represent the main synthetic methods to prepare chemically crosslinked GL-based hydrogels.

The insertion of natural polymers in a macromolecular network by radical polymerization processes, can be achieved employing two different synthetic approaches. The first method involves the chemical modification of the protein to introduce functionality able to undergo radical polymerization reactions [7, 9, 12]. In addition, GL without any functional changes, can take part in graft radical polymerization reactions that involve the heteroatoms of the substrates [24, 25]. Both synthetic approaches allows to prepare biocompatible bioconjugates showing improved physico-chemical and mechanical properties respect to the starting natural species. Condensation reactions with chemical crosslinking agents such as bi-functional or polyfunctional molecules produce polymeric networks by bridging free carboxylic acid residues or amine groups between adjacent protein molecules (i.e. glutaraldehyde, genipin, dextran dialdehyde, citric and polylactic acid derivatives [26].

Methacrylated gelatin (MA-GL) was prepared by reaction of the GL primary amines groups with an excess of MA in phosphate buffer at 45°C (Figure 6).

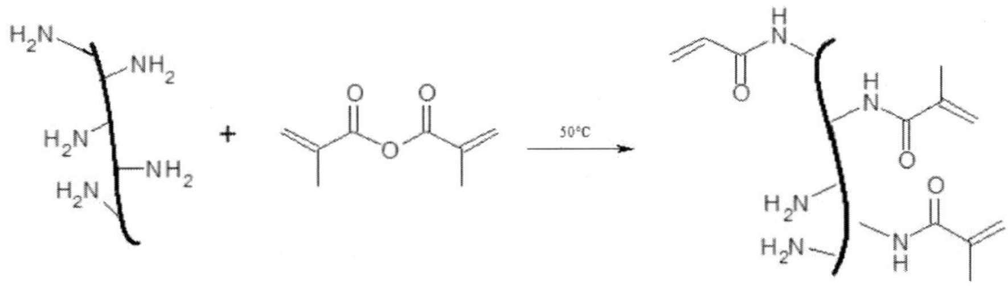


Figure 6.

Hydrogels based on methacrylamide-modified gelatin are attractive materials for biomedical applications and their chemical cross-linkage leads to well controllable chemical networks. The hydrogels were prepared by radical cross-linking of MA-GL in aqueous medium in the presence of a water-soluble photoinitiator, Irgacure 2959 [27]. The rheological properties of the GL-based hydrogels can be controlled by the degree of substitution, polymer concentration, initiator concentration, and UV irradiation conditions producing materials with different mechanical properties and potential applications in different fields. MA-GL was proposed as hydrogel platform for creating cell-laden microtissues and microfluidic devices because of cells readily bound to, proliferated, elongated, and migrated when encapsulated in microfabricated MA-GL hydrogels [28]. The authors suggest that these hydrogels could be useful for creating complex, cell-responsive microtissues, such as endothelialized microvasculature, or for other applications that require cell-responsive microengineered hydrogels. Methacrylated gelatin was successful employed in the polymerization reaction with poly(ethylene glycol)dimethacrylated (PEGDMA) to produce hydrogels for cell culture and tissue-engineering applications that exhibits both features of each reagent [29]. Poly(ethylene glycol) hydrogels are popular in the biomedical fields because they are nontoxic and exhibit favorable hydration and nutrient transport properties. However, cells cannot adhere to, remodel, proliferate within, or degrade PEG hydrogels. The insertion in the polymeric network of GL moieties yields an enzymatically degradable hydrogel that cells can degrade, adhere to and spread within. The mechanical and biological properties of PEG-MA-GL composite hydrogels can be controlled by altering the concentration of PEGDMA, MA-GL, or both, to produce a biomaterial that can be customized to meet the wide range of biological and mechanical profiles required for tissue engineering. Thermal polymerization of MA-GL, using AIBN as initiator system, in a concentrated oil-in-water emulsion, of the continuous phase gave rise to a polyHIPE, a porous material possessing a highly interconnected, trabecular morphology, useful as scaffolds for tissue engineering applications

[30]. The obtained scaffold by radical polymerization possessed a morphology characterized by relatively large voids and interconnects, and as a consequence, it was more suitable for hepatocytes colonization [31].

Koepff et al. reported the derivatization of enzymatic gelatin hydrolisate (HGL) with glycidilmethacrylate (GMA) that reacts with both amino and acidic groups and leads to polymerizable gelatin (GMA-HGL) (Figure 7) [32]. Comparing to the native protein, the hydrolysates are characterized not only by enhanced water solubility, but also by a greater number of nucleophilic groups disposable for the reaction with the acylating agent. Unfortunately, solutions of pure GMA-HGL in water form brittle hydrogels when exposed to light in the presence of a photoinitiator. Hence, the use of flexibilizing reactive diluents is necessary to break the intermolecular H-bridges, to lower network density and increase the mechanical stability. Therefore, further modifications of HGL were attempted to increase the monomer tolerance and lower the amount of water or even avoid it entirely. HGL was modified with an PEG-300-monomethacrylate previously derivatized with maleic anhydride and further activated with N-hydroxysuccinimide (PMA-HGL) (Figure 7) [33].

Because of the two different reactive sites on HGL-amino and carboxylic groups-introduction of two different moieties is possible with selective reagents. PEG-monomethylethers with molecular weights of 1000 and 4000 were primarily modified with maleic anhydride and subsequently activated with NHS to give the reactive intermediates. HGL could be converted with these products (Figure 7) under mild conditions in very good yields providing HGL derivatives. The reaction of the remaining carboxylic groups with GMA resulted in polymerizable gelatin derivatives, labelled $PGMA_1$-HGL, $PGMA_2$-HGL [34]. The authors photocopolymerized the gelatin derivatives with HEMA and poly(ethylene glycol 400) dimethacrylate to provide hydrogels useful as scaffold for tissue engineering field.

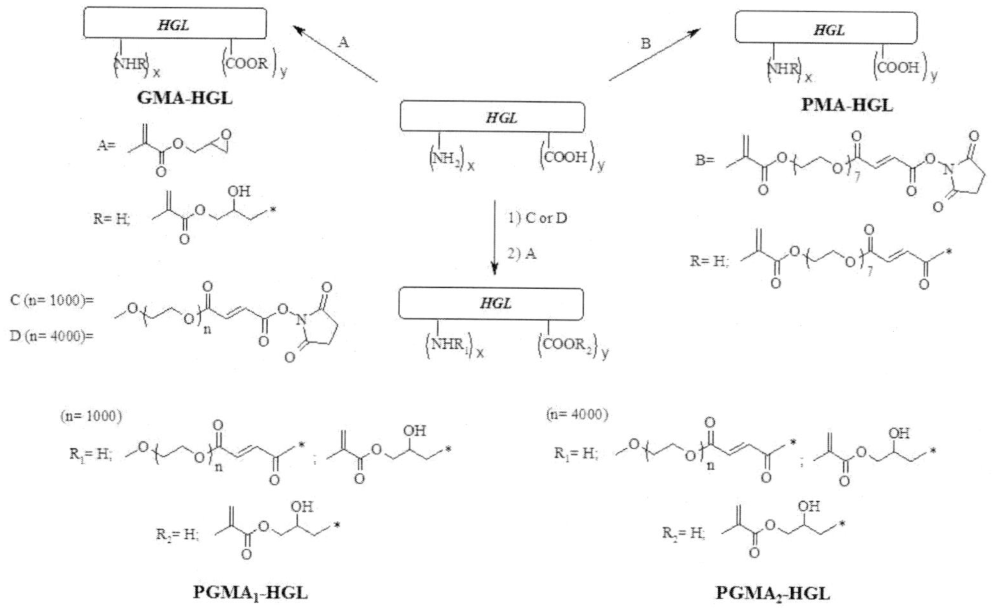


Figure 7.

Curcio et al. employed methacrylated gelatin hydrolysates (MA-HGL) as pro-hydrophilic macromer in the synthesis of thermo-responsive microspheres by free-radical suspension polymerization using NIPAAm and MEBA, as thermo-responsive functional monomer and crosslinking agent, respectively (Figure 8) [35]. The potential application of these materials as drug carriers was demonstrated by performing Diclofenac release experiments at temperatures around the LCST. Depending on the temperature of the surrounding environment, the release of the drug across the hydrogels takes place by rapid and reversible modification of volume hydrogels and by diffusion of the therapeutic through the polymeric network.

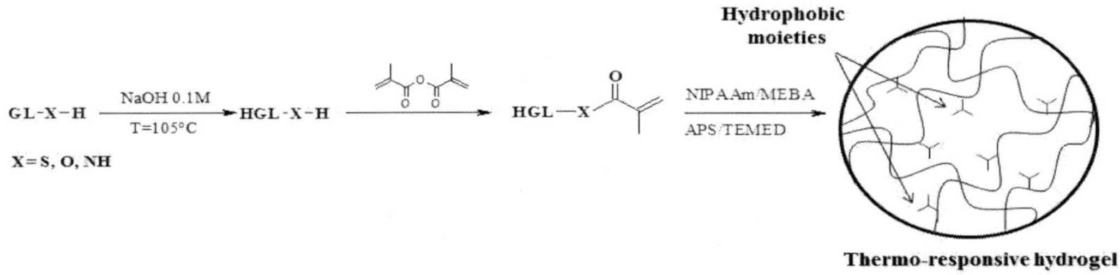


Figure 8.

The reaction beetwen the gelatin and a specific functional monomer was proposed to prepare, by radical polymerization, GL-based stimuli responsive hydrogels. The graft copolymerization of GL with various monomers is an effective method to improve its properties. The possibility to insert commercial gelatin in a crosslinked structure bearing thermo-sensitive moieties by reverse phase suspension radical polymerization represents an interesting innovation that significantly improves device performance, opening new applications in biomedical and pharmaceutical fields. Curcio et al. proposed the synthesis of thermo-responsive microspheres based on commercial GL as potential drug delivery systems. In this work, NIPAAm and MEBA were covalently inserted in the GL structure by radical grafting, using APS as initiator [36]. The thermal dissociation of the initiator formed anionic radicals that attack H-atoms in hydroxyl, thiolic or amino groups in the side chain of gelatin, forming a macroradical with several active sites. At those sites, polymer chain of NIPAAm starts and propagates as regular radical polymerization of polyacrylates (Figure 9).

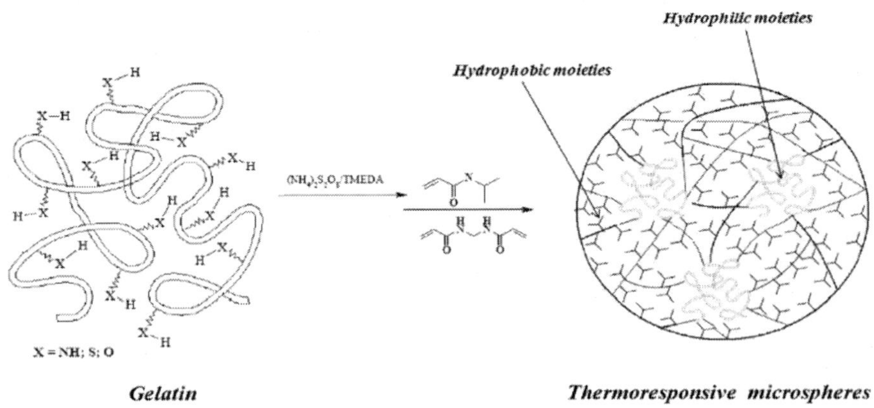


Figure 9.

In order to prepare biocompatible hydrogel magnetic nanoparticles acrylamide, GL, and MEBA, as a cross-linker, were polymerized [37]. The combination of iron complexation ability of amide groups in the main chains and the water sorption characteristics of the GL allow the synthesis of hydrogel network templates for in situ deposition of magnetite iron oxide particles by utilizing the reduction process induced by ammonia solution (Figure 10). In addition, the presence of the GL in the network ensures the in vitro blood compatibility of magnetic nanocomposites.

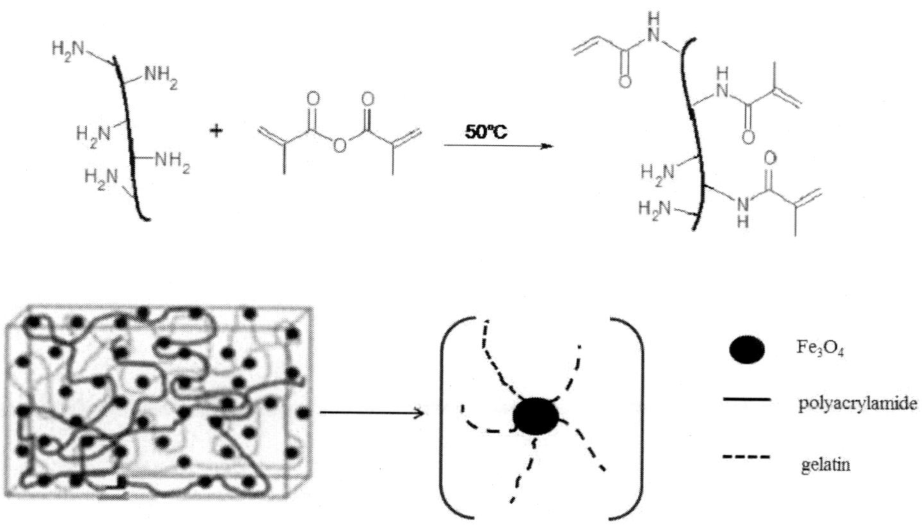


Figure 10.

The device was proposed as DDS and the drug releasing studies were performed under external magnetic field, using Doxorubicin, as model drug. The results indicate that the prepared hydrogels may be useful for potential applications for magnetically controlled drug release-systems, magnetic-sensitive sensors, and pseudomuscular actuators.

Recently, a novel pH-sensitive hydrogel based on GL and acrylic acid (AA), using γ-radiation as super clean source for polymerization and crosslinking, was synthesized [38]. Ionizing radiations are able to generate radicals on the monomer and polymer in addition to the production of $OH^{\cdot}$ and $H^{\cdot}$ radicals, as the primary products of water radiolysis. On exposure to γ-ray, as ionizing radiation, monomers radicals combine and propagate to form linear or branched but soluble polymers. In the same time, GL and the formed pAA are also radiolyzed forming macro radicals that contribute to chain initiation and crosslinking formation. Ketoprofen, an anti-inflammatory and analgesic agent, was used as model drug to evaluate the prepared hydrogel as drug carrier. *In vitro* release studies in different pH similar to the gastrointestinal fluids have been made to show the influence of the environmental pH and the preparation conditions on the release profiles. Because of the use of protein as a natural backbone, it is expected that the resulted pH-sensitive hydrogel show more compatibility with body when they use as drug delivery systems.

Sadeghi et al. report the study of a novel GL-based hydrogel synthesized through graft copolymerization of methacrylic acid (MAA) onto gelatin, using APS and MEBA, as free radical initiator and crosslinker respectively [39]. The macroradicals produced from thermal decomposition of APS, initiate graft copolymerization of MAA, leading to the graft copolymer. GL-*g*-pMAA hydrogel exhibited a pH-responsiveness character so that a swelling-deswelling pulsatile behavior was recorded at pH 2.0 and 8.0. The on-off switching behavior makes the hydrogel as a good candidate for controlled delivery of bioactive agents.

The disadvantage related to the use of GL for the development of hydrogels is their structural and thermal lability, which may be improved by using crosslinkers like glutaraldehyde (GA) and diisocyanates, as well as carbodiimides, polyepoxy compounds and acyl azide methods [40]. GA is by far the most widely used agent, due to its efficiency to stabilize protein-based biomaterials and in spite of some local cytotoxicity and calcification of long-term implants [41] (Figure 11).

Figure 11.

The mechanical, thermal, swelling and release properties of GA crosslinked gelatin films have been investigated in order to verify the influence of GA concentration (from 0.05 to 2.5 wt %) on the stability of the films. The data suggest that the use of GA at low concentration, which is desiderable to prevent toxicity, allows to modulate the physico-chemical properties of gelatin films, in order to obtain stable materials with a wide range of possible biomedical applications [42].

Einerson et al. explored modification of the gelatin backbone with PEG-dialdehyde and/or ethylenediaminetetraacetic dianhydride (EDTAD) to alter the physicochemical properties of the gelatin, and to affect the subsequent release, degradation and solubility of model drugs from and within the hydrogel (Figure 12) [43]. PEG and EDTAD are widely accepted as having low toxicity [44]. Furthermore, modification of GL with EDTAD introduces polyanionic molecules into the gelatin chain, increasing the hydrophilicity of the gelatin backbone with the addition of charged groups, and thereby potentially improving the swelling capability of the resulting hydrogel.

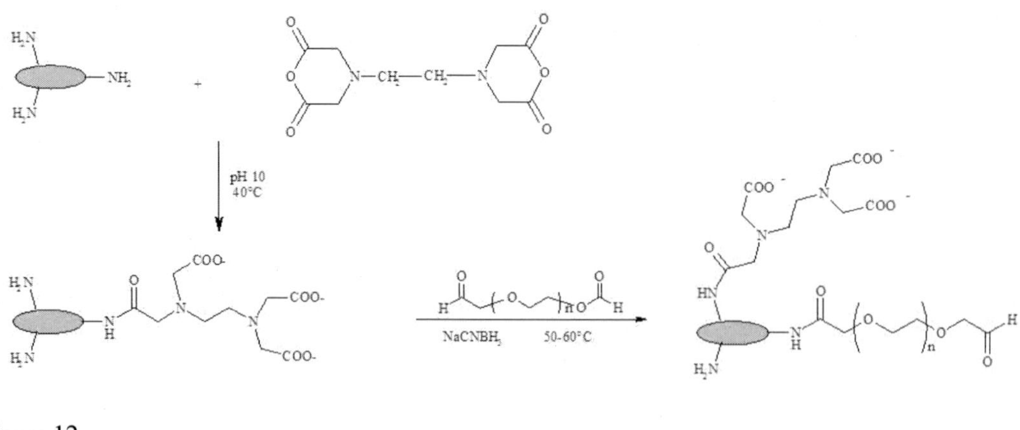


Figure 12.

The modulation of crosslinking modality (i.e. percent GA or self-crosslinking via exposure to dry heat) of unmodified and modified gelatin affect the solubility and density of the resulting matrix, varying the swelling/degradation and the release mechanism of therapeutic agents.

Magnetic micro- and nanoparticles based on ferrofluid (maghemite) were prepared by inverse emulsion crosslinking of sodium salt of carboxymethylcellulose and GL (Figure 13). Crosslinking was carried out with GA within aqueous droplets dispersed into toluene in presence of surfactants [45].

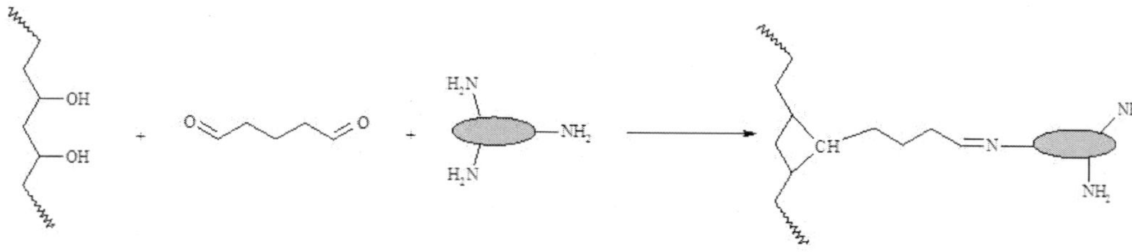

Figure 13.

The composition, the morphology, swelling in water characteristics, the capacity to include and release water soluble drugs of these particles depend on elaboration process parameters such as the polymer ratio, the polymer mixture/crosslinking agent (respectively maghemite) ratio, or the nature and the concentration of the surfactant. The ability to include drugs (Cefotaxime and Methotrexate) is directly correlated with the swelling degree and the gelatin content within the particles. The absence of toxicity, associated with the bactericide activity (for at least 48 h) make these systems potential drug carriers.

Vandenbulcke et al. have reported on the preparation of gelatine hydrogels crosslinked with partially oxidised dextrans (GL-DO) [45]. Reaction of dextran with sodium periodate leads to the formation of oxidized dextran containing different types of dialdehyde structure. Crosslinking is predominantly due to Schiff's base formation in reaction between the ε-amino groups of lysine or hydroxylysine side groups of gelatine and the available aldehyde (Figure 14).

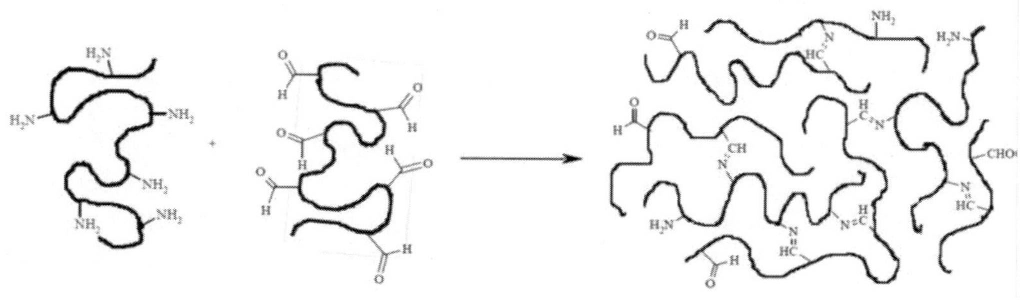

Figure 14.

It was found that the rate of gelation and the physicochemical properties of GL-DO depends on the molecular weight and on the degree of oxidation of dextran, the type of gelatin and on the reaction conditions. Rheological measurements demonstrate that the gel strength is governed by two factors: chemical crosslinking by reaction with polyaldehyde and physical structuring of the gelatin [46]. In order to evaluate the in vitro release kinetics of bioactive factors from GL-DO hydrogel films to formulate a bioactive dressing capable of delivering growth factors in the wound site in order to assist and improve wound healing [47]. Fluorescein isothiocyanate-dextran or polypeptides were evaluated in terms of their release characteristics and mechanical properties. Cross-linked gelatin hydrogels were found to be appropriate release systems for medium-term sustained delivery of biologically active epidermal growth factor, but release characteristics were strongly dependent on the nature of the protein which was incorporated. The biosafety of GL-DO as wound dressing material was evaluated *in vitro* in cultures of dermal fibroblasts, epidermal keratinocytes, and endothelial cells, three cell types which play a major role in the process of cutaneous wound healing, and *in vivo* by subcutaneous implantation studies in mice [48]. These findings suggest that dextran dialdehyde cross-linked GL hydrogel has acceptable biocompatibility which justifies further testing in fullthickness wounds in animal models.

Gelatin gels were successfully obtained in the presence of CAD (Figure 15) and residual amino group content, swelling ratio and enzymatic degradability of the hydrogel were evaluated [49].

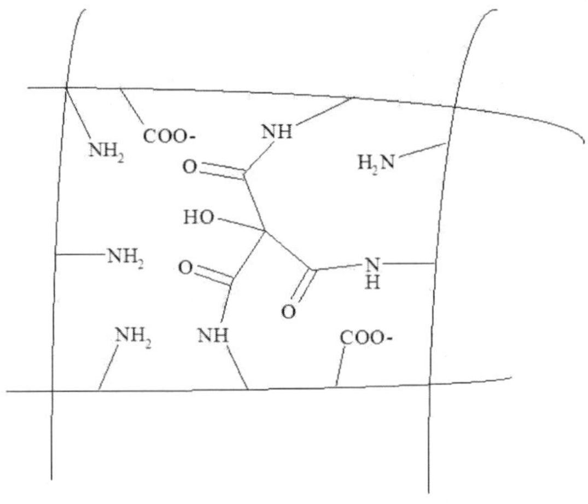


Figure 15.

The biomaterial based on GL and CAD was proposed as tissue adhesive by adding divalent anions due to electrostatic attraction and salting out to help shorten gelation time of the adhesive [50]. Furthermore, gel strength was increased with the addition of DMSO.

Hydrogels consisting of GL and poly(L-glutamic acid) (PLGA) were proposed as a surgical glue that can possibly replace fibrin glues prepared from human blood components. This polymer pair promptly forms a gel and can firmly bond to soft tissues when a water-soluble carbodiimide (WSC) is added to a mixed solution triggering the crosslinking [51, 52]. However, the toxicity of WSC suggested to design a new GL-PLGA glue without using WSC for gelation of this polymer combination. N-hydroxysuccinimide-esters was employed to synthesize a NHS derivative of PLGA which induces prompt gelation when mixed with an aqueous solution of gelatin (Figure 16) [53].

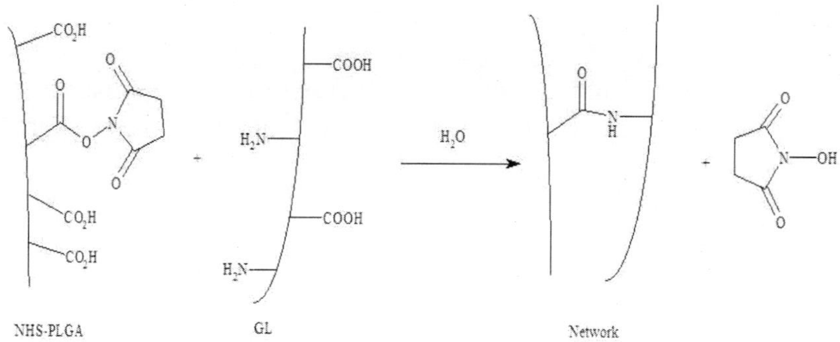


Figure 16.

The main factor limiting the use of common chemical crosslinking is their cytotoxicity that may initiate calcification of implants, there remains the need for crosslinkers from natural sources with improved biocompatibility. Genipin (GP) is one such crosslinker, obtained from the fruits of *Gardenia jasminoides*, which is ~10,000 times less toxic and degrades more lowly when compared with glutaraldehyde [54, 55]. Apart from improved stability, crosslinking also alters swelling or degradation-induced diffusion and thus the release of incorporated bioactive agents from hydrogels [56]. GL-based hydrogels crosslinked with GP were designed and the impact of crosslinking temperature and crosslinker amount on gel strength, microstructure, cytocompatibility, swelling and drug release were studied [57, 58]. The mechanism of the reaction of amino acids or proteins with GP is still not well understood at present. The mechanism proposed by Touyama's group [59] for the formation of the GP-methylamine monomer suggest that GP may form intramolecular and intermolecular cross-links in biological tissue [60] (Figure 17).

Biocompatibility, *in vitro* cell viability, cell cycle experiments indicated normal cell proliferation without any effect on cell cycle demonstrating the appropriateness of GP as a crosslinker and GP-crosslinked hydrogels as drug delivery systems for transdermal, and other, applications.

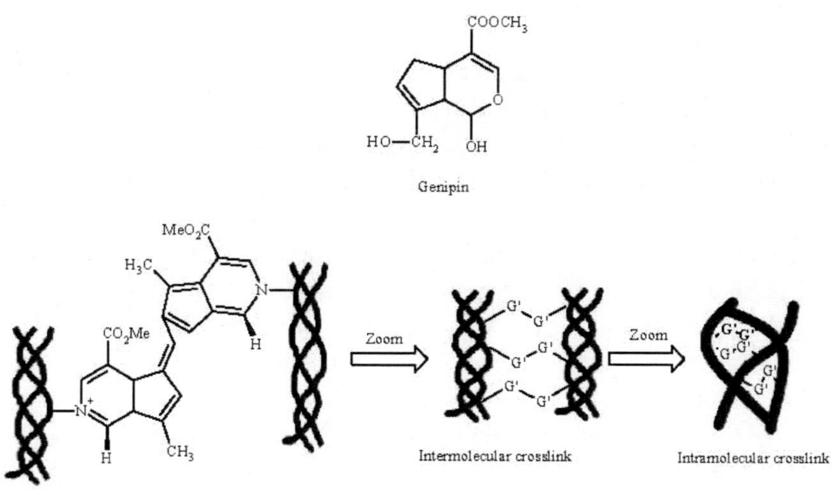


Figure 17.

## 2.3. Casein

Casein (CA), the major milk protein, is inexpensive, readily available, non-toxic and highly stable. CA comprises about 94% protein and 6% low molecular weight compounds collectively called colloidal calcium phosphate. Mainly four CA phosphoproteins, αS1-, αS2-, β-, and κ-CA, exist approximately in proportions of 4:1:4:1 by weight respectively in cow milk [61,62]. All of the four CAs are amphiphilic and have ill-defined structures. Many of the structural and physicochemical properties of CAs facilitate their functionality in drug delivery systems [63]. CA possesses a number of interesting properties such as high hydrophilicity, good biocompatibility, particularly in oral delivery applications, lack of toxicity and availability of reactive sites for chemical modification that make it a good candidate for conventional and novel drug delivery systems. However, the limitations of CA may include its possible immunogenicity/allergenicity. In the following will be reported on the systems that make use of CA as a drug carrier with a special emphasis on CA based micro and nanoparticles as promising drug delivery vehicles [64].

Polymeric hydrogels composed of polyacrylamide and CA have been synthesized by dissolving the protein in NaOH solution and then AAm and MEBA were added (Figure 18) [65].

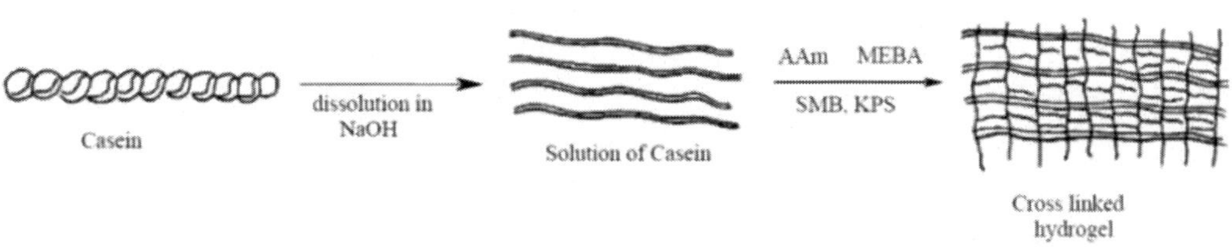


Figure 18.

Due to presence of protein in the polymeric network, the swelling behaviour of CA cross-linked polyacrylamide hydrogels has been found to be pH-dependent and it favours the medium with neutral pH 7.0. This finding makes the device suitable as carrier of drugs.

CA-based hydrogel for the controlled release of BSA has also been prepared using GP to crosslink CA protein in an aqueous system [66]. The mechanical strength of the crosslinked CA hydrogel could be tuned by the amount of GP. The recorded release behavior could be related to crosslinking and swelling degrees of the hydrogel networks formed by various amounts of GP. It is suggested that the GP-crosslinked CA hydrogels might be a suitable polymeric carrier for protein drug delivery in the intestine. Lately the enzymatic crosslinking of milk proteins, using transglutaminase (TGase), a natural enzyme, has attracted considerable attention [67]. The enzyme-induced gelation take place under mild conditions, thus offering a more biocompatible approach for the in situ generation of the three-dimensional polymeric network required for biomedical applications and for entrapment of heat sensitive drugs without using chemical crosslinkers. TGase catalyzes covalent inter-molecular cross-linking through an acyl-transfer reaction, between the γ-carboxyamide group of a peptide-bound glutamine residue (acyl donor) and the primary amino group of an amine (acyl acceptor) (Figure 19).

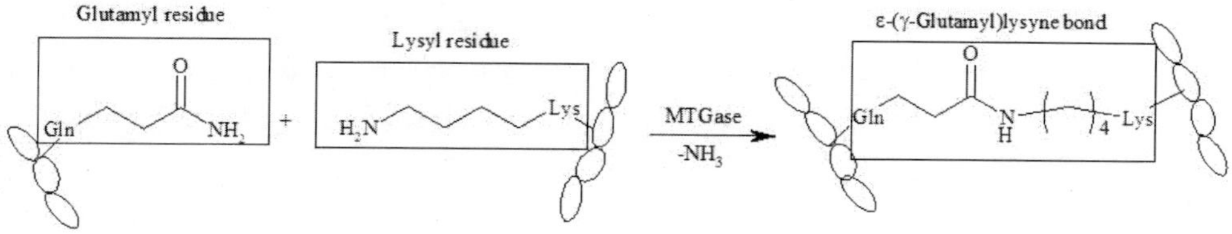


Figure 19.

Moreover, this crosslinking method could offer the potential for kinetic modulation of the hydrogel formation and strength via simple control of the enzyme concentration.

CA has been used to prepare biodegradable microspheres by an emulsification technique [68]. Fluorescein sodium-loaded CA–gelatin floating beads have been prepared by an emulsification extraction method and cross-linked with D, L-glyceraldehyde [69]. CA emulsifying properties caused air bubble incorporation and formation of large holes in the beads. The high porosity of the matrices influences the bead properties such as drug loading, drug release and floatation. CA microspheres may be advantageous to be used as an alternative to albumin as a matrix for microsphere drug carriers. CA microspheres are relatively inexpensive; they have better amphiphilicity and good dispersibility in aqueous systems, and they form uniform spherical structures [70]. Crosslinked CA microparticles were prepared through interfacial crosslinking of CA with glutaraldehyde or terephthaloyl chloride following w/o emulsion formation. Crosslinked CA microspheres could be promising parenteral biodegradable carriers for sustained delivery of cytotoxic drugs when administered via intramuscular, intraperitoneal, direct intratumoral or intraarterially for embolization in solid tumor deposits [71-78].

In recent years, CA-based nanovehicles were utilized for drug and nutraceutical delivery applications varying from nanosized micelles to enzymatically crosslinked nano-gel particles, or nanoparticles prepared by graft copolymerization, heat gelation, or polyelectrolyte ionic complexation. Zhu and Li [79], prepared poly(methyl methacrylate) (PMMA)/CA core-shell nanoparticles via a graft copolymerization of methyl methacrylate (MMA) with CA micelles induced by a small amount of copper ions (Figure 20).

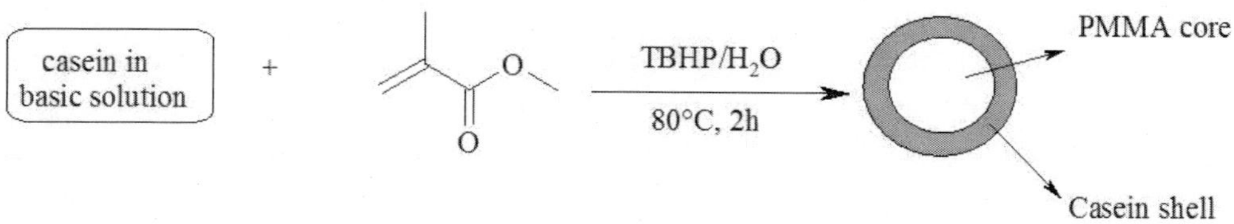


Figure 20.

The presence of CA micelles promoted the emulsion polymerization of the monomer and provided particle stability. Nanoparticles, ranging from 80 to 130 nm in diameter, showed a well-defined core-shell morphology, where PMMA cores were coated with CA shells. Thus, this process is considered commercially viable and could be used as a drug delivery vehicle. In another study, Pan et al. [80], grafted dextran to CA to increase its hydrophilicity through the Maillard reaction, which is a nontoxic reaction happening naturally during the processing, cooking, and storage of foods [81]. Nanoparticles encapsulating extremely insoluble β-carotene were fabricated via self-assembly of CA-graft-dextran copolymer. The encapsulated β-carotene can be released by pepsin or trypsin hydrolysis. This is a green process of simultaneous nanoparticle formation and encapsulation driven by hydrophobic interaction between CA and β-carotene. The amphiphilic character of the particles provides a possibility for practical applications of the particles to deliver unstable and hydrophobic nutrients and drugs. Finally, nanogel CA particles were prepared by cross-linking the CA micelles with the enzyme transglutaminase [82]. These particles consist of a covalently linked CA network from which micellar calcium phosphate can be removed without compromising structural integrity.

## 2.4. Fibrin

Fibrin (FB) is a natural polymer, which is formed by the enzymatic polymerization of fibrinogen (FBN). Soluble fibrinogen is converted to fibrin monomers by thrombin-catalyzed removal of fibrinopeptides A and B on the amino-termini of the Aα- and Bβ-chains [83].

Fibrin is known to stimulate regenerative and remodeling responses in various cell types and it plays an important role in angiogenesis [84]. This is advantageous for tissue engineering, since one of the major problems in engineered tissues is the limited blood supply. During wound healing the fibrin matrix naturally provides a suitable environment for angiogenesis and thereby recovery of the blood supply [85]. In addition, fibrin is also used in stem cell research. Due to its cell adhesive properties, biocompatibility and biodegradation fibrin can be used in the isolation and the delivery of stem cells.

Usually, FB-based hydrogels were designed by insertion in the crosslinked structure of the precursor FBN and its subsequently thrombin activation. Combinations of biodegradable polyethylene glycol and fibrinogen can be used for smooth muscle, cartilage, cardiac, endothelial [86] and embryonic stem cell culturing [87]. As a provisional scaffold, PEGylated fibrinogen functions as a bio-mimetic fibrin clot regulating the cell invasion, and maintains its stability *in vivo* for several months. PEGylated FB was used for mesenchymal stem cell delivery onto injured myocardium and facilitated the regeneration of infarcted myocardium [88]. PEGylated FB patch for mesenchymal transplantation was synthesized by modifying FBN with the benzotriazole carbonate derivative of PEG to create secondary crosslinking. Dikovsky et al. reports on the synthesis of a hybrid scaffold comprised of synthetic polyethylene glycol and endogenous FBN precursor molecules [89]. The PEGylated fibrinogen, obtained by reaction of FBN with $PEGDA_1$, is cross-linked using a photoinitator in the presence of cells to form a dense cellularized hydrogel network. The FB-like scaffold material maintains its bio-functionality through the fibrinogen backbone, while changes in the molecular architecture of the synthetic precursor are used to alter the nanostructrual properties of the scaffold, including mesh size and permeability. Experiments using smooth muscle cells cultured inside the PEG-fibrinogen scaffold demonstrates a qualitative relationship between the molecular architecture of the matrix and the cellular morphology. PEGylated fibrinogen hydrogel can maintain cell adhesion and support proteolytic degradability based on the specific characteristics of the protein backbone. The biological backbone of these materials also ensures adequate bioactivity and sufficient biocompatibility [15]. The synthetic PEG constituent provides the possibility of controlling the enzymatic biodegradation and structural properties of the materials without introducing additional cross-linking steps. This combination of synthetic polymers and reconstituted proteins for scaffold design may help broaden the usefulness of natural protein-based scaffolds in tissue engineering and regeneration.

It is believed that using fibrin as a cell-delivery vehicle is limited by fibrin's low mechanical properties and low hydration capacity [90]. For this reason, efforts were made to add biomacromolecules to the fibrin gel to improve its biocompatibility [91,92]. Yang et al. report the synthesis and potential applications of hybrid molecules composed of fibrinogen coupled to the reducing ends of short-chain hyaluronic acids (sHAs) by reductive amination (Figure 21) [93].

The sHA–fibrinogen gel was examined for its ability to encapsulate and support the differentiation of chondrocyte-like cells. Experimental data suggest that the three-dimensional gel is a better support than the fibrin gel for chondrogenesis induction.

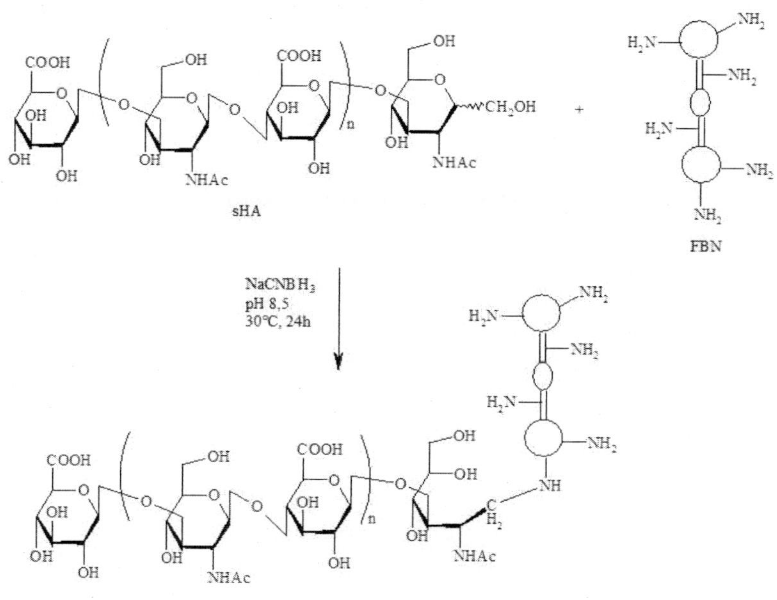


Figure 21.

## 2.5. Elastin

Elastin is an insoluble, polymeric, extracellular matrix protein [94] that provides the connective tissue in the body (skin, lungs, ligaments, arteries, and specialized cartilages) with the unique physical characteristics properties of strength, extensibility and elastic recoil [95]. Elastin is a 68 kDa protein comprised of approximately 800 amino acid residues [96], containing two main domains: highly hydrophobic domains and crosslinking domains, which give the tissue its mechanical properties, and at the same time, allow it to form a crosslinked network with neighboring molecules [97].

In recent years, elastin-based biomaterials are increasingly investigated due to their remarkable properties such as elasticity, self-assembly, long-term stability, and biological activity [98].

Moreover, elastin is highly insoluble as a consequence of extensive lysine mediated cross-linking and therefore difficult to process into new biomaterials [99]. Consequently, soluble forms of elastin including α-elastin (EL) [100,101], recombinant tropoelastin (rTE) [102,103], and engineered recombinant elastin-like polypeptides (ELPs) [104] are frequently used to form cross-linked hydrogels.

Specifically, tropoelastin is the soluble precursor of elastin, while EL is an oxalic acid-solubilised derivation of elastin [105]. ELPs, a class of polypeptides, inspired by the amino acid sequence of natural elastin, which are composed of oligomeric repeats of the pentapeptide sequence Val-Pro-Gly-X-Gly (VPGXG) (X is any amino acid except Pro) [106], represent the major class of newly synthesized polymers for tissue engineering using human elastin sequences for their building blocks. These repetitive polypeptides are synthesized with the goal of precise control of the physical and functional properties of the polypeptide at the genetic or chemical synthesis level [96].

Various crosslinking methods including chemical [107-115], enzymatic [116], physical [117-119], and γ-irradiation [120,121] have been used to fabricate elastin-based hydrogels. The reactions may be carried out in an aqueous or organic phase depending on the type of crosslinker and desired properties of hydrogel [122]. Regarding the chemical cross-linking, various cross-linkers such as GA [107,108], disuccinimidyl glutarate [108], bis(sulfosuccinimidyl) suberate [103,109-111], copper sulfate and pyrroloquinoline quinone [108], ethylene glycol diglycidyl ether [100], hexamethylene diisocyanate [113], trissuccinimidyl aminotriacetate [106], disuccinimidyl suberate [110,111], and b-[tris(hydroxymethyl) phosphino] propionic acid [114] have been used to crosslink genetically engineered ELPs, rTE and EL. Deficiencies associated with these crosslinking approaches include the persistence of small pore sizes and inadequately communicating channels [101].

### *2.5.1. α-Elastin*

As reported, α-elastin is an oxalic acid-solubilised derivative of elastin and it is often used to produce elastin-based hydrogels. In the production of EL, bovine ligament elastin is treated with a mild acid hydrolysis to yield a high molecular weight digest that retains the amino acid composition of native elastin. Despite structural heterogeneities resulting from hydrolysis, EL retains several key physicochemical properties of nascent elastin. Similar to the formation of elastin from its precursor rTE, EL hydrogel fabrication involves two steps: coacervation and crosslinking. EL molecules coacervate in an aqueous solution by intermolecular hydrophobic associations and the protein molecules are then chemically crosslinked to increase mechanical integrity and form hydrogels for biomedical and tissue engineering applications [123].

In a work of 2005 [109] to fabricate elastin-based biomaterials easy to synthesize and which maintain the biological and material properties of native elastin, a system composed of two commercially available and water-soluble components like EL and ethylene glycol diglycidyl ether (EGDE), a diepoxy crosslinker. A pH-dependent cross-linking scheme to yield insoluble EL biomaterials was provided by the authors who hypothesized that reaction pH could be used to modulate the physical properties of the materials that are controlled by the degree of crosslinking. Specifically, they stated that the crosslinking could be controlled by forming three types of crosslinks (i.e. ether, amide, ester) between the elastin fragments in a pH-dependent manner (Figure 22). Under alkaline pH, EGDE may predominately react with the hydroxyl groups of EL's serine, threonine and tyrosine residues to yield stable ether bonds. At neutral pH, EGDE reacts with the amines found on lysine as well as the fragment N-termini to form stable amide bonds. Finally, under acidic conditions, EGDE reacts predominately with the carboxylic acids of aspartic acid, glutamic acid and also the fragment C-termini to form hydrolytically degradable ester bonds [124]. The obtained crosslinked EL hydrogels have been suggested as a platform for complex composite materials with both mechanical and bioactive properties that closely mimic native vascular tissue.

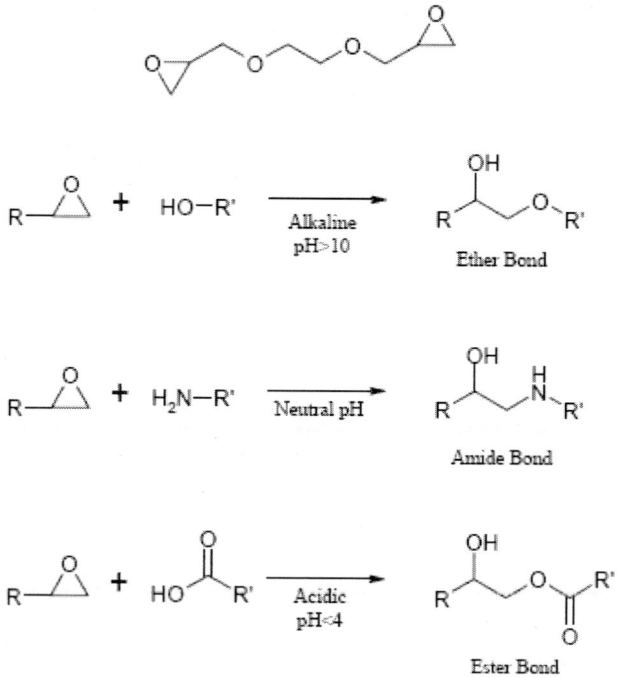


Figure 22.

The fabricated hydrogels had strength, elasticity, and biocompatibility properties similar to those of naturally occurring human elastic tissue. In vitro and in vivo experiments revealed that these hydrogels could support cellular growth; however, some deficiencies have been remarked. In particular, the existence of a discrete discontinuous pore network (non-homogenous and limited porosity) prohibited adequate cellular migration deep into the hydrogels [95].

In recent years, dense gases (DGs) have been become particularly useful for biomaterial processing as an alternative to organic solvent to fabricate porosity in polymeric matrices [125], and remarkable progress has been made in this application field over the last decade [126,127]. DGs are fluids with their pressure and temperature close to the critical point. Beyond the critical point, the substance is called a supercritical fluid and has solvation strength and diffusivity approaching that of liquids and gases, respectively. The most commonly used DG is carbon dioxide because it is inexpensive and nontoxic and has low critical parameters ($T_c$ = 31.1 °C, $P_c$ = 73.8 bar) [128]. Gas foaming technology [129,130], crosslinking reactions at high pressure $CO_2$ [131,132], and supercritical $CO_2$-water emulsion techniques [133,134] have been used to fabricate porous hydrogels or scaffolds from different kinds of polymers. However, limited studies have been carried out on the preparation of porous hydrophilic biomaterials using biopolymers in combination with supercritical $CO_2$ without using organic solvents or any other additives such as surfactant [135]. Annabi et al. [101] demonstrated that EL hydrogels produced by cross-linking with GA fabricated at high pressure $CO_2$ are more useful than those fabricated at atmospheric pressure. The main differences between the two hydrogels are: EL hydrogels produced at high $CO_2$ pressure were easily handled, highly porous and robust while those synthesized at atmospheric pressure are very fragile. Comparison at the microscopic level gave information about the porous structure. High pressure $CO_2$ reduced the pore size of the fabricated hydrogels. The organization of the matrix produced under high pressure is similar to that seen in some natural elastin microstructures within the body [136]. The 10-fold reduction in pore wall thickness of EL hydrogels fabricated at high pressure $CO_2$ compared with the samples produced at atmospheric pressure can be explained by reference to the mechanism of pore formation in porous membrane [137]. Furthermore, a highly homogenous interconnected-porous structure was detected for EL hydrogels acquired using high pressure $CO_2$. In comparison, EL hydrogels produced at atmospheric pressure showed an internal non-uniform porous structure covered by an impermeable layer. For these reasons, cells were able to attach and proliferate on the surface of EL hydrogels fabricated at atmospheric pressure but could not penetrate into the structure due to the skin formation on the top surface, lack of channels and undesirable pore size. However, encouragingly cells were able to colonise both the top surface and the internal surfaces of channels within EL hydrogels produced under high pressure $CO_2$. The same authors prepared rTE/EL hybrid hydrogels to overcome the limitations due to the low number of lysine residues (less than 1%) in EL resulted in limited crosslinking with GA and consequently poor mechanical integrity [138]. The addition of rTE containing 35 lysine residues per molecule to the protein solution was expected to increase the cross-linking density and promote the mechanical properties of elastin-based hydrogels.

### *2.5.2. Elastin-Like Polypeptides*

Elastin-like polypeptides (ELPs) have been explored as hydrogels due to their capacity for self-assembly and phase transition behavior; they also mimic many features of the extracellular matrix and have the potential to guide the migration, growth and organization of cells during regeneration processes [98,139]. ELPs can be genetically encoded, their synthesis from a synthetic gene in a heterologous host (e.g., bacteria or eukaryotic cell) provides complete control over the amino acid sequence and molecular weight, two variables that are not easy to precisely control in synthetic polymers.

ELPs are readily expressed from a plasmid gene in Escherichia coli to relatively high yields (~500 mg/L growth), can be readily purified in a batch process by exploiting their inverse temperature phase transition without the need for chromatography [140], can be engineered to approach the viscoelastic properties of native elastin upon crosslinking, and, finally, they are biocompatible, biodegradable, and non-immunogenic [141].

These repetitive biopolymers exhibit conformational and aggregation sensitivity to environmental conditions such as temperature, ionic strenth, pH, and redox state. Below a critical temperature, monomeric ELPs are soluble in aqueous solution. When the temperature exceeds this critical limit, ELPs undergo a sharp (2-3°C range) phase transition leading to contraction, desolvation, and aggregation of the polypeptide. The critical temperature at which this transition occurs is altered by amino acid sequence, molecular weight, ELPs concentration, ionic strength, pH, and redox state, and the transition is completely reversible upon cooling [142].

The temperature sensitivity of ELPs may be exploited in tissue engineering for those applications that may benefit from biomaterial formulations that are injectable and may be triggered in some way to form a solid matrix after defect filling. The inherent thermal transition properties of ELPs provide a natural trigger for coacervation, and reversibility of the phase transition enables recovery of ELPs from applications that desire a scaffold-free outcome. ELPs may also be designed to be mixed with a biocompatible crosslinker that is also triggered by temperature or other external stimulus in order to form a mechanically robust scaffold after crosslinking. These features allow ELPs solutions to be mixed with cells or other bioactive factors prior to gelation (thermal or crosslinking) and to be injected into a defect site to form a mechanically functional scaffold upon implantation [142]. Furthermore, the phase transition behavior of ELPs is also maintained in their cross-linked state, which provides a secondary variable to tune their mechanical behaviors by modulating the degree of solvation of the cross-linked ELP hydrogel [106].

The two predominant crosslinks of native elastin are desmosine and isodesmosine, each involving four lysine residues. When brought into sufficient proximity and proper alignment, these lysine residues permit covalent crosslink formation [143]. The synthesis of chemically cross-linked elastin-mimetic hydrogels is based on selective reaction between the ε-amino groups of the regularly spaced lysine residues in ELPs and a bifunctional electrophilic cross-linker. These reactions can be conducted under mild and controllable conditions with negligible degradation of the protein polymer. Elastin mimetic networks should contain cross-links at well-defined positions in the limit of complete substitution of the amino groups, although the formation of intramolecular loops and monosubstituted side chains can occur as side reactions during this process [111].

Conticello and co-workers showed that hydrogels could be fabricated from ELPs with engineered chemoselectivity for cross-link formation at precisely spaced periodicities along the polypeptide backbone [144]. The gene sequence for a single ELPs was designed with lysine residues spaced every six pentapeptides, and the protein was microbially expressed. Purified ELPs was chemically cross-linked to produce ELPs hydrogels that, similar to the behavior of γ-radiated gels, also exhibited reversible temperature dependent expansion and contraction. Additionally, they demonstrated that cross-linking in organic solvent, where ELPs molecules exhibit no inverse phase transition, leads to the formation of hydrogels that are more uniform in structure.

Different cross-linking methods of ELPs [106,109,111] have been investigated; chemical methods include cross-linking by radiation [120-122], photoinitiation [145], chemical crosslinking [106,108-111,113] and enzymatic cross-linking by tissue transglutaminase [116], while the formation of physically cross-linked networks has also been demonstrated for ELPs block copolymers [118,119,146,147].

The choice of cross-linker has a critical impact on the properties of the polymer gel because it becomes incorporated into the network. Selective cross-linking of lysine residues in biomolecules has been accomplished using a variety of reagents including GA, bis(imidoesters), and NHS esters of bifunctional carboxylic acids.

GA has a long history as a cross-linking agent for biomaterials, and has been shown to react primarily with the ε-amino groups of lysine to give mixtures of products that are best described as complex [148,149]: neither GA reaction chemistry nor the chemical identity of the resulting cross-links is, indeed, well-defined.

In a study of Tirrell's group [107], GA was used both in solution and in the vapor phase to form hydrogel film. Before exposing protein films to GA solutions in PBS (pH 7.4), it was necessary to anneal the transparent films, which were cast below the LCST from aqueous solutions, in a high humidity environment (98%) for several hours, after which they became hydrated and opaque. Otherwise, it was observed that the optically clear films readily dissolved in the GA solutions, even above the LCST. However, the results of endothelial cell culture on GA crosslinked films were variable, perhaps because of non-specific interactions between the proteins and endothelial cells or because of the cytotoxicity of GA [150].

Tirrell et al. employed hexamethylene diisocyanate as an alternative lysine-targeted crosslinker, because its cytotoxicity is considerably lower than that of GA [151]. Films prepared from proteins with hexamethylene diisocyanate crosslinking were transparent, had low water contents, were viscoelastic and highly extensible, and retained the temperature-dependent swelling behavior of the component proteins [113].

Bis-(imidoesters) are also widely employed for chemical fixation of biological specimens; however, these reagents form positively charged amidinium ions upon reaction with amino groups. The responsive properties of these polyelectrolyte gels can be dominated by electrostatic repulsion between the charged amidinium groups rather than by hydrophobic assembly of the elastin segments. Since most bioelastomeric materials, including elastin, do not have an appreciable charge on the polymer network, this cross-linking chemistry was avoided for the preparation of the hydrogels [152].

NHS esters of bifunctional carboxylic acids were chosen for the preparation of synthetic polypeptide networks on the basis of the following criteria: (a) well-defined reaction chemistry, (b) production of stable, uncharged amide cross-links, and (c) commercial availability of reagents with different spacer compositions and lengths between reactive end groups [153]. The cross-linking reaction involves nucleophilic displacement of the NHS leaving group from the cross-linker by the εamino group of a lysine residue on the polypeptide to form an amide bond [111] (Figure 23).

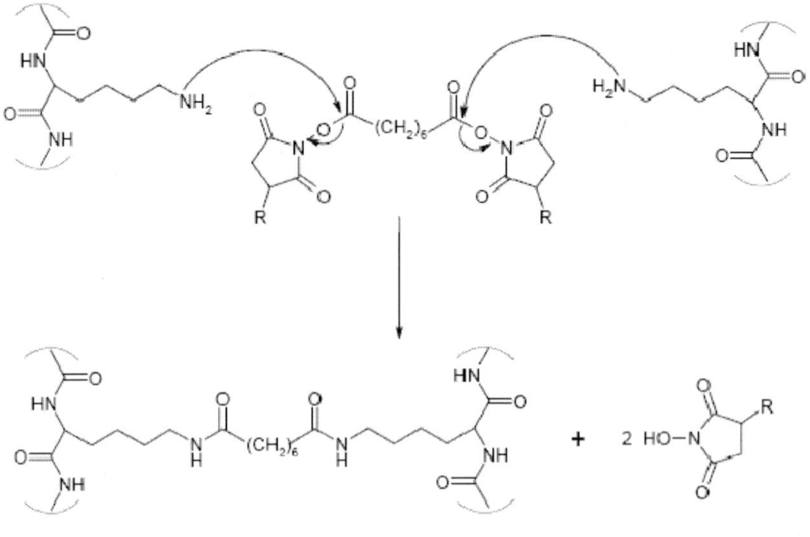


Figure 23.

This reaction is rapid and irreversible at room temperature in neutral to slightly basic aqueous solution (pH 7-9) or in polar organic solvents such as dimethyl sulfoxide. When the bifunctional molecule reacts with lysine residues from two different polypeptide chains, an intermolecular cross-link is formed between the chains. As the reaction proceeds, the effective degree of intermolecular cross-linking between polypeptide chains approaches the gel point of the mixture. This cross-linking process more closely emulates the biological assembly of elastin in that charged amino groups of lysine residues within the soluble precursor are converted to neutral cross-links during formation of the network [154].

Different succimidyl derivative such as Bis(sulfosuccinimidyl) suberate and disuccinimidyl suberate are proposed for the same purpose [108], and the reaction scheme for protein cross-linking is shown in Figure 24.

The advantage of the proposed protocol are related to the specificity of the reaction of NHS-esters with amines, and thus the cross-linking should be directed to the lysine residues and should leave the cell binding domains unperturbed.

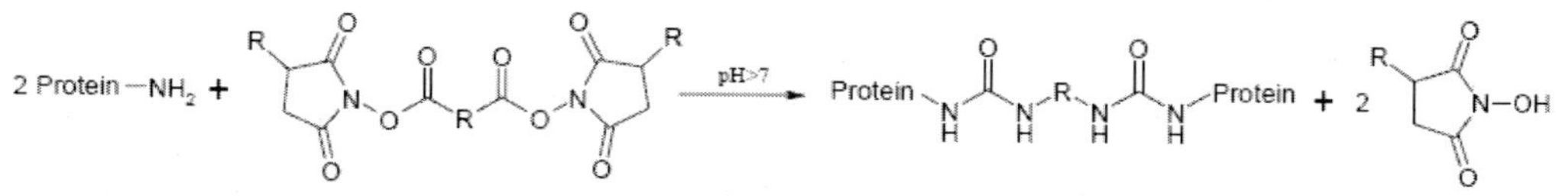


For disuccinimidyl suberate, R=H; for Bis(sulfosuccinimidyl) suberate, R= $SO_3^-Na^+$.

Figure 24.

In a different work, tris-succinimidyl aminotriacetate was chosen as the cross-linker for gelation experiments on ELPs [106]. Authors found that hydrogels are formed at lower protein concentrations than a chemically similar bifunctional cross-linking agent. This is not unexpected because a gel network can be formed if only two of the three available reactive groups are involved in intermolecular cross-links. The third reactive group can further facilitate the formation of a more densely cross-linked network.

The zero-length crosslinker pyrroloquinoline quinone (PQQ) in the presence of divalent copper, was shown to crosslink elastin polypeptides, yielding a biomaterial with an elastic modulus similar to that of native elastin [155]. Although aldehyde fixatives have been commonly used in biomaterial constructs due to their efficient formation of crosslinks between amino acid side chains, low antigenicity and sufficient mechanical strength, they exhibit high cytotoxicity [156,157].

Genipin (GP) a naturally occurring covalent crosslinking agent, offers comparable crosslinking efficacy to aldehydes, but with greatly reduced cytotoxicity [158,159]. GP fixation with amino group-containing molecules is thought to begin with the formation of the GP-amino complex resulting from a GP mediated nucleophilic attack [160]. Subsequently, an aldehyde group is formed upon opening of the GP ring. Crosslinking is completed when this complex attacks a second amino group. An inherent phenomenon of GP fixation not present in zero length covalent fixation (such as with PQQ) is the self-polymerization, which occurs by radical reaction of two amino-attached open GP rings [161]. PQQ acts by oxidative deamination of the ε-amino groups of lysine residues. Cross-linking depends on the correct short-range configuration of these residues, which limits PQQ's capacity to cross-link and alters the tensile properties of the ELPs sheets. GP fixation can be considered more ''reactive,'' assuming sufficient amino-bearing substrate is present, given that it can introduce both short-range and long-range crosslinks. GP self-polymerization has been shown to yield an increase in molecular weight from 226 (monomer) to 3900–7900.

Despite the variety of cross-linking approaches that have been proposed in the literature, many of these methods cannot be used to create injectable ELPs scaffolds in which the liquid precursors can be readily injected into a defect site, followed by in situ formation of a conformal hydrogel, because of the cytotoxicity of the reactants or byproducts, the need for organic solvents, or the suboptimal kinetics of the cross-linking reactions [116]. Motivated by these considerations, the objective of the study of Chilkoti and co-workers [104,114] was to develop a cross-linking strategy that permits a mixture of soluble ELPs and cells to be injected and cross-linked in vivo with optimal kinetics (<5 min gelation time), with minimal cytotoxicity to provide a hydrogel whose mechanical properties match those of cartilaginous tissues. They introduced the Mannich-type condensation of (hydroxymethyl)phosphines with primary and secondary amines of amino acids as a new cross-linking method for polypeptide-based biomaterials. Chemically cross-linked ELP hydrogels were formed rapidly in an aqueous solution by reaction of ELPs containing periodic lysine residues with β-[tris(hydroxymethyl) phosphino]propionic acid (THPP) (Figure 25).

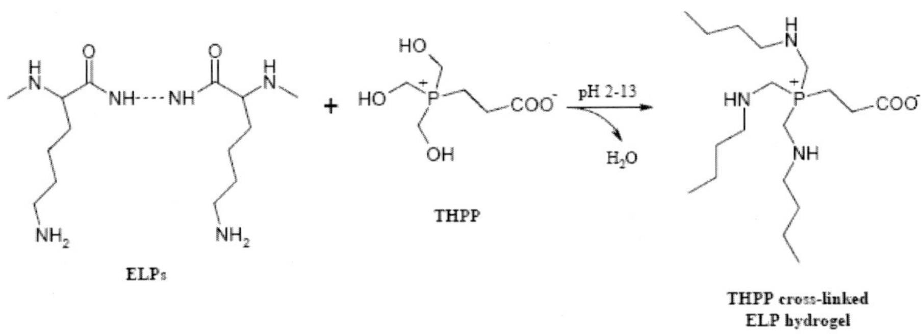


Figure 25.

This reaction is of interest for in situ hydrogel formation for tissue engineering applications because: (a) it can be carried out under physiological conditions without an additional reduction procedure [114,162]; (b) the chemical linkages, aminomethylphosphines, are stable in a physiological milieu and the cross-linking reaction releases water as the sole byproduct [163]; (c) gelation proceeds rapidly with initial stabilization of the hydrogel in several minutes and continued evolution of the hydrogel properties over an hour [164]; and (d) the cross-linking agent can be mixed with cells without compromising their viability [165]. In addition, the formation of cross-linking sites presents phosphine centers to potentially coordinate transition metals as well as reactive carboxylic acids for the installation of bioactive ligands into the ELPs hydrogels [166].

Urry and co-workers have formed ELPs hydrogels through nonspecific radical cross-linking of ELPs in the coacervate state using γ-irradiation and have demonstrated preservation of the inverse temperature phase transition [167].

In another report [145], Chaikif's group describe the incorporation of functional groups into the protein polymer backbone that facilitate site-specific solid-state photo-cross-linking using either UV or visible light actived photoinitiators. ELPs were modified to incorporate an acrylate moiety with different functionalization degree. Acrylate modified elastomeric proteins were associated with lower inverse transition temperatures than the unmodified recombinant protein. This approach provides for both spatial and temporal control over the cross-linking reaction.

In another study, McHale group designed and synthesized ELPs capable of undergoing enzyme- initiated gelation via tissue transglutaminase, with the ultimate goal of creating an injectable, *in situ* crosslinking scaffold to promote functional cartilage repair. Addition of the enzyme promoted ELPs gel formation and chondrocyte encapsulation in a biocompatible process, which resulted in cartilage matrix synthesis *in vitro* and the potential to contribute to cartilage mechanical function *in vivo* [116].

#### *2.5.3. Tropoelastin*

Tropoelastin is the soluble precursor of elastin [94], secreted from cells as a 60–72 kDa protein in humans. The monomer has two major types of domains, hydrophobic and hydrophilic [168,169]. *In vivo*, rTE molecules are cross-linked in the extracellular space through the action of lysyl oxidase on the exposed lysines of rTE [170].

Due to the lack of availability of active lysyl oxidase, chemical cross-linkers such as bis(sulfosuccinimidyl)suberate are commonly used to model elastin cross-linking *in vitro* [103,171].

The availability of recombinant forms of rTE has allowed for further research into the formation of rTE-based scaffolds. Synthetic elastin scaffolds based on the polymerization of rTE have strength, elasticity, and biocompatibility that approach that of human elastic tissue and are amenable to molding into a variety of shaped tissue substrate [172].

Glycosaminoglycans such as heparin and dermatan sulfate are complex polysaccharides that play important roles in cell growth, differentiation, morphogenesis, and cell migration [173,174]. Glycosaminoglycans have also been shown to promote rTE self-assembly in vitro by binding rTE through lysine side chains and lowering the critical concentration for the coacervation [175].

In another report, elastic biomaterials are generated from rTE in the absence of the classically accepted cross-linking pathway. Under alkaline conditions rTE proceeds through a sol–gel transition leading to the formation of an irreversible hydrogel [102].

### 2.6. Lysozyme

Lysozyme (LZM) is an enzyme of 129 amino acids found in tears, nasal secretions, animal tissues, organs, and serum as well as milk and cervical mucus. Lysozyme attacks the cell wall polysaccharides of different bacterial species, especially Gram-positive infectious organisms such as Staphylococci and Streptococci species and kills them, even if some Gram-negative bacterial species such as E.coli and Salmonella are also destroyed by LZM. This protein is a small globular protein and contains both α-helix and β-sheet in its secondary structure and has high solubility in water [176].

Miller et al. focused on the gelation behavior of the model protein hen egg white lysozyme in the presence of a reductant, dithiothreitol, in aqueous solution.

In these reducing conditions, the protein denatures and self-assembles to form β-sheet-rich fibrils that become entangled to form a three-dimensional hydrogel with an average pore size of 25.6 nm and mechanical properties comparable to natural biopolymers. The formation of the β-sheet -rich fibrils is an irreversible process; however, their aggregation and formation of physical cross-links that produce the gel is thermally reversible. The obtained gel is transparent, viscoelastic, shows a high water content (97% w/w), and exhibited good biocompatibility when cultured with fibroblast cells [177]

The objective of a study of Van Nostrum group [178] was to develop a method to functionalize LZM with methacrylamide functions, for subsequent immobilization and controlled release from a hydrogel network.

LZM has six amine groups in the side chains of the lysine residues and one amine at its N-terminus. The reactivity of these seven groups is different and well documented [179]. Three of them, localized on the surface of the macromolecule, are most reactive and readily available to be functionalized, the other three are buried inside the ternary protein structure (Figure 26).

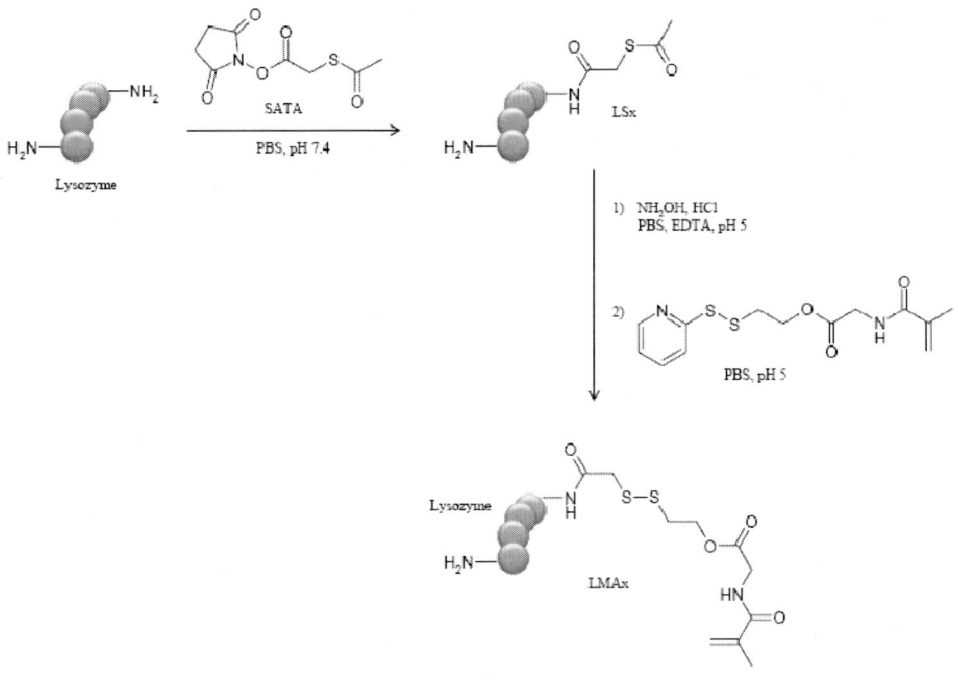


Figure 26.

The modification reaction is well controlled and the number of linkers introduced per protein molecule can be tailored by changing the reaction conditions. LZM can be modified with up to three methacrylamide moieties without major conformational changes. The lytic activity was still 60% after introducing an average of 2.3 methacrylamide units. The modified LZM was successfully immobilized into a hydrogel network and subsequently released by reduction of the degradable linker.

In another work [180], Romanovskaya et al. proposed the development of new polymeric non-adhesive hydrogel coatings with immobilized hydrolytic enzymes. Authors state that a promising carrier for fabricating such materials with proteolytic and bacteriolytic activity is poly-*N*-vinylpyrrolidone, which is widely used in medicine, and characteristically has its own physiological activity, is nontoxic, and can be modified. Authors prepared and studied the physicochemical properties of lysozyme immobilized in poly-*N*-vinylpyrrolidone modified by silicic acid sol. In this way, elastic polymeric hydrogels that were insoluble under physiological conditions and had highly retained respective activities that were prolonged for 24 h were prepared. Authors found that the lytic and proteolytic activities of free and immobilized enzymes as functions of temperature did not undero significant changes. The hydrogels were stable under conditions modeling the pH of wounds and on storage for one year without losing their proteolytic and bacteriolytic activity.

## 2.7. Keratin

Keratins (KRs) are a family of fibrous proteins that can be found in several sources including hair, nails, wool, horns, hooves, feathers and other epithelial coverings.

Based on their structure, function and regulation, KRs can be classified into two distinct classes: soft KRs and hard KRs. Soft KRs are cytoskeletal elements found in epithelial tissues, they form loosely-packed bundles of cytoplasmic intermediate filaments and endow mechanical resilience to epithelial cells [181]; hard KRs form ordered arrays of intermediate filaments embedded in a matrix of cysteine-rich proteins, contribute to the tough structure of epidermal appendages and protective tissues such as nails, hooves and hair.

The structural subunits of both epithelial and hair KRs consist of two chains of differing molecular weight and composition, designated as types I and II, that each contain non-helical end-terminal domains and a highly-conserved, central alpha-helical domain. Type I (acidic) and type II (neutral-basic) keratin chains interact to form heterodimers, which in turn further polymerize to form 10-nm intermediate filaments [182]. The proteic chains are composed of several repeating sequences of amino acids and these sequences define the possibility of intermolecular links and the access of amino acids to chemical reactions. The presence of cysteine (Figure 32), a sulfur-containing amino acid, in keratin chain leads to characteristic inter- and intra-molecular disulfide bonds which determine the properties of this class of proteins [183]. The most distinctive feature of KRs, indeed, is the high concentration of cystine residues. Cystine is a dimeric amino acid formed by two molecules of cysteine that covalently link to make a disulfide bond; most of these residues are localized at the terminal regions of the protein. Disulfide bonds between cysteine residues within polypeptide support the protein's secondary structure [184] and KRs can be considered as three-dimensional polymers interlinked by S-S bonds between reduced keratin-monomeric units [185].

Like other naturally derived biomaterials, KRs are characterized by intrinsic biological activity and remarkable biocompatibility. Extracted KRs are able to self-assemble, forming structures that regulate cellular recognition and participate in some regulatory functions that mediate cellular behavior [186,187]. In addition, keratin biomaterials derived from wool and human hair possess cell binding motifs, such as leucine-aspartic acid-valine (LDV) and glutamic acid-aspartic acid-serine (EDS) binding residues, which are capable of supporting cellular attachment [188,189]. Together, these properties create a favorable three dimensional matrix that allows for cellular infiltration, attachment and proliferation.

KRs, due to these features and their ability to support also the sustained release of therapeutic agents, have attracted both academic and industrial interest. Keratin-based biomaterials, indeed, could find applications in several fields, such as tissue engineering, regenerative medicine, drug delivery, trauma and medical devices.

Many approaches are employed to extract KRs from hair or wool fibers. These methods are based on oxidative and reductive chemistries and have been published since the early 1900s. Keratoses and kerateines represent oxidized and reduced keratin derivatives, respectively, and they have been used to prepare materials for medical applications, such as wound healing, bone regeneration, hemostasis, and peripheral nerve repair [190]. The use of either oxidative or reductive extraction techniques can lead to different mechanical properties due to the absence or presence of disulfide bonds in the material, respectively [182].

The protocols described for KRs extraction have been reviewed over the last century, most notably by Crewther et al. in 1965 [191]. These protocols have been most often applied to wool, but in many cases also to human hair with only slight modifications [192]. Soluble proteins extracts are then processed to obtain keratin-based hydrogels. Due to KRs capability for self-assembly, the formation of hydrogels occurs spontaneously upon addition of water to extracted freeze-dried proteins.

In 2008 Sierpinski et al. [193] reported a chemical process to oxidize the keratin crosslinks within human hair fibers, extract the cortical contents and used this keratin-based material to form hydrogels. Human hair was treated with a solution of peracetic acid. Free proteins were extracted into excess Tris base, followed by deionized water. Then extracts were purified by dialysis, and concentrated and isolated by lyophilization. Due to the intrinsic capacity of KRs for self-assembly, a keratin-based hydrogel was formed by re-hydration of the obtained lyophilized material. The synthesized keratin-based hydrogels can mediate regeneration response in a peripheral nerve injury model in mice because they enhance the *in vitro* activity of Schwann cells by inducing cellular attachment, proliferation and migration, and up-regulate expression of specific genes required for important neuronal functions.

Keratin hydrogels derived from human hair have been also shown their efficacy in arresting hemorrhage and improving survivability in a rabbit model of liver injury similar to that of two other commercial hemostats, QuickClot® and HemCon® bandage [184]. In this study keratin was extracted from human hair according to a long-established protocol with slight modifications [194]. After the extraction procedure, the obtained keratin solution was exposed to air forming a cross-linked hydrogel that was unable to be re-dissolved in water or to be re-reduced to the point of solubility. The keratin hydrogel promoted thrombus formation in the injury site forming a physical seal of the wound site that acted as a porous scaffold to allow for cellular infiltration and granulose tissue formation.

Keratin-based hydrogels for the sustained release of ciprofloxacin were synthesized using keratin proteins obtained by oxidative extraction. Ciprofloxacin is a fluoroquinolone antibiotic indicated for treatment of urinary tract infections, skin infections and lower respiratory tract infections [195,196] but also in some cases for bone, joint and soft tissue infections [197]. Keratin was extracted from human hair fibers by an oxidative method involving the use of peracetic acid [193] and hydrogels were formed by adding phosphate-buffered saline to dry powder of extracted proteins which do not contain disulfide bonds because they are broken and converted to sulfonic acid. Therefore, hydrogels are likely held together through hydrophobic interactions and chain entanglement, but not through covalent disulfide cross-linking. These keratose proteins spontaneously form hydrogels at 15%

weight/volume and the role of hydrophobic interactions in maintaining the assembly of keratose proteins is necessary for gel formation [198,199]. The obtained hydrogel was able to release 60% of the loaded drug over the first 10 days and that continued release was detectable over the course of 3 weeks. The results of this work indicate the utility of keratin-based materials for the sustained release of therapeutic agents, such as ciprofloxacin, reducing acute bacterial infection.

Feather proteins are almost pure keratin (90% or more) and contain more cysteine than most tissue proteins and food proteins [200]. Hydrogels made from feather protein are opaque and do not swell sufficiently to be useful for biocompatible applications. However, feather proteins present several reactive side groups, including amino, hydroxyl, sulfhydryl and carboxyl moieties, which can be used as sites of chemical modification and crosslinking to produce novel polymeric hydrogels with improved functionalities.

Lin et al. [184] described the synthesis of pH- and salt-sensitive hydrogels employing poultry feather protein by crosslinking ethylene diamine tetraacetic dianhydride (EDTAD)-modified feather protein isolate (FPI) with GA. In order to obtain FPI characterized by a different degree of acylate modification, different molar ratios of EDTAD/FPI were used. The first step of the synthetic procedure was the isolation of the feather proteins. Feather fiber was dissolved in $NaHSO_3$ solution to open the double sulfide linkage in the cysteine and break the original three-dimensional network structure in the feather protein. Then it was hydrolyzed with NaOH solution, filtered and treated with HCl to precipitate the dissolved protein to yield FPI. After the isolation, cysteine residues of FPI were chemically modified using EDTAD (Figure 27) to obtain acylated proteins which were cross-linked with GA. The hydrogel formation was based on the hydrogen bond between EDTAD-modified FPI segments.

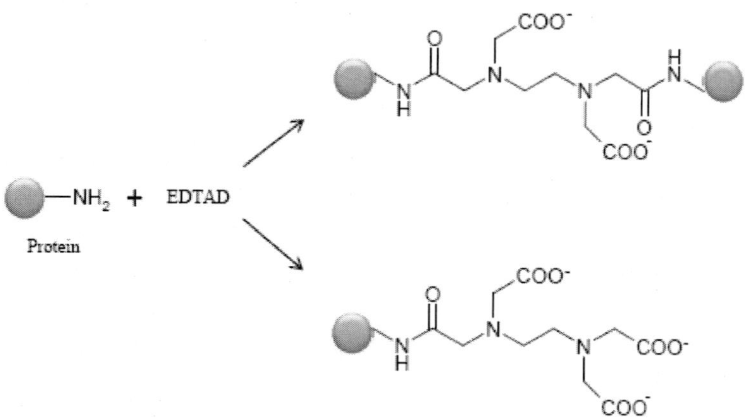


Figure 27.

Keratin-based hydrogels were investigated for their hydrolytic stability *in vitro* [190]. In this study kerateine was extracted from human hair by performing a treatment with thioglycolic acid based on a protocol first published in 1934 [201]. The obtained kerateine solution was allowed to crosslink; crosslinking was affected simply by allowing the oxygen entrapped in the gel to catalyze the re-oxidation of cysteine to cystine. General cell compatibility of the obtained materials was demonstrated by marked cell adhesion and proliferation.

## 2.8. Collagen

Collagen (CL) represents the most abundant protein in animals, where it provides the principal structural and mechanical support [202]. It is a major structural protein, forming molecular strands that strengthen the tendons and resilient sheets that support the skin and internal organs. Bones and teeth are also made of CL and this protein plays an important role in protecting and supporting the softer tissues and connecting them with the skeleton.

CL consists of three different chains which are held in a triple-helical structure by weaker bonds [203,204], such as hydrogen bonds, dipole-dipole bonds, ionic bonds, van der Waals interactions. Each chain contains 1000 amino acids and a sturdy structure is formed by a repeated sequence of three amino acids. Every third amino acid is glycine, a small amino acid that fits perfectly inside the helix. Many of the remaining positions in the chain are filled by two amino acids such as proline and hydroxyproline, a modified version of proline which is responsible for CL stability. Triple-helical molecules are packed together into the staggered patterns called fibrils and CL, within a fibril, is stabilized by the numerous intra- and intermolecular forces and each molecule undergoes a strong molecular connection with neighboring CL molecules. The role of naturally occurring CL cross-links is important in the formation of a stable fibrillar structure.

Collagen is an excellent biocompatible candidate for biomedical applications, presents several advantages as a biomaterial and is widely used in reconstructive medicine, pharmacy and cosmetics [205-207] as carrier systems for delivery of drugs, proteins and genes.

The properties of CL-based materials are influenced by the source of CL and by the method of preparation involving purification, fibril formation or casting and subsequent crosslinking. Purified CL does not exhibit the necessary mechanical strength required for *in vivo* applications [208], thus, chemically crosslinking CL subunits with biocompatible molecules is often employed to strengthen this kind of materials for tissue engineering applications.

Chemical cross-linking involves the formation of either intra- or intermolecular ionic or covalent bonds between the amino acid residues. CL crosslinking has been achieved through condensation of amino acid residues by both dehydrothermal treatment and photochemical crosslinking with ultraviolet light [209,210] or employing polyvalent cations, such as trivalent chromium [211] and aluminium [212], in the aim to obtain ionic cross-links. Other approaches involve reagents that covalently crosslink CL to form a hydrogel matrix. Aldehydes, such as formaldehyde and glutaraldehyde, covalently bridge lysine and hydroxylysine residues with tyrosine, asparagine or glutamine to form a cross-linked network [213]. A resistant CL-based material was obtained by employing HMDI, but this crosslinking results in cytotoxic effects that inhibit cell proliferation [214]. Carbodiimides, such as EDC, represent the most promise chemical crosslinking agents because they facilitate amidation and esterification reactions between carboxylic acid and either amine or alcohol residues without becoming incorporated within the resulting CL matrix [215].

Polyamines, such as second-generation polypropylene imine dendrimers [208] and polyethyleneimine have been used as crosslinkers increasing the crosslink density above the level obtained with CL alone. Polyethyleneimine-functionalized Single Walled Carbon Nanotubes (SWNTs) have been also used as macromolecular crosslinking agents [216].

Hwang et al. [217] described the crosslinking of CL hydrogels with GP. The reaction between GP and CL induces the formation of cyclic structures, which serve as the intramolecular and intermolecular crosslinks between CL molecules within fibers [150]. Two potential molecular mechanisms for GP modification of biopolymers containing primary amine groups are described in literature [160,218] and for CL, they are reported in Figure 28.

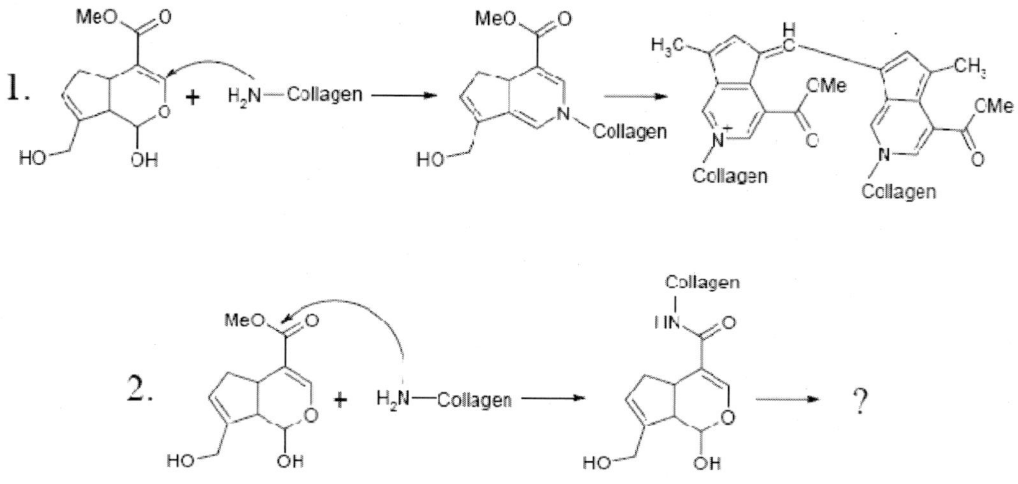


Figure 28.

The first cross-linking pathway begins with a nucleophilic attack of the CL's primary amine of lysine, hydroxylysine and arginine residues on the GP C3 carbon to form a nitrogen iridoid [156]. Subsequently, the tertiary nitrogen of CL replaces an oxygen atom in the GP's six-membered ring. The intramolecular and intermolecular cross-links formed are the linked cyclic structures of nitrogen substituted GP attached to CL [150]. Two iridoids subsequently dimerize through a radical reaction (Figure 36). The second proposed pathway begins with a nucleophilic substitution reaction that replaces the ester group of GP by a secondary amide linkage. The final product in the case of this pathway is unknown [218].

Zhang et el. [219] developed a novel CL-chondroitin sulfate-hyaluronic acid hybrid hydrogel scaffold, through CL self-assembling, that partly mimicked the composition and structure of cartilage extracellular matrix. In order to improve physical and chemical performances, hydrogels were cross-linked with different concentrations of GP using gel-crosslinking method [220]. Chondroitin sulfate and hyaluronic acid were firstly cross-linked by adipic dihydrazide (ADH) and EDC to reduce their solubility in water. Covalent attachment of ADH to the carboxylic acid groups results in the availability of pendant hydrazine amino functionalities. Then, the obtained ADH-derivatives [221,222] were blended with neutral CL solution to form composite CL-chondroitin sulfate-hyaluronic acid hybrid hydrogels (CCH). As the composite hydrogels are characterized by poor biodegradability, a chemical cross-linking agent with good biocompatibility was adopted to improve the performance. For cross-linking, a gel-crosslinking method was adopted, in which CCH hydrogels after physical gel formation were immersed directly into GP solution. GP is able to react with the amino groups of the synthesized adipic dihydrazide derivatives and CL to form intermolecular cross-link inside the hybrid hydrogel system. Though the reaction of GP cross-linking occurs mainly on the surface of hybrid hydrogel scaffold, the physicochemical properties of scaffold vary with the dosage of cross-linking agent. The prepared hydrogels exhibited excellent cellular compatibility and might have a promising application in cartilage tissue engineering.

A series of pluronic samples, triblock copolymers consisting of poly(ethylene oxide)-*block*-poly(propylene oxide)-*block*poly(ethylene oxide) (PEO-*block*-PPO-*block*-PEO), were also investigated as bifunctional macromolecular cross-linking agents while simultaneously serving as surfactants for SWNTs [223]. The cross-linked CL matrices were prepared by blending type I bovine CL with either pluronics or SWNTs dispersed in an aqueous pluronic solution and crosslinked utilizing carbodiimide chemistry (Figure 29).

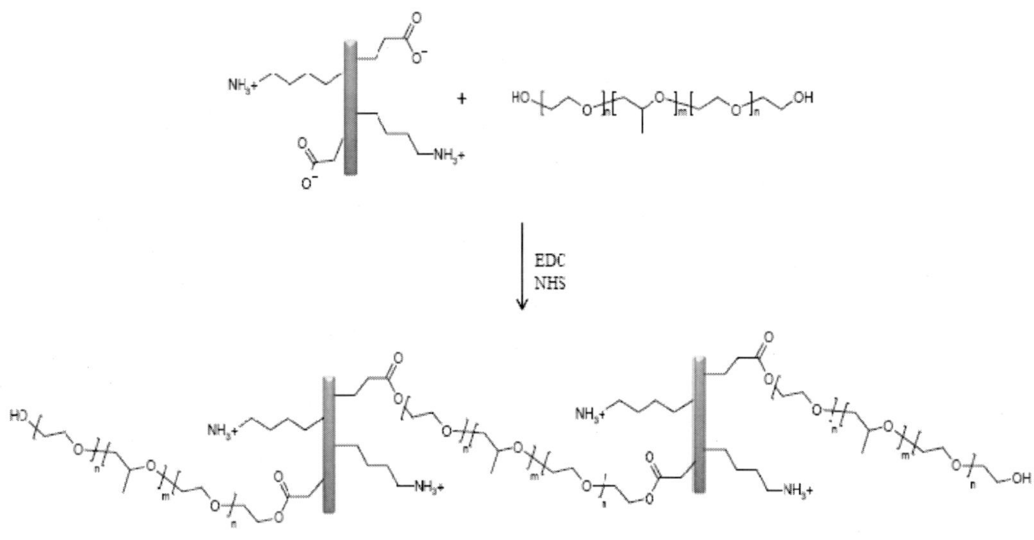


Figure 29.

Collagen was mixed with HCl in the aim to dissolve the fibrils and was then blended either with the pluronic crosslinker. The pH of the reaction mixture was adjusted to 5.5 to obtain the optimal pH for EDC activation. The mixture was then activated with EDC and NHS and blended until the solution viscosity began to increase. Subsequently, the CL reaction mixture was compressed between two glass plates. The resulting material was a cross-linked CL hydrogel with sufficient mechanical strength to be manipulated and transferred without damaging the matrix.

Sulea et al. [224] described the preparation of chlorhexidine digluconate (CHXD) containing CL hydrogels to combine the healing capacity of CL with the high bactericidal activity against a wide spectrum of both Gram-positive and Gram-negative bacteria in view of controlling the wound infection. In this study, three series of drug delivery systems consisting of type I CL hydrogel containing 0, 0.02, 0.05 and 0.10% CHXD and having pHs 3.8 and 7.4, the last non-cross-linked and cross-linked with 0.15% GA, were prepared to be used for wound healing.

Albu et al. [225] reported the synthesis of CL-based hydrogels for the *in vitro* release of doxycycline hyclate employing GA as crosslinking agent. The kinetics of *in vitro* release of the therapeutic agent from the CL hydrogels demonstrated that the amount of released drug decrease with increasing CL concentration. GA plays an important role in increasing of the CL hydrogel viscosity and the cross-linking has the same effect as increasing of CL concentration. Thus, the rate of doxycycline release decreases both with the increasing of CL concentration in the hydrogel and the presence of the cross-linking agent.

Photopolymerizable biomimetic hydrogel scaffolds made from protein-polymer conjugates of poly(ethylene glycol) (PEG) and CL were also synthesized for tissue engineering applications [226]. The PEGylation of CL was accomplished by a Michael-type addition reaction in which free thiols present in unpaired cysteine residues in the denatured protein were covalently conjugated to unsaturated double bonds on functionalized PEG-diacrylate [227].

For this purpose, the CL molecule required modification using a succinimidylacetyl-thioacetate (SATA) (Figure 30) reaction scheme, which introduced acetylated SH groups on the lysine residues [228].

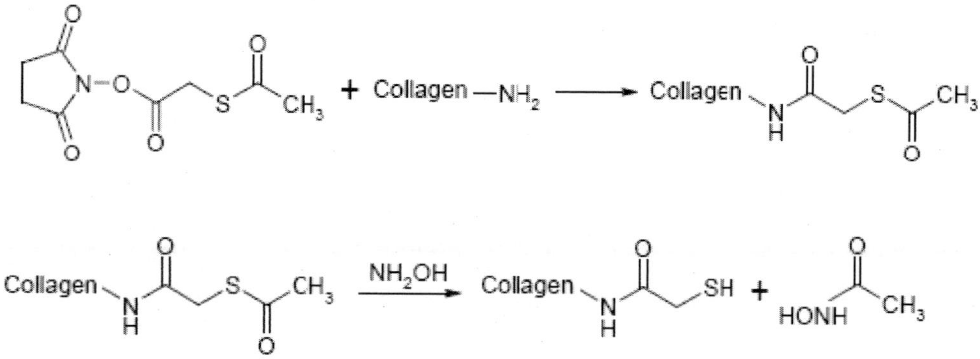


Figure 30.

In all of the PEGylation reactions, molar excess of PEGDA relative to the thiols was used in order to ensure that only a single acrylate on the PEGDA molecule would react with a single protein thiol. The other acrylate group on the protein-bound PEG was left unreacted in order to allow for free-radical polymerization of the hydrogel in the presence of photoinitiator and UV light. The synthesized protein-polymer conjugates are characterized by remarkable biocompatibility, cell adhesion ability and proteolytic degradability based on the specific characteristics of the protein backbone.

CL-polymer conjugates can be also prepared by reaction of acrylamide groups with thiol groups. Peptides containing a single thiol group were coupled to the PEG-diacrylamide in aqueous solution at room temperature in about 2 h [227]. If only a portion of the acrylamide groups were targeted for reaction with peptide, sufficient amounts of PEG-diacrylamide remained to be polymerized by free-radical mechanisms via photoinitiation. The photopolymerization step can be performed in contact with cells, providing a means to produce bioactive scaffolds for tissue engineering. After polymerization, the peptide will thus be pendantly attached into the network, meaning that the peptide will be attached to the cross-linked network via a PEG spacer. Peptides were attached to acrylate or acrylamide groups on PEG molecules via a conjugate addition reaction (Figure 2).

In recent works, H-CL, manufactured as an inexpensive byproduct of the leather industries and containing 25% insoluble phosphate salts, has been employed in the aim to prepare superabsorbent hybrid hydrogels. The commercial H-CL was purified by filtration and used in the synthesis of several hydrogels modified via crosslinking graft copolymerization with different acrylic monomers [229-231].

Pourjavadi et al. have been prepared a novel hydrolyzed CL-based hydrogel by grafting the binary mixture of AAm and 2-acrylamido-2-methylpropanesulfonic acid (AMPSA) onto the CL backbone in the presence of a crosslinking agent, MEBA, and using as free-radical initiators, APS (Figure 31) [232].

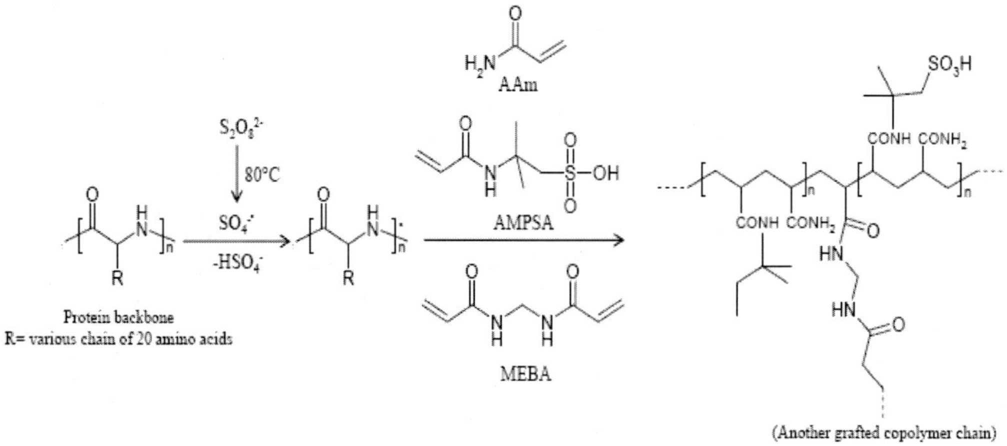


Figure 31.

The macroradicals that are generated by APS initiate simultaneous graft copolymerization of AAm and AMPSA onto CL backbone. The reversible swelling/de-swelling behavior in solutions with acidic and basic pH makes the hydrogels a suitable candidate for various applications.

H-collagen was also graft copolymerized with binary mixtures of acrylic acid and AMPSA, AA and hydroxyethyl acrylate (HEA) (Figure 32) and MAA and AAm to prepare hybrid hydrogel materials with super-swelling properties [233, 234].

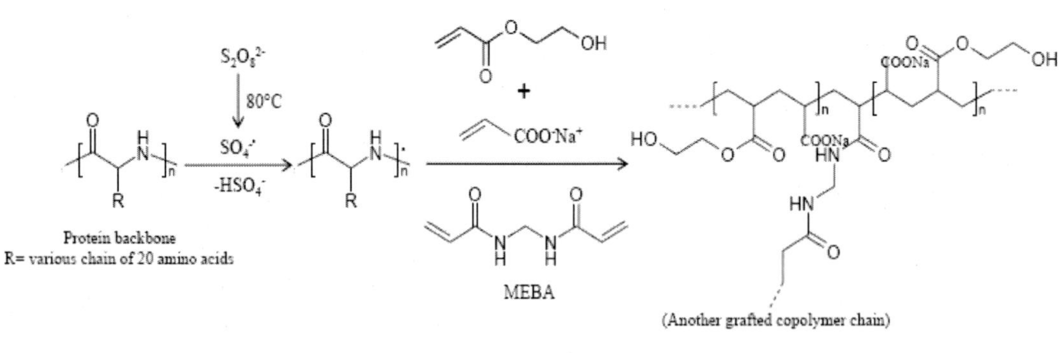


Figure 32.

## REFERENCES

[1] Ohm, D.K & Kristi L.K. (2010). Protein- and peptide-modified synthetic polymeric biomaterials. *Peptide Science, 94,* 32-48.

[2] Lutz, J.F. & Borner, H.G. (2008). Modern trends in polymer bioconjugates design. *Progress Polymer Sciences, 33,* 1–39.

[3] Klouda, L. & Mikos, A.G. (2008). Thermoresponsive hydrogels in biomedical applications. *European Journal of Pharmaceutics and Biopharmaceutics*, 68, 34–45.

[4] Tabata, Y. & Ikada, Y. (1998). Protein release from gelatin matrices. *Advanced Drug Delivery Reviews, 31,* 287–301.

[5] Young, S.; Wong, M.; Tabata, Y. & Mikos, A.G. (2005). Gelatin as a delivery vehicle for the controlled release of bioactive molecules. *Journal6 of Controlled Release, 109,* 256– 274.

[6] Desponds, A. & Freitag, R. (2005). Light-responsive bioconjugates as novel tools for specific capture of biologicals by photoaffinity. *Biotechnology Bioengeneering, 91,* 583–591.

[7] Iemma, F.; Spizzirri, U.G.; Muzzalupo, R.; Puoci, F.; Trombino, S. & Picci, N. (2004). Spherical hydrophilic microparticles obtained by the radical copolymerisation of functionalised bovine serum albumin. *Colloid of Polymer Sciences, 283,* 250–256.

[8] Iemma, F.; Spizzirri, U.G.; Puoci, F.; Muzzalupo, R.; Trombino, S. & Picci, N. (2005). Radical crosslinked albumin microspheres as potential drug delivery systems: preparation and in vitro studies. *Drug Delivery, 12,* 179–184.

[9] Iemma, F.; Spizzirri, U.G.; Puoci, F.; Muzzalupo, R.; Trombino, S.; Cassano, R.; Leta, S. & Picci, N. (2006). pH-Sensitive hydrogels based on bovine serum albumin for oral drug delivery. *International Journal of Pharmaceutics, 312,* 151–157.

[10] Spizzirri, U.G.; Iemma, F.; Puoci, F.; Cirillo, G.; Curcio, M.; Parise, O.I. & Picci, N. (2009). Synthesis and release profile analysis of thermo-sensitive albumin hydrogels. *Colloid of Polymer Sciences, 287,* 779–787.

[11] Spizzirri, U.G.; Cirillo, G.; Iemma, F.; Puoci, F.; Curcio, M.; Parise, O.I. & Picci, N. (2011). Thermo-responsive albumin hydrogels with LCST near the physiological temperature. *Journal of Applied Polymer Science, 121,* 342–351.

[12] Cirillo, G.; Iemma, F.; Spizzirri, U.G.; Puoci, F.; Curcio, M.; Parisi, O.I. & Picci, N. (2011). Synthesis of stimuli-responsive microgels for in vitro release of diclofenac diethyl ammonium. *Journal of Biomaterial Science Polymer Edition, 22,* 823–844.

[13] Elbert, D.L. & Hubbell J.A. (2001). Conjugate addition reactions combined with free-radical cross-linking for the design of materials for tissue engineering. *Biomacromolecules, 2,* 430-441.

[14] Halstenberg, S.; Panitch, A.; Rizzi, S.; Hall, H. & Hubbell J.A. (2002). Biologically engineered protein-graft-poly(ethylene glycol) hydrogels: a cell adhesive and plasmin-degradable biosynthetic material for tissue repair. *Biomacromolecules, 3,* 710-723.

[15] Gonen-Wadmany, M.; Oss-Ronen, L. & Seliktar, D. (2007). Protein–polymer conjugates for forming photopolymerizable biomimetic hydrogels for tissue engineering. *Biomaterials, 28,* 3876–3886.

[16] Oss-Ronen, L. & Seliktar D. (2010). Photopolymerizable hydrogels made from polymerconjugated albumin for affinity-based drug delivery. *Advanced Biomaterials Advanced Engeneering Materials, 12,* 45–52.

[17] Oss-Ronen, L. & Seliktar D. (2011). Polymer-conjugated albumin and fibrinogen composite hydrogels as cell scaffolds designed for affinity-based drug delivery. *Acta Biomaterialia, 7,* 163-170.

[18] Tada, D.; Tanabe, T.; Tachibana, A. & Yamauchi, K. (2005). Drug release from hydrogel containing albumin as crosslinker. *Journal of Bioscience and Bioengineering, 100,* 551–555.

[19] Tada, D.; Tanabe, T.; Tachibana, A. & Yamauchi, K. (2007). Albumin-crosslinked alginate hydrogels as sustained drug release carrier. *Materials Science and Engineering C, 27,* 870–874.

[20] Taguchi, T.; Saito, H.; Iwasashi, M.; Sakane, M.; Kakinoki, S.; Ochiai, N. & Tanaka, J. (2007). Development of a novel glue consisting of naturally-derived biomolecules: citric acid and human serum albumin. *Journal of Nanoscience and Nanotechnology, 7,* 742-747.

[21] Kakinoki, S.; Taguchi, T.; Saito, H.; Tanaka, J. & Tateishi T. (2007). Injectable in situ forming drug delivery system for cancer chemotherapy using a novel tissue adhesive: Characterization and in vitro evaluation. *European Journal of Pharmaceutics and Biopharmaceutics, 66,* 383–390.

[22] Hirose, M.; Tachibana, A. & Tanabe, T. (2010). Recombinant human serum albumin hydrogel as a novel drug delivery vehicle. *Materials Science and Engineering C, 30,* 664–669.

[23] Tabata, Y.; Nagano, A. & Ikada, Y. (1999). Biodegradation of hydrogel carrier incorporating fibroblast growth factor. *Tissue Engeneering, 5,* 127–138.

[24] Spizzirri, U.G.; Iemma, F.; Puoci, F.; Cirillo, G.; Curcio, M.; Parisi, O.I. & Picci, N. (2009). Synthesis of antioxidant polymers by grafting of Gallic acid and Catechin on gelatine. *Biomacromolecules, 10,* 1923-1930.

[25] Cirillo, G.; Kraemer, K.; Fuessel, S.; Puoci, F.; Curcio, M.; Spizzirri, U.G.; Altimari, I. & Iemma, F. (2010). The biological activity of a gallic acid-gelatin conjugate. *Biomacromolecules, 11,* 3309-3315.

[26] Kuijpers, A.J.; Engbers, G.H.; Krijgsveld, J.; Zaat, S.A.; Dankert, J. & Feijen, J. (2000). Cross-linking and characterisation of gelatin matrices for biomedical applications. *Journal Biomaterials Science, Polymer Edition, 11,* 225– 243.

[27] Van Den Bulcke, A.I.; Bogdanov, B.; De Rooze, N.; Schacht, E.H.; Cornelissen, M. & Berghmans, H. (2000). Structural and rheological properties of methacrylamide modified gelatin hydrogels. *Biomacromolecules, 1,* 31-38.

[28] Nichol, J.W.; Koshya, S.T.; Baea, H.; Hwanga, C.M.; Yamanlara, S. & Khademhosseinia, A. (2010). Cell-laden microengineered gelatin methacrylate hydrogels. *Biomaterials, 31,* 5536-5544.

[29] Hutson, C.B.; Nichol, J.W.; Aubin, H.; Baea, H.; Yamanlara, S.; Al-Haque, S.; Koshya, S.T.; & Khademhosseinia, A. Synthesis and characterization of tunable poly(ethylene glycol): gelatin methacrylate composite hydrogels. Tissue Engineering: Part A, DOI: 10.1089/ten.tea.2010.0666.

[30] Barbetta, A.; Dentini, M.; Zannoni, E.M. & De Stefano M.E. (2005). Tailoring the porosity and morphology of gelatin-methacrylate polyHYPE scaffolds for tissue engineering applications. *Langmuir, 21,* 12333-12341.

[31] Barbetta, A.; Massimi, M.; Conti Devirgiliis, L. & Dentini, M. (2006). Enzymatic cross-linking versus radical polymerization in the preparation of gelatin polyHYPEs and their performance as scaffolds in the culture of hepatocytes. *Biomacromolecules, 7,* 3059-3068.

[32] Koepff, P.; Braumer, K. & Babel, W. (1998). U.S. Pat. Appl US 5,733,994; 3.

[33] Pochlauer, P. & Hendel, W. (1998). One-pot formation of succinimidyl esters by the system chlorophosphate/hydroxysuccinimide/base. *Tetrahedron, 54,* 3489–3494.

[34] Schuster, M.; Turecek, C.; Weigel, G.; Saf, R.; Stampfl, J.; Varga, F.; Liska, R. (2009). Gelatin-based photopolymers for bone replacement materials. *Journal of Polymer Science: Part A: Polymer Chemistry, 47,* 7078–7089.

[35] Curcio, M.; Puoci, F.; Spizzirri, U.G.; Iemma, F.; Cirillo, G.; Parisi, O.I. & Picci, N. (2010). Negative thermo-responsive microspheres based on hydrolyzed gelatin as drug delivery device. *AAPS Pharmaceutical Science Technologies, 11,* 652-662.

[36] Curcio, M.; Spizzirri, U.G.; Iemma, F.; Puoci, F.; Cirillo, G.; Parisi, O.I. & Picci N. (2010). Grafted thermo-responsive gelatin microspheres as delivery systems in triggered drug release. *European Journal of Pharmaceutics and Biopharmaceutics, 76,* 48-55.

[37] Narayana Reddy, N.; Varaprasada, K.; Ravindraa, S.; Subba Reddy, G.V.; Reddy, K.M.S.; Mohan Reddy K.M. & K. Rajua, M. (2011). Evaluation of blood compatibility and drug release studies of gelatin based magnetic hydrogel nanocomposites. Colloids and Surfaces A: Physicochemical and Engineering Aspects, 385, 20-27.

[38] Amany, I.R. (2010). Gelatin based pH-sensitive hydrogels for colon-specific oral drug delivery: Synthesis, characterization, and in vitro release study. *Journal of Applied Polymer Science, 118,* 2642-2649.

[39] Sadeghi, M. & Heidari, B. (2011). Crosslinked graft copolymer of methacrylic acid and gelatin as a novel hydrogel with pH-responsiveness properties. *Materials, 4*, 543-552.

[40] Bigi, A.; Bracci, B.; Cojazzi, G.; Panzavolta, S. & Roveri, N. (1998). Drawn gelatin films with improved mechanical properties. *Biomaterials, 19,* 2335-2340.

[41] Khor E. (1997). Methods for the treatment of collagenous tissues for bioprostheses. *Biomaterials, 18,* 95-103.

[42] Bigi, A.; Cojazzi, G.; Panzavolta, S.; Rubini, K. & Roveri, N. (2001). Mechanical and thermal properties of gelatin films at different degrees of glutaraldehyde crosslinking. *Biomaterials, 22,* 763-768.

[43] Einersona, N.J.; Stevensa, K.R. & Kao, W.J. (2002). Synthesis and physicochemical analysis of gelatin-based hydrogels for drug carrier matrices. *Biomaterials, 24,* 509–523.

[44] Hwang, D.C. & Damadoran, S. (1996). Chemical modification strategies for synthesis of protein-based hydrogel. *Journal of Agricultural and Food Chemistry, 44,* 751–758.

[45] Schacht, E.; Bogdanov, B.; Van Den Bulcke, A. & De Rooze, N. (1997). Hydrogels prepared by crosslinking of gelatin with dextran dialdehyde. *Reactive & Functional Polymers, 33,* 109-116.

[46] Bogdanov, B.; Schacht, E. & Van Den Bulcke, A. (1997). Thermal and rheological properties of gelatin-dextran hydrogels. *Journal of Thermal Analysis, 49,* 847-856.

[47] Draye, J.-P.; Delaey, B.; Van de Voorde, A.; Van Den Bulcke, A.; Bogdanov, B. & Schacht, E. (1998). In vitro release characteristics of bioactive molecules from dextran dialdehyde cross-linked gelatin hydrogel films. *Biomaterials, 19,* 99-107.

[48] Draye, J.-P.; Delaey, B.; Van de Voorde, A.; Van Den Bulcke, A.; De Reu, B. & Schacht, E. (1998). In vitro and in vivo biocompatibility of dextran dialdehyde cross-linked gelatin hydrogel films. *Biomaterials, 19,* 1677-1687.

[49] Saito, H.; Taguchi, T.; Kobayashi, H.; Kataoka, K.; Tanaka, J.; Murabayashi, S. & Mitamura, Y. (2004). Physicochemical properties of gelatin gels prepared using citric acid derivative. *Materials Science and Engineering C, 24,* 781–785.

[50] Aoki, H.; Taguchi, T.; Saito, H.; Kobayashi, H.; Kataoka, K. & Tanaka, J. (2004). *Materials Science and Engineering C, 24,* 787–790.

[51] Otani, Y.; Tabata, Y. & Ikada, Y. (1996). A new biological glue from gelatin and poly(L-glutamic acid). *Journal Biomedical Material Research, 31,* 157-166.

[52] Otani, Y.; Tabata, Y. & Ikada, Y. (1996). Rapidly curable biological glue composed of gelatin and poly(L-glutamic acid). *Biomaterials, 17,* 1387-1391.

[53] Iwata, H.; Matsuda, S.; Mitsuhashi, K.; Itoh, E. & Ikada, Y. (1998). A novel surgical glue composed of gelatin and *N*-hydroxysuccinimide activated poly(L-glutamic acid): Part 1. Synthesis of activated poly(L-glutamic acid) and its gelation with gelatin. Biomaterials, 19, 1869-1876.

[54] Liang, H.C.; Chang, W.H.; Lin, K.J. & Sung HW. (2003). Genipin-crosslinked gelatin microspheres as a drug carrier for intramuscular administration: in vitro and in vivo studies. *Journal Biomedical Material Research A,* 65, 271–282.

[55] Sung, H.-W.; Chen, C.-N.; Huang, R.-N.; Hsu, J.-C. & Chang, W.-H. (2000). In vitro surface characterization of a biological patch fixed with a naturally occurring crosslinking agent. *Biomaterials, 21,* 1353-1362.

[56] Gupta, P.; Vermani, K. & Garg S. (2002). Hydrogels from controlled release to pH-responsive drug delivery. *Drug Discovery Today, 7,* 569–579.

[57] Yao, C.H.; Liu, B.S.; Chang, C.J.; Hsu, S.H. & Chen, Y.S. (2004). Preparation of networks of gelatin and genipin as degradable biomaterials. Materials Chemistry and Physics, 83, 204–208.

[58] Thakur, G.; Mitra, A.; Rousseau, D.; Basak, A.; Sarkar, S. & Pal, K. (2011). Crosslinking of gelatin-based drug carriers by genipin induces changes in drug kinetic profiles in vitro. *Journal Material Science: Materials in Medicine, 22,* 115–123.

[59] Touyama, R.; Inoue, K.; Takeda, Y.; Yatsuzuka, M.; Ikumoto, T.; Moritome, N.; Shingu, T.; Yokoi, T. & Inouye, H. (1994). Studies on the blue pigments produced from genipin and methylamine. II. On the formation mechanisms of brownish-red intermediates leading to the blue pigment formation. *Chemical & Pharmaceutical Bulletin, 42,* 1571-1578.

[60] Sung, H.W.; Chen, C.N.; Huang, R.N.; Hsu, J.C. & Chang, W.H. (2000). In vitro surface characterization of a biological patch fixed with a naturally occurring crosslinking agent. *Biomaterials, 21,* 1353–1362.

[61] Fox, P.F. (2003). Milk proteins: general and historical aspects, in: P.F. Fox, P.L.H. McSweeney (Eds.), Part A: Advanced dairy Chemistry-1 Proteins, Kluwer Academic/Plenum Publishers, New York, 1–48.

[62] Thompson, A.; Boland, M.J. & Singh, H. (2009). Milk proteins, from expression to food, first ed., S.L. Taylor, Elsevier, Academic Press, Amsterdam.

[63] Livney, Y.D. (2010). Milk proteins as vehicles for bioactives. *Current Opinion in Colloid Interface Sciences, 15,* 73–83.
[64] Elzoghby, A.O.; Abo El-Fotoh, W.S. & Elgindy, N.A. (2011). Casein-based formulations as promising controlled release drug delivery systems. *Journal of Controlled Release, 153,* 206–216.
[65] Bajpai, S.K. (1999). Casein cross-linked polyacrylamide hydrogels : study of swelling and drug release behaviour. *Iranian Polymer Journal, 8,* 231-239.
[66] Songa, F.; Zhang, L.-M.; Yang, C. & Yan, L. (2009). Genipin-crosslinked casein hydrogels for controlled drug delivery. *International Journal of Pharmaceutics, 373,* 41–47.
[67] Lorenzen, P.C.; Schlimme, E. & Roos, N. (1998). Crosslinking of sodium caseinate by a microbial transglutaminase. *Nahrung, 42,* 151–154.
[68] Latha, M.S. & Jayakrishnan, A. (1994). Glutaraldheyde crosslinked bovine casein microspheres as a matrix for the controlled release of theophylline: In-vitro Studies. *Journal Pharmacy and Pharmacology, 46,* 8–13.
[69] Bulgarelli, E.; Forni, F. & Bernabei, M.T. (2000). Effect of matrix composition and process conditions on casein–gelatin beads floating properties. *International Journal of Pharmaceutics, 198,* 157–165.
[70] Knepp, W.A.; Jayakrishnan, A.; Quigg, J.M.; Sitren, H.S.; Bagnall, J.J. & Goldberg, E.P. (1993). Synthesis, properties, and intratumoral evaluation of mitoxantrone-loaded casein microspheres in Lewis lung carcinoma. *Journal Pharmacy and Pharmacology, 45,* 887–891.
[71] Chen, Y.; Willmott, N.; Anderson, J. & Florence, A.T. (1987). Comparison of albumin and casein microspheres as a carrier for doxorubicin. *Journal Pharmacy and Pharmacology, 39,* 978–985.
[72] Knepp, W.A.; Jayakrishnan, A.; Quigg, J.M.; Sitren, H.S.; Bagnall, J.J. & Goldberg, E.P. (1993). Synthesis, properties, and intratumoral evaluation of mitoxantrone-loaded casein microspheres in Lewis lung carcinoma. *Journal Pharmacy and Pharmacology, 45,* 887–891.
[73] Knepp, W.A.; Jayakrishnan, A. & Goldberg, E.P. (1994). Casein microspheres: preparation and evaluation as a carrier for controlled drug delivery, *International Journal of Pharmaceutics, 106,* 221–228.
[74] Latha, M.S.; Jayakrishnan, A.; Rathinam, K. & Mohanty, M. (1994). Casein as a carrier matrix for 5-fluorouracil: drug release from microspheres, drug-protein conjugates and in-vivo degradation of microspheres in rat muscle. *Journal Pharmacy and Pharmacology, 46,* 858–862.
[75] Latha, M.S.; Lal, A.V.; Kumary, T.V.; Sreekumar, R. & Jayakrishnan, A. (2000). Progesterone release from glutaraldehyde cross-linked casein microspheres: in vitro studies and in vivo response in rabbits. *Contraception, 61,* 329–334.
[76] Puthli, S. & Vavia, P. (2008). Gamma irradiated micro system for long-term parenteral contraception: an alternative to synthetic polymers. *European Journal of Pharmaceutical Sciences, 35,* 307–317.
[77] Latha, M.S. & Jayakrishnan, (1994). A. Glutaraldheyde crosslinked bovine casein microspheres as a matrix for the controlled release of theophylline: In-vitro Studies. *Journal Pharmacy and Pharmacology, 46,* 8–13.

[78] Heelan, B.A. & Corrigan, O.I. (1997). In vitro analysis of the release of incorporated agents from sodium caseinate microspheres. *Journal of Microencapsulation, 14,* 63–78.

[79] Zhu, J. & Li, P. (2003).Synthesis and characterization of Poly(methyl methacrylate)/Casein nanoparticles with a well-defined core-shell structure. Journal of Polymer Science: Part A: Polymer Chemistry, 41, 3346–3353.

[80] Pan, X.; Yao, P. & Jiang, M. (2007). Simultaneous nanoparticle formation and encapsulation driven by hydrophobic interaction of casein-graft-dextran and β-carotene. *Journal of Colloid Interface Sciences, 315,* 456–463.

[81] Pan, X.; Mu, M.F.; Hu, B.; Yao, P. & Jiang, M. (2006). Acidic solution properties ofβ-casein-graftdextran copolymer prepared throughmaillard reaction. *Biopolymers, 81,* 29.

[82] Huppertz, T. & de Kruif, C.G. (2008). Structure and stability of nanogel particles prepared by internal cross-linking of casein micelles. *International Dairy Journal, 18,* 556–565.

[83] Herrick, S.; Blanc-Brude, O.; Gray A. & Laurent, G. (1999). Fibrinogen. *International Journal of Biochemistry and Cell Biology, 31,* 741–746.

[84] Sahni, A. & Francis, C.W. (2000). Vascular endothelial growth factor binds to fibrinogen and fibrin and stimulates endothelial cell proliferation. *Blood, 96,* 3772-3778.

[85] Zisch, A.H.; Schenk, U.; Schense, J.C.; Sakiyama-Elbert, S.E. & Hubbell J.A. (2001). Covalently conjugated VEGF—fibrin matrices for endothelialization. *Journal of Controlled Release, 72,* 101-113.

[86] Almany, L. & Seliktar, D. (2005). Biosynthetic hydrogel scaffolds made from fibrinogen and polyethylene glycol for 3D cell cultures. *Biomaterials, 26,* 2467-2477.

[87] Griffith, L.G. & Naughton, G. (2002). Tissue engineering-current challenges and expanding opportunities. *Science, 295,* 1009-1014.

[88] Zhang, G.; Wang, X.; Wang, Z.; Zhang, J. & Suggs L. (2006). A PEGylated fibrin patch for mesenchymal stem cell delivery. *Tissue Engeneering, 12,* 9-19.

[89] Dikovskya, D.; Bianco-Peleda, H. & Seliktar, D. (2006). The effect of structural alterations of PEG-fibrinogen hydrogel scaffolds on 3-D cellular morphology and cellular migration. *Biomaterials, 27,* 1496–1506.

[90] Lee, K.Y. & Mooney, D.J. (2001) Hydrogels for tissue engineering. *Chemical Review, 101,* 1869–1879.

[91] Chou, C.H.; Cheng, W.T.; Kuo, T.F.; Sun, J.S.; Lin, F.H. & Tsai, J.C. (2007). Fibrin glue mixed with gelatin/hyaluronic acid/chondroitin-6-sulfate tri-copolymer for articular cartilage tissue engineering: the results of real-time polymerase chain reaction. *Journal of Biomedical Material Research A, 82,* 757–67.

[92] Stern, S.; Lindenhayn, K.; Schultz, O. & Perka, C. (2000). Cultivation of porcine cells from the nucleus pulposus in a fibrin/hyaluronic acid matrix. *Acta Orthopaedica Scandinavica, 71,* 496–502.

[93] Yang, C.L.; Chen, H.W.; Wang, T.C. & Wang, Y.J. (2011). A novel fibrin gel derived from hyaluronic acid-grafted fibrinogen. *Biomedical Material, 6,* 1-11.

[94] Mithieux, S.M. & Weiss, A.S. (2005). Elastin. Fibrous Proteins: Coiled-Coils, Collagen and Elastomers. *Advances in Protein Chemistry, 70*, 437-461.

[95] Annabi, N.; Mithieux, S.M.; Weiss, A.S. & Dehghani, F. (2010). Cross-linked open-pore elastic hydrogels based on tropoelastin, elastin and high pressure $CO_2$. *Biomaterials, 31*, 1655-1665.

[96] Sandberg, L.B.; Soskel, N.T. & Leslie, J.G. (1981). Elastin structure, biosynthesis, and relation to disease states. *New England Journal of Medicine, 304,* 566–579.

[97] Nettles, D.L.; Chilkoti, A. & Setton, L.A. (2010). Applications of elastin-like polypeptides in tissue engineering. *Advanced Drug Delivery Reviews, 62*, 1479–1485.

[98] Daamen, W.F.; Veerkamp, J.H.; van Hest, J.C.M. & van Kuppevelt, T.H. (2007). Elastin as a biomaterial for tissue engineering. *Biomaterials, 28*, 4378–98.

[99] Annabi, N.; Mithieux, S.M.; Boughton, E.A.; Ruys, A.J.; Weiss, A.S. & Dehghani, F. (2009). Synthesis of highly porous crosslinked elastin hydrogels and their interaction with fibroblasts in vitro. *Biomaterials, 30,* 4550-4557.

[100] Leach, J.B.; Wolinsky, J.B.; Stone, P.J. & Wong, J.Y. (2005). Crosslinked alpha-elastin biomaterials: towards a processable elastin mimetic scaffold. *Acta Biomaterialia, 1,* 155-164.

[101] Annabi, N.; Mithieux, S.M.; Weiss, A.S. & Dehghani, F. (2009). The fabrication of elastin-based hydrogels using high pressure $CO_2$. *Biomaterials, 30*, 1–7.

[102] Mithieux, S.M.; Tu, Y.; Korkmaz, E.; Braet, F. & Weiss, A.S. (2009). In situ polymerization of tropoelastin in the absence of chemical cross-linking. *Biomaterials, 30*, 431-435.

[103] Mithieux, S.M.; Rasko, J.E.J. & Weiss, A.S. (2004). Synthetic elastin hydrogels derived from massive elastic assemblies of self-organized human protein monomers. *Biomaterials, 25*, 4921-4927.

[104] Lim, D.W.; Nettles, D.L.; Setton, L.A. & Chilkoti, A. (2008). In situ cross-linking of elastin-like polypeptide block copolymers for tissue repair. *Biomacromolecules, 9*, 222-230.

[105] Vrhovski, B. & Weiss, A.S. (1998). Biochemistry of tropoelastin. *European Journal of Biochemistry, 258*, 1-18.

[106] Trabbic-Carlson, K.; Setton, L.A. & Chilkoti, A. (2003). Swelling and Mechanical Behaviors of Chemically Cross-Linked Hydrogels of Elastin-like Polypeptides. *Biomacromolecules, 4,* 572-580.

[107] Welsh, E.R. & Tirrell, D.A. (2000). Engineering the extracellular matrix: a novel approach to polymeric biomaterials. I. Control of the physical properties of artificial protein matrices designed to support adhesion of vascular endothelial cells. *Biomacromolecules, 1,* 23-30.

[108] Martino, M. & Tamburro, A.M. (2001). Chemical synthesis of cross-linked poly(KGGVG), an elastin-like biopolymer. *Biopolymers, 59*, 29-37.

[109] McMillan, R.A.; Caran, K.L.; Apkarian, R.P. & Conticello, V.P. (1999). High-resolution topographic imaging of environmentally responsive, elastin-mimetic hydrogels. *Macromolecules, 32*, 9067-9070.

[110] Di Zio, K. & Tirrell, D.A. (2003). Mechanical properties of artificial proteinmatrixes engineered for control of cell and tissue behavior. *Macromolecules, 36*, 1553-1558.

[111] McMillan, R.A. & Conticello, V.P. (2000). Synthesis and characterization of elastin–mimetic protein gels derived from a well-defined polypeptide precursor. *Macromolecules, 33*, 4809-4821.

[112] Bellingham, C.M., Lillie, M.A.; Gosline, J.M.; Wright, G.M.; Starcher, B.C., Bailey, A.J., Woodhouse, K.A. & Keeley, F.W. (2003). Recombinant human elastin polypeptides self-assemble into biomaterials with elastin-like properties. *Biopolymers, 70,* 445-455.

[113] Nowatzki, P.J. & Tirrell, D.A. (2003). Physical properties of artificial extracellular matrix protein films prepared by isocyanate crosslinking. *Biomaterials, 25,* 1261-1267.

[114] Lim, D.W.; Nettles, D.L.; Setton, L.A. & Chilkoti, A. (2007). Rapid cross-linking of elastin-like polypeptides with (hydroxymethyl)phosphines in aqueous solution. *Biomacromolecules, 8*, 1463-1470.

[115] Vieth, S.; Bellingham, C.M.; Keeley, F.W.; Hodge, S.M. & Rousseau, D. (2007). Microstructural and tensile properties of elastin-based polypeptides crosslinked with genipin and pyrroloquinoline quinone. *Biopolymers, 85,* 199-206.

[116] McHale, M.K.; Setton, L.A. & Chilkoti, A. (2005). Synthesis and in vitro evaluation of enzymatically cross-linked elastin-like polypeptide gels for cartilaginous tissue repair. *Tissue Engineering, 11,* 1768-1779.

[117] Mithieux, S.M.; Tu, Y.; Korkmaz, E.; Braet, F. & Weiss, A.S. (2009). In situ polymerization of tropoelastin in the absence of chemical cross-linking. *Biomaterials, 30,* 431-435.

[118] Nagapudi, K.; Brinkman, W.T.; Thomas, B.S.; Park, J.O.; Srinivasarao, M.; Wright, E.; Conticello, V.P. & Chaikof, E.L. (2005). Viscoelastic and mechanical behavior of recombinant protein elastomers. *Biomaterials, 26,* 4695-4706.

[119] Nagapudi, K.; Brinkman, W.T.; Leisen, J.; Thomas, B.S.; Wright, E.R.; Haller, C.; Wu, X.; Apkarian, R.P.; Conticello, V.P. & Chaikof, E.L. (2005). Protein-based thermoplastic elastomers. *Macromolecules, 38,* 345-354.

[120] Lee, J.; Macosko, C.W. & Urry, D.W. (2001). Phase transition and elasticity of protein-based hydrogels. *Journal of Biomaterials Science Polymer Edition, 12,* 229-242.

[121] Lee, J.; Macosko, C.W. & Urry D.W. (2001). Mechanical properties of cross-linked synthetic elastomeric polypentapeptides. *Macromolecules, 34,* 5968-5974.

[122] Lee, J.; Macosko, C.W. & Urry, D.W. (2001). Swelling behavior of g-irradiation cross-linked elastomeric polypentapeptide-based hydrogels. *Macromolecules, 34,* 4114-4123.

[123] Cox, B.A.; Starcher, B.C. & Urry, D.W. (1973). Coacervation of α-elastin results in fiber formation. *Biochimica et Biophysica Acta, 317,* 209-213.

[124] Ito, S.; Ishimaru, S. & Wilson, S.E. (1998). Effect of coacervated alpha-elastin on proliferation of vascular smooth muscle and endothelial cells. *Angiology, 49,* 289-297.

[125] Tai, H.; Popov, V.K.; Shakesheff, K.M. & Howdle, S.M. (2007). Putting the fizz into chemistry: applications of supercritical carbon dioxide in tissue engineering, drug delivery and synthesis of novel block copolymers. *Biochemical Society Transactions, 35*, 516-521.

[126] Kazarian, S.G. (2000). Polymer processing with supercritical fluids. *Polymer Science, 42,* 78-101.

[127] Cooper, A.I. (2000). Polymer synthesis and processing using supercritical carbon dioxide. *Journal of Materials Chemistry, 10,* 207-234.

[128] McHugh MA. Supercritical fluid extraction: principles and practice. Boston: Butterworths; 1986.

[129] Quirk, R.A.; France, R.M.; Shakesheff, K.M. & Howdle, S.M. (2005). Supercritical fluid technologies and tissue engineering scaffolds. *Current Opinion in Solid State & Materials Science, 8,* 313-321.

[130] Cansell, F.; Aymonier, C. & Loppinet-Serani, A. (2003). Review on materials science and supercritical fluids. *Current Opinion in Solid State & Materials Science, 7,* 331-340.

[131] Cooper, A.I.; Wood, C.D. & Holmes, A.B. (2000). Synthesis of well-defined macroporous polymer monoliths by sol-gel polymerization in supercritical $CO_2$. *Industrial & Engineering Chemistry Research, 39,* 4741-4744.

[132] Cooper, A.I. & Holmes, A.B. (1999). Synthesis of molded monolithic porous polymers using supercritical carbon dioxide as the porogenic solvent. *Advanced Materials, 11,* 1270-1274.

[133] Partap, S.; Rehman, I.; Jones, J.R. & Darr, J.A. (2006). Supercritical carbon dioxide in water emulsion-templated synthesis of porous calcium alginate hydrogels. *Advanced Materials, 18,* 501-504.

[134] Lee, J.Y.; Tan, B. & Cooper, A.I. (2007). $CO_2$-in-water emulsion-templated poly(vinyl alcohol) hydrogels using poly(vinyl acetate)-based surfactants. *Macromolecules, 40,* 1955-1961.

[135] Palocci, C.; Barbetta, A.; La Grotta, A. & Dentini, M. (2007) Porous biomaterials obtained using supercritical $CO_2$-water emulsions. *Langmuir, 23,* 8243-8451.

[136] Vesely I. (1998). The role of elastin in aortic valve mechanics. *Journal of Biomechanics, 31,* 115-123.

[137] van de Witte, P.; Dijkstra, P.J.; van den Berg, J.W.A. & Feijen, J. (1996). Phase separation processes in polymer solutions in relation to membrane formation. *Journal of Membrane Science, 117,* 1-31.

[138] Annabi, N.; Mithieux, S.M.; Anthony, S.W. & Dehghani, F. (2009). Development and characterization of a novel elastin hydrogel. *Materials Research Society Symposium Proceedings*, 1140.

[139] Simnick, A.J.; Lim, D.W.; Chow, D. & Chilkoti, A. (2007). Biomedical and biotechnological applications of elastin-like polypeptides. *Polymer Reviews, 47*, 121-154.

[140] Meyer, D.E. & Chilkoti, A. (1999). Purification of recombinant proteins by fusion with thermally responsive polypeptides, *Nature Biotechnology, 17,* 1112-1115.

[141] Urry, D.W.; Parker, T.M.; Reid, M.C. & Gowda, D.C. (1991). Biocompatibility of the bioelastic materials, poly(GVGVP) and its gamma-irradiation cross-linked matrix - summary of generic biological test-results. *Journal of Bioactive and Compatible Polymers, 6*, 263-282.

[142] Urry, D.W. (1997). Physical Chemistry of Biological Free Energy Transduction As Demonstrated by Elastic Protein-Based Polymers. *Journal of Physical Chemistry B, 101,* 11007-11028.

[143] Urry, D.W. & Parker, T.M. (2002). Section: Extracellular matrix proteins; Mechanics of elastin: molecular mechanism of biological elasticity and its relationship to contraction. *Journal of Muscle Research and Cell Motility, 23,* 543-559.

[144] McMillan, R.A.; Caran, K.L.; Apkarian, R.P. & Conticello, V.P. (1999). Anisotropic Coarsening of Two-Dimensional Surface Domains in Copolymer Thin Films. *Macromolecules, 32*, 9007-9010.

[145] Nagapudi, K.; Brinkman, W.T.; Leisen, J.E.; Huang, L.; McMillan, R.A.; Apkarian, R.P.; Conticello, V.P. & Chaikof, E.L. (2002). Photomediated Solid-State Cross-Linking of an Elastin−Mimetic Recombinant Protein Polymer. *Macromolecules, 35,* 1730-1737.

[146] Wright, E.R. & Conticello, V.P. (2002). Self-assembly of block copolymers derived from elastin-mimetic polypeptide sequences. *Advanced Drug Delivery Reviews, 54*, 1057-1073.

[147] Wright, E.R.; McMillan, R.A.; Cooper, A.; Apkarian, R.P. & Conticello, V.P. (2002). Thermoplastic Elastomer Hydrogels via Self-Assembly of an Elastin-Mimetic Triblock Polypeptide. *Advanced Functional Materials, 12*, 149-154.

[148] Nimni, M.E.; Cheung, D.T.; Strates, B.; Kodama, M. & Skeikh, K. (1987). Chemically modified collagen: A natural biomaterial for tissue replacement. *Journal of Biomedical Materials Research, 21,* 741-771.

[149] Chvapil, M.; Speer, D.; Mora, W. & Eskelson, C. (1983). fect of tanning agent on tissue reaction to tissue implanted collagen sponge. *Journal of Surgery Research, 35,* 402-409.

[150] Sung, H.W.; Liang, I.L.; Chen, C.N.; Huang, R.N. & Liang, H.F. (2001). Stability of a biological tissue fixed with a naturally occurring crosslinking agent (genipin). *Journal of Biomedical Materials Research, 55,* 538-546.

[151] Van Luyn, M.J.A.; Van Wachem, P.B.; Olde Damink, L.H.H.; Dijkstra, P.J.; Feijen, J. & Nieuwenhuis, P. (1992). Relations between in vitro cytotoxicity and crosslinked dermal sheep collagens. *Journal of Biomedical Materials Research,* 26, 1091-1110.

[152] Gosline, J.M. (1980). The elastic properties of rubber-like proteins and highly extensible tissues. *Symposia of the Society for Experimental Biology, 34,* 332-357.

[153] Staros, J.V. (1982). N-hydroxysulfosuccinimide active esters: bis(N-hydroxysulfosuccinimide) esters of two dicarboxylic acids are hydrophilic, membrane-impermeant, protein cross-linkers. *Biochemistry, 21*, 3950-3955.

[154] Matsuo, E.S.; Orkisz, M.; Sun, S.T.; Li, Y. & Tanaka, T. (1994). Origin of Structural Inhomogeneities in Polymer Gels. *Macromolecules, 27,* 6791-6796.

[155] Shah, M.A.; Bergethon, P.R.; Boak, A.M.; Gallop, P.M. & Kagan, H.M. (1992). Oxidation of peptidyl lysine by copper complexes of pyrroloquinoline quinone and other quinones: A model for oxidative pathochemitry. *Biochimica et Biophysica Acta, 1159*, 311–318.

[156] Sung, H.W.; Huang, D.M.; Chang, W.H.; Huang, R.N. & Hsu, J.C. (1999). Evaluation of gelatin hydrogel crosslinked with various crosslinking agents as bioadhesives: In vitro study. *Journal of Biomedical Materials Research, 46,* 520-530.

[157] Chang, Y.; Tsai, C.C.; Liang, H.C. & Sung, H.W. (2002). In vivo evaluation of cellular and acellular bovine pericardia fixed with a naturally occurring crosslinking agent (genipin). *Biomaterials, 23,* 2447-2457.

[158] Liu, B.S.; Yao, C.H.; Hsu, S.H.; Yeh, T.S.; Chen, Y.S. & Kao, S.T. (2004). A Novel Use of Genipin-Fixed Gelatin as Extracellular Matrix for Peripheral Nerve Regeneration. *Journal of Biomaterials Application, 19,* 21-34.

[159] Nickerson, M.T.; Farnworth, R.; Wagar, E.; Paulson, A.T.; Hodge, S.M. & Rousseau, D. (2006). Some physical and microstructural properties of genipin-crosslinked gelatin–maltodextrin hydrogels . *International Journal of Biological Macromolecules, 38,* 40–44.

[160] Butler, M.F.; Ng, Y.F. & Pudney, P.D.A. (2003). Mechanism and kinetics of the crosslinking reaction between biopolymers containing primary amine groups and genipin. *Journal of Polymer Science Part A: Polymer Chemistry, 41,* 3941–3953.

[161] Liang, H.C.; Chang, W.H.; Liang, H.F.; Lee, M.H. & Sung, H.W. (2004). Crosslinking structures of gelatin hydrogels crosslinked with genipin or a water-soluble carbodiimide. *Journal of Applied Polymer Science, 91,* 4017-4026.

[162] Berning, D.E.; Katti, K.V.; Barnes, C.L. & Volkert, W.A. (1999). Chemical and Biomedical Motifs of the Reactions of Hydroxymethylphosphines with Amines, Amino Acids and Model Peptides. *Journal of the American Chemical Society, 121*, 1658–1664.

[163] Henderson, W.; Petach, H.H. & Sarfo, K. (1994). A novel polymeric phosphine oxide-derived support for enzyme immobilization. *Journal of the Chemical Society, Chemical Communications, 3,* 245–246.

[164] Ong, S.R.; Trabbic-Carlson, K.A.; Nettles, D.L.; Lim, D.W.; Chilkoti, A. & Setton, L.A. (2006). Epitope tagging for tracking elastin-like polypeptides. *Biomaterials, 27,* 1930–1935.

[165] Hanson, N.A.; Nettles, D.L.; Vail, T.P.; Lim, D.W.; Chilkoti, A. & Setton, L.A. (2006). Epitope tagging for tracking Elastin-like polypeptides. *Transactions of the Annual Meeting - Orthopaedic Research Society.*

[166] Volkert, W.A. & Hoffman, T.J. (1999). Therapeutic Radiopharmaceuticals. *Chemical Reviews*, 99, 2269–2292

[167] Urry, D.W.; Haynes, B.; Zhang, Z.; Harris, R.D. & Prasad, K.U. (1988). Mechanochemical coupling in synthetic polypeptides by modulation of an inverse temperature transition. *Proceedings of the National Academy of Sciences USA, 85,* 3407-3411.

[168] Wise, S.G. & Weiss, A.S. (2009). Tropoelastin. The International Journal of Biochemistry & Cell Biology, 41, 494–497.

[169] Clarke, A.W.; Arnspang, E.C.; Mithieux, S.M.; Korkmaz, E.; Braet, F. & Weiss, A.S. (2006). Tropoelastin massively associates during coacervation to form quantized protein spheres. *Biochemistry, 45,* 9989–9996.

[170] Kagan, H.M. & Sullivan, K.A. (1982). Lysyl oxidase: preparation and role in elastin biosynthesis. *Methods in Enzymology, 82 (Pt A)*, 637–650.

[171] Naumann, C.; Mithieux, S.M.; Szekely, D.; Tu, Y.; Weiss, A.S. & Kuchel, P.W. (2009). Setting paint'' analogy for the hydrophobic self-association of tropoelastin into elastin-like hydrogel. *Biopolymers, 91,* 321–330.

[172] Mithieux, S.M.; Wise, S.G.; Raftery, M.J.; Starcher, B. & Weiss, A.S. (2005). A model two-component system for studying the architecture of elastin assembly in vitro. *Journal of Structural Biology, 149,* 282–289.

[173] Yamada, S. &, Sugahara, K. (2008). Potential therapeutic application of chondroitin sulfate/dermatan sulfate. *Current Drug Discovery Technologies, 5,* 289–301.

[174] Capila, I. & Linhardt, R.J. (2002). Heparin-protein interactions. *Angewandte Chemie International Edition, 41,* 391–412.

[175] Tu, Y. & Weiss, A.S. (2008). Glycosaminoglycan-mediated coacervation of tropoelastin abolishes the critical concentration, accelerates coacervate formation, and facilitates spherule fusion: Implications for tropoelastin microassembly. *Biomacromolecules, 9,* 1739–1744.

[176] Yan, H.; Saiani, A.; Gough, J.E. & Miller, A.F. (2006). Thermoreversible Protein Hydrogel as Cell Scaffold. *Biomacromolecules, 7,* 2776-2782.
[177] Yan, H.; Frielinghaus, H.; Nykanen, A.; Ruokolainen, J. & Miller, A. (2008). Thermoreversible lysozyme hydrogels: properties and an insight into the gelation pathway *Soft Matter, 4*, 1313-1325.
[178] Verheyen, E.; Delain-Bioton, L.; van der Wal, S.; El Morabit, N.; Barendregt, A.; Hennink, W.E. & van Nostrum, C.F. (2010). Conjugation of Methacrylamide Groups to a Model Protein via a Reducible Linker for Immobilization and Subsequent Triggered Release from Hydrogels. *Macromolecular Bioscience, 10*, 1517–1526.
[179] Kleopina, G.V.; Kravchenko, N.A. & Kaverzneva, E.D. (1965). Role of the 6-amino groups of lysine in lysozyme. *Russian Chemical Bulletin, 14,* 806-812.
[180] Romanovskaya, I.I.; Dekina, S.S. & Pashkin, I.I. (2009). Drug synthesis methods and manufacturing technology. *Pharmaceutical Chemistry Journal, 43,* 615-618.
[181] Coulombe, P.A.; Bousquet, O.; Ma, L.; Yamada, S. & Wirtz, D. (2000). The 'ins' and 'outs' of intermediate filament organization. *Trends in Cell Biology, 10*, 420-428.
[182] Rouse, J.G. & Van Dyke, M.E. (2010). A Review of Keratin-Based Biomaterials for Biomedical Applications. *Materials, 3,* 999-1014.
[183] Sionkowska, A. (2011). Current research on the blends of natural and synthetic polymers as new biomaterials: Review. *Progress in Polymer Science, 36.* 1254-1276.
[184] Lin, H.; Sritham, E.; Lim, S.; Cui, Y. & Gunasekaran, S. (2010). Synthesis and Characterization of PH-and Salt-Sensitive Hydrogel Based on Chemically Modified Poultry Feather Protein Isolate. *Journal of Applied Polymer Science, 116,* 602-609.
[185] Yamauchi, K.; Yamauchi, A.; Kusunoki, T.; Kohda, A. & Konishi, Y. (1996). Preparation of stable aqueous solution of keratins, and physiochemical and biodegradational properties of films. *Journal of Biomedical Materials Research, 31,* 439-444.
[186] Magin, T.M.; Vijayaraj, P. & Leube, R.E. (2007). Structural and regulatory functions of keratins. *Experimental Cell Research*, *313*, 2021−2032.
[187] Izawa, I. & Inagaki, M. (2006). Regulatory mechanisms and functions of intermediate filaments: A study using site- and phosphorylation state-specific antibodies. *Cancer Science, 97*, 167-174.
[188] Tachibana, A.; Furuta, Y.; Takeshima, H.; Tanabe, T. & Yamauchi, K. (2002). Fabrication of wool keratin sponge scaffolds for long-term cell cultivation. *Journal of Biotechnology, 93,* 165-170.
[189] Verma, V.; Verma, P.; Ray, P. & Ray, A.R. (2008). Preparation of scaffolds from human hair proteins for tissue-engineering applications. Biomedical Materials, 3, 25007.
[190] Hill, P.; Brantley, H. & Van Dyke, M. (2010). Some properties of keratin biomaterials: Kerateines. *Biomaterials, 31,* 585-593.
[191] Crewther, W.G.; Fraser, R.D.B.; Lennox, F.G. & Lindley, H. (1965). The chemistry of keratins. In: Anfinsen, C.B. Jr; Anson, M.L.; Edsall, J.T. &, Richards, F.M. editors. Advances in Protein Chemistry. New York: Academic Press; 191–346.
[192] Aboushwareb, T.; Eberli, D.; Ward, C.; Broda, C.; Holcomb, J.; Atala, A. & Van Dyke, M. (2009). A Keratin Biomaterial Gel Hemostat Derived from Human Hair: Evaluation in a Rabbit Model of Lethal Liver Injury. *Journal of Biomedical Materials Research Part B: Applied Biomaterials, 90B,* 45-54.

[193] Sierpinski, P.; Garrett, G.; Ma, J.; Apel, P.; Klorig, D.; Smith, T.; Koman, A.; Atala, A. & Van Dyke, M. (2008). The use of keratin biomaterials derived from human hair for the promotion of rapid regeneration of peripheral nerves. *Biomaterials, 29,* 118-128.

[194] Ju, H.K.; Kim, S.Y. & Lee, Y.M. (2001). pH/temperature-responsive behaviors of semi-IPN and comb-type graft hydrogels composed of alginate and poly(N-isopropylacrylamide). *Polymer, 42,* 6851-6859.

[195] Schaeffer, A.J. (2002). The expanding role of fluoroquinolones. *American Journal of Medicine, 113Suppl 1A,* 45S–54S.

[196] Segev, S.; Yaniv, I.; Haverstock, D. & Reinhart, H. (1999). Safety of long-term therapy with ciprofloxacin: data analysis of controlled clinical trials and review. *Clinical Infectious Diseases, 28,* 299-308.

[197] Petri, W.A. Jr. (2001). Antimicrobial Agents (Cont.). In: Hardman, J.G. & Limbird, L.E. editors. The Pharmacological Basis of Therapeutics. New York: McGraw-Hill; 1171–1188.

[198] Bernot, K.M.; Lee, C.H. & Coulombe, P.A. (2005). A small surface hydrophobic stripe in the coiled-coil domain of type I keratins mediates tetramer stability. *Journal of Cell Biology, 168,* 965-974.

[199] Ikkai, F. & Naito, S. (2002). Dynamic light scattering and circular dichroism studies on heat-induced gelation of hard-keratin protein aqueous solutions. *Biomacromolecules, 3,* 482-487.

[200] Lenzi, F.; Sannino, A.; Borriello, A.; Porro, F.; Capitani, D. & Mensitieri, G. (2003). Probing the degree of crosslinking of a cellulose based superabsorbing hydrogel through traditional and NMR techniques. *Polymer, 44,* 1577-1588.

[201] Goddard, D.R. & Michaelis, L. (1934). A study on keratin. *Journal of Biological Chemistry, 106,* 605-614.

[202] Bailey, A.J. & Paul, R.G. (1998). Collagen is not so simple protein. *Journal of the society of leather technologist and chemists, 82,* 104-118.

[203] Goh, K.L.; Hiller, J.; Haston, J.L.; Holmes, D.F.; Kadler, K.E.; Murdoch, A.; Meakin, J.R. & Wess, T.J. (2005). Analysis of collagen fibril diameter distribution in connective tissues using small-angle X-ray scattering. *Biochimica et Biophysica Acta, 1722,* 183-188.

[204] Bella, J. (2010). A new method for describing the helical conformation of collagen: dependence of the triple helical twist on amino acid sequence. *Journal of Structural Biology, 170,* 377-391.

[205] Deng, C.; Zhang, P.; Vulesevic, B.; Kuraitis, D.; Li, F.; Yang, A.F.; Griffith, M. & Suuronen, E.J. (2010). A collagen–chitosan hydrogel for endothelial differentiation and angiogenesis. *Tissue Engineering Part A, 16,* 3099-3109.

[206] Lee, C.H.; Singla, A. & Lee, Y. (2001). Biomedical applications of collagen. *International Journal of Pharmaceutics, 221,* 1-22.

[207] Seal, B.L.; Otero, T.C. & Panitch, A. (2001). Polymeric biomaterials for tissue and organ regeneration. *Materials Science and Engineering: R: Reports, 34,* 147-230.

[208] Duan, X. & Sheardown, H. (2005). Crosslinking of collagen with dendrimers. *Journal of Biomedical Materials Research Part A. 75A*, 510-518.

[209] Friess, W. (1998). Collagen – biomaterial for drug delivery. *European Journal of Pharmaceutics and Biopharmaceutics, 45,* 113-136.

[210] Khor, E. (1997). Methods for the treatment of collagenous tissues for bioprostheses. *Biomaterials, 18,* 95-105.

[211] Gorham, S.D. (1991). Biomaterials: Novel Materials from Biological Sources. Stockton Press, New York.

[212] Heidemann, E. (1988), The Chemistry of Tanning. CRC Press, Boca Raton, FL.

[213] Lee, C.R.; Grodzinsky, A.J. & Spector, M. (2001). The effects of cross-linking of collagen-glycosaminoglycan scaffolds on compressive stiffness, chondrocyte-mediated contraction, proliferation and biosynthesis. *Biomaterials, 22,* 3145-3154.

[214] Rault, I.; Frei, V.; Herbage, D.; Abdul Malak, N. & Huc, A. (1996). Evaluation of different chemical methods for cros-linking collagen gel, films and sponges. *Journal of Materials Science: Materials in Medicine, 7,* 215-221.

[215] Gratzer, P.F. & Lee, J.M. (2001). Control of pH alters the type of cross-linking produced by 1-ethyl-3-(3-dimethylaminopropyl)-carbodiimide (EDC) treatment of acellular matrix vascular grafts. *Journal of Biomedical Materials Research, 58,* 172-179.

[216] Homenick, C.M.; Sheardown, H. & Adronov, A. (2010). Reinforcement of collagen with covalently-functionalized single-walled carbon nanotube crosslinkers. *Journal of Materials Chemistry, 20,* 2887-2894.

[217] Hwang, Y.J.; Larsen, J.; Krasieva, T.B. & Lyubovitsky J.G. (2011). Effect of Genipin Crosslinking on the Optical Spectral Properties and Structures of Collagen Hydrogels. *ACS Applied Materials and Interfaces, 3,* 2579-2584.

[218] Sundararaghavan, H.G.; Monteior, G.A.; Lapin, N.A.; Chabal, Y.J.; Miksan, J.R. & Shreiber, D.I.J. (2008). Genipin-induced changes in collagen gels: Correlation of mechanical properties to fluorescence. *Journal of Biomedical Materials Research Part A, 87A*, 308–320.

[219] Zhang, L.; Li, K.; Xiao, W.; Zheng, L.; Xiao, Y.; Fan, H. & Zhang, X. (2011). Preparation of collagen–chondroitin sulfate–hyaluronic acid hybrid hydrogel scaffolds and cell compatibility in vitro. *Carbohydrate Polymers, 84,* 118-125.

[220] Lien, S.M.; Li, W.T. & Huang, T.J. (2008). Genipin-crosslinked gelatin scaffolds for articular cartilage tissue engineering with a novel crosslinking method. *Materials Science & Engineering C, 28*, 36-43.

[221] Kirker, K.R.; Luo, Y.; Nielson, J.H.; Shelby, J. & Prestwich, G.D. (2002). Glycosaminoglycan hydrogel films as bio-interactive dressings for wound healing. *Biomaterials, 23*, 3661-3671.

[222] Luo, Y.; Kirker, K.R. & Prestwich, G.D. (2000). Cross-linked hyaluronic acid hydrogel films: New biomaterials for drug delivery. *Journal of Controlled Release, 69*,169-184.

[223] Homenick, C.M.; de Silveira, G.; Sheardown, H. & Adronov, A. (2011). Pluronics as crosslinking agents for collagen: novel amphiphilic hydrogels. *Polymer International, 60,* 458-465.

[224] Sulea, D.; A Ghica, M.V.; Micutz, M.; Albu, M.G.; Brazdaru, L.; Staicu, T., Leca, M. & Popa, L. (2010). Characterization and in vitro release of chlorhexidine digluconate comprised in type I collagen hydrogels. *Revue Roumaine de Chimie, 55,* 543-551.

[225] Albu, M.G.; Ghica, M.V.; Popa, L.; Leca, M. & Trandafir, V. (2009). Kinetics of in vitro release of doxycycline hyclate from collagen hydrogels. *Revue Roumaine de Chimie, 54,* 373-379.

[226] Gonen-Wadmany, M.; Oss-Ronen, L. & Seliktar, D. (2007). Protein-polymer conjugates for forming photopolymerizable biomimetic hydrogels for tissue engineering. *Biomaterials, 28,* 3876-3886.

[227] Elbert, D.L. & Hubbell, J.A. (2001). Conjugate Addition Reactions Combined with Free-Radical Cross-Linking for the Design of Materials for Tissue Engineering. *Biomacromolecules, 2,* 430-441.

[228] Chen, J.S.; Noah, E.M.; Pallua, N. & Steffens, G.C. (2002). The use of bifunctional polyethyleneglycol derivatives for coupling of proteins to and crosslinking of collagen matrices. *Journal of Materials Science: Materials in Medicine, 13,* 1029-1035.

[229] Pourjavadi, A.; Kurdtabar, M., Mahdavinia, G.R. & Hosseinzadeh, H. (2006). Synthesis and super-swelling behavior of a novel protein-based superabsorbent hydrogel. *Polymer Bulletin, 57,* 813-824.

[230] Pourjavadi, A.; Salimi, H.; Amini-Fazl, M.S.; Kurdtabar, M.; Amini-Fazl, A.R. (2006). Optimization of synthetic conditions of a novel collagen-based superabsorbent hydrogel by Taguchi method and investigation of its metal ions adsorption. *Journal of Applied Polymer Science, 102,* 4878-4885.

[231] Pourjavadi, A. & Kurdtabar, M. (2007). Collagen-based highly porous hydrogel without any porogen: Synthesis and characteristics. *European Polymer Journal, 43,* 877–889.

[232] Pourjavadi, A.; Salimi, H. & Kurdtabar, M. (2007). Hydrolyzed collagen-based hydrogel with salt and pH-responsiveness properties. *Journal of Applied Polymer Science, 106,* 2371-2379.

[233] Pourjavadi, A.; Kurdtabar, A. & Ghasemzadeh, H. (2008). Salt- and pH-Resisting Collagen-based Highly Porous Hydrogel. *Polymer Journal, 40,* 94-103.

[234] Pourjavadi, A. & Salimi, H. (2008). New Protein-Based Hydrogel with Superabsorbing Properties: Effect of Monomer Ratio on Swelling Behavior and Kinetics. *Industrial & Engineering Chemistry Research, 47,* 9206-9213.

Hydrogels: Synthesis, Characterization and Applications ISBN: 978-1-61942-842-3
Editors: F. V. Câmara and L. J. Ferreira 

*Chapter 2*

# THE ROLE OF HYDROGELS IN THE DEVELOPMENT OF ARTIFICIAL BONE SUBSTITUTES

***Luke M. Geever*** [*1], ***Martin O. Cloonan*** [‡2], ***Declan M. Devine*** [≠1], ***John A. Killion*** [±1], ***Laura Grehan*** [£3], ***Sinead M. Devery*** [¥3] ***and Clement L. Higginbotham*** [#1]

[1] Materials Research Group, Research Hub, Athlone Institute of Technology, Athlone, Co. Westmeath, Ireland
[2] Department of Chemistry, School of Science, Galway-Mayo Institute of Technology, Galway, Ireland
[3] Department of Life and Physical Science, Toxicology Unit, Athlone Institute of Technology, Athlone, Co. Westmeath, Ireland

## ABSTRACT

Biomedical hydrogels for tissue engineering and more specifically for bone regeneration purposes are investigated throughout this book chapter. The most commonly utilised natural and synthetic polymer materials researched in the development of such scaffolds is initially documented. Synthesis techniques and in particular the role of photopolymerisation in the design of these devices is also explored. The importance of the physical and mechanical performance of the hydrogel scaffolds is examined, while current mechanisms of improving mechanical performance including the use of composite materials are considered. The importance of porosity, scaffold surface modification, degradability and degradation rate is also highlighted. Biological considerations in terms of preliminary toxicological analysis, cell isolation and loading, growth factor incorporation and release, as well as other advanced clinical requirements

* Tel: 00353 90 6428054; fax: 00353 90 6424493: e-mail address: lgeever@ait.ie
‡ Tel: 00353 91 742049; fax: 00353 91 758412; e-mail address: Martin.Cloonan@gmit.ie
≠ Tel: 00353 90 6428054; fax: 00353 90 6424493: e-mail address: ddevine@ait.ie
± Tel: 00353 90 6468059; fax: 00353 90 6424493; e-mail address: jkillion@research.ait.ie
£ Tel: 00353 90 6468059; fax: 00353 90 6424493; e-mail address: lgrehan@research.ait.ie
¥ Tel: 00353 90 6468050; fax: 00353 90 6424493: e-mail address:sdevery@ait.ie
# Tel: 00353 90 6424463; fax: 00353 90 6424493: e-mail address: chigginbotham@ait.ie

necessary for bone regeneration scaffolds are evaluated. While it is clear following critical appraisal that hydrogels alone will not suffice as load bearing bone tissue engineering devices, hydrogel composites look set to play a vital role in the development of the next generation of bone substitute materials.

## 1.1. Introduction

Tissue engineering is a multi-disciplined field involving knowledge in both life sciences and engineering to create artificial tissues and organs to regenerate damaged tissue. Langer and Vacanti describe tissue engineering as "an interdisciplinary field that applies the principles of engineering and the life sciences towards the development of biological substitutions that restore, maintain or improve tissue or organ functions" (Langer & Vacanti 1993). Tissue regeneration works on the concept of replicating the original tissue through cellular migration, inhibitory interactions, proliferation/apoptosis, differentiation and tissue patterning/morphogenesis. This includes the seeding of specific cells onto the carrier, i.e. scaffold or hydrogel. When the scaffold is synthesised it is then cultured *in vitro* and ultimately implanted into the appropriate tissue. Once the scaffold is in the body, natural tissue regeneration processes then take place, where the new tissue regenerates as the scaffold degrades. The general process of tissue engineering is illustrated in Figure 1.

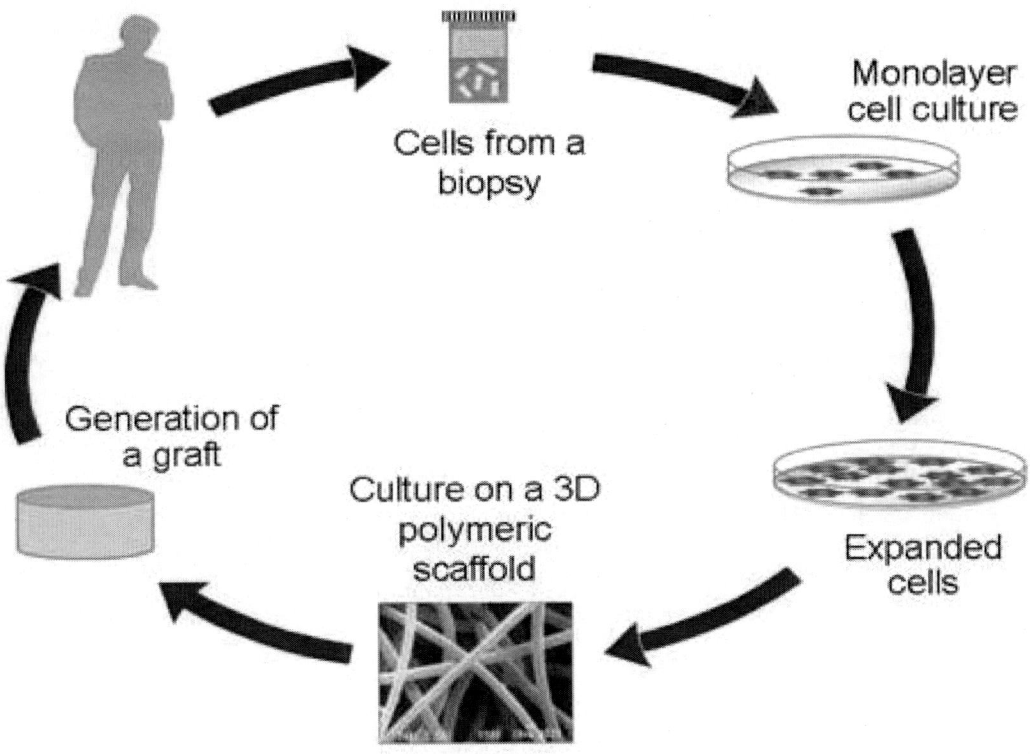


Figure 1.Basic illustration of tissue engineering process (Mayorga et al. 2010).

## 1.2. Bone Tissue Engineering

Bone is a very complex tissue with multiple cell phenotypes, distinct tissue types and high vascularisation. It plays a very demanding mechanical role in the structural support of the body. Bone has a highly anisotropic structure and remodels itself along local stress field lines in order to optimise its properties (Bertram & Swartz 1991). The main elements of bone tissue are cells, minerals, organic matrix and water (Oyen 2008). The mineral phase (approximately 69% of bone weight) is a biological analogue of the mineral hydroxyapatite (Jing *et al.* 2001). Bone also contains a significant amount of protein in the form of collagen which is a highly cross-linked material, due to both intermolecular and intramolecular crosslinks causing it to be highly insoluble (Barone *et al.* 2011). Collagen orientates the deposition of minerals and promotes cell attachment to the bone. The mechanical properties of different bone types can be seen in Table 1.

**Table 1. Mechanical properties of cortical and trabecular bone (Boyan et al. 1999)**

| Bone Type | Compressive Strength (MPa) | Elastic Modulus (GPa) | Flexural Strength (MPa) | Flexural Modulus (GPa) |
|---|---|---|---|---|
| Cortical Bone | 170 | 7-30 | 180 | 20 |
| Trabecular Bone | 40-60 | 0.05-0.5 | - | 2 |
| Adult mandible | 90-140 | 2-10 | - | 5-12 |
| Adult vertebra | - | 0.1-0.25 | - | - |

### 1.2.1. Bone Repair

In 2006 it was estimated that 36 million Americans suffered from bone related diseases while 3 million were diagnosed with fracture related injuries. Bone grafting is a common surgical procedure; it has been estimated that 2.2 million grafting procedures are performed worldwide each year (Lewandrowski *et al.* 2000). In addition studies have approximated that for every fracture, 5–10% result in impaired healing, which can cause a delayed union or non-union (Calori *et al.* 2007; Tzioupis & Giannoudis 2007). A critical size defect (CSD) is the diameter of a bone beyond which complete calcification of the defect will not occur during the lifetime of the injured animal (Schmitz & Hollinger 1986). For example, CSD values in some long bones include: 3 cm for the forearm, 5 cm in the femur and tibia and 6 cm in the humerus. Conventional treatment of fractures usually results in full restoration of skeletal integrity. However, in some situations such as severe trauma, defects in specific parts of the skeleton, or in elderly patients and/or in the presence of metabolic diseases, repair of bone defects may require additional stimulations. In general, biological activities attributed to bone repair enhancers are classified as either osteogenic, osteoconductive or osteoinductive (Bauer & Muschler 2000). The most frequently used strategies to stimulate bone healing are bone-grafts, biomaterial implants, growth factors and cell therapy (Albrektsson 2001). The combinations of different methods will be discussed in the following sections.

### 1.2.2. Biological Bone Grafts

Currently there are three main types of biological grafts: autografts, allografts and xenografts.

1. Autografts: An autograft involves harvesting bone from one location in the patient's body and transplanting it into the defect of the same patient, i.e. a self-transplant of tissue. This biological graft represents 58% of the current bone substitutes. Autografts are the gold standard because they avoid most issues related to transfer, or infection and rejection. The bone is often harvested from the iliac crest and can be either trabecular or cortical tissue (Barone *et al.* 2011). The advantages of using autografts include clinical experience, their osteoconductivity, osteoinductivity and osteogenicity. However, it has some limitations including being painful to harvest, the possibility of donor site morbidity, chronic donor shortage and dependency on donor quality.
2. Allograft: This type of grafting procedure differs from autografts in that the bone is harvested from one human and implanted into another. The harvesting comes from processing bone from a cadaver and transplanting it into the patient's defect. Therefore, allografts run the risk of immune response and possible disease transmission. In order to minimise this risk, prior to placing the allograft into the patient, the graft must be thoroughly sterilised. The quality of allografts is inferior to that of autologous grafts. Allografts have a poor degree of cellularity, less revascularisation, and a higher resorption rate compared to autologous grafts resulting in a slower rate of new bone tissue formation.
3. Xenografts: The bone tissue from this type of biological graft is harvested from a different species prior to implantation into the patient's defect. One advantage of these grafts is their availability in various shapes and sizes. However, it does run the risk of immunological reactions and infections, handling issues and poor initial strength.

As highlighted by the disadvantages of each type of existing biological graft, to date there is still no ideal graft to help repair critical size bone defects. Therefore, an extensive body of research has been carried out to repair such bone defects with various types of biomaterials.

## 1.3. Scaffold Design Requirements

Scaffolds effectively perform as a platform for tissue formation where various cell types and growth factors are incorporated into the scaffold prior to implantation. To date, countless scaffolds have been synthesised with an array of biomaterials in an effort to regenerate tissues that have been damaged for one reason or another. However, for every tissue type, there are a number of design requirements when deciding the appropriateness of any particular scaffold for the use in tissue engineering. Below, the defined scaffold requirements for tissue engineering are briefly summarised in terms of biocompatibility, mechanical properties, vascularisation, degradation and manufacturing technology. These scaffold requirements are examined in greater detail throughout the remainder of the book chapter, with a key focus on the area of bone regeneration.

### 1.3.1. Biocompatibility

The first, and one of the most important scaffold requirements, is that of biocompatibility. Williams in 1987 described biocompatibility as "the ability of a material to perform with an appropriate host response in a specific application". Hydrogels and other scaffolds for bone regeneration require comprehensive biocompatibility testing as they are intended for implantation and subsequent interaction with host cells and tissue. Previous investigations into the biocompatibility of hydrogels have shown that common sources of toxicity include photoinitiators, chemical crosslinkers, unreacted monomers, stabilisers and organic solvents. However, improvements can be made by optimum selection of photoinitiators as well as photoinitiator concentration and use of macromolecular monomers. Karaca *et al.* (2010) showed that minimisation of oxygen inhibition by photoinitiator selection promotes biocompatibility of the hydrogel while Studer *et al.* (2003) showed that minimisation of photoinitiator concentration leads to improved biocompatibility. Hydrogels are also structurally similar to the natural extracellular matrix of many tissues, providing a three dimensional, highly hydrated environment comparable to soft tissue for encapsulated cells. The porous structure of hydrogels also allows diffusion of nutrients, oxygen, metabolites and cellular wastes. Due to the importance of biocompatibility in tissue engineering, this topic will be further discussed in Section 1.8.1.

### 1.3.2. Mechanical Properties

The mechanical properties play a vital role in the success of the scaffold. Scaffolds must provide mechanical stability within the defect until full regeneration of tissue has occurred. The mechanical forces affect the remodelling and homeostasis for tissues that are load-bearing tissues, i.e. bone and cartilage (Horan *et al.* 2005; Puppi *et al.* 2010). In the case of hydrogels, the mechanical properties can be controlled by alternating the methods of crosslinking and the length of the polymer chains (See also Section 1.5.1).

### 1.3.3. Vascularisation

Nearly all tissues in the human body are vascularised. Vascularisation (formation of new blood vessels in adult tissue) is a key requirement for scaffold design of these tissues as it supplies a base for elimination of waste by perfusion and also nutrient exchange (Slaughter *et al.* 2009). Therefore, a scaffold should contain a porous surface and microstructure, allowing ingrowth, *in vitro* cell adhesion and reorganisation for adequate vascularisation *in vivo* to occur (Puppi *et al.* 2010). Pore interconnectivity directly influences the diffusion of physiological nutrients and gases to cells as well as the removal of metabolic waste and by-products from cells (Vacanti *et al.* 1988; Mikos *et al.* 1993). However, if a scaffold has a high degree of porosity, this will have a detrimental effect on the mechanical properties of the scaffold. Therefore, there is a need for a balance between the porous structure and the mechanical properties of the scaffold to have sufficient mechanical strength while at the same time allowing adequate cell infiltration and vascularisation in designing a successful scaffold (O'Brien 2011) (See also Section 1.8.4.2).

### 1.3.4. Degradation

Degradation is necessary in tissue regeneration applications. Ideally, in bone regeneration the rate at which a scaffold degrades in the defect, should mimic the rate of new bone formation allowing cells to produce their own extracellular matrix. As the scaffold degrades, the by-products should be non-toxic. In the case of hydrogels, there are three methods in which degradation can occur: hydrolysis, dissolution and enzymatic cleavage. In most cases, synthetic hydrogels degrade through hydrolysis of ester linkages (Drury & Mooney 2003; Chwalek *et al.* 2011). Scaffolds can also be made degradable by incorporating cleavable groups into the backbone of the polymer.

### 1.3.5. Manufacturing Technology

A major challenge is presented to manufacturers in creating a porous architecture, i.e. scaffolds for tissue engineering. The challenges involved in the manufacturing of scaffolds involve the synthesis of the material and the process by which that material can be formed into a three-dimensional device (Hollister 2009). This is a process by which the scaffold can be taken from the laboratory and scaled up to manufacturing conditions, while still maintaining its material parameters and its cost effectiveness. Successful bone regeneration scaffolds which have made it to the market include:

- INFUSE Bone Graft, Medtronic 2002.
- Vitoss Scaffold FOAM, Orthovita 2004.
- FortrOss, Pioneer Surgical 2008 (Place *et al.* 2009).

## 1.4. Fabrication Techniques

A vast number of fabrication techniques have been designed to produce scaffolds for tissue engineering. Various techniques are used for different material and tissue engineering applications for optimum scaffold properties. The main methods of scaffold fabrication are described below.

### 1.4.1. Photopolymerisation

Photochemical initiated polymerisation reactions are a well-established technology for many industrial applications and are currently finding use for biomedical applications and tissue engineering (Williams *et al.* 2005). A photoinduced polymerisation reaction is usually considered as a reaction where the initiation step is produced by a photochemical event. Thus, on exposure to UV or visible light the liquid monomer formulation is converted into a solid polymer. However, if the monomer contains more than one reactive function, a crosslinked polymer network is readily produced. Photopolymerisation is different from photocrosslinking where every chain propagation step requires the absorption of a photon of light. Initiating polymerisation by radiation curing is beneficial as it reduces the need for volatile organic solvents, allows temporal control of initiation and proceeds more rapidly than thermal polymerisation. Most monomers, oligomers or pre-polymers commonly employed in photopolymerisation do not readily produce sufficient initiating species upon light exposure. Therefore, it is usually necessary to incorporate a photoinitiator that will begin the polymerisation process (Fouassier 1995). As UV light energy is absorbed by the photoinitiator, it fragments into free radical or cationic reactive species. There are also a few cases of initiators (e.g. iodonium and sulphonium salts, arene complexes), which are able to initiate polymerisations via both cationic and free radical processes. The free radical class of initiators represent greater than 90% of commercially used initiator chemistry due to their availability, high quantum yields of radical generation and monomer radical formations, while cationic curing remains more limited in scope (Moon *et al.* 2005). The free radicals initiate the polymerisation process, attacking reactive double bonds on monomer molecules and converting them to a polymer.

Hydrogels are a key group of biomaterials that are often synthesised using photopolymerisation. Many hydrogels are biocompatible in nature, resulting in minimal inflammatory responses and tissue damage. However it is widely accepted that a variety of hydrogels from monomeric precursors require a number of time consuming washing steps to ensure the cytocompatibility of such scaffolds, mainly due to the presence of minute levels of unreacted monomer present following photopolymerisation. Therefore macromolecular monomers are being used more often, as hydrogels synthesised using these materials have been reported to exert a minimal toxicological response, and do not necessitate washing steps. Poly(ethylene glycol) derivatives have been widely investigated and reviewed as scaffold materials for bone regeneration (Zhu 2010). Therefore in this book chapter alternative chemically modified photopolymerisable derivatives such as chitosan and polyvinylalcohol are discussed.

Chitosan has been chemically modified to obtain photopolymerisable derivatives (Qui 2008). The majority of the curable groups attached to polymers and biopolymers have been applied to the polysaccharide chitosan. Chitosan contains primary alcohol and amine functional groups that allow facile functionalisation with curable groups. In general an amino group is more reactive as long as the alcohol is un-ionised. The predominant curable groups are the acrylates and methacrylates. Reagents for the synthesis of the photopolymerisable chitosan's include glycidyl methacrylate **I** (Han *et al.* 2009; Ngamviriyavong *et al.* 2010), poly(ethylene glycol) diacrylate **II** (Ma *et al.* 2009), ethylene glycol acrylate **III** (Amsden *et al.* 2007; Li *et al.* 2009), ethylene glycol acrylate methacrylate **IV** (Gao *et al.* 2010), ethoxylated trimethylolpropane triacrylate **V** and tris(2-hydroxyethyl) isocyanurate triacrylate

**VI** (Correa *et al.* 2009), Figure 2. Some of the polymers formed from the photopolymerisation of the acrylate and methacrylate chitosans have been evaluated as scaffolding materials for bone regeneration (Amsden *et al.* 2007; Ma *et al.* 2010).

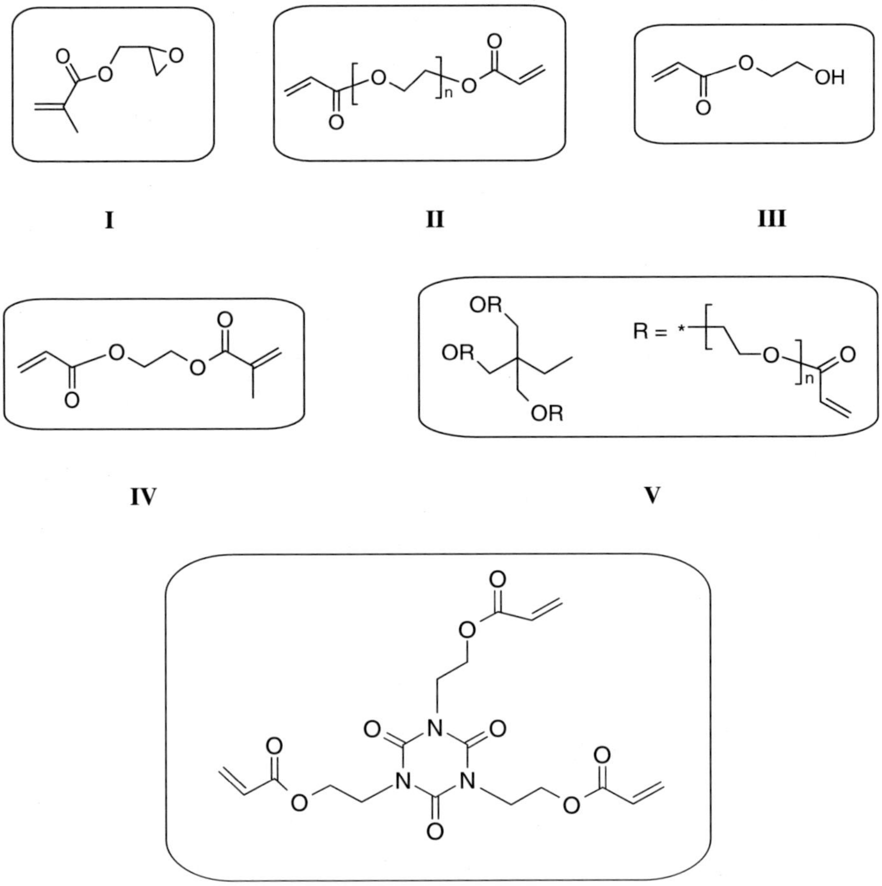


Figure 2. Reagents for attaching curable acrylate and methacrylate groups to chitosan.

*O*-Carboxymethylchitosan has also been methacrylated in the synthesis of a photocrosslinkable macromolecule (Poon *et al.* 2007). Acrylates are more reactive than methacrylates in radical photopolymerisation reactions but they also give rise to a greater degree of 1,4-conjugate addition side reactions (Michael addition), which results in the formation of aliphatic groups. The aliphatic groups can have toxic effects when used as biomaterials. *N,O*-Maleoyl chitosan **VIII** has been synthesised by reacting chitosan with maleic anhydride **VII** in the presence of an acid or base catalyst, Scheme 1 (Dong *et al.* 2006; Fan *et al.* 2009; Zhong *et al.* 2010). These derivatives are less sensitive to Michael addition reactions due to the presence of the 1,2-disubstituted olefin structure. Under the reaction conditions the double bond of the maleic group isomerises to the more stable *trans* diastereomer. Both organic and water soluble photo-curable chitosan versions have been prepared. The water soluble versions are achieved by converting the carboxylic acid group of the macromer **VIII** to its sodium salt. A photoinitiator is required in the photopolymerisation of the acrylates, methacrylates and the *N,O*-maleoyl photopolymers. Cell cytotoxicity tests on some of the chitosan-based biodegradable biomaterials derived from the *N,O*-maleoyl chitosan photopolymer have shown that they are relatively non-toxic. The *N,O*-maleoyl photopolymer has also proven to be relatively non-toxic (Zhong *et al.* 2010). Thus there is ample scope for these new materials in bone regeneration research.

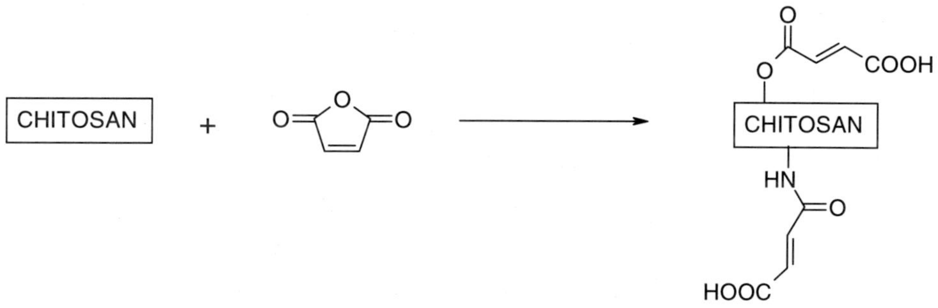


**VII** **VIII**

Scheme 1. Synthesis of N,O-Maleoyl Chitosan VIII.

The curable cinnamate group has been covalently bonded to chitosan via acylation with cinnamoyl chloride **IV** in the presence of base, Scheme 2, and photopolymerised with and without a photoinitiator (Ma *et al.* 2011). The cinnamoyl group is nontoxic. Application of the *N,O*-cinnamoyl chitosan photopolymer in bone regeneration has yet to be explored.

Other aromatic systems which were used as a photopolymerisable group include the azidobenzene functional group (Obara *et al.* 2003; Fukuda 2006; Ono *et al.* 2009; Rickett *et al.* 2011). Synthesis involved coupling chitosan to *p*-azidobenzoic acid **V** using a dehydrating agent such as 1-ethyl-3-[3-dimethylaminopropyl] carbodiimide (EDC), Scheme 3. A photoinitiator is not required for the photopolymerisation of the *p*-azidobenzene macromer. Some of the polymers produced from the *p*-azidobenzene chitosan macromer have been tested and shown to be cyctocompatible (Rickett *et al.* 2011) and thus the macromer has the potential of producing scaffolding materials for bone regeneration.

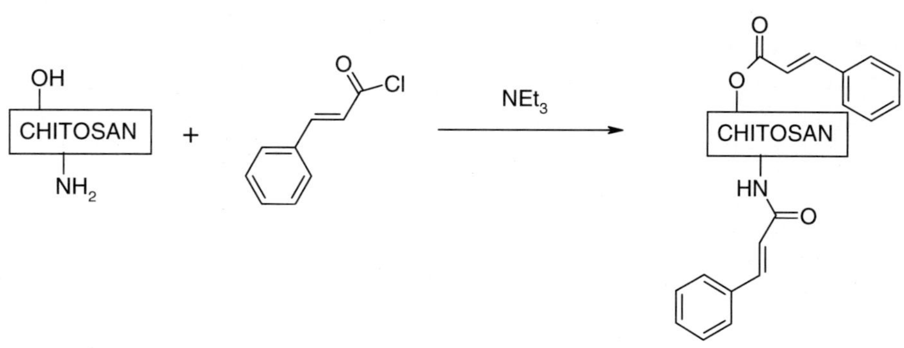


**IV**

Scheme 2. Synthesis of the photopolymerisable N,O-Cinnamoyl Chitosan.

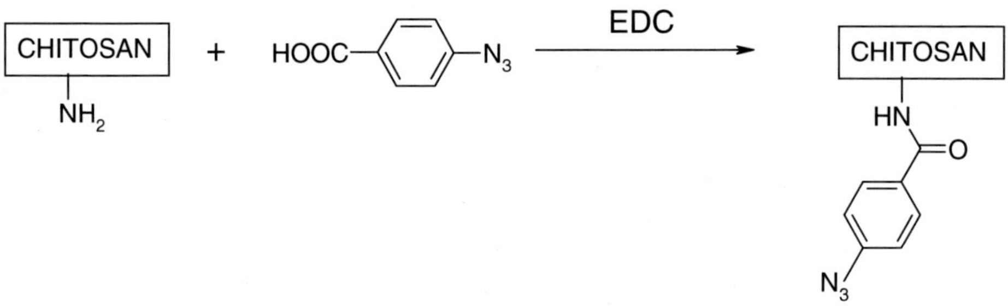


**V**

Scheme 3. Synthesis of the photocrosslinkable N-Azido Chitosan.

The curable furan heterocycle was covalently bonded to chitosan by reacting with furfuryl isocyanate **VI**, Scheme 4 (Kim *et al.* 2011). This photocrosslinkable chitosan derivative invokes a photo-oxidation-induced cross-linking mechanism. The synthesised furfuryl isocyanate chitosan derivative showed no significant toxicity. The scope of this photopolymerisable chitosan derivative in bone regeneration has not been explored at present.

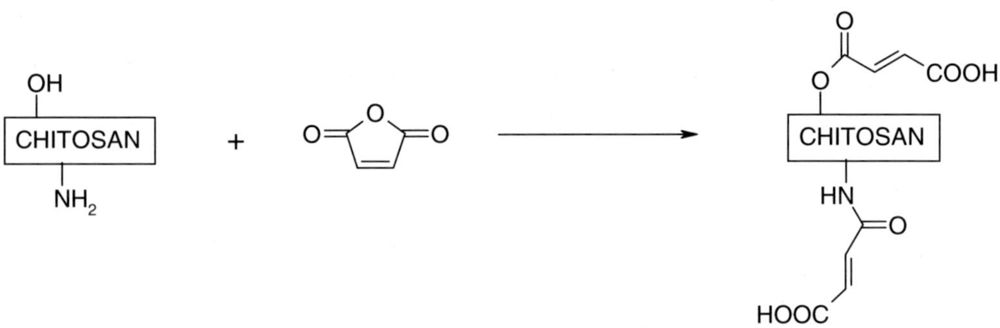


**VI**

Scheme 4. Synthesis of furfuryl isocyanate Chitosan.

Polyvinylalcohol (PVA) has been functionalized extensively to produce photopolymerisable derivatives (Pande *et al.* 1973; Sperry *et al.* 1984; Dickinson *et al.* 1994; Kityk *et al.* 1997; Martens *et al.* 2000; Blaya *et al.* 2002; Schmedlen *et al.* 2002; Ito *et al.* 2005; Ibara-Escutia *et al.* 2010). All the reagents used for chitosan are also applicable to PVA due to the presence of pendant nucleophilic hydroxyl groups in PVA, Scheme **5**. In the literature the majority of these reagents have been used. PVA has also been chemically modified with the curable allyl group using allyl isocyanate as a reagent (Pande *et al.* 1973). The presence of the 1,3-dihydroxy (diol) functional group in PVA allows an extra dimension to the synthesis of photopolymerisable PVA compared to chitosan. Acetals or aldehydes can be used to link the curable group to PVA via reacting with the diol group, Scheme 5.

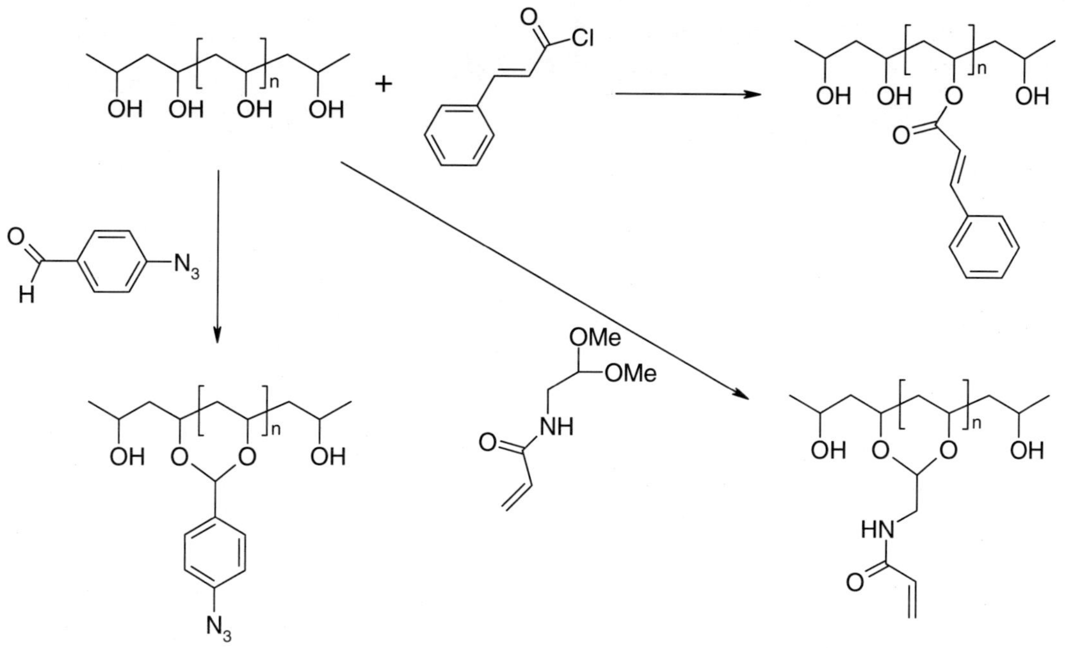

Scheme 5. Synthesis of photopolymerisable PVA via reactions involving the hydroxyl group or the 1,3-diol group of PVA.

## 1.4.2. Solvent Casting and Particulate Leaching Techniques

Particulate leaching is very simple, easy and inexpensive (Dietmar 2000). This process involves dissolving a polymer in a solvent with the addition of particles, i.e. porogens. The procedure for creating these scaffolds involves casting a solution into a mould. Once the solvent has evaporated, the particulates are leached out and the empty voids left behind create a scaffold with a porous structure with uniform porosity (Liu *et al.* 2007). Solvent casting works very well with thin membranes; however, if the sample cross-section is too thick, it is very difficult to remove the particles from within the polymer matrix (Mikos *et al.* 1994). The advantages of this technique include cost effectiveness, ease of fabrication and greater control of porosity (Subia *et al.* 2010). The disadvantages include a necessary step of removing the porogen, limited mechanical properties and limited thickness (Puppi *et al.* 2010).

## 1.4.3. Gas Foaming

This method uses supercritical fluid technology to form a porous scaffold. The gas foaming technique is carried out by placing the polymer in a pressure chamber and inputting high pressure carbon dioxide into the scaffold. The carbon dioxide dissolves into the polymer making it become unstable. This causes phase separation and produces the pores within the scaffold. The advantage of using gas forming as a fabrication technique over solvent casting and particulate leaching is that it does not use any organic solvents, which enhances biocompatibility. However, scaffolds produced using this technique have limited mechanical properties and poor pore interconnectivity (Ikada 2006).

### 1.4.4. Phase Separation

The phase separation technique involves the thermodynamic instability of a polymer, in which a porous scaffold is created by the division between the polymer rich and polymer poor phases. Currently, there are a number of processes of inducing phase separation in mixtures of such polymers. An example of this would be liquid-liquid phase separation. This process creates polymer rich and polymer poor phases in the polymer liquid. Once the polymer poor phase has been removed, the porous scaffold is formed (Lee *et al*. 2004). The structure of the scaffold is controlled by phase separation conditions (Liu *et al*. 2007). However, it can be difficult to regulate the scaffold morphology (Subia *et al*. 2010). Liu *et al*. (2000) produced macroporous poly(2-hydroxyethylene) methacrylate hydrogels using this technique.

### 1.4.5. Electrospinning

This technique designs nonwoven fabrics consisting of extremely fine polymer fibers. The method involves applying a high electrostatic field to a capillary connected to a syringe containing a polymer solution. One electrode is placed in the collector and the other in the polymer solution (Subia *et al*. 2010). Electrospinning can produce fibers of varying lengths: from a few nanometers to hundreds of microns (Xu *et al*. 2004). Yang *et al*. (2007) produced gelatine/PVA nanofiber scaffolds for controlled release of drugs. The main advantage of electrospinning is that it can produce scaffolds which are appropriate for growth of tissues and cells by controlling the pore size and fiber length. *N*-methylene phosphonic chitosan hydrogels and electrospun nanofibrous hydrogels have been synthesised and results show that they can be both osteoconductive and osteoinductive (Datta *et al*. 2012).

### 1.4.6. Rapid Prototyping (RP)

Rapid prototyping Technologies (RP) or Solid Free Form fabrication (SFF) is a computer controlled technique. It is designed as a set of manufacturing processes capable of creating a part from a computer-aided design (CAD) package. Once the 3-D image of the tissue defect (through MRI or CT scan) has been taken and designed in CAD, the machine is then ready to form the scaffold starting with the bottom of the scaffold and working its way up in layers (Puppi *et al*. 2010). Within rapid prototyping there are several techniques which include: stereolithography, selective laser sintering (SLS), fused deposition modeling (FDM) and 3D printing (3D-P). The advantages of RP are control on pore size, pore shape, interconnectivity, branching, geometry and orientation. However, the equipment used to fabricate these hydrogels is very expensive and due to the process it is limited to the polymers which can be used.

### 1.4.7. Melt Moulding

The process of melt moulding includes mixing a polymer powder and porogen component in a mould and heating the mould above the glass-transition temperature ($T_g$) or melt temperature ($T_m$) of the polymer, while at the same time applying pressure on the blend where the polymer is bound together. Once blended, the moulding is removed and the porogen is leached out. Scaffolds produced using this technique assume the shape of the mould. This process allows independent control over the pore size and porosity, however it does require high temperatures for non-amorphous polymers (Puppi *et al.* 2010).

### 1.4.8. Freeze Drying

Freeze drying is based on the principle of sublimation involving the removal of a solvent (usually water). Firstly, the polymer is dissolved in solvent to its required concentration. The mixture is cooled down below its glass transition temperature where the polymer chains are in a frozen state resulting in the formation of ice crystals. Once frozen the solvent is removed by applying a vacuum pressure, where the sublimation cycle produces a highly porous sponge. The porosity can be controlled by the freezing rate (a faster freezing rate produces smaller pores) and pH.

## 1.5. Hydrogels

By definition, hydrogels are polymeric networks with three-dimensional configuration capable of imbibing large amounts of water or biological fluids. Since the first publication in the 1960s (Wichterle & Lim 1960) based on hydroxylethyl methacrylate (HEMA) hydrogels, the number of papers published has increased exponentially. Their ability to absorb water is due to the presence of hydrophilic groups such as -OH, -CONH-, -$CONH_2$, -COOH, and -$SO_3H$ (Peppas & Khare 1993). Due to their hydrophilic nature, they can absorb from 10-20 % wt. up to thousands of times their dry weight in aqueous media. Hydrogels can be tailored by varying porosity, pore size, surface to volume ratio, compressive modulus, topography and wettability which makes them promising candidate materials for bone regeneration.

### 1.5.1. Crosslinking

Hydrogels maintain their structure due to the crosslinks present within the network. There are two main types of crosslinking found in hydrogels; chemical crosslinks and physical crosslinks as shown in Figure 3.

#### *1.5.1.1. Chemical Crosslinking*

Chemical crosslinks are three-dimensional molecular networks formed from the introduction of primary covalent bonds. These hydrogels will not dissolve in aqueous solution unless the covalent bond can be broken by chemical or physical cleavage processes. The main methods of chemically crosslinking hydrogels include:

1. *Radical polymerisation* through monomers in the presence of crosslinking agent (Geever *et al*. 2008).
2. *Chemical reaction of complementary groups* where covalent linkages between chains can be established by the reaction of functional groups with complementary reactivity, i.e. crosslinking by condensation reaction (Crescenzi *et al*. 2003).
3. *High energy radiation* of gamma and electron beam to polymerise unsaturated compounds (Zhao *et al*. 2003).

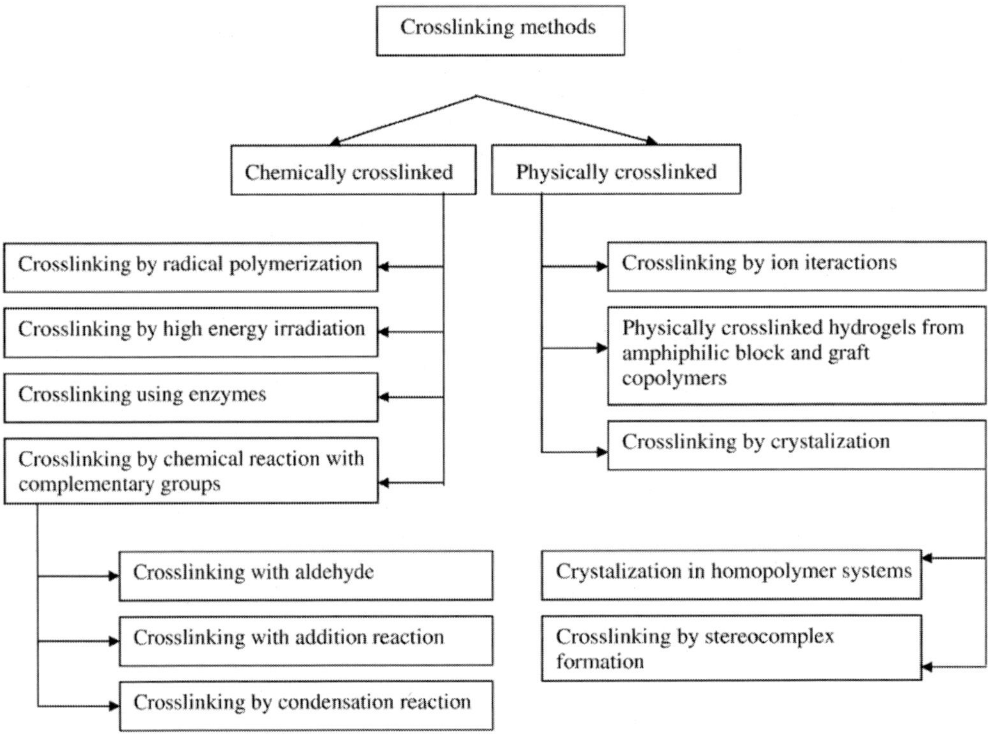


Figure 3. Methods of crosslinking hydrogels (Hamidi et al. 2008).

### *1.5.1.2. Physical Crosslinking*

Alternatively, physically crosslinked hydrogels are held together by molecular entanglement, and/or secondary forces. They do not require the introduction of external crosslinking agents and this will therefore help promote their biocompatibility. Physically crosslinked hydrogels can be formed by:

1. *Ionic interactions* such as with a polysaccharide with mannuronic and glucuronic acid residues which are crosslinked by calcium ions (Hu & Chou 1996).
2. *Crystallisation* through the process of numerous cycles of freeze thawing (Smith *et al*. 2009).
3. *Amphiphilic block and graft copolymers* which assemble in organised structures such as micelles (Förster & Antonietti 1998).

4. *Hydrogen bonding* for example between the oxygen in PEG and the carboxylic of PMAA (Fu *et al*. 2007).
5. *Protein interactions*.

#### *1.5.1.3. Interpenetrating Networks*

Hydrogels can also contain crosslinked networks consisting of both chemical and physical crosslinks. These hydrogels are called interpenetrating networks (IPNs). In general IPNs have properties that either retain the characteristics of the single networks (chemical or physical crosslinks) or are an average of the two independent networks (chemical and physical) (Elisseeff *et al*. 2000). Gong *et al*. (2003) synthesised IPNs consisting of different synthetic and biological polymer combinations with mechanical properties much enhanced when compared to their distinctive components.

### 1.5.2. Types of Hydrogels

#### *1.5.2.1. Synthetic Polymers*

Synthetic hydrogels are appealing for tissue engineering because their chemistry and properties are controllable and reproducible. Synthetic polymers can be reproducibly fabricated with specific molecular weights, block structures, degradable linkages and crosslinking modes. These properties in turn, determine gel formation dynamics, crosslinking density and material mechanical and degradation properties (Drury & Mooney 2003). Two of the most commonly used synthetic hydrogels include Poly (ethylene glycol) and Poly(vinyl alcohol):

- Poly (ethylene glycol) (PEG) is US Food and Drug Administration (FDA) approved due to its biocompatibility and low toxicity. It has been used in several tissue engineering applications including, bone (Ma *et al*. 2010) and cartilage (Bodugoz-Senturk *et al*. 2009). PEG has excellent antifouling properties which causes resistance to protein adsorption (Csucs *et al*. 2003). PEG has been synthesised to contain in excess of 90% water. The advantage of such a high water content is the ability to create an environment suitable for cells, as it can replicate natural soft tissues (Lin & Anseth 2009). This helps enhance the biocompatibility and also vascularisation which are two of the key requirements for scaffold design.
- Poly(vinyl alcohol) (PVA) is obtained from poly(vinyl acetate) through aminolysis, hydrolysis or alcoholysis (Jones 1973). PVA is a hydrophilic synthetic polymer and has been successfully used in tissue engineering due to its biocompatibility and biodegradation properties. These tissue engineering applications include: wound healing (Yang *et al*. 2010), heart valves (Jiang *et al*. 2004), cartilage (Ma *et al*. 2009) and bone (Ma *et al*. 2010). PVA hydrogels may be synthesised through various techniques including, incorporation of a chemical crosslinker agent, freeze/thaw synthesis and electron beam fabrication.

#### *1.5.2.2. Natural Polymers*

Natural polymeric hydrogels have been frequently used in tissue engineering applications because they are either components of, or have macro-molecular properties similar to the natural extracellular matrix (ECM) (Drury & Mooney 2003). Natural hydrogels in general have been shown to have weak mechanical properties and potentially exert immunogenicity or evoke inflammatory responses due to the presence of immunogen/pathogen moieties. However, they do offer various advantageous properties such as usually being non-toxic and biocompatible. Two of the main natural polymers include collagen and chitosan:

- Collagen is the most abundant protein and is a main component of the extracellular matrices of bone, skin, cartilage, ligament and tendon. In total there are approximately 20 members of the collagen family. Of primary interest to tissue engineering are the classes, Type I (bone and skin), Type II (cartilage) and Type III (vasculature). Collagen can be physically crosslinked or chemically crosslinked. Physically crosslinked hydrogels have limited mechanical properties and are thermally reversible. However, if they are chemically crosslinked with the addition of a crosslinker, i.e. glutaraldehyde (Veríssimo *et al.* 2010), this can significantly improve the mechanical properties of the hydrogels. Furthermore, the mechanical properties of collagen materials are controlled by the process of purification, source of collagen, fibril formation and crosslinking type (Alina 2011).
- Chitosan is a natural polysaccharide and is derived by the deacetylation of shell-fish-derived chitin using alkaline solution. Chitosan has been abundantly used in tissue engineering due to its biocompatibility, low toxicity, biodegradability, antimicrobial activity and its approved status by the FDA. Chitosan has been used in numerous tissue engineering applications such as nerve regeneration (Lu *et al.* 2007), drug delivery for cancer therapy (Azab *et al.* 2007), wound healing (Obara *et al.* 2003), cartilage (Jin *et al.* 2009) and bone repair (Wang & Stegemann 2010). Similarly to collagen, chitosan can be crosslinked physically or chemically with glutaraldehyde (Kandile *et al.* 2009), though it does need to overcome its low solubility issues in the majority solvents (Di Martino *et al.* 2005). One specific area of tissue engineering that chitosan is targeted towards is bone tissue regeneration. Chitosan has been shown to promote new tissue growth by osteoblasts in culture (Fakhry *et al.* 2004). Also, Ma *et al.* (2010) blended injectable hydrogels of chitosan derivatives with PEGDA and N,N-dimethylacrylamide through photopolymerisation as bone tissue engineering matrix (Ma *et al.* 2010).

## *1.6. Existing Materials for Bone Regeneration*

As mentioned previously, hard tissues such as bone and tooth are intended to support loads, being stiffer (higher elastic modulus) and stronger (higher tensile strength) than soft tissues. Bioactive ceramics, bioactive glasses and metals, can be used to replicate these properties. Inorganic materials which are similar in composition to the mineral phase of bone, either synthetic bioactive ceramics, such as calcium phosphates, β –tricalcium phosphate (β-TCP) and bioactive glasses (BAGs), or natural coralline hydroxyapatite (cHA) and their combinations have also been investigated due to their interesting properties, mainly osteoconductivity and osteoinductivity (Puppi *et al.* 2010).

### 1.6.1. Bioceramics

Bioceramics can be defined as biomaterials of the ceramic origin. Bioceramics are considered biocompatible, hard, brittle materials with exceptional compressive strength; however they have moderately weak tensile properties. Additionally, they can be soluble, biodegradable and resorbable depending on the type of bioceramic (Jayaswal *et al.* 2010). Bioceramics have numerous applications throughout the body (Mastrogiacomo *et al.* 2007; Zhai *et al.* 2012) and can be classified as either bioinert (zirconia and alumina) or bioactive (Bioglass) (Best *et al.* 2008). Within bioactive ceramics, they can act as resorbable or nonresorbable. Bioceramics can be used as porous networks, microspheres and as the reinforcing phase of polymer composites (Jayaswal *et al.* 2010). Types of bioceramics include bioactive glass, calcium phosphate and hydroxyapatite.

- A description of a bioactive material would be the capability of a material to form an apatite-like layer on the surface of the scaffold when in contact with tissue (Vallet-Regí *et al.* 2008). Hench & West (2006) identified two classes of bioactive materials (A and B) for bone regeneration. The first one (Class A) in which the scaffold materials has osteoproduction and osteoconduction at the implant surface. The second category (Class B) is where bioactivity of osteoconduction only. Bioactive ceramics include calcium phosphates, glasses and glass ceramics.
- Calcium phosphate is a resorbable bioactive glass and has been commercially available in tissue engineering since the 1980s (Jayaswal *et al.* 2010). Calcium phosphate has been used in bone tissue engineering as these scaffolds improve bone apposition to implants. Calcium phosphate is the main mineral component of bone. Calcium phosphates include both hydroxyapatite and tricalcium phosphate (TCP). The properties of each material are dependent on the amount of calcium to phosphorus ions within the structure (Buchs *et al.* 1991; Gauthier *et al.* 2005).
- Hydroxyapatite is a bioresorbable type of bioceramic. Hydroxyapatite has the capacity to support the formation of new bone as the chemical structure is comparable to bone. Hydroxyapatite can be used as powders or porous blocks to fill bone defects. This allows the bone to regrow in the defect and is used as an alternative to biological grafts, i.e. autografts (Jayaswal *et al.* 2010). Hydroxyapatite has the capabilities of integrating rapidly with the human body, especially with bone where it can bond with near identical unions. However, it does have limited mechanical strength, so therefore it cannot be used for applications of a load bearing nature.

## 1.7. Hydrogel Composites

Hydrogel composites are based on the fabrication of a hydrogel and inorganic reinforcement materials bound together. The hydrogel can consist of a polymer-ceramic or polymer-metallic matrix with interpenetrating network microstructures consisting of both bioceramic particles blended in with the polymer phase. The combination of hydrogels and bioceramics into hydrogel composite materials give a synergistic effect in overcoming the limitation of using either component individually. These limitations include:

- Unreinforced hydrogels are typically more ductile but not stiff enough to be used to replace hard tissues in load bearing applications.
- Metals or bioceramics processibility is very limited, metals are much stiffer than human hard tissues and can fail due to infection, while bioceramics are not only more brittle but also stiffer than natural mineralised tissues and present a low mechanical stability and fracture toughness preventing their use in the complete regeneration of large bone defects (Grimm 2004).

Nevertheless, hydrogel based composites can be designed to meet stiffness and strength requirements for hard tissue substitution. A study by Gaharwar *et al.* (2010) incorporated silicate nanoparticles prior to photopolymerisation. Results showed a significant improvement in mechanical properties. The silicate particles also promoted bioadhesion of osteoblasts on the scaffold surface. A similar study by Gaharwar *et al.* (2011) developed nanohydroxyapatite (nHA) hydrogels for orthopaedic tissue engineering applications. The properties of the PEG-nHA hydrogels had higher compressive strengths, fracture stresses, and toughness when compared with PEG-based hydrogels. The improvement in the mechanical properties is a result of the interactions between the polymer and nanoparticles which interfere with the permanent cross-linking of PEG. Bioglasses can also be incorporated into hydrogels. Couto *et al.* (2009) synthesised a novel injectable temperature-responsive hydrogel with bioactive glass ($SiO_2$: $CaO$:$P_2O_5$) and chitosan for orthopaedic and tissue regenerative applications.

## 1.8. Biological Considerations

One of the key issues in tissue engineering is the development of suitable biodegradable materials and scaffolds for the subsequent growth of tissues. A biomaterial is needed that has the appropriate composition and three-dimensional structure to promote cellular proliferation and differentiation as well as the strength to provide a scaffold for tissue regeneration (Yoshikawa *et al.* 2001). While clinical problems are extremely diverse, and no one approach will likely suit all potential applications an ideal bone graft should mimic the living tissue from a mechanical, chemical, biological and functional point of view (Murugan & Ramakrishna 2005). For bone tissue engineering, major issues surrounding tissue engineering scaffolds include biocompatibility and bioactivity or inclusion of factors that signal the induction of new bone formation (an osteoinductive environment) and supply of a scaffold conducive to new bone formation (an osteoconductive environment). With hydrogels, the use of appropriate matrix materials, control of porosity and pore characteristics, mechanical strength, degradation and surface properties can be optimised to meet the required needs for individual tissues by variation of composition, processing parameters and fabrication processes (Park 2009).

### 1.8.1. Biocompatibility

As outlined in Section 1.3.1, biocompatibility of implantable devices is an important prerequisite for pre-clinical evaluation of biomaterials, concerning the ability of the material to perform effectively with an appropriate host response for the desired application (Gomes *et al.* 2000; Onuki *et al.* 2008). Materials which are considered to be biocompatible may have extensive applications in human health care, improving quality of life and elongating life expectancy while non-biocompatibility of an implant can result in failure and subsequent necessity for removal of the implant, which can lead to serious medical consequences (Gallagher *et al.* 2006; Tienhoven *et al.* 2006). Hence a biocompatible material must perform its intended function with the absence of undesirable local or systemic effects. The materials and their degradation products along with any unreacted compounds or additives should not demonstrate immunogenicity or cytotoxicity, should not cause irritancy or carcinogenicity, should not induce an inflammatory response and should promote normal differentiation in the surrounding tissues. For implantable devices, bio adhesive contact must be established between the materials and living tissue allowing it to support cell- biomaterial interactions at the site of application (Onuki *et al.* 2008; Basu & Nath 2009). Interestingly biocompatibility requirements vary with the site and application. For bone tissue regeneration degradable materials are required, the scaffold degradation products should be processed through normal metabolic processes and eliminated from the body without any negative influence on the organism. In addition, the material should not be negatively influenced by the surrounding environment while performing its particular function (Silva *et al.* 2004; Basu & Nath 2009). All newly developed biomaterials must comply with a series of standard testing criteria, recommended by the international organisation for standardisation (ISO) before consideration for clinical applications. These guidelines allow for the selection of relevant cell based toxicity assays and provide clear descriptions of test selection, design and data interpretation. The ISO10993 series describes methods for the assessment of toxicity of medical devices and describes various possibilities for endpoint measurements and sample selection (Tienhoven *et al.* 2006). These assays can be performed *in vitro* and allow for the removal of high risk materials before more comprehensive toxicological evaluation of selected samples. The US food and drug administration incorporated such ISO guidelines into their biocompatibility testing methodology in 1995 and they remain as the current testing approach (Pizzoferrato *et al.* 1994). The ISO 10993 guidance documents have categorised medical devices as: non-contact devices; surface contacting devices; external communicating devices and implant devices. Medical devices have also been categorised by duration of contact, namely; limited exposure, prolonged exposure and permanent contact (ISO 10993-1, 2009). As bone tissue engineering scaffolds are intended for permanent implantation, the relevant guidelines include ISO 10993-3 (2003) for genotoxicity and carcinogenicity testing, ISO 10993-5 (2009) for cytotoxicity testing, ISO 10993-6 (1994) for local implantation effects, ISO 10993-9 (1999) for potential degradation product toxicity and ISO 10993-11 for systemic toxicity.

#### *1.8.1.1. Cytotoxicity Testing*

Cytotoxicity tests are performed to act as reliable and reproducible screening methods and represent the initial phase in testing biocompatibility of biomaterials (Gomes *et al.* 2000). They allow for large numbers of potential biomaterials and their modifications to be screened with comparative cost effectiveness and speed (Silva *et al.* 2004). The tests incorporate various endpoint analyses although it is generally accepted that the cytotoxic effect of a material can be assessed by evaluation of cell death (in the form of apoptosis or necrosis), loss of membrane integrity, reduced cell adhesion or proliferation, altered cell morphology and reduced biosynthetic activity (Kirkpatrick 1997). Since biodegradable materials are preferred for bone regeneration, cytotoxicity data deals with direct surface toxicity and with substances which leach out of the biomaterials which may exhibit activity responsible for cell toxicity or alteration of metabolic balances. Materials intended for implantation such as hydrogels for bone regeneration will first exert their effect at a cellular level, results from *in vitro* cytotoxicity tests can therefore predict immediate toxic response which can then be extrapolated for prediction of potential systemic response (Garle *et al.* 1994). The contribution of *in vitro* cytotoxicity tests to preliminary hydrogel and biomaterial evaluation has been extensively reported in the literature (Gomes *et al.* 2000; Fini *et al.* 2005; Yin *et al.* 2009; Kirf *et al.* 2010). However, *in vitro* cytotoxicity tests exceed the sensitivity of *in vivo* tests as metabolism, distribution and absorption variables are minimised, allowing maximisation of dosage per cell. The clinical acceptance of a material is dependent on many different factors, of which target cellular toxicity is only the primary screening test (Hanson *et al.* 1996), comprehensive assessment of toxicity requires a combination of assays which are sensitive to various stages of cell- material interactions.

#### *1.8.1.2. Genotoxicity Testing*

Genotoxicity is defined as any toxic modification of the structure or function of the hereditary material, DNA. Chemical, physical and biological agents which have the ability to cause alterations to the genetic material (i.e. nucleotide sequence of genes, chromosomal structure or number), resulting in permanent changes in cell function are called mutagens. The application of genotoxicity tests within biomaterial evaluation is of growing scientific interest due to the high correlation between mutagenicity and carcinogenicity. The ISO standards stipulate that all materials that will be in contact with mucous, bone or dentinal tissue exceeding 30 days, as well as all implantable devices who's contact exceeds 24 h, must undergo genotoxicity testing (Chauvel-Lebret *et al.* 2000; Silva *et al.* 2004). Testing regimens are undertaken using a multi-tiered test battery approach, based on the Organisation of Economic Cooperation and Development (OECD) test designs, as no single assay is capable of detecting all types of DNA damage. Hence assays should cover the three levels of genotoxic effects: point mutations (changes in the base sequence of DNA); structural (insertions, inversions, duplications, deficiencies and translocations); or numerical (gain or loss of whole or sets) chromosomal aberrations. The most commonly performed genotoxicity tests for regulatory purposes include a test for gene mutation in bacteria, the Ames *Salmonella typhimurium* reverse mutation assay, an *in vitro* mammalian cell gene mutation (mouse lymphoma thymidine kinase assay) or chromosome aberration assay and an *in vivo* test for chromosomal damage using rodent haematopoietic cells (OECD 471, 1997; OECD 476, 1997; OECD 473, 1997; ICH S2(R1), 2008; ICH S2(R1), 2008). However, if the results from two of these tests are negative, then a third test is usually not required. Interestingly the contribution of additional genotox tests to the evaluation of biomaterial/hydrogel toxicity has also been reported in the literature. Due to its highly sensitive and rapid nature, and its ability to quantitatively report DNA damage in individual eukaryotic cells, the comet assay is being applied to assessing the DNA damaging potential of biomaterials including hydrogels (Lebret *et al.* 2001; Devine *et al.* 2006; Noushad *et al.* 2009; Yin *et al.* 2009; Kirf *et al.*, 2010; Parry *et al.*, 2010).

## 1.8.2. In Vitro Evaluation of Biomaterials

### *1.8.2.1. Mammalian Cell Culture Systems*

The need to restrict the use of animals has enhanced the necessity to use *in vitro* systems for preliminary testing of biomaterials, allowing the exclusion of those unsuitable for human application and selection of those with potential usefulness. Continuous cell lines are most frequently used at the first stage of screening and cellular models of toxicity are now in common usage for biomaterials evaluation. Continuous cell lines can be obtained from primary cultures which have been phenotypically adapted via spontaneous, chemical or viral induced transformation which allows them to be cultured indefinitely under appropriate conditions. As bone forming cells, osteoblasts or osteoblast like cells are used as standards for *in vitro* studies to evaluate bone responses to various biological or chemical stimulations from underlying substrates. Immortalised cells from mouse (MC3T3-E1, C3H10T 1/2), rat (UMR106, ROB-C26) and human (SaOS-2, MG63) are widely used to assess various reactions and differentiation events in response to biomaterials (Zhao *et al.* 2007). However, these transformed cells do not retain complete functional activity or responses to biological cues and may undergo dedifferentiation with increased culture time (Kirf *et al.* 2010). Consequently, primary cells relevant to the purpose for which the biomaterial has been developed are preferred for biofunctionalilty testing as these cell cultures closely mimic the *in vivo* state and generate more physiologically relevant data. Primary osteoblasts can be isolated from fresh bone tissue samples by using enzymatic or explant methods and treated with osteogenic media containing β-glycerophosphate or glucocorticoids which stimulate osteoblastic phenotype. In the enzymatic method, bone tissue is treated with trypsin and collagenase for removal of bone matrix and release of osteoblasts while in the explant method, small bone fragments are directly cultured in media which allows osteoblast spontaneous migration from the explant (Wurtz & Berdal 2003; Silva 2004). Mesenchymal stem cells (MSCs), found in many adult tissues are an attractive stem cell source for the regeneration of damaged tissues in clinical applications because they are characterised as undifferentiated cells that have the potential to differentiate to lineages of mesenchymal tissues, including bone, cartilage, fat, tendon, muscle and marrow stroma (Kern *et al.* 2006). MSCs have shown great potential for the replacement of damaged tissue such as bone as they are susceptible to manipulation by biological molecules, driving them down a specific lineage. For example MSCs cultured with dexamethasone, ascorbate-2-phosphate, and beta-glycerophosphate in collagen hydrogels displayed a distinct expression pattern of the osteogenic markers alkaline phosphatase (ALP), type I collagen, osteocalcin (OC), and cbfa-1 suggesting MSCs in combination with hydrogels may be suitable for investigation into bone tissue engineering applications (Reichert *et al.* 2009)

### *1.8.2.2. Cell Encapsulation*

Although scaffolds for tissue engineering may be designed as purely structural supports providing passive or biological cues to guide cell and tissue growth, utilisation of 3D scaffolds for cell encapsulation during fabrication which serve as temporary supports offers several advantages over cell seeding on to prefabricated porous scaffolds. Hydrogel attributes (high water content and tissue-like elastic properties) make them ideal candidates for growing cells and tissues, recreating the *in vivo* environment by allowing permeation of biologic nutrients and signals for promotion of proliferation or induction of differentiation while protecting the entrapped cells from the host immune system. Encapsulation can be performed under mild gelation conditions using biocompatible macromolecular precursors which can be employed as injectable systems, delivered and cured at the site of interest in a minimally invasive manner (Nicodemus & Bryant 2008; Park 2009). The swelling, mechanical and degradation properties of hydrogel scaffolds are closely linked with the crosslinked structure and can be controlled through a variety of processing conditions. This crosslinked structure also controls ECM diffusion so gel structure and degradation rate can be closely matched to ECM secretion for optimisation of biocompatibility. A variety of cell types have been encapsulated in hydrogels for a wide range of tissue engineering applications. For example, hydrogels fabricated using biomaterials such as 5-ethyl-5-(hydroxymethyl)-beta,beta-dimethyl-1,3-dioxane-2-ethanol diacrylate (EHD) and poly(ethylene glycol) diacrylate (PEGDA) were used for encapsulation of BMSCs. It was found that metabolic activity, viability and osteodifferentiation of BMSCs were minimally affected by the gel and the processing conditions, demonstrating that EH-PEG hydrogels may be a viable option for the encapsulation of bone marrow stromal cells (BMSCs) for bone tissue regeneration (Betz *et al.* 2008).

## 1.8.3. Osteoinductivity

### *1.8.3.1. Natural Bone Growth And Repair*

Bone is a specialised connective tissue with a diverse set of physiological roles including protection, structural support, haematopoiesis and calcium homeostasis. From a structural standpoint it is composed of both mineral and organic phases, mainly hydroxyapatite, type 1 collagen and small amounts of non-collagenous proteoglycans such as osteopontin, ostocalcin, oseonectin and bone sialoprotein (Hill 1998; Wurtz & Berdal 2003). It is a dynamic tissue that is continually being remodelled according to physiological circumstances through bone formation and bone resorption, ensuring the conservation and renewal of the bone matrix. The principal cell types found in bone are osteoblasts, osteoclasts and osteocytes. Osteoblasts, of mesenchymal origin are responsible for the synthesis of the bone matrix, osteoid, a substance made predominantly of collagen which forms a scaffold in which minerals including calcium and phosphate begin to crystallise, forming hydroxyapatite, the major inorganic constituent of bone. Some active osteoblasts become trapped within the matrix they secrete and differentiate into osteocytes (Hill *et al.* 1998). Others will undergo apoptosis or revert back to lining cells which cover the surface of bone. Osteoclasts are multinucleated cells, derived from monocytes, they are hematopoietic in origin and function to resorb mineralised bone tissue by anchoring to the surface of bone via podosomes, creating a seal with the underlying bone matrix. Ruffling of the plasma membrane within this sealed zone allows the osteoclasts to increase the surface area of the cell and release proteolytic enzymes, cathepsin K, and hydrochloric acid (HCl) that dissolves the bones mineral content and degrades the remaining collaganeous bone matrix while protecting neighbouring cells (Price *et al.* 1994)

### *1.8.3.2. Regulation of Bone Formation*

The complex process of bone formation, maintenance and regeneration outlined above involves the interaction of cellular elements with many local and systemic regulators including hormones, growth and differentiation factors, cytokines and ECM components. (Baylink *et al.* 2009). Osteoblasts, derived from MSCs, need to undergo several transitional steps before becoming mature osteoblasts, with each step requiring the activation or suppression of the above fore mentioned cellular elements for progression of differentiation to occur (Shah *et al.* 1999). Among these systemic regulators, members of the TGF-β superfamily, multifunctional cytokines including TGF-βs, activins and bone morphogenetic proteins (BMPs), specifically BMP-2 and BMP-7 have been shown to have immense osteoinductive potential. Members of the TGF-β family interact with two different types of serine/threonine kinase receptors (type I and type II). Type II receptor kinases are constitutively active and transactivate the type I receptor kinases (Miyazono 2000). BMPs bind to activin receptor like kinase ALK-3 or ALK-6 and BMP-2 receptors while TGF-β binds ALK-5 and TGFβRII. Upon activation of the type I receptors by BMPs, the receptor then phosphorylates the receptor regulated signalling mothers against decapentaplegic (Smad) proteins (R-Smads). This is followed by association with common mediator Smads (co-Smad or Smad 4) which then translocate to the nucleus for transcriptional activation of specific target genes such as *Runx*-2 (or *cbfa*-1), initiating expression of Runx-2 protein, which binds to osteocalcin transcription promoter, allowing for osteoblast differentiation. Members of this superfamily can also activate mitogen activated protein (MAP) kinase pathways, resulting in p38, extracellular regulated kinase (ERK) and mitogen activated protein kinase (MAPK) activation, independent or in cooperation with smads (Canalis 2009; Biggs & Dalby 2010; Juarez & Guise 2011), pathways which have also been shown to be responsible for differentiation toward an osteoblastic phenotype (Khatiwala *et al.* 2008). BMPs potential to induce osteogenic cell differentiation *in vitro* and bone defect healing *in vivo* has been extensively reported in the literature (Yamamoto *et al.* 1999; Cancedda *et al.* 2007; Fischer *et al.* 2011). The osteoinductive potency of BMPs makes them clinically valuable as an alternative to bone grafts and they have been approved by the FDA for use in non-union fractures and spinal fusions (Zimmermann *et al.* 2009). TGF-β also plays a key role in osteoblast differentiation and bone development and remodelling, coordinating the activity of osteoblasts and osteoclasts (Alliston *et al.* 2001), however, its effects depend on the differentiation stage of the cells. It has been shown to stimulate the proliferation and early osteoblast differentiation, while inhibiting terminal differentiation (Juarez & Guise 2011). An important parameter in the success of using these factors in treating bone defects is the use of an appropriate carrier which allows vascular and mesenchymal cell invasion and subsequent osteogenesis. An ideal carrier scaffold should be capable of sustained release, protect the cytokines from protolithic deactivation, be biocompatible, bioresorbable and have appropriate porosity to allow cell infiltration (Seeherman & Wozney 2005; Fischer *et al.* 2011). With hydrogels, these properties can be tailored to meet the required needs and therefore may represent an attractive alternative for current approaches for the delivery of osteoinductive growth factors. *In vivo* evaluation of a hyaluronic acid (HA) hydrogel showed that injection of controlled release gels containing BMP-2 over the rat calvarium resulted in newly formed bone with high expression of bone markers, osteocalcin and osteonectin and angiogenic bone (Martinez *et al.* 2011). Another study has shown that gelatin hydrogels incorporating BMP-2 allowed optimisation of the retention period by alteration in water content (BMP-2 retention prolonged with decrease in water content). The same study found that alkaline phosphatase activity and osteocalcin content surrounding the implant site were significantly higher than BMP-2 solution injection or collagen sponge carriers, highlighting the potential of hydrogels as versatile osteoinductive carriers which can promote bone augmentation for potential orthopaedic applications (Yamamoto *et al.* 2003).

### 1.8.4. Osteoconductivity

The overall biocompatibility of materials at an implant site involves a complex interaction of cells and tissues with the biomaterial. Elucidation of the mechanisms involved in these interactions is a prerequisite for the rational development of medical devices with optimal osteoconductive properties. Irrespective of the application, implantation of a synthetic device causes a reaction at the interface between the implant and the host organism. The biological reaction which takes place is termed interfacial biology and plays a crucial role in determining the fate of the implants (Peluso *et al.* 1996; Kirkpatrick *et al.* 1997). The interaction involves attachment, adhesion, contact, and spreading of the cells on the surface of the scaffold. Tissue cell adhesion to the surface of the implanted biomaterial can be enhanced for facilitation of integration at the interface, while reduction of immune cell adhesion is desirable for limiting unwanted cell mediated events and immunological reactions. Generally, a complex ECM protein layer will be acquired (from plasmatic or tissue fluid origin) on material surfaces prior to interaction with host cells, converting the biomaterial into a biologically recognisable material. Studies have shown that these endogenous proteins (fibronectin, vitronectin, complement etc.) adsorb in response to surface free energy, providing a framework for initiation of cellular adhesion and constituting the major cellular recognition system for biomaterials (Anderson *et al.* 2008). This adsorption to the surface is influenced by the surface properties of the biomaterial such as hydrophobicity, charge, roughness and chemical composition and the composition of this layer at the interface determines the biological response and therefore the long-term compatibility of the implanted biomaterial (Gallagher *et al.* 2006). Following this initial attachment phase, host cells synthesise their own ECM. In bone, osteoblast ECM secretion consists primarily of collagen with small amounts of non- collagenous proteins such as osteocalcin, osteonecin, bone sialoproteins, fibronectin, proteoglycans and bone morphogenetic proteins (Anselme 2000). Cellular adhesion to these proteins is mediated by adhesion receptors such as α and β-chain transmembrane proteins, integrins (with cell adhesive activity centering around the RGD tripeptide in the central domain of ECM molecules such as fibronectin). Osteospecific adhesion is regulated by the orientation, spacing and composition of this protein layer as integrin binding is dependent on accessibility of the motif sequences (Biggs & Dalby 2010). Initial cellular adhesion is followed by the formation of supramolecular complexes, containing structural adaptor proteins such as vinculin, talin and paxillin colocalized with molecules involved in signal transduction such as protein kinases and phosphatases (Anselme 2000). Anchoring focal complexes are then formed by increased integrin recruitment and clustering (Shah *et al.* 1999). These are then reinforced intracellulary to form focal adhesion plaques, diverse protein networks which link the ECM to intracellular actin filaments, enabling extracellular transmission of forces to the cell interior which are translated into biochemical signals (mechanotransduction), initiating signalling pathways, regulating nuclear transcription factor activity and allowing for the facilitation of cell migration, proliferation and differentiation.

### *1.8.4.1. Modification with Biologically Active Molecules*

Since cell attachment and adhesion has been shown to have an influence on signal transduction and regulation of gene expression (Shah *et al.* 1999). The degree of maturation and mineralisation of osteoblastic cells may be highly dependent on the nature of the initial attachment and therefore on the surface properties of the biomaterial. Materials which are engineered to promote osteoblast specific adhesion and prevent inflammatory cell and bacterial adhesion may therefore enhance functional differentiation resulting in the formation of mineralised matrix and bony tissue formation (Kalajzic *et al.* 2004). Hydrogels have been widely studied as tissue scaffolds because they are biocompatible and non-adhesive to anchorage dependant cells and can therefore positively influence biocompatibility *in vitro*. Cell specific adhesion can be programmed via modification with biologically active adhesion molecules allowing for desirable effects on host tissue and optimisation of biocompatibility (Dettin *et al.* 2005). Many studies have concentrated on modification via incorporation of cell adhesive peptides isolated from adhesive proteins, such as Arg-Gly-Asp (RGD). Reyes and Garcia (2004) have shown enhanced osteospecific adhesive potential and reduction in inflammatory cell adhesion of orthopaedic materials by utilisation of the specificity of this motif (relative to surfaces modified with full length type 1 collagen). The RGD sequence is an ubiquitous adhesive motif found in many proteins including several bone ECM components such as fibronectin, osteopontin, type 1 collagen, osteonectin and bone sialoprotein. Its effects on cell adhesion and spreading have been extensively studied on a variety of hydrogel surfaces. Alsberg *et al.* (2001) showed increased attachment and spreading of osteoblasts and increased *in vivo* bone formation on RGD modified alginate hydrogels when compared to unmodified controls while Burdick *et al.* (2002) demonstrated increased osteoblast attachment, proliferation and mineralisation on PEG-RGD hydrogels. It has also been shown that MSC differentiation on RGD modified PEGDA hydrogels promotes osteogenic differentiation, stimulating an increase in bone marker proteins such as alkaline phosphatase and osteocalcin and highlighting the potential of hydrogels to positively influence cell specific adhesion and differentiation (Yang *et al.* 2005).

### *1.8.4.2. Pore Size and Porosity*

Scaffold porosity, pore size and overall pore structure all affect adequate tissue formation and infiltration into biomaterial constructs. Scaffolds for bone regeneration should possess an interconnected and spread porosity and highly porous surface and microstructure, creating an increased internal surface area for cell adhesion and spreading. By allowing transfer of nutrients, oxygen and growth factors through the material as well as removal of metabolic wastes and by products and space for neovascularisation, an open network structure permits cell viability and proliferation and allows for the deposition of cell secreted proteins so that as the scaffold degrades, ECM deposition will fill the void space, creating a new viable tissue. (Bryant & Anseth 2003; Lum & Elisseeff 2003; Puppi *et al.* 2010). As pore size and interconnectivity determines degree of transport and vascularisation and influences mechanical properties of the scaffold, optimisation can allow for increased biocompatibility (Bryant & Anseth 2003). For example, pore structures which are not fully interconnected may retard revascularisation and formation of mineralised matrix throughout the scaffold

(Murugan & Ramakrishna 2005). If pores are too large there is a decrease in specific surface area available, reducing ligand density available for tissue cells to bind to while smaller pores can inhibit cellular migration resulting in the formation of a cellular capsule at the edges of the scaffold, followed by reduction of waste product removal and distribution of nutrients or resulting in the ingrowth of unmineralised or fibrous tissue (Murphy & O'Brien 2010). Although there are conflicting reports in the literature for optimal pore size for osteoblast activity, a balance must be maintained which provides sufficient surface area for attachment while still allowing for cellular migration. A study into porous implants by Klawitter *et al.* 1976 demonstrated an optimal range of 100-135μm with a minimum size of 75-100μm for significant bone growth. However, since these early reports, other studies have suggested pore sizes ranging from 100-400um for optimum bone formation and vascularisation whereby the increased surface area provided by smaller pores may have a beneficial effect on cellular adhesion, while the improved cellular infiltration with larger pores may outweigh this effect, providing optimal conditions for bone tissue repair (Itala *et al.* 2000; Murphy *et al.* 2009). With hydrogels, pore size, porosity, surface to volume ratio and interconnectivity can be tailored to meet required needs for individual tissues. Oestogenic expression and differentiation of hMSCs in response to varying pore sizes and porosities was investigated in EH-PEG hydrogels by Betz *et al.* 2010. It was found that a macroporous environment promoted a dramatic increase in BMP-2 expression by hMSCs due to rapid osteoblastic differentiation in these scaffolds.

#### *1.8.4.3. Compressive Modulus*

Stiffness of the underlying surface has been shown to affect cellular behaviour; it is believed that when the compressive modulus of the underlying substrate matches the tissue modulus *in vivo*, cells exhibit behaviour which more closely mimics physiological behaviour. A study by Chung *et al.* (2006) showed that in 2D culture, proliferation of osteoblasts was enhanced on stiffer poly (N-isopropylacrylamide-co-acrylic acid) gels compared to softer gels while gel stiffness was also shown to influence the differentiation of cardioprogenitor 3D cell constructs in PEG hydrogels (Kraehenbuehl *et al.* 2008). Chatterjee *et al.* (2010) studied the response of osteoblasts to variation in 3D hydrogel modulus (accomplished by variation in PEGDA composition, without the use of bioactive factors). It was found that osteoblasts preferentially differentiated on scaffolds whereby the compressive modulus matched that of mineralised bone tissue *in vivo* and that hydrogels of modulus ≈ 225kPa or higher were required for induction of significant mineralisation. The intercellular pathways that direct cell response to scaffold stiffness have not been fully elucidated, however, it is thought that the extracellular signal-regulated kinase/mitogen activated protein kinase (ERK/MAPK) pathway may play an important role (Khatiwala *et al.* 2008). Changes in compressive modulus are transduced by integrin binding, stimulating downstream protein tyrosine kinases such as FAK (focal adhesion kinase). FAK activity leads to activation of RhoA, a GTPase which has been widely implicated in integrin mediated signalling and mechanotransduction. Acting through a downstream effector, ROCK (Rho-associated protein kinase), RhoA in association with ERK/MAPK stimulates the phosphorylation of ERK1/2, regulating the activity of RUNX2, an osteogenic transcription factor which controls the expression of osteogenic genes and drives differentiation toward an ostoblastic phenotype. With the use of PEG hydrogel substrates, utilised as they permit the control of elastic modulus independent of changes in ligand density (Peyton *et al.* 2006), it has been shown that on stiffer matrices, FAK activation is enhanced leading to higher RhoA and ERK activity and higher expression levels of RUNX2 (Khatiwala *et al.* 2008). These studies suggest that hydrogels are ideal materials to direct cell function due to the ease with which their mechanical properties can be controlled.

#### *1.8.4.4. Topography*

The influence of surface features or topography of an implantable material has long been recognised as a significant parameter in the appreciation of biocompatibility. Basement membranes are covered with grooves, ridges, pits, pores, and the fibrillar meshwork of the ECM. These topographical features provide tensile strength and mechanical rigidity as well as providing binding sites for cell adhesion molecules. The fibers of the ECM are arranged in an ordered fashion with cell adhesion and response depending on confirmation, dimensions and symmetry. The biological response to these topographical features has been shown to be cell type specific and control of these surface features can be exploited to influence specific protein interactions, allowing fabrication of TE scaffolds which can influence the adhesion, migration and differentiation of specific cell types. For example, it has been shown that fibroblasts attach to and proliferate on smooth surfaces while osteoblastic cells prefer rougher surfaces, with osteoblast proliferation increasing in close correlation with increasing surface roughness (Kunzler *et al.* 2007). However, the use of different surface modification techniques and methods for determining roughness, along with the use of osteoblasts in different maturation stages makes comparison between literature results difficult. Investigation into the effect of surface topography on changes in gene expression may provide information into the regulatory mechanisms behind bone formation and further elucidate elements of osteoblast response. A study by Hamilton *et al.* in (2007) suggests that rougher surfaces result in significant increases in phosphorylation of FAK and ERK 1/2 as well as increased nuclear translocation of ERK 1/2 and transcription factor RUNX2, driving differentiation toward on ostoblastic phenotype. Due to the critical influence of surface topography on tissue organisation and response, the study of material topography on intracellular signalling has the potential to provide information to optimise the design of materials that promote bone formation. The surface topography of hydrogels can be easily altered by changing the fabrication process, allowing optimisation of conditions to allow for increased cell specific attachment and providing a versatile scaffold for investigation of cellular response.

#### *1.8.4.5. Surface Wettability*

Lee *et al.* (1998) created polyethylene surfaces with increasing wettability gradients in order to investigate the interaction of various cell types and serum proteins on hydrophilic/hydrophobic surfaces. It was observed that serum proteins and cells adhered and proliferated better on surfaces with moderate hydrophillicity (with water contact angle of approx. 57°) regardless of the cell type. The use of either extreme in surface wettability may lead to a decrease in biocompatibility as it has been demonstrated that extremely hydrophilic materials inhibit protein adsorption while extremely hydrophobic materials may promote irreversible non-specific binding of proteins. However, it has been shown that hydrophobic surfaces strongly denature adsorbed proteins while on hydrophilic surfaces protein binding domains were found to be conserved. Protein adhesion to hydrogels could therefore result in enhanced functional presentation of ligands and subsequent improvement in cellular adhesion and subsequent proliferation when compared to more hydrophobic scaffolds (Keselowsky *et al.* 1995).

## 1.9. Summary and Clinical Requirements Necessary for Bone Regeneration Scaffolds

In this section of the book chapter, some of the information most pertinent to bone regeneration is summarised, and the clinical requirements necessary for bone regeneration scaffolds are further described. As stated previously, bone has a unique ability to heal following injury. However in certain cases, bones' healing ability is interrupted by the size of the defect (critical sized defects) or through infection or other ailment which prevents the bone fragments from joining (non-unions). Hence there is an urgent clinical need to develop an alternative tissue engineering construct for the treatment of these defects. Bone replacement/grafting product sales were valued at €588 million in 2007 and are projected to reach €807 million in 2014 (BCC Market Report 2009). According to a recent Market Report, the worldwide tissue engineering and regenerative medicine market approached €1.25 billion in 2008 and is projected to reach €2.6 billion by 2013 (Market Report, Life Science Intelligence 2009). Therefore, the potential return of investment made in the research field of using hydrogels for bone regeneration is tremendous.

Tissue Engineering is an emerging research area that offers outstanding opportunities for Regenerative Medicine. In these research endeavours researchers combine biological compounds such as cells or growth factors to a scaffold or delivery device which protects the biological component until it has fulfilled its function. A major development in tissue engineering to enhance bone regenerative medicine has been the identification of the bone morphogenetic proteins (BMPs), a family of protein initiators. BMPs play a pivotal role in inducing bone, cartilage, ligament and tendon formation (Xiao *et al.* 2007). The BMPs however need an appropriate carrier matrix so that they can be delivered to the desired site of action, without which the BMPs regenerating action is lost (Pluhar *et al.* 2006). There are hydrogel based tissue engineering solutions clinically available which utilise BMPs for bone regeneration sold as Infuse ® bone graft (Medtronic Inc.) and OP-1® (Stryker Inc.). However, it has been shown that when using the collagen scaffold supplied with these BMPs, the half-life of the initial burst release is less than 10 minutes (Lieberman *et al.* 2002). There are also a number of published patents relating to hydrogels for bone regeneration (Rapaport 2010). Rapaport, (2010) describes peptide hydrogels that are a matrix for calcium phosphate mineral which produces osteoinductive effects. Wack *et al. (*2007) also reports hydrogel scaffolds for bone regeneration application. The diamond process has been coined by Giannoudis *et al.* (2007) to describe the interaction of osteogenic cells, osteoconductive scaffolds, the mechanical environment and the growth factors in the bone regeneration. This diamond concept has been modified to also include the vascularity in the process (Giannoudis *et al.* 2008), and the role the surgeon has in creating a biological chamber into which these treatments may be placed (Calori 2011). The diamond concept neatly describes the current consensus and understanding of the physiological processes governing the fracture healing process, involves a number of mediators and cellular elements which interact at the molecular level and exert an effect in association with physiological and biomechanical principles. It is believed that the interaction of all of these parameters generates complex pathways facilitating the bone healing process (Giannoudis *et al.* 2008). Therefore, an ideal tissue engineering construct should be designed to incorporate the principles of the diamond concept to assist/improve bone healing in harmony with the bones natural ability to heal.

## CONCLUSION

Bone substitute materials are a possible alternative to biological grafts. There are a number of crucial requirements when designing tissue engineered scaffolds. The scaffold should: be biocompatible; possess interconnecting pores; be made from material with controlled biodegradability or bioresorbability; have appropriate surface chemistry to favour cellular attachment; possess adequate mechanical properties to match the intended site of implantation and intraoperative handling. Designing scaffolds that meet all of these requirements is a goal that continues to challenge researchers. Current bone substitute materials are in general ceramic, polymer, or a combination of both. Ceramic scaffolds have a number of inherent drawbacks including slow degradation rate, brittleness, high susceptibility to fatigue fracture and do little to retain biological factors released by the body at the site of the injury. It can be concluded that while many natural and synthetic polymers exhibit one or more of the key requirements for tissue engineering applications, an ideal bone substitute scaffold does not currently exist. Hydrogels have recently come to the forefront of musculoskeletal regenerative research and are considered by many the most likely material to overcome the limitations exhibited by current artificial bone substitutes.

Due to the unlimited ability to alter polymeric or monomeric feed ratios, it is conceivable that a hydrogel which meets the majority of the criteria outlined above could be developed. Hydrogels have been widely used in biomedical applications because of their biocompatibility with the human body and also because they resemble natural living tissue more than any other class of synthetic biomaterials. This is due to their high water content and soft consistency that is similar to natural tissue. The mechanical properties of hydrogels can be altered by varying the polymeric feed ratios or components. This enables hydrogels to be utilised in non-load bearing areas such as guided bone regeneration. In this area, a barrier membrane is used to prevent soft tissue ingrowth into defects typically created in the Alveolar bone following tooth abstraction. In doing so, this permits bone to grow uninterrupted into the defect space. Nevertheless, due to the high water content of hydrogels, it is unlikely that they would be capable of supported a long bone fracture or defect without the aid of a reinforcing filler or the support of a fixation device. While the merits of using composite hydrogels and fillers are discussed through this chapter, it is noteworthy that fixation devices (either internal or external) are routinely used in the treatment of fractures and large bone defects and therefore this should not prevent the use of hydrogels in the treatment of these aliments.

## REFERENCES

Albrektsson T & C., J. (2001) Osteoinduction, osteoconduction and osseointegration. *Eur Spine J*, pp. 96-101.

Alina, S. (2011) Current research on the blends of natural and synthetic polymers as new biomaterials: Review. *Progress in Polymer Science*, 36 (9) pp. 1254-1276.

Alliston, T., Choy, L., Ducy, P., Karsenty, G., Derynck, R. (2001) TGF-β induced repression of CBFA1 by Smad3 decreases cbfa1 and osteocalcin expression and inhibits osteoblast differentiation. *The EMBO journal* vol.20, pp. 2254-2272

Alsberg, E., Anderson, K.W., Albeiruti, A., Francceschi., Mooney, D.J. (2001) Cell-interactive alginate hydrogels for bone tissue engineering, *J Dent Res* vol, vol.11, pp. 2025-2029.

Amsden, B.G., Sukarto, A., Knight, D.K., Shapka, S.N. (2007). Methacrylated Glycol Chitosan as a Photopolymerizable Biomaterial. *Biomacromolecular*, vol 8, pp. 3758-3766.

Anderson, J.M., Rodriguez, A., Chang, D.T. (2008) Foreign body reaction to biomaterials, *Semin Immunol* vol.20, pp.86-100.

Anselme, K. (2000) Osteoblast adhesion on biomaterials, *Biomaterials* , vol.21 pp.667-681.

Azab, A.K., Kleinstern, J., Doviner, V., Orkin, B., Srebnik, M., Nissan, A. & Rubinstein, A. (2007) Prevention of tumor recurrence and distant metastasis formation in a breast cancer mouse model by biodegradable implant of 131I-norcholesterol. *Journal of Controlled Release*, 123 (2) pp. 116-122.

Barone, D.T.J., Raquez, J.M. & Dubois, P. (2011) Bone-guided regeneration: from inert biomaterials to bioactive polymer (nano)composites. *Polymers for Advanced Technologies,* 22 (5) pp. 463-475.

Basu, B. and Nath, S. (2009) Fundamentals of biomaterials and biocompatibility, *Advanced biomaterials: Fundamentals, Processing and Applications. Review,* pp. 3-18.

Bauer, T.W. & Muschler, G.F. (2000) Bone graft materials. An overview of the basic science. *Clinical Orthopaedics and Related Research*, 371 pp. 10-27.

Baylink, D.J., Finkelman, R.D. (2009) Growth factors to stimulate bone formation, *Journal of bone and mineral research*, vol.8, pp.565-572.

BCC Research Report. (2009). *Regenerative Medicines: Bone and Joint Applications. Market Report.*

Bertram, J. & Swartz, S. (1991) The 'Law of bone transformation': A case of crying wolff? *Biological Reviews,* 66 (3) pp. 245-273.

Best, S.M., Porter, A.E., Thian, E.S. & Huang, J. (2008) Bioceramics: Past, present and for the future. *Journal of the European Ceramic Society*, 28 (7) pp. 1319-1327.

Betz,M.W., Yeatts, A.B., Richbourg, W.J., Caccamese, J.F., Coletti, D.P., Falco, E.E. (2010) Macroporous hydrogels upregulate osteogenic signal expression and promote bone regeneration. *Biomacromolecules*, vol.11, pp.1160-1168.

Biggs, M.J.P., Dalby, M.J. (2010) Focal adhesions in osteogenesis, *Proc Inst Mech Eng H*, vol.224, pp.1441-1453.

Blaya, S., Carretero, L., Madrigal, R. F., Ulibarrena, M., Fimia, A. (2002). New photopolymerizable holographic recording material based on polyvinylalcohol and 2-

hydroxyethylmethacrylate (HEMA). *Applied Physics B: Laser and Optics,* vol 74, pp. 603-605.

Bodugoz-Senturk, H., Macias, C.E., Kung, J.H. & Muratoglu, O.K. (2009) Poly(vinyl alcohol)–acrylamide hydrogels as load-bearing cartilage substitute. *Biomaterials*, 30 (4) pp. 589-596.

Boyan, B.D., Lohmann, C.H., Romero, J. & Schwartz, Z. (1999) Bone and cartilage tissue engineering. *Clinics in Plastic Surgery*, 26 (4) pp. 629-645.

Bryant, S.J., Anseth, K.S. (2003) Controlling the spatial distribution of ECM components in degradable PEG hydrogels for tissue engineering cartilage. *Journal of Biomedical Materials Research,* 64A, 70-79.

Buchs, B., Rizzoli, R. & Bonjour, J.P. (1991) Evaluation of bone resorption and renal tubular reabsorption of calcium and phosphate in malignant and nonmalignant hypercalcemia. *Bone*, 12 (1) pp. 47-56.

Burdick, J.A., Mason, M.N., Hinman, A.D., Thorne, K., Anseth, K.S. (2002) Delivery of osteoinductive growth factors from degradable PEG hydrogels influences osteoblast differentiation and mineralization. *Journal of controlled release,* vol.83, pp.53-63.

Calori, G.M., Albisetti, W., Agus, A., Iori, S. & Tagliabue, L. (2007) Risk factors contributing to fracture non-unions. *Injury*, 38, Supplement 2 (0) pp. S11-S18.

Calori, G.M., Mazza, E., Colombo, M. & Ripamonti. C. (2011). The use of bone-graft substitutes in large bone defects: Any specific needs? *Injury*, Vol. 42(S2), pp. S56-S63

Canalis, E., (2009) Prospect-Growth factor control of bone mass, *J Cell Biochem ,* vol.108, pp.769-777.

Cancedda, R., Giannoni, P., Mastrogiacomo, M. (2007) A tissue engineering approach to bone repair in large animal models and in clinical practice, *Biomaterials*, vol.28, pp.4240-4250.

Chatterjee, K., Gibson, S.L., Wallace, W.E., Parekh, S.H., Lee, Y.J., Cicerone, M.T.,, Young, M.F., Simon, C.J. (2010) The effect of 3D hydrogel scaffold modulus on osteoblast differentiation and mineralization revealed by combinatorial screening, *Biomaterials*, vol.31, pp.5051-5062.

Chauvel-Lebret, D.J., Auroy, P., Tricot-Doleux, S., Bonnaure-Mallet, M. (2000) Evaluation of the capacity of SCGE assay to assess the genotoxicity of biomaterials, *Biomaterials*, vol.22, pp.1795-1801.

Chung, E.H., Gilbert, M., Virdi, A.S., Sena, K., Sumner, D.R., Healy, K.E. (2006) Biomimetic artificial ECMs stimulate bone regeneration, *J Biomed Mater Res* A, vol.15, pp. 815-826.

Chwalek, K., Levental, K.R., Tsurkan, M.V., Zieris, A., Freudenberg, U. & Werner, C. (2011) Two-tier hydrogel degradation to boost endothelial cell morphogenesis. *Biomaterials*, vol.32, pp. 9649-9657.

Correa, D., Tayalia, P., Cosendey, G., Santos, D.D., Aroca, R., Mazur, E., Mendonca, C., (2009). Two-Photon Polymerization for Fabricating Structures Containing the Biopolymer Chitosan J. *Nanoscience and Nanotechnology*, vol 9, pp. 5845-5849.

Couto, D.S., Hong, Z. & Mano, J.F. (2009) Development of bioactive and biodegradable chitosan-based injectable systems containing bioactive glass nanoparticles. *Acta Biomaterialia*, 5 (1) pp. 115-123.

Crescenzi, V., Francescangeli, A., Capitani, D., Mannina, L., Renier, D. & Bellini, D. (2003) Hyaluronan networking via Ugi's condensation using lysine as cross-linker diamine. *Carbohydrate Polymers*, 53 (3) pp. 311-316.

Csucs, G., Michel, R., Lussi, J.W., Textor, M. & Danuser, G. (2003) Microcontact printing of novel co-polymers in combination with proteins for cell-biological applications. *Biomaterials*, 24 (10) pp. 1713-1720.

Datta, P., Dhara, S. & Chatterjee, J. (2012) Hydrogels and electrospun nanofibrous scaffolds of N-methylene phosphonic chitosan as bioinspired osteoconductive materials for bone grafting. *Carbohydrate Polymers*, 87 (2) pp. 1354-1362.

Dettin, M., Conconi, M.T., Gambaretto, R., Bagno, A., Bello, C.D., Menti, A.M., Grandi, C., Parnigotto, P.P. (2005) Effect of synthetic peptides on osteoblast adhesion, *Biomaterials*, vol.26, pp.4507-4515.

Devine, D.M., Devery, S.M., Lyons, J.G., Geever, L.M., Kennedy, J.E., Higginbotham, C.L. (2006) Multifunctional polyvinylpyrrolidinone-polyacrylic acid copolymer hydrogels for biomedical applications. *International Journal of Pharmaceutics* 326: 50-59.

Di Martino, A., Sittinger, M. & Risbud, M.V. (2005) Chitosan: A versatile biopolymer for orthopaedic tissue-engineering. *Biomaterials*, 26 (30) pp. 5983-5990.

Dickinson, P., Pratt, J. E., (1994) *Photocurable compositions comprising grafted polyvinyl alcohol derivates. US patent* 5,332,651.

Dietmar W, H. (2000) Scaffolds in tissue engineering bone and cartilage. *Biomaterials*, 21 (24) pp. 2529-2543.

Dong Y M, Mao W, Wang H W, Zhao Y Q, Li X J, Bi D X,Yang L L, Ge Q and Fang X M. (2006). Measurement of critical concentration for mesophase formation of chitosan derivatives in both aqueous and organic solutions. *Polymer Int.*, vol 55, pp. 1444.

Drury, J.L. & Mooney, D.J. (2003) Hydrogels for tissue engineering: scaffold design variables and applications. *Biomaterials*, 24 (24) pp. 4337-4351.

Elisseeff, J., Mcintosh, W., Anseth, K., Riley, S., Ragan, P. & Langer, R. (2000) Photoencapsulation of chondrocytes in poly(ethylene oxide)-based semi-interpenetrating networks. *Journal of Biomedical Materials Research*, 51 (2) pp. 164-171.

Fakhry, A., Schneider, G.B., Zaharias, R. & Şenel, S. (2004) Chitosan supports the initial attachment and spreading of osteoblasts preferentially over fibroblasts. *Biomaterials*, 25 (11) pp. 2075-2079.

Fan, J., Chen, J., Yang, L., Lin H., Cao, F., (2009). Preparation of dual-sensitive graft copolymer hydrogel based on N-maleoyl-chitosan and poly(N isopropylacrylamide) by electron beam radiation. *Bull. Mater. Sci.,* vol 32, pp. 521-526.

Fini, M., Motta, A., Torricelli, P., Giavaresi, G., Aldini, N.N., Tschon, M., Giardino, R., Migliaresi, C. (2005) The healing of confined critical size cancellous defects in the presence of a silk fibrin hydrogel, *Biomaterials*, vol.26, pp.3527-3536

Fischer, J., Kolk, A., Wolfart, S., Pautke, P.H., Plank, C., Smeets, R. (2011) Future of local bone regeneration- Protein verses gene therapy, *Journal of Cranio-Maxillo-Facial Surgery*, vol.39, pp.54-64

Förster, S. & Antonietti, M. (1998) Amphiphilic Block Copolymers in Structure-Controlled Nanomaterial Hybrids. *Advanced Materials*, 10 (3) pp. 195-217.

Fouassier, J.P. Photoinitiation, (1995) *Photopolymerization and Photocuring; fundamentals and applications.* Hanser Publishers.

Fu, X.J., Cao, S.Q., Wang, N.X., Zhang, S.Z., Wang, H. & Yang, Y.J. (2007) Effect of hydrogen bonding and hydrophobic interaction on the formation of supramolecular hydrogels formed by l-phenylalanine derivative hydrogelator. *Chinese Chemical Letters*, 18 (8) pp. 1001-1004.

Fukuda, J., Khademhosseini, A., Yeo, Y., Yang, X., Yeh, J., Eng, G., Blumling, J., Wang, C-F., Kohane, D. S., Langer. R., (2006). Micromolding of photocrosslinkable chitosan hydrogel for spheroid microarray and co-cultures. *Biomaterials*, vol 27, pp. 5259-5267.

Gaharwar, A.K., Rivera, C.P., Wu, C.-J. & Schmidt, G. (2011) Transparent, elastomeric and tough hydrogels from poly(ethylene glycol) and silicate nanoparticles. *Acta Biomaterialia,* 7 pp 4139–4148.

Gaharwar, A.K., Schexnailder, P.J., Jin, Q., Wu, C.-J., Schmidt, G., 2010. Addition of Chitosan to Silicate Cross-Linked PEO for Tuning Osteoblast Cell Adhesion and Mineralization. *ACS Applied Materials & Interfaces* 2, 3119-3127.

Gallagher, W., (2006). Molecular basis of cell-biomaterial interaction: Insights gained from transcriptomic and proteomic studies, *Biomaterials*, vol. 27, pp. 5871-5882

Gao, X., Zhou, Y., Ma, G., Shi, S., Yang, D., Lu, F., Nie., J. A. (2010) Water-soluble photocrosslinkable chitosan derivative prepared by Michael-addition reaction as a precursor for injectable hydrogel. *Carbohydrate Polymers*, vol 79, pp. 507-512.

Garcia, A.J., Reyes, C.D. (2004) Bio-adhesive surfaces to promote osteoblast differentiation and bone formation, *J Dent Res*, vol. 84, pp.407-413

Garle, M.J., Fentem, J.H., Fry, J.R. (1994) In vitro cytotoxicity tests for the prediction of acute toxicity in vivo, *Toxic. in vitro*, vol.8, pp.1303-1312

Gauthier, O., Müller, R., Von Stechow, D., Lamy, B., Weiss, P., Bouler, J.-M., Aguado, E. & Daculsi, G. (2005) In vivo bone regeneration with injectable calcium phosphate biomaterial: A three-dimensional micro-computed tomographic, biomechanical and SEM study. *Biomaterials*, 26 (27) pp. 5444-5453.

Geever, L.M., Cooney, C.C., Lyons, J.G., Kennedy, J.E., Nugent, M.J.D., Devery, S. & Higginbotham, C.L. (2008) Characterisation and controlled drug release from novel drug-loaded hydrogels. *European Journal of Pharmaceutics and Biopharmaceutics*, 69 (3) pp. 1147-1159.

Giannoudis, P.V., Einhorn, T.A., Schmidmaier G. & Marsh D (2008). The diamond concept: open questions. *Injury*, Vol. 39(S2), pp S5–8.

Giannoudis, P.V., Einhorn, T.A. & Marsh, D (2007). Fracture healing: the diamond concept. *Injury*, Vol. 38(S4), pp. S3–6.

Gomes, M.E., Reis, R.L., Cunha, A.M., Blitterswijk, C.A., de Bruijn, J.D. (2000). Cytocompatability and response of osteoblastic-like cells to starch based polymers:effect of several additives and processing conditions, *Biomaterials*, vol.22, pp.1911-1917.

Gong, J.P., Katsuyama, Y., Kurokawa, T. & Osada, Y. (2003) Double-Network Hydrogels with Extremely High Mechanical Strength. *Advanced Materials,* 15 (14) pp. 1155-1158.

Grimm, B., Heyligers, I.C., Miles, A.W., Turner, I.G., Haaren, E.H., Smit, T.H., Boshuizen, K. (2004). Subsidence and implant stability in impaction bone grafting using bone allografts and ceramic bone substitutes. Comparison between three experimental methods. *European Hip Society, Innsbruck, Austria* 14, 69.

Hamidi, M., Azadi, A. & Rafiei, P. (2008) Hydrogel nanoparticles in drug delivery. *Advanced Drug Delivery Reviews,* 60 (15) pp. 1638-1649.

Hamilton, D.W., Brunette, D.M. (2007) The effect of substratum topography on osteoblast adhesion mediated signal transduction and phosphorylation. *Biomaterials*, vol.28, pp.1806-1819

Han, J., Wang, K., Yang, D., Nie. J. (2009). Photopolymerization of methacrylatedchitosan/PNIPAAm hybrid dual-sensitive hydrogels as carrier for drug delivery. *Int. J. Biological Macromolecules*, vol 44, pp. 229-235.

Hanson, S., Lalor, P.A., Niemi, S.M., Northup, S.J., Ratner, B.D., Spector, M., Vale, B.H., Willson, J.E. (1996) *Testing biomaterials, Biomaterials*, review, pp.215-220

Hench, L.L. & West, J.K. (2006) Biological applications of bioactive glasses. *Life Chemistry Reports*, 13 pp. 187-241.

Hill, P.A., Orth, M. (1998) Bone remodelling. Review , *British journal of orthodontics,* vol.25, pp.101-107.

Hollister, S.J. (2009) Scaffold Design and Manufacturing: From Concept to Clinic. *Advanced Materials*, 21 (32-33) pp. 3330-3342.

Horan, R.L., Antle, K., Collette, A.L., Wang, Y., Huang, J., Moreau, J.E., Volloch, V., Kaplan, D.L. & Altman, G.H. (2005) In vitro degradation of silk fibroin. *Biomaterials*, 26 (17) pp. 3385-3393.

Hu, D.S.-G. & Chou, K.J.-N. (1996) Kinetics of water swelling and development of porous structure in ionic poly(acrylonitrile-acrylamide-acrylic acid) hydrogels. *Polymer*, 37 (6) pp. 1019-1025.

Ibarra-Escutia, P., Gómez, J.J., Calas-Blanchard, C. Marty, J.L., Ramírez-Silva, M.T. (2010). *Amperometric biosensor based on a high resolution photopolymer deposited onto a screen-printed electrode for phenolic compounds monitoring in tea infusions*. vol 81, pp. 1636-1642.

ICH S2(R1) (2008) *Guidance on Genotoxicity Testing and Data Interpretation for Pharmaceuticals Intended for Human Use.*

Ikada, Y. (2006) *Tissue engineering: fundamentals and applications*. UK: Elsevier Ltd

Itala, A.I., Ylanen, H.O., Ekholm, C., Karlsson, K.H., Aro, H.T. (2000) Pore diameter of more than 100μm is not requisite for bone ingrowth in rabbits, *J Biomed Mater*, vol.12, pp.679-683

Ito, Y., Nogawa, M., Takeda, M., Shibuya, T. (2005). Photo-reactive polyvinylalcohol for photo-immobilized microarray. *Biomaterials*, vol 26, pp. 211-216

Jayaswal, G., Dange, S. & Khalikar, A. (2010) Bioceramic in dental implants: A review. *The Journal of Indian Prosthodontic Society,* 10 (1) pp. 8-12.

Jiang, H., Campbell, G., Boughner, D., Wan, W.-K. & Quantz, M. (2004) Design and manufacture of a polyvinyl alcohol (PVA) cryogel tri-leaflet heart valve prosthesis. *Medical Engineering & Physics*, 26 (4) pp. 269-277.

Jin, R., Moreira Teixeira, L.S., Dijkstra, P.J., Karperien, M., Van Blitterswijk, C.A., Zhong, Z.Y. & Feijen, J. (2009) Injectable chitosan-based hydrogels for cartilage tissue engineering. *Biomaterials*, 30 (13) pp. 2544-2551.

Jing, R., Yanqun, Z., Jiuqiang, L. & Hongfei, H. (2001) Radiation synthesis and characteristic of IPN hydrogels composed of poly(diallyldimethylammonium chloride) and Kappa-Carrageenan. *Radiation Physics and Chemistry*, 62 (2-3) pp. 277-281.

Jones, J.I. (1973) Polyvinyl alcohol. Properties and applications. Edited by C. A. Finch. John Wiley, Chichester. 1973. Pp. xviii + 622. Price: £14.00. *British Polymer Journal*, 5 (6) pp. 493-494.

Juarez, P., Guise, T.A. (2011) TGF-β in cancer and bone: Implications for treatment of bone metastases, *Bone*, vol.48, pp.23-29

Kalajzic, I., Stall, A., Yang, W.P., Wu, Y., Johnson, S.E. (2004) Expression profile of osteoblast lineage at defined stages of differentiation, *Biological chemistry,* vol.280, pp.24618-24626

Kandile, N.G., Mohamed, M.I., Zaky, H.T., Nasr, A.S. & Abdel-Bary, E.M. (2009) Synthesis and properties of chitosan hydrogels modified with heterocycles. *Carbohydrate Polymers*, 75 (4) pp. 580-585.

Karaca, N., Temel, G., Karaca Balta, D., Aydin, M. & Arsu, N. (2010) Preparation of hydrogels by photopolymerization of acrylates in the presence of Type I and one-component Type II photoinitiators. *Journal of Photochemistry and Photobiology A: Chemistry*, 209 (1) pp. 1-6.

Kern, S., Eichler, H., Stoeve, J., Kluter, H., Bieback, K. (2006) Comparative analysis of mesenchymal stem cells from bone marrow, umbilical cord or adipose tissue, *STEM CELLS*, vol.22, pp.625-634

Keselowsky, B.G., Collard, D.M., Garcia, A.J. (1995) Surface chemistry modulates fibronectin confirmation and directs integrin binding and specificity to control cell adhesion, *J. Biomedical Mater. Res,* vol.29, pp. 1545-1555

Khatiwala, C.B., Kim, P.D., Peyton, S.R., Puntam, A.J. (2008) ECM compliance regulates osteogenesis by influencing MAPK signalling downstream of RhoA and ROCK, *Journal of bone and mineral research,* vol.24, pp. 886-698.

Kim, K.I., Na, H.-N., Ito, Y., Son, T.I. (2011). Synthesis of visible light-induced cross-linkable chitosan as an anti-adhesive agent. *Macromolecular Research,* vol 19, pp. 216-220

Kirf, D., Higginbotham, C.L., Rowan, N.J., Devery, S.M. (2010) Cyto-and genotoxicological assessment and functional characterization of *N*-vinyl-2-pyrrolidone-acrylic acid-based copolymeric hydrogels with potential for future wound healing applications. *Biomedical Materials* 5:1-11.

Kirkpatrick, C.J., Wagner, M., Kohler, F., Bittinger, F., Otto, M., Klien, C.L. (1997) *The cell and molecular biological approach to biomaterial research: a perspective,* vol.8, pp.131-141

Kityk, I.V., Kasperczyk, J., Sahraoui, B., Yasinskii, M.F., Holan B. (1997). Low temperature anomalies in polyvinyl alcohol photopolymers. *Polymer*, vol. 38, pp. 4803-4806.

Klawitter, J.J., Bagwell, J.G., Weinstein, A.M., Sauer, B.W. (1976) An evaluation of bone growth into porous high density polyethylene, *Journal of biomed Mater Res*, vol.10 pp.311-323.

Kraehenbuehl, T.P., Zammaretti, P., Van der viles, A.J., Schoenmakers, R.G., Lutolf, M.P., Jaconi, M.E., Hubbell, J.A. (2008) Three dimensional extracellular matrix directed cardioprogenitor differentiation: Systematic modulation of a synthetic cell-responsive PEG hydrogel, *Biomaterials* , vol.29, pp.2757-2766.

Kunzler, T.P., Drobek, T., Schuler, M., Spencer, N.D. (2007) Systematic study of osteoblast and fibroblast response to roughness by means of surface morphology gradients, *Biomaterials*, vol.28, pp.2175-2182

Langer, R. & Vacanti, J. (1993) *Tussue Engineering. Science*, 260 pp. 920-926.

Lebret, D.J.C., Auroy, P., Tricot-Doleux, S., Bonnaure-Mallet, M. (2001) Evaluation of the capacity of the SCGE assay to assess the genotoxicity of biomaterials. *Biomaterials* 22: 1795-1801.

Lee, J.H., Khang, G., Lee, J.W., Lee, H.B. (1998) Interaction of different types of cells on polymer surfaces with wettability gradient, *Journal of colloid and interface science,* vol. 205, pp.323-330

Lee, K.-W.D., Chan, P.K., Feng, X., (2004). Morphology development and characterization of the phase-separated structure resulting from the thermal-induced phase separation phenomenon in polymer solutions under a temperature gradient. *Chemical Engineering Science* 59, 1491-1504

Lewandrowski, K.-U., D. Gresser, J., Wise, D.L. & Trantolo, D.J. (2000) Bioresorbable bone graft substitutes of different osteoconductivities: a histologic evaluation of osteointegration of poly(propylene glycol-co-fumaric acid)-based cement implants in rats. *Biomaterials*, 21 (8) pp. 757-764.

Li, Q., Yang, D., Ma, G., Xu, Q., Chen, X., Lu, F., Nie. J. (2009) Synthesis and characterization of chitosan-based hydrogels. *Int. J. Biological Macromolecules,* vol 44, pp. 121-127.

Lieberman, J.R., Daluiski, A. & Einhorn, T.A. (2002). The role of growth factors in the repair of bone. Biology and clinical applications. *J Bone Joint Surg Am.* Vol. 84, pp. 1032–44.

Life Science Intelligence, (2009). *Worldwide Markets and Emerging Technologies for Tissue Engineering and Regenerative Medicine.* Market Report,

Lin, C.-C. & Anseth, K. (2009) PEG Hydrogels for the Controlled Release of Biomolecules in Regenerative Medicine. *Pharmaceutical Research,* 26 (3) pp. 631-643.

Liu, C., Xia, Z. & Czernuszka, J.T. (2007) Design and Development of Three-Dimensional Scaffolds for Tissue Engineering. Chemical Engineering Research and Design, 85 (7) pp. 1051-1064.

Liu, Q., Hedberg, E.L., Liu, Z., Bahulekar, R., Meszlenyi, R.K. & Mikos, A.G. (2000) Preparation of macroporous poly(2-hydroxyethyl methacrylate) hydrogels by enhanced phase separation. Biomaterials, 21 (21) pp. 2163-2169.

Lu, G., Kong, L., Sheng, B., Wang, G., Gong, Y. & Zhang, X. (2007) Degradation of covalently cross-linked carboxymethyl chitosan and its potential application for peripheral nerve regeneration. *European Polymer Journal*, 43 (9) pp. 3807-3818.

Lum, L., Elisseeff, J. (2003) Injectable Hydrogels for cartilage tissue engineering, *Topics in tissue engineering*, vol.376, pp.401-410, June.

Ma, G., Liu, Y., Kennedy, JF., Nie, J. (2011). Synthesize and properties of photosensitive organic solvent soluble acylated chitosan derivatives (2). *Carbohydrate Polymers*, vol 84, pp. 681-685.

Ma, G., Yang, D., Li, Q., Wang, K., Chen, B., Kennedy, J.F. & Nie, J. (2010) Injectable hydrogels based on chitosan derivative/polyethylene glycol dimethacrylate/N,N-dimethylacrylamide as bone tissue engineering matrix. *Carbohydrate Polymers*, 79 (3) pp. 620-627.

Ma, G., Zhang, X., Han, J., Song, G., Nie, J. (2009). Photo-polymeriable chitosan derivative prepared by Michael reaction of chitosan and polyethylene glycol diacrylate (PEGDA*). Int. J. Biological Macromolecules* vol 45, pp. 499-503.

Ma, R., Xiong, D., Miao, F., Zhang, J. & Peng, Y. (2009) Novel PVP/PVA hydrogels for articular cartilage replacement. Materials *Science and Engineering:* C, 29 (6) pp. 1979-1983.

Martens, P., Anseth, K.S. (2000). Characterization of hydrogels formed from acrylate modified poly(vinyl alcohol) macromers. *Polymer*, vol 41, pp. 7715-7722

Martinez Sanz, E., Ossipov, D.A., Hilborn, J., Larsson, S., Jonsson, K.B., Varghese, O.P. (2011) Bone reservoir: Injectable hyaluronic acid hydrogel for minimal invasive bone augmentation, *Journal of controlled release*, vol.152, pp. 232-240

Mastrogiacomo, M., Papadimitropoulos, A., Cedola, A., Peyrin, F., Giannoni, P., Pearce, S.G., Alini, M., Giannini, C., Guagliardi, A. & Cancedda, R. (2007) Engineering of bone using bone marrow stromal cells and a silicon-stabilized tricalcium phosphate bioceramic: Evidence for a coupling between bone formation and scaffold resorption. *Biomaterials*, 28 (7) pp. 1376-1384.

Mayorga, M., Oyalowo, A., Rementer, C., Soucy, M. & Weng, L. (2010) *Biomaterials based tissue engineering*. Thesis. Imperial College London, Department of Materials.

Mikos, A.G., Thorsen, A.J., Czerwonka, L.A., Bao, Y., Langer, R., Winslow, D.N. & Vacanti, J.P. (1994) Preparation and characterization of poly(l-lactic acid) foams. *Polymer*, 35 (5) pp. 1068-1077.

Mikos, A.G., Sarakinos, G., Lyman, M.D., Ingber, D.E., Vacanti, J.P. & Langer, R. (1993) Prevascularization of porous biodegradable polymers. *Biotechnology and Bioengineering*, 42 (6) pp. 716-723.

Miyazono, K. (2000) TGF-β signalling by Smad proteins, *Cytokine and growth factor reviews* vol.11, pp.15-22.

Moon, J.H; Shul, Y.G; Han, H.S; Hong, S.Y; Choi, Y.S; Kim, H.T. (2005) A study on UV-curable adhesives for optical pick-up: I. Photo-initiator effects. *International Journal of Adhesion and Adhesives,* 25, 301-312.

Murphy, C.M., O'Brien, F.J. (2010) Understanding the effect of mean pore size on cell activity in collagen-glycoaminoglycan scaffolds, *Cell adhesion and migration*, pp.377-381.

Murphy, C.M., Haugh, M.G., O'Brien, F.J. (2009) The effect of mean pore size on cell attachment, proliferation and migration in collagen-glycoaminoglycan scaffolds for bone tissue engineering, *Biomaterials*, vol.31, pp.461-486.

Murugan, R., Ramakrishna, S. (2005) Development of nanocomposites for bone grafting, *Composites science and technology,* vol.65, pp.2385-2406.

Ngamviriyavong, P., Thananuson, A., Pankongadisak, P., Tanjak, P., Janvikul, W., (2010). Antibacterial Hydrogels from Chitosan Derivatives. *J. Metals, Materials and Minerals*, vol 20, pp. 113-117.

Nicodemus, G.D., Bryant, S.J. (2008) Cell encapsulation in biodegradable hydrogels for tissue engineering applications, *Tissue engineering,* vol.14, pp. 149-165.

Noushad, M., Kannan, T.P., Husein, A., Abdullah, H., Ishmail, A.R. (2009) Genotoxicity evaluation of locally produced dental porcelain-an *in vitro* study using the Ames and comet assays. *Toxicology in vitro* 23: 1145-1150.

Obara, K., Ishihara, M., Ishizuka, T., Fujita, M., Ozeki, Y., Maehara, T., Saito, Y., Yura, H., Matsui, T., Hattori, H., Kikuchi, M. & Kurita, A. (2003) Photocrosslinkable chitosan hydrogel containing fibroblast growth factor-2 stimulates wound healing in healing-impaired db/db mice. *Biomaterials*, 24 (20) pp. 3437-3444.

O'brien, F. (2011) Biomaterials & scaffolds for tissue engineering. *Materials Today,* 14 (3) pp. 88-95.

OECD 471 (1997) *Bacterial Reverse Mutation Test.*

OECD 473 (1997) *In vitro Mammalian Chromosome Aberration Test.*

OECD 476 (1997) *In vitro Mammalian Cell Gene Mutation Test.*

Ono, K., Saito, Y., Yura, H., Ishikawa, K., Kurita, A., Akaike, T., Ishihara, M., (2009). Photocrosslinkable chitosan as a biological adhesive. *J. Biomed Material Research*, vol 49, pp. 289-295.

Onuki, Y., Bhardwaj,U., Papadimitrakopoulos, W., Burgess, D.J. (2008) A review of the biocompatibility of implantable devices: current challenges to overcome foreign body response, *Journal of Diabetes science and technology,* vol.2, pp.1003-1015.

Oyen, M.L. (2008) The materials science of bone: lessons from nature for biomimetic materials synthesis. *MRS Bulletin* (Materials Research Society), 33 (1) pp. 49-55.

Pande, K.C., Kallenbach, S.E. (1973). *Ally carbamate esters of hydroxy-containing polymers. US Patent 3,*776,889.

Park, J.B. (2009) The use of hydrogels in bone tissue engineering. *Biomaterials and Bioengineering in Dentistry*, vol.16, pp. 115-118.

Parry, M.C., Bhabra, G., Sood, A., Machado, F., Cartwright, L., Saunders, M., Newson, I.R., Blom, A.W., Case, C.P. (2010) Thresholds for indirect DNA damage across cellular barriers for orthopaedic biomaterials. *Biomaterials* 31: 4477-4483.

Peluso, G., Petillo, O., Mazzarella, L., Cara, F.L., Sada, A., Melone, M.A.B., Davies, P.J.A., Gentile, V. (1996) Cell-biomaterial interactions:role of transglutaminase enzyme, *Journal of materials science*, vol.7, pp.707-711.

Peppas, N.A. & Khare, A.R. (1993) Preparation, structure and diffusional behavior of hydrogels in controlled release. *Advanced Drug Delivery Reviews*, 11 (1-2) pp. 1-35.

Peyton, S.R., Raub, C.B., Keschrumrus, V.P., Putnam, A.J. (2006) The use of poly (ethylene glycol) hydrogels to investigate the impact of ECM chemistry and mechanics on smooth muscle cells, *Biomaterials*, vol.27, pp.4881-4839

Pizzoferrato, A., Ciapetti, G., Stea, S., Cenni, E., Arciola, R., Granchi, D. (1994) Cell culture methods for testing biocompatibility, *Clinical materials*, vol.15, pp.170-180

Place, E.S., Evans, N.D. & Stevens, M.M. (2009) Complexity in biomaterials for tissue engineering. *Nat Mater*, 8 (6) pp. 457-470.

Pluhar G.E., Turner, A.S., Pierce, A.R., Toth, C.A. & Wheeler D.L. (2006). A comparison of two biomaterial carriers for osteogenic protein-1 (BMP-7) in an ovine critical defect model. *J Bone Joint Surg Br. Vol. 88*(7), pp. 960-6.

Poon, Y.F., Zhu, Y.B., Shen, J.Y., Chan-Park M.B., Ng. S.C. (2007). Cytocompatible Hydrogels Based on Photocrosslinkable MethacrylatedO-Carboxymethylchitosan with Tunable Charge: Synthesis and Characterization. *Adv. Funct. Mater.*, vol 17, pp. 2139-2150.

Price, J.S., Oyajobi, B.O., Russell, R.G. (1994) The cell biology of bone growth, *Eur J Clin Nutr*, vol.48, pp.131-149

Puppi, D., Chiellini, F., Piras, A.M. & Chiellini, E. (2010) Polymeric materials for bone and cartilage repair. *Progress in Polymer Science*, 35 (4) pp. 403-440.

Qui., Y. (2008). *Chitosan derivatives for tissue engineering*. Ph.D. Thesis, Clemson University.

Rapaport, H. (2010). *Amphiphilic peptides and hydrogel matrices thereof for bone repair. US Patent* 0015197.

Reichert, J.C., Heymer, A., Berner, A., Nöth, U. (2009) Fabrication of polycaprolactone collagen hydrogel constructs seeded with mesenchymal stem cells for bone regeneration, *Biomedical materials*, vol.4, pp.1748-6041

Rickett, TA., Amoozgar, Z., Tuchek, C.A., Park, J., Yeo, Y., Shi, R. (2011). *Rapidly Photo-Cross-Linkable Chitosan Hydrogel for Peripheral Neurosurgeries biomacromolecules*, vol 12, pp. 57-65.

Schmedlen, R.H., Masters, K.S., West, J.L. (2002). Photocrosslinkable polyvinyl alcohol hydrogels that can be modified with cell adhesion peptides for use in tissue engineering. *Biomaterials*, vol 23, pp. 4325-4332.

Schmitz, J.P. & Hollinger, J.O. (1986) The crictical size defect as an experimental model for craniomandibulofacial nonunions. *Clin Orthop*, 205 pp. 299-308.

Seeherman, H., Wozney, J.M. Delivery of bone morphogenetic proteins for orthopaedic tissue regeneration, *Cytokine and growth factor reviews,* vol.16, pp.329-345 (2005)

Shah, A.K., Lazatin, J., Sinha, R.K., Lennox, T., Hickok, N.J., Tuan, R.S. (1999) Mechanism of BMP-2 stimulated adhesion of osteoblastic cells to titanium alloy, *Biology of the cell*, vol.91, pp.131-142

Silva, G.A., Marques, A.P., Gomes, M.E., Coutinho, O.P., Reis, R.L. (2004) *Cytotoxicity screening of biodegradable polymeric systems*, pp. 339-353.

Slaughter, B.V., Khurshid, S.S., Fisher, O.Z., Khademhosseini, A. & Peppas, N.A. (2009) Hydrogels in Regenerative Medicine. *Advanced Materials,* 21 (32-33) pp. 3307-3329.

Smith, T.J., Kennedy, J.E. & Higginbotham, C.L. (2009) The rheological and thermal characteristics of freeze-thawed hydrogels containing hydrogen peroxide for potential wound healing applications. *Journal of the Mechanical Behavior of Biomedical Materials,* 2 (3) pp. 264-271.

Sperry, J.A., Maguire, D.D. (1984). *Photopolymer derivatives of polyvinyl alcohol. US Patent 4*,478,977.

Studer, K., Decker, C., Beck, E. & Schwalm, R. (2003) Overcoming oxygen inhibition in UV-curing of acrylate coatings by carbon dioxide inerting, *Part I. Progress in Organic Coatings,* 48 (1) pp. 92-100.

Subia, B., Kundu, J. & Kundu, S. (2010). Biomaterial Scaffold Fabrication Techniques for Potential Tissue Engineering Applications. In D. Eberli (Ed.), *Tissue Engineering.*

Tienhoven, E.A.E., Korbee, D., Schipper, L., Verharen, H.V., De Jong, W.H. (2006) In Vitro and in vivo (cyto)toxicity assays using PVC and LDPE as model materials, *Journal of biomedical materials research*, vol.78, pp.175-182.

Tzioupis, C. & Giannoudis, P.V. (2007*) Prevalence of long-bone non-unions. Injury, 38, Supplement 2* (0) pp. S3-S9.

Vacanti, J.P., Morse, M.A., Saltzman, W.M., Domb, A.J., Perez-Atayde, A. & Langer, R. (1988) Selective cell transplantation using bioabsorbable artificial polymers as matrices. *Journal of Pediatric Surgery*, 23 (1) pp. 3-9.

Vallet-Regí, M., Balas, F., Colilla, M. & Manzano, M. (2008) Bone-regenerative bioceramic implants with drug and protein controlled delivery capability. *Progress in Solid State Chemistry,* 36 (3) pp. 163-191.

Veríssimo, D.M., Leitão, R.F.C., Ribeiro, R.A., Figueiró, S.D., Sombra, A.S.B., Góes, J.C. & Brito, G.A.C. (2010) Polyanionic collagen membranes for guided tissue regeneration:

Effect of progressive glutaraldehyde cross-linking on biocompatibility and degradation. *Acta Biomaterialia*, 6 (10) pp. 4011-4018.

Wack, M.A., Schroder, L.K. & Brinkerhuff H.E. (2007). *Hydrogel bone plate spacer. US Patent 0191848.*

Wang, L. & Stegemann, J.P. (2010) Thermogelling chitosan and collagen composite hydrogels initiated with β-glycerophosphate for bone tissue engineering. *Biomaterials*, 31 (14) pp. 3976-3985.

Wichterle, O. & Lim, D. (1960) Hydrophilic Gels for Biological Use. *Nature*, 185 (4706) pp. 117-118.

Williams, C.G., Malik, A.N., Kim, T.K., Manson, P.N., Elisseff, J.H. (2005) Variable cytocompatability of six cell lines with photoinitiators used for polymerizing hydrogels and cell encapsulation, *Biomaterials*, vol.26, pp. 1211-1288

Williams, D.F. (1987) *Definitions in biomaterials*. Amsterdam: Elsevier

Wurtz, T., Berdal, A. (2003) Osteoblast precursors at different anatomic sites, *Critical reviews in eukaryotic gene expression*, vol.13, pp.147-162.

Xiao Y.-T., Xiang L.-X. & Shao J.-Z. (2007) Bone morphogenetic protein. *Biochem Biophys Res Commun*. Vol. 362(3), pp. 550-3

Xu, C.Y., Inai, R., Kotaki, M. & Ramakrishna, S. (2004) Aligned biodegradable nanofibrous structure: a potential scaffold for blood vessel engineering. *Biomaterials*, 25 (5) pp. 877-886.

Yamamoto, M., Tabata, Y., Hong, L., Miyamoto, S., Hashimoto, N., Ikada, Y. (1999) Bone regeneration by transforming growth factor β1 released from a biodegradable hydrogel, *Journal of controlled release*, vol.64, pp.133-142.

Yamamoto, M., Takahashi, Y., Tabata, Y. (2003) Controlled release by biodegradable hydrogels enhances the ectopic bone formation of bone morphogenetic protein, *Biomaterials*, vol.24, pp.4375-4383

Yang, D., Li, Y. & Nie, J. (2007) Preparation of gelatin/PVA nanofibers and their potential application in controlled release of drugs. *Carbohydrate Polymers,* 69 (3) pp. 538-543.

Yang, F., Williams, C.G., Wang, D., Lee, H., Manson, P.N., Elisseeff. (2005) The effect of incorporating RGD adhesive peptide in polyethylene glycol diacrylate hydrogel on osteogenesis of bone marrow stromal cells, *Biomaterials*, vol.26, pp.5991-5998

Yang, X., Yang, K., Wu, S., Chen, X., Yu, F., Li, J., Ma, M. & Zhu, Z. (2010) Cytotoxicity and wound healing properties of PVA/ws-chitosan/glycerol hydrogels made by irradiation followed by freeze–thawing. *Radiation Physics and Chemistry*, 79 (5) pp. 606-611.

Yin, L., Zhao, X., Cui, L., Ding, J., He, M., Tang, C., Yin, C. (2009) Cytotoxicity and genotoxicity of superporous hydrogel containing interpenetrating polymer networks, *Food and chemical toxicology*, vol. 47, pp.1139-1145

Yoshikawa, T., Noshi, T., Mitsuno, H., Hattori, K., Ichijima, K., Takakura, Y. (2001) Bone and soft tissue regeneration by bone marrow mesenchymal cells, *Materials science and engineering*, vol.17, pp.19-26

Zhai, W., Lu, H., Chen, L., Lin, X., Huang, Y., Dai, K., Naoki, K., Chen, G. & Chang, J. (2012) Silicate bioceramics induce angiogenesis during bone regeneration. *Acta Biomaterialia,* 8 (1) pp. 341-349.

Zhao, G. Interaction of surface energy and microarchitecture in determining cell and tissue response to biomaterials. (2007) Available from:*http://smartech.gatech.edu/bitstream/handle/1853/24656/Zhao_Ge_200708_PhD.pdf.*

Zhao, L., Mitomo, H., Zhai, M., Yoshii, F., Nagasawa, N. & Kume, T. (2003) Synthesis of antibacterial PVA/CM-chitosan blend hydrogels with electron beam irradiation. *Carbohydrate Polymers*, 53 (4) pp. 439-446.

Zhong, C., Wub, J., Reinhart-King, C.A., Chu, C.C. (2010). Synthesis, characterization and cytotoxicity of photo-crosslinked maleic chitosan–polyethylene glycol diacrylate hybrid hydrogels. *Acta Biomaterialia,* vol 6, pp. 3908-3918.

Zhu, J. (2010). Bioactive modification of poly(ethylene glycol) hydrogels for tissue engineering. *Biomaterials*, vol. 31, pp. 4639-4656.

Zimmerman, G., Wagner, C., Schmeckenbecher, K., Wentzensen, A., Moghaddam, A. (2009) Treatment of tibial shaft non-unions: bone morphogenetic proteins verses autologous bone graft, *Int. J. Care Injured,* vol.40, pp.550-553.

Hydrogels: Synthesis, Characterization and Applications ISBN: 978-1-61942-842-3
Editors: F. V. Câmara and L. J. Ferreira 

*Chapter 3*

# Hydrogel as a Controlled Release System

***Roseane Maria Ribeiro Costa, Anivaldo Pereira Duarte Junior, Reagan Mampuya Matondo and José Otávio Carréra Silva Júnior***
Federal University of Pará, Institute of Health Sciences
Faculty of Pharmacy, Laboratory of Quality Control,
Cidade Universitária Professor José da Silveira Neto, Belém-PA, Brazil

## Abstract

As a great challenge for any country today, billions are being invested in the development of technology to obtain more efficient and competitive products that may bear fruits and boost the economy of the country. The application of technological innovation is employed in several areas aimed at obtaining products, structures, devices and / or systems that help have more sophisticated control of specific properties such as use, size and shape. The application of hydrogels technology is being widely used in pharmaceutical industry as drug delivery systems that have physical, chemical and biological characteristics which make them very interesting vehicles as carriers of drugs, medicines and cosmetics. The hydrogels can be used in all routes of administration for presenting compatibility with the tissues, low toxicity and generally for being biodegradable. They are made of polymeric materials in order to aggregate the different properties between the constituent polymers and possess the convenience of having characteristics that may be shaped according to their target application. Hydrogels synthesized through blends of synthetic and natural polymers have been an alternative of interest to merge interesting properties. The response to stimuli in polymeric hydrogels may increase or decrease according to the three-dimensional shape of the polymer chain and it also may change in environmental conditions. In order to guarantee the properties that a perfect carrier can hold, the quality control of these systems from their pre-formulation to the final product it is crucial to assure their efficiency as a controlled release system.

# 1. INTRODUCTION

## 1.1. Hydrogels

The research on hydrogels began in the 60's with gels of poly (hydroxyethyl methacrylate ) along with ethylene diacrylate by Wichterle and Lim [1] which was the first material used for the production of contact lenses. From this date, studies with hydrogels have become major routes in the advances of the technology of acquisition and application of these materials [2].

Hydrogels are defined as polymeric networks with three-dimensional configurations capable of imbibing high amounts of water or biological fluids [3-6].Therefore, they are able to absorb/retain large amounts of water, reaching thousands of times the mass of the polymers involved in their matrix [7]. This affinity to absorb water is attributed to the presence of hydrophilic groups such as -OH,-CONH-,-$CONH_2$-, and -$SO_3H$ in hydrogel-forming polymers structures[8].

They originate a three-dimensional network held together by covalent bonds (crosslinking) or electrostatic forces [9, 10]. The polymeric networks can be formed by one or more monomers, homopolymers or copolymers, so-called respectively, whereas at least one of them must be hydrophilic. They can be constituted of synthetic monomers: anionic, such as the family derived from acrylic acid, the crotonic acid; cationic, such as vinylpyridine, derived from amino ethyl methacrylate; neutral, such as the hydroxyalkyl methacrylates, derived from acrylamide, N-vinyl pyrrolidone, hydrophobic acrylic; and crosslinking, such as ethylene-bis-acrylamide, ethylene glycol dimethacrylate [10].

Hydrogels can be synthesized through chemical and physical methods. According to their physical structure, hydrogels can be classified into semi-crystalline or amorphous. And in relation to their electric charge, they can be neutral or ionic [11].

These systems have several advantages: low toxicity, ability to swell in water and biological fluids (which make them similar to living tissues); elastomeric consistency (which minimizes friction between tissues and the hydrogel); high permeability (which allows the flow of body fluids by the hydrogel due to its high porosity); ease of obtaining them in different forms; incorporation and controlled release of drugs of different polarities, making them attractive as promising drug carriers for controlled release systems [2, 12].

## 1.2. Controlled Release Systems

Controlled release systems (CRS) are very promising systems with regard to increased therapeutic efficacy of a drug through the maintenance of drug levels within the therapeutic range, a significant decrease in toxicity while minimizing side effects, thus reducing the need for administration of various doses and therefore, adherence to treatment [13-17].

The fact that most of the hydrogels are biocompatible and / or biodegradable contributes for a continuous investigation for application as a controlled release system. These systems have been employed successfully in a variety of drugs and proteins, among them: bovine serum albumin [18], antihypertensives [19, 20], antihyperglycemic [21], antibacterial [22], ophthalmic [23].

In developing these systems several factors must be taken into account, including: properties of the active ingredient, especially solubility; type of therapy; route of administration; protection of the active principle; stability and release mechanism; the nature of the polymers (properties and concentration) as well as process conditions. Polymers are among the most commonly used excipients for obtaining pharmaceutical forms in therapies of controlled release of the drug due to their various applications and features [24].

## 1.3. Polymeric Excipients

Polymers are one of the most versatile classes of materials and have changed our daily lives for decades with important applications in the medical field [19, 25]. The combination of polymer science with pharmaceutical sciences led to a spectacular advance in terms of "innovation" (flexibility in the physical state, shape, size and surface) in the design and development of new drug delivery systems [25, 26]. Polymeric delivery systems are mainly applied to achieve a temporal or spatial control of drug release and are very important tools for pharmaceutical technology [26, 27].

The physicochemical properties of polymers are very important factors that determine their application. These properties are dependent on the chemical nature of monomers, process and technique of preparing the polymer, molecular weight, which depends on polymerization conditions for each monomer, and the macromolecular structure [28].

Yet the physical properties of polymers are related to the strength of covalent bonds, the rigidity of the polymer chain segments and the strength of the intermolecular forces [28]. These properties are also related to their behavior in front of variations in temperature. When reaching glass transition temperature (Tg), their molecular chains undergo a rearrangement, when reaching melting temperature (Tf) they change their physical state and there still is the transition temperature phase (Tm), in which the polymer is above this total mobility [18]. Due to the inherent diversity of structures and the requirements of fully understanding the surface and properties of the polymer which can provide its chemical, interfacial, mechanical and biological desired functions, the selection and design of a polymer becomes a very challenging task [25].

In developing CRS, some properties of the polymers show more relevancy in their preparation, such as permeability [29], surface properties such as hydrophilicity, lubricity, smoothness, durability, degradability, surface energy [30], adhesion, solubility [25] and glass transition temperature [31]. These information are also very relevant in the choice of polymeric material that will form the hydrogel.

The excipients used in obtaining hydrogels are mostly natural, synthetic or combinations of natural and synthetic polymers (Table 1). Synthetic polymers are available in a wide variety of compositions with easy adjustment of the properties [19]. Hydrophilic synthetic polymers usually own clusters of polar atoms in the chain. Such groups can be hydroxyl, amine, carboxylic, ionic or otherwise, which make the polymer interact more strongly with water, by dissolving, dispersing or swelling itself. The increase of ionic charge in a polymer increases the repulsion between its molecules and consequently leads to an increase of its hydrophilicity and swelling capacity, an important characteristic in polymer matrices in the transmission of controlled release drugs [20].

The application of the hydrogels when directed to a particular organ or tissue may need an increase of the hydrophobic character in the polymer matrix. In this case, hydrophobic monomers or oligomers are grafted in the polymer chain or, if there are hydrophilic functional groups, these can be modified through reactions with compounds that increase the hydrophobicity [21].

**Table 1. Natural, synthetic or combinations of natural and synthetic polymers used in preparation of hydrogels**

| Natural Origin | Abbreviation | Reference |
|---|---|---|
| Chitosan | CH | [22, 23] |
| Chondroitin Sulfate | CS | [24] |
| Arabic Gum | GA | [25, 26] |
| Dextran | D | [27, 28] |
| Galactomannan | G | [29, 30] |
| **Synthetic origin** | | |
| Poly(vinyl alcohol) | PVA | [31-33] |
| Polyacrylamide | PAAm | [34] |
| Poly (N-isopropylacrylamide) | PNIPAAm | [35, 36] |
| Poly [(N-isopropylacrylamide)-co-methacrylic acid | P[(N-iPAAm)-co-(MAA)] | [37] |
| Poly(methacrylic acid-co-poly (ethylene glycol) | P(MA-co-PEG) | [38] |
| Poly (ethylene oxide) | PEO | [39] |
| **Combinations of natural and synthetic polymers** | | |
| Polyacrylamide-co-methylcellulose | PAAM-co-MC | [40] |
| Cashew gum-co-acrylamide | CGMA-co-AAm | [41] |
| Chitosan-poly (hidroxyethyl methacrylate) | pHEMA | [42] |
| Alginate-Ca +2- co-poly (N-isopropylacrylamide) | CA/PNIPAAm | [43] |
| Chitosan-co-poly (acrylic acid) | CH-co-PAA | [44] |

Natural polymers (or biopolymers) are found in abundance and in a biodegradable way in nature, and offer excellent biocompatibility, but may vary from lot to lot due to difficulties in purification [19]. They have good mechanical and functional properties. Biopolymers are widely used in preparations such as hydrogels. There are some examples listed in Table 1 [45, 46].

A variety of non-biodegradable polymers, cellulose derivatives (natural origin) and acrylic (synthetic origin), is used in the preparation of hydrogels [4, 6, 47].

Factors affecting the biodegradation of polymers are: structure and chemical composition, distribution of repeated units in monomers; presence of ionic groups, presence of unexpected units or chain defects, configuration of the structure, molecular weight and molecular weight distribution, morphology; process conditions: the implant site, physical-chemical factors and mechanism of hydrolysis. Biodegradable polymers of natural or synthetic origin are degraded *in vivo* by nonenzymatic or enzymatic hydrolysis or both processes, producing biocompatibility and metabolites which are eliminated through normal channels [48, 49].

The biocompatibility and biodegradability of a polymer also depend on a fine-tuning of its hydrophilicity. This can be done by modifying the surface energy of the polymer by graft polymerization of some hydrophilic monomer chain, conducted by high-energy radiation [19, 50].

## 1.4. Drug- Hydrogel Adsorption

The main inherent properties of the drug to be incorporated into a CRS are: stability, solubility (e.g., hydrophobic drugs tend to precipitate in aqueous media), pharmacokinetics (drugs having a very fast clearance require an increase in the number of doses), biodistribution (drugs with wide distribution in the body are more likely to cause toxic effects) and the molecular weight (the smaller the molecular weight, the better is the absorption) [2].

The phenomenon of adsorption is an example of drug and excipient interaction and through this study it is possible to evaluate the ability of interaction between adsorbent and adsorbate [51]. The physical adsorption is due to electrostatic interactions, hydrogen bonding or van der Waals forces and it is usually reversible, whereas in chemical adsorption, the adsorbed is tied to the adsorbent by primary chemical bonds, including ion exchange, protonation and complexation, and it is irreversible [52]. The extent of adsorption depends on the physicochemical properties of the drug and excipient [53].

The adsorption of the drug to the hydrogel can be accomplished through two methods. In the first one, the drug is added to the reaction system at the time of the synthesis of the hydrogel. In the second method, the already synthesized hydrogel is immersed in an aqueous solution containing the drug at a known concentration until reaching balance. The latter method is most often used, as there is little interference of the drug at the time of the synthesis and so, little possibility of polymerization of the hydrogel [37].

To illustrate how the drug binds to the hydrogel, and how the adsorption of the drug to the polymer happens, we can apply mathematical models of isotherms: Langmuir and Freundlich. According to the Langmuir model, the drug will be adsorbed in the hydrogel in a homogeneous manner, since it is available in monolayers. This model is expressed in the equations 1 and 2. In the Freundlich model, the drug is adsorbed in a heterogeneous way, because it is available in multilayers as in equations 3 and 4 [54-58].

$$qeq = KL \times \frac{Ceq}{1} + (\alpha L \times Ceq) \quad (1)$$

$$\frac{Ceq}{qeq} = \frac{1}{KL} + \left(\frac{\alpha L}{KL}\right) \times Ceq \quad (2)$$

$$qeq = Kf \times Ceq^{1/n} \quad (3)$$

$$\ln qeq = \ln Kf + {}^{1}/_{n} \times \ln Ceq \quad (4)$$

Where: qeq is the total amount of the drug adsorbed by the hydrogel ($\mu g.g^{-1}$); Ceq is the concentration of the drug solution in equilibrium ($\mu g.mL^{-1}$), KL and αL are Langmuir constants. Kf and n are Freundlich constants. These constants are calculated through the intercept and slope of the line. If n> 1, adsorption was favorable [59].

## 1.5. Release Kinetics

The studies on the *in vitro* release of drugs, determine the fraction of drug released as a function of time. The characterization of the drug release profile can provide information on the structure of the system, the mechanism involved in the release process, the interaction of the drug with the carrier and it can still be used as quality control for the proposed formulation [60].

The profile of drug release is directly correlated with the drug-polymer interaction. The drug release depends on how much the polymer matrix absorbs water; on its dissolubility in the medium; on the migration and diffusion of the drug molecules as a function of drug solubility and diffusion coefficient; on the physical structure of the polymer, the crystalline structure has a higher degree of molecular packaging and therefore tends to resist more to the release of a drug compared to amorphous polymers with the same chemical structure; on the pH of the polymer, when the drug is released in a specific pH, i.e., in the gastrointestinal tract or colon, non-ionic polymers should not be used because they are pH-independent [53, 61, 62]. However, the method of adsorption of the drug, the partition of drugs, the hydrophilic/lipophilic balance, the osmotic effect and the elasticity of the polymer are also factors that affect the controlled release of drugs [63].

The most common mechanism of drug release from hydrogels is the passive diffusion, molecules of different sizes and characteristics would freely diffuse into/out of the hydrogel matrix during loading and storage periods. From various behavior studies on the release mechanisms of an active compound from a hydrogel device, focused on the rate-limiting step of the release phenomena, drug release mechanisms from hydrogels can be categorized as: diffusion-controlled, erosion, relaxation of polymer chains and dissolution [6, 20].

The kinetics of *in vitro* release aims at clarifying how the drug is released by the system. Analysis of the kinetics of drug release by polymer matrix may result from one or more release mechanisms. Mathematical models have been developed in order to describe the drug release from the dosage form that contains it. Several articles that describe these models are available in the literature [18, 20, 34, 40].

## 2. SYNTHESIS OF HYDROGELS

Several methods for obtaining hydrogels are found in the literature. Due to the flexibility of these materials, they can be molded according to the destination of the application and characteristics of the polymer. Especially hydrogels for applications in drug delivery systems are required to have acceptable biodegradability and biocompatibility which necessitate the development of novel synthesis and crosslinking methods to design the desired products [6, 63].

The polymeric networks can be formed by one or more monomers, homopolymers or copolymers, so-called respectively, whereas at least one of them must be hydrophilic [4, 64]. They can be constituted of synthetic monomers: anionic, such as the family derived from acrylic acid, the crotonic acid; cationic, such as vinylpyridine, derived from amino ethyl methacrylate; neutral, such as the hydroxyalkyl methacrylates, derived from acrylamide, N-vinyl pyrrolidone, hydrophobic acrylic; and crosslinking, such as ethylene-bis-acrylamide, ethylene glycol dimethacrylate [9, 41, 42] .

They can be formed by physical or chemical crosslinks (Figure 1) of homopolymers or copolymers, being appropriately used to give the three-dimensional structures specific mechanical and chemical characteristics. The crosslink can take place after or at the same time as the copolymerization [2, 10]. Physical gels are three-dimensional networks where the polymer chains bond through non-covalent interactions. Junction zones form when the separate polymer chains interact over a certain length and not point to point (point crosslinks). One of the ways to form physical crosslinking is hydrophobic interaction in which hydrophobic blocks are coupled to hydrophilic blocks creating a polymer amphiphile [43]. When the bonds between the polymers are of covalent nature, the crosslinking is of chemical character. Covalent interactions are much stronger than non-covalent ones, providing excellent mechanical stability. Chemical crosslinking methods include radical polymerization, chemical reaction of complementary groups, high energy irradiation and enzyme usage [44].

A polymer widely used in obtaining hydrogels is PAAm (polyacrylamide), which has a synthetic origin, and whose main property is hydrophilicity. It has good mechanical strength; it is biocompatible and absorbs large amounts of water. On the other hand, chemically modified polysaccharides, cellulose derivatives are widely used in preparations such as hydrogels. The methylcellulose (MC), resulting from a methyl substitution in a cellulose molecule, can be transformed into gel through chemical crosslinking using di-aldehydes in the presence of a strong acid [46-48]. Hydrogels obtained from the combination of these two polymers, polyacrylamide-co-methylcellulose (PAAma-co-MC), were considered potential drug carrier candidates in controlled release systems. They were synthesized by free radical polymerization, in which N-N'-methylene-bisacrylamide (MBAAm), N,N,N',N' tetramethylethylenediamine (TEMED) and sodium persulfate (PS) were used, as cross-linker, initiator and catalytic agent for the reaction, respectively [39, 45].

Semi-interpenetrating network hydrogels using silk fibroin/polyacrylamide were prepared by free radical method, using ammonium persulfate as initiator and TEMED, as activator. The authors related that the intense bonding and 3D networking between polyacrylamide, the stable and cytocompatible matrices were obtained with natural fibroin protein. This hidrogel was synthesized as drug releasing matrices for potential tissue engineering and biomedical applications [65].

Hassan and co-workers have studied the incorporation and release profile of the bovine serum albumine (BSA) in Poly(vinyl alcohol) hydrogels by freezing and thawing technique [66]. This "freeze-drying" technique is useful in the preparation of porous hydrogels from water-soluble polymers such as polysaccharides. Monomers can be polymerized in the frozen state around aqueous crystals, and then water can be subsequently removed by thawing to result in a macroporous hydrogel. In this technique, the release can be optimized by controlling both the conditions and the number of cycles of freezing and thawing.

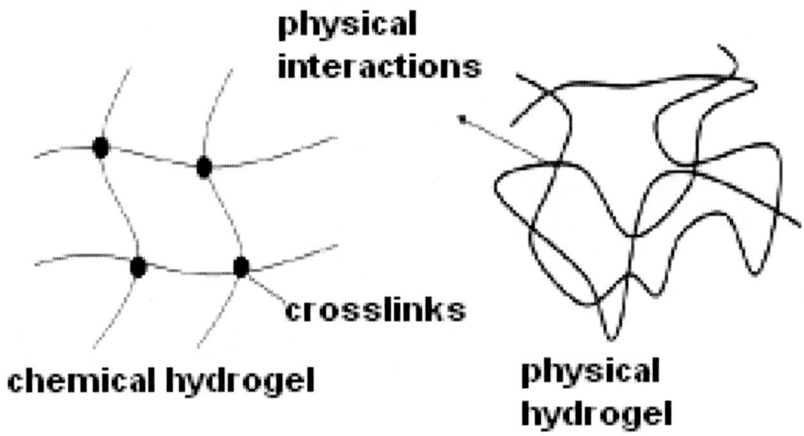


Figure 1. Structures representing chemical and physical hydrogels [45].

A physically cross-linked palmitoyl glycol chitosan hydrogel was also formed by freeze-drying erodibility, an aqueous dispersion of the polymer either in the presence or absence of a macromolecule fluorescein isothiocyanatedextran model as a drug delivery system [67].

Synthetic polymers sensitive to temperature changes are being used in systems for drug delivery. Thermo-sensitive hydrogels demonstrate a large change of equilibrium swelling in response to small pH and/or temperature variations [68, 69].

Park and Hoffam [70] prepared and characterized positively charged hydrogels by copolymerizing varying ratios of N-isopropylacrylamide and NN′-dimethylaminopropylmethacrylamide. These hydrogels exhibit different lower critical solution temperature (LCST) ranges depending on the environmental pH values. They observed that the pH-dependent phase transition is strongly affected by temperature, while the temperature-dependent transition is, in turn, largely influenced by the pH. As the temperature is raised, the transitional degree of gel swelling change becomes sharper and larger, and the pH value of the phase transition shifts to a lower pH. The authors also found that swelling is faster than deswelling for these cationic hydrogels, and that, the swelling kinetics of initially dry and glassy gels were strongly dependent on both the pH value and temperature.

Thermo sensitive semi-IPN hydrogel constituted of PNIPAAm supported in alginate-$Ca2^{+}$ matrices in different alginate and PNIPAAm concentrations were prepared and characterized by Moura *et al.* [71]. Results showed that controlling these properties makes these semi-IPN hydrogels, thermo sensitive and potentially viable for different applications such as controlled and/or prolonged release drug carriers and substrates for cell growth.

A self-folding miniature device has been developed to provide enhanced mucoadhesion, drug protection, and targeted unidirectional delivery. This device has been obtained from pH-sensitive hydrogel by free-radical photopolymerization based on crosslinked poly(methyacrylic acid) (PMAA) that swells significantly when in contact with body fluids [72].

Rosiak [7] developed a hydrogel for wound healing in which the radiation formation of hydrogels using N-vinylpyrrolidone as polymer in aqueous solutions proceeds according to the radical mechanism. In both processes of polymerization and crosslinking, the main role is played by indirect effect. These processes are initiated by intermediate products of radiolysis of water forming reactive $H^+$ and especially $OH^-$ free radicals. In the presence of oxygen in the irradiated solutions, the process of gel formation is accompanied by the reactions leading to the scission of polymer main chain. These methods are called clean as a sterilized product is obtained.

But recently, Brant [21] synthesized stable hydrogens from mixtures of aqueous solutions of chitosan (CH) and poly (N-vinyl-2-pyrrolidone) (PVP) as well as films of blends of these polymers in different proportions through ultraviolet radiation at 254 nm. The author noted that the strong interaction between PVP and CH led to an increase of blend insolubility in relation to pure hidrophilicity of the blend, reducing the crystallinity of the chitosan. Another aspect seen as important by the author was the increase of the thermal stability of chitosan, influenced by the increase of PVP portions in the blends.

Several advances have been made in hydrogels technology in drug delivery systems, such as the development of nanoparticle formulations, due to their physical, chemical and biological properties. The nanogels, which are hydrogels obtained at a nanoscale present characteristics such as hydrophilicity, flexibility, versatility, high water absorptivity and biocompatibility of these particles and all the advantages of the nanoparticles. Mainly long life span in circulation and the possibility of being actively or passively targeted to the desired biophase, e.g. tumor sites. Different methods have been adopted to prepare nanogels of different consistency. Beside the commonly used synthetic polymers, active research is focused on the preparation of nanogels using naturally occurring hydrophilic polymers [6].

## 3. Characterization of Hydrogels

The hydrogels can be characterized in different ways, for example, according to their chemical composition, size and structure [73]. Some methods for characterization of hydrogels can be seen in Table 2.

Through the application of these methodologies it is possible to know the physico-chemical and microbiological features of these systems. The evaluation of such characteristics is of great importance for understanding the behavior of the hydrogel and checking the intended application of the material studied. From the results obtained one can predict the behavior *in vivo* as well as to promote the optimization of the system, the development of the formulation and the obtaining process [73, 74].

**Table 2. Methods used for characterization of hydrogels**

| Characterization Parameter | Analytical Tool | References |
|---|---|---|
| Crystallinity | Differential scanning calorimetry (DSC)<br>Scanning electron microscoscopy (SEM) | [73] |
| Amount of Water | Differential scanning calorimetry (DSC) | [75] |
| Composition | Liquid chromatography, (HPLC) or (SLC) – HPLC<br>UV – visible spectrophotometry<br>Fourier transform infrared spectrometry<br>Mass spectrometry<br>X - ray fluorescence<br>Fluorescence spectroscopy<br>$^{1}$H-NMR spectroscopy | [73, 75-88] |
| Particle diameter and size distribution | Photon correlation spectroscopy<br>Transmision electron microscopy<br>Scanning electorn microscopy | [73, 74, 77, 89] |
| Surface area | BET method (Brunauer, Emmett, and Teller method) | [90, 91] |
| Surface structure and morphology | Proton nuclear magnetic resonance (1H NMR) spectroscopy<br>Scanning electron microscoscopy (SEM)<br>Field emission scanning electron microscopy (FESEM)<br>Transmission electron microscopy (TEM)<br>X - ray diffraction (XRD)<br>Contact Angle Measuring | [73-77, 79, 81, 92, 93] |

## Density Measurement

The density of the hydrogel can be measured using a picnometric method by adding a defined mass of the sample in a pycnometer of known mass and then filling it with an inert solvent with known density [94].

## Brunauer-Emmett-Teller

The aim of the Brunauer-Emmett-Teller (BET) technique is to obtain information about the specific area and the porous structure of a solid, being represented by an adsorption isotherm.Through the interpretation of the curve represented by the isotherm it can be concluded that the porous feature is related to the material under study.The type of isotherm is a function of the effect of pore size on the adsorption phenomenon.

Generally, the isotherms can be decomposed into three characteristic regions.The first occurs at low values of relative pressure, and concerns the adsorption of $N_2$ molecules in energetically favorable regions of the particle surface.In this region the information on the particle surface area lies and where BET is determined.The second region occurs for intermediate values of relative pressure and it is related to the overlapping layers of gas on the particle surface.The third and last region of the isotherm occurs for high values of relative pressure and it is characteristic of the condensation phenomenon of gas volume.

The sample preparation to BET measurement begins with the lyophilization of the sample. Then the sample is "degassed" for a certain period of time and right afterwards the process of measuring the surface area starts [90].

According to the pore size, the hydrogels can be classified in microporus or macroporus hydrogels.Due to the presence of pores with larger size and volume than those found in traditional hydrogels and also greater high specific surface area [91].

## Nuclear Magnetic Resonance Spectroscopy

Nuclear magnetic resonance (NMR) is an analytical method that allows the rapid structural characterization of liquid and solid samples, being perfectly applied to the study of macromolecular systems, such as hydrogels [79].

It is very safe technique as ionizing radiation is not used. Signs can be produced from hydrogen $^{1}H$ and carbon $^{13}C$ atoms, which are the most studied by NMR.

The NMR can be used to determinethe size and composition of hydrogels and also to elucidate the structure and composition of copolymers [73, 80-83].Through this technique the free chemical groups in the structure of the polymer or copolymers which belong to the hydrogel structure may be known as well as study the chemical bonds that occur in the formation of this system [84-86, 93].

The dynamic behaviorof the water in thehydrogelcan be characterized by $^{1}H$ NMR. The water relaxation times values T1 and T2, generally exhibit a linear relationship with the cube root of the swelling degree [95].The concentration and type of polymer used to obtain the hydrogel also change the T1 and T2 values of the water, showing the fraction of the water bound to the polymer [96].

Using$^{13}C$ NMR, the spectra obtained may present low resolution, even when the spinning rate and decoupling field strength vary.This occurs when the sample is amorphous and low-resolution peaks are due to the large distribution of isotropic chemical shifts occurring indisordered systems. Generally, the carbon 1C gives rise to a very large and asymmetric signal, ascribable to the presence of different conformational situations. This phenomenon is known as γ-gaucheeffect [97-99].

## Fourier Transform Infrared Spectroscopy

Fourier transform infrared spectroscopy (FTIR) is used to characterize hydrogels because it can identify chemical groups present in the sample showing interaction between the monomers used in the synthesis and the formation of copolymers, as well as confirm if there was crosslinking of the substance to the polymer network of the hydrogel [86]

The FTIR spectra of the components of the synthesis reaction of the hydrogel can be obtained separately and the characteristic absorbance bands of each one marked down. Then the FTIR spectra of the synthesized hydrogel will point out the disappearance of these characteristic peaks, showing that there was an interaction between these components during the synthesis of hydrogel. The comparison of theFTIR spectra between the same hydrogel prepared by different techniques or by the same technique, with a few changes in the methodology may be useful to see if there was any change in the chemical structure of the hydrogel [82, 87, 88].

To obtain porous hydrogels, some characteristic substances are used in the synthesis, such as silica particles [100]. The use of FTIR technique is important to confirm the extraction of the substance that was used in the synthesis with the intention of obtaining porous hydrogels and ensure that this substance is no longer part of the hydrogel.

## Optic and Electron Microscopy

Optical microscopy has less capacity to increase when compared to electron microscopy. However, the optical microscopy can be used when you want to visualize the presence or absence of pores in the structure of the hydrogel, with the use of swollen wet gel samples directly without further treatment [101].

The maximum possible magnitude increase with optical microscopy is approximately 2000 times, so some very small structural elements can not be viewed using optical microscopy. In these circumstances the electron microscopy is used, since it reaches a much larger magnitude increase, around 1.000.000 times [102].

In electron microscopy the image of the structure under investigation consists of an electron beam. The high magnitude and resolution increase are achieved due to the small wavelength of the electron beam as a result of the acceleration caused by the application of high voltages [102].

The scanning electron and transmission microscopy (SEM and TEM, respectively) are able to visualize the hydrogels directly.Therefore details of the hydrogel morphology are known, and the scanning microscopy is the best for this function [73].

The image obtained by TEM is formed by an electron beam that passes through the sample, so the sample must be prepared in the form of a very thin sheet. The TEM has a lower size limit of detection, provides structural information by electron diffraction; however it is necessary to color the material with some contrast and keep in mind that the small sample size may be influenced by the vacuum [73, 103].

The SEM is a newer technique than the transmission microscopy, consisting primarily of an electron beam that travels through the sample surface and the reflected or backscattered electrons are captured and transformed into an image on a monitor.The sample surface must be covered by a thin layer of electrically conductive material [102].

The hydrogel sample that will be investigated through the SEM should be dry, and in order to keep the pores of the material intact, the hydrogel should be placed dry in liquid nitrogen and then break it down for better visualization of the polymer network [93].Another option is to give the hydrogel swell to equilibrium at room temperature and freeze-dry in a freeze dry system to remove water completely and finally sputter-coat with gold [90, 104-107].

## Atomic Force Microscopy

The atomic force microscopy (AFM) has the versatility of obtaining the resolution of the scanning electron and transmission microscopy with the ability of imaging in an aqueous medium, typical of optical microscopes. Besides being able to produce images of conductive and nonconductive surfaces, it does not use lenses and does not require a light source or an electron beam to obtain the images [108].

In the AFM technique a probe with a tip at the atomic scale scans the surface of the sample and produces a topological map based on the forces acting between the probe tip and sample surface, providing high spatial resolution in the visualization of the surfaces of different nature at atomic level and in various media [73, 108].

The probe may be in contact with the sample or not, depending on the technique used [73].The three major subtypes of AFM are: contact mode, non-contact mode and tapping mode [108, 109].

This technique has the feature of not only determining the particle size, but also determining the colloidal attraction or resistance to deformation, making it a very interesting technique. However, size and shape are the main applications today. The study of polymer structure from the atomic level with the three-dimensional visualization and quantitative precision measurements are of great importance for the characterization of hydrogels [73, 110-112].

Studies with atomic force microscopy can be divided into topographic applications and non-topographic. In the first application, the goal is to obtain images of the surface of the sample for a structural characterization and dynamic [108, 113].

The surface roughness of hydrogels is another feature that can be achieved by processing the images captured by the topographic AFM. Mean surface roughness, mean-square-roughness and maximum roughness are processed from the roughness analysis using an image processing software[109].

The advantages of this technique are the possibility of observing structures at atomic scale in real space and the ability to operate at room temperature and in the presence of air, and therefore, reducing the cost of construction and operation [108]. Other features that differentiate the AFM from the other electron microscopy techniques are the largest increase magnitude, reaching an increase of $10^9$ times better resolution and three-dimensional imaging [102].

## Zeta Potential

The zeta potential determines the charge of the particle surface. It is measured from the signal oscillation of the light backscattered from particles subjected to an electric field [73]. The zeta potential reflects the surface potential of the particles, which is influenced by changes in the interface with the dispersing medium, due to the dissociation of functional groups on the particle surface or adsorption of ionic species present in aqueous dispersion [114].

## X-ray Diffraction

X-ray scattering studies is used to verify the morphology of the sample, especially when you can not see from scanning electron microscopy [100]. The diffraction pattern can be obtained from the isolated polymer and of the synthetized hydrogel.

The crystallinityof the material analyzed is calculated according to the intensity of the peaks shown by XRD pattern. These data can, for example, suggest an increase or decrease the level of crosslinking of the hydrogel, and evaluate the porous structure of porous hydrogels [115-117].

## Water Content or Swelling

A fundamental characteristic of the hydrogel is the ability to absorb a large amount of water through the interaction with the polymer chains, promoting the expansion of the hydrogel volume. The structure formed by the binding of the hydrogel polymer forms a three-dimensional network that acts as a retraction force which maintains the hydrogel structure and determines the maximum amount of water absorbed and therefore the maximum swelling of the hydrogel. This inter-relationship between expansion and contraction reaches balance at a particular solvent and temperature and is explained by the Flory rubber elasticity theory [118-121].

The swelling characteristics of the hydrogel are very important to determine its application in medical and pharmaceutical products. The swelling reached at the equilibrium-state influences on the solute diffusion coefficient, surface properties, mobility, and optical and mechanical properties of hydrogels. The equilibrium state is reached after the immersion time in which there is no more swelling of the hydrogel, and this time varies according to the composition of the hydrogel and the degree of crosslinking [86, 118, 122-126].

Some parameters related to the swelling of a hydrogel are presented below. The determination of the swelling degree is performed from a dry hydrogel with a known mass which is immersed in a liquid medium, usually water. In predetermined intervals of time the sample is removed from the medium, dried at the outer end with paper to remove water excess and then weighed [75, 78].

The water content and degree of swelling is expressed in most cases in percentage in the following equations 5 and 6:

$$Water\ content = \frac{water\ mass}{water\ mass + dry\ hydrogel\ mass} \times 100 \qquad (5)$$

Another way to express swelling is through the swelling ratio, represented by the ratio of mass of the swollen sample and the mass of the dry sample, as in the equation below:

$$swelling\ ratio = \frac{weight\ of\ swollen\ gel}{weight\ of\ dry\ gel} \qquad (6)$$

When the water content absorbed by the hydrogel reaches more than 90%, the hydrogel is considered superadsorvent [118].

Some factors that may cause changes in the swelling of the hydrogel are: pH, temperature, solvent and ionic strength of the medium [80, 86]. The molecular weight, the presence of hydrophilic chemical groups in the polymer chains, the degree of crosslinking and cristanility of the polymer also cause changes in swelling [81, 93, 102]. Figure 2 shows the degree of swelling of a hydrogel composed of polyacrylamide and methylcellulose.

The kinetics of swelling can be calculated by the following equation 7:

$$F = \frac{M_t}{M_{eq}} = kt^n \quad (7)$$

Where $M_{eq}$ is the hydrogel mass at swelling equilibrium, t is time and k is the diffusion constant and depends on the type of hydrogel and swelling medium. The parameters obtained when calculating the degree of swelling as a function of time (F vs T) are the diffusional exponent (n) and diffusion constant (k). For the calculation of n and k a graph of ln F vs. ln t is used. The value of the diffusional exponent n is the slope of the curve and the linear coefficient is ln k [54, 86, 127].

The value of n=0.5 indicates that the kinetics of swelling was by Fickian diffusion, n=1 shows anomalous transport and n values between 0.5 and 1 demonstrate that there have been two types of transport phenomena where diffusion and relaxation of the polymer chains were simultaneously responsible for the swelling kinetics[86, 128].

Equation 7 can be applied from the early stages up to 60% of the swelling curve (F vs t).The increase of the swelling degree with time is practically linear, and after the 60% stage, the swelling no longer follows this trend and since then hardly suffers more variation over time to reach the equilibrium state. The curves obtained from the relation F vs $R^2$.t which have $R^2$ values inferior to 0.9994 indicate that this difference in linear behavior in the first 60% of swelling and the non-linear behavior of the rest of the process still reaching equilibrium are more pronounced, thus demonstrating that equation 7 is valid only for the first 60% of the hydrogel swelling [54, 129].

Figure 2 below shows a curve of *lnF vs ln (t)* of a polyacrylamide and methylcellulose hydrogel immersed in a pH 7 solution. From the curve the two stages of the hydrogel swelling can be seen.

The diffusion of the solvent into the hydrogel can be obtained through the analysis of D values, obtained from equation 8: [54]

$$D = \pi r^2 \left(\frac{k}{4}\right)^{\frac{1}{n}} \quad (8)$$

Where D is given in $cm^2\ s^{-1}$, r is the radius of the dry hydrogel (cm) in a cylindrical shape. The values of n and k are obtained from equation 7.

Another way to calculate the diffusion coefficients for thin films of water in hidrogels utilizing diffusion equations was first developed by Crank and Park [130]. A sample of thin film was suspended in the sorption apparatus and the kinetics of sorption for a single step in humidity was recorded in the usual manner. For a single step in controlled humidity, with a double-sided thin film of thickness $d$ the initial kinetics of sorption into the bulk may be described by the following equation 9[128]:

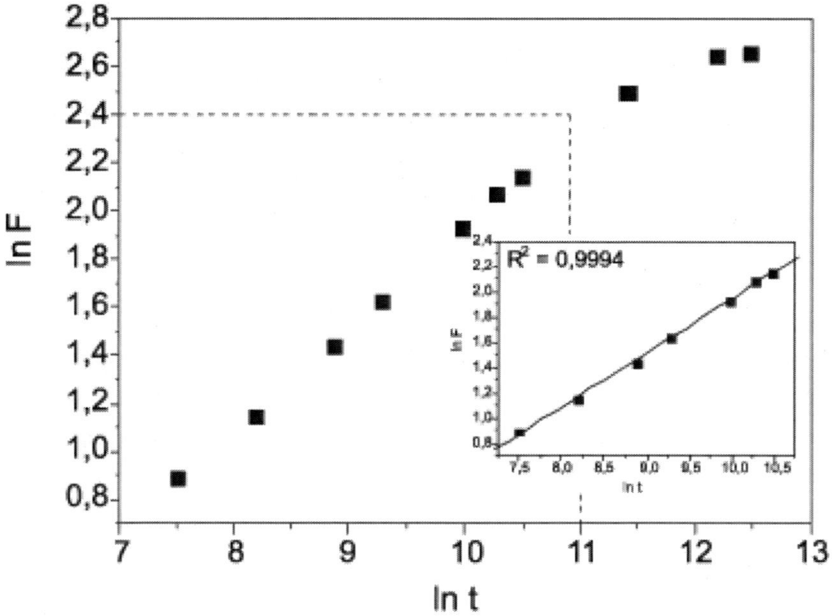


Figure 2. Relation of ln F vs. ln (t) of a hydrogel demonstrating the linear and nonlinear swelling stage.

$$\frac{M_t}{M_\infty} = \frac{4}{d}\sqrt{\frac{Dt}{\pi}} \qquad (9)$$

Where $M_t$ is the amount adsorbed at time $t$, $M_\infty$ is the amount adsorbed at thermodynamic equilibrium, and $D$ is the diffusion coefficient.

The average molar mass of waste ($M_c$), according to the theory of Flory and Rehn for a perfect network, can be obtained from equation 10: [54, 78, 106, 131, 132].

$$M_c = -V_1 d_P \frac{\left(V_s^{1/2} - V_{s/2}\right)}{\ln(1) - V_S + V_S + xV_S^2} \qquad (10)$$

Where $V_1$ is the molar volume of the solvent (mL mol$^{-1}$), $d_p$ is the polymer density, $V_s$ is the volume fraction of the swollen gel and $x$ is the interaction parameter between solvent and polymer [54]. The parameters presented above are the most often used to characterize a three-dimensional polymer network, however, other parameters such ascrosslink density (q), which is proportional to the molar fraction of the crosslinking agent used, and number of effective elastic chains per unit volume $V_e$ also are important for the structural characterization of the three-dimensional network of a hydrogel.

The values of q and $V_e$ are calculated from equations 11 and 12[54]:

$$q = \frac{M_0}{M_c} \qquad (11)$$

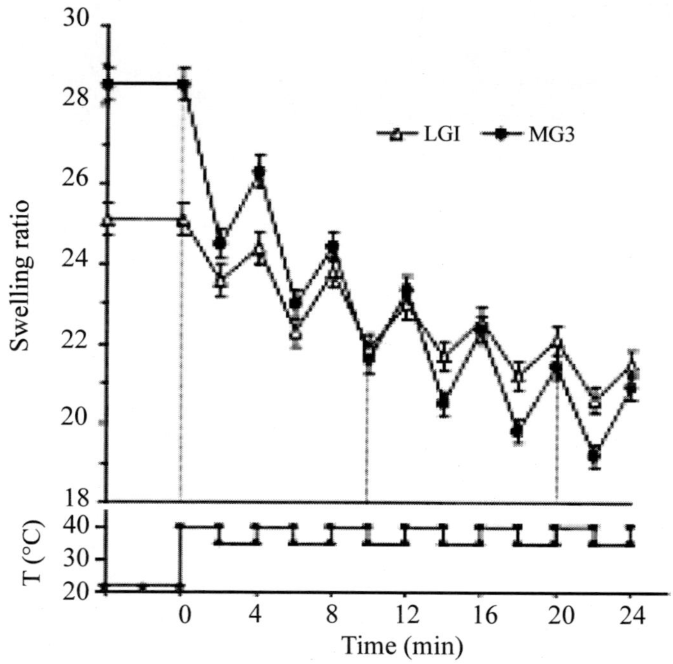


Figure 4. Oscillating shrinking–swelling kinetics of the semi-IPN–like PNIPAAm hydrogels over the 2-min temperature cycles between 35 and 40°C [133].

Where $M_0$ is the molar mass of the repeating unit.

$$V_e = \frac{d_P N_A}{M_C} \tag{12}$$

Where $N_A$ is Avogadro's number.

An important feature to be assessed in hydrogels according to the intended application, is the shrinking kinetics. In the development of hydrogels that respond quickly to external stimuli like temperature of the medium, this parameteris of great importance. The rapid response to external stimulus (temperature) is due to the presence of macropores in the hydrogel structure [101, 107].

A rapid response to temperature change and consequently a fast shrinking kinetics was described by Zhang & Jandt, 2008 that evaluated a porous fast-responsive poly(N-isopropylacrylamide) hydrogel using a hydrophobic liquid template method with kinetic shrinking significantly higher than the observed in conventional hydrogels, resulting in a loss of 97% water in 5 minutes, reaching equilibrium state within 10 minutes [104].

The study of cycles of de-swelling and swelling of hydrogels are performed with the rapid increase in the medium temperature where the hydrogel is immersed across the volume phase transition temperature. The time required for the swelling and deswelling process of the hydrogel to reach equilibrium is then observed. One of the objectives is to verify if the cycles of swelling and deswelling will change the mechanical strength or modulus properties of hydrogels [100]. In the Figure 4 it can be observed an example of swelling and deswelling over the 2 minutes temperature cycles between 35 and 40 °C [133].

## Mechanical Properties

The mechanical properties of hydrogels are determined by obtaining the values of maximum compressive stress (σmax) and modulus of elasticity (E). To represent these properties force-deformation curves are used. Through them one can determine whether the deformation of the hydrogel is elastic type, i.e., the applied force and deformation generated are linear and when the force is removed, the hydrogel returns to its original conformation [102, 134, 135].

The resistance to compression is held in a device that exerts a compressive force on the sample at a given speed until decreasing the size of the hydrogel at a certain height or until the rupture of the sample. Tension is measured in a resistive extender that verifies the tensile resistance of the material [86, 136, 137].

Stress-strain measurements can also be performed with the aid of a digital scale that registers the load transmitted vertically to the gel through a rod. The force acting is calculated from a reading of the balance. The deformation of the hydrogel is calculated using a digital comparator which is sensitive to displacements of $10^{-3}$ mm and is based in uniaxial deformation and the statistical theories of rubber elasticity yield for Gaussian chains [138].

The viscoelastic properties are obtained by dynamic mechanical analysis (DMA), and can be carried out in dry or wet material. The mechanical strength of the swollen gel network can be influenced by two independent factors: the crosslinking density and the polymer volume fraction [64, 139]. Sliding frictional behavior is performed on the hydrogel surface when it is wet [140].

## Thermal Analysis

The thermal analysis techniques that can be applied to hydrogels are: thermogravimetry (TGA), differential scanning calorimetry (DSC) and differential thermal analysis (DTA). Such assessments measure the physical properties of a sample as a function of temperature while the sample is subjected to a temperature programming. Thermogravimetry records the mass loss, the differential scanning calorimetry registers the enthalpy change and DTA the temperature difference of the sample and a reference, according to temperature variation.

These techniques help to determine the amount of water incorporated into the hydrogel and determine the glass transition temperature [86]. The phenomenon of deswelling and swelling in hydrogels possessing stimuli-response when subjected to temperatures above the "lower critical solution temperature" by DSC may be observed [88, 90, 100, 139, 141-143].

Measurements of bound (nonfreezing) and free water can be determined by DSC. The sealed pan is quickly frozen and its temperature reduced to −30 °C inside the DSC chamber and the system is allowed for several minutes to come to equilibrium. The sample holder assembly is then heated under $N_2$ flow. The heat of melting the freezing water is determined from the area under the endothermic curve and is calibrated using pure distilled water as a standard. The fraction of free water in total water is approximately calculated as the ratio of the endothermic peak area for water-swollen hydrogel to melting endothermic heat of fusion (79.9 cal/g) for pure water [116, 128].

The stability and thermal decomposition of the hydrogel is verified by TGA tests. Where the heating of the sample is measured from room temperature to about 700 ° C at a heating rate that can vary between 5 and 20 °C/min and under a dynamic atmosphere of nitrogen or synthetic air, depending on the sample [144].

## 4. Applications as Drug Delivery Systems

Hydrogels are being widely used for various purposes, such as in delivery systems for pesticides [145], anti-inflammatories [146], antimicrobials [147], hypoglycemic [148], and antihypertensives. Immobilization of biologically active species in hydrogel matrices, their use as drug delivery systems and enzyme traps as well as the modification of material surfaces to improve biocompatibility and their ability to bond antigens and antibodies had been undertaken by many researchers in the pharmaceutical area. The network structure and the thermodynamic nature of the components of these networks play a key role in their diffusional behavior, molecular mesh size changes [149]. However, for the controlled delivery of low molecular weight hydrophilic compounds, hydrogels are generally not suitable since their high water content and the presence of large pores in relation to the drug size results in rapid drug release [150].

This ability of undergoing structure alterations enables hydrogels to be used as carriers of drugs, proteins and cells overcoming several drawbacks of others drug delivery systems. Drugs, proteins and cells can be released from hydrogels according to their target tissue or organ and administration route as determined by the molecular designer. In this way, it is possible to provide the drug or others compounds of interest at a specific predetermined temporal and/or spatial manner within the body to fulfill the specific therapeutic needs [151, 152]. So they are used in various routes of administration as drug carriers (Figure 4).

Materials in the form of hydrogel matrices enable sustained and/or controlled release of embedded medicines to body fluids after their implantation, injection or other introduction into the organism [153]. Hydrogel biomaterials for intelligent delivery can be used for epidermal, oral, rectal, ocular, and subcutaneous application. Hamidi *et al.* [154] reported that the pharmaceutical hydrogels can be classified as: i) oral hydrogel systems, ii) transdermal and implantable hydrogel systems, iii) topical and transdermal hydrogel systems, iv) hydrogel devices for gastrointestinal (GI) drug delivery, and v) hydrogel-based ocular delivery systems.

For an effective therapy and drug bioavailability, hurdles of each route of administrations have to be taken into consideration. Peppas *et al.* [152], reported main administration route problems which are: severe acidic environment of the stomach; discomfort and uncontrolled diffusion of drug in the rectum for suppositories; effective tear drainage, blinking and low permeability of the cornea in the eye, leading to a frequent dosing regimen; etc. In the past five years, there have been numerous new or improved pharmaceutical applications of hydrogels in order to overcome these obstacles. Effective drug delivery depends on the balance of drug loading and the effect of polymer on its release profile.

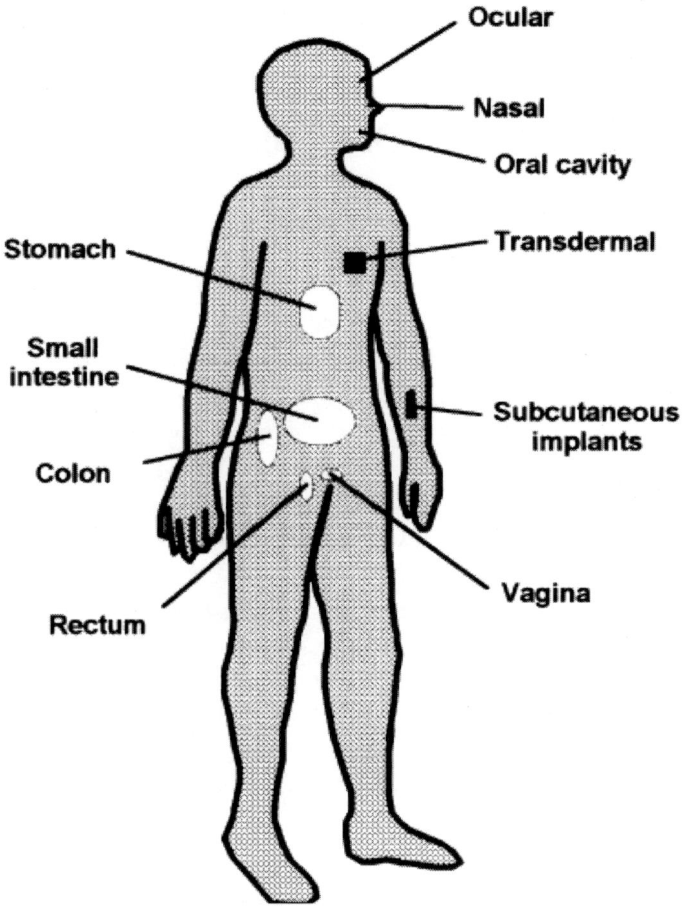


Figure 4. Schematic different routes of administration of hydrogels in controlled release systems of drugs [152].

Significant progress has been made in improving the properties of hydrogels used for drug delivery and expanding the range of drugs and kinetics which can be achieved using a hydrogel-based delivery vehicle. Human-friendly hydrogel systems, due to the rising trend to prolong life span and improve results of medical care, seem to be one of the most expected and required products. However, several challenges remain to improve the clinical applicability of hydrogels for drug delivery.

## Nasal Hydrogel Systems

Jie wu *et al.* [155] reported in 2007 a thermosentive hydrogel for a nasal delivery, especially for peptide or protein drug. Nasal mucociliary clearance rate is a great hurdle in delivering drug in this part of the body. This controlled delivery system was formed by mixing N-[(2-hydroxy-3-trimethylammonium) propyl] chitosan chloride, poly(ethylene glycol) with a small amount of α-β-glycerophosphate. It was verified that this system was a flowing solution at/or below the room temperature but a gel at the body temperature enhancing the contact time between drug and the mucosa providing a sustained drug delivery.

For being non-toxic, this system has the potential application as nasal drug delivery system especially for peptide or protein drugs used chronically. In their work, the model drug was insulin. It was verified that the blood glucose level decreased 40–50% of initial concentration for at least 4–5 h after administration.

## Transdermal and Implantable Hydrogel Systems

Local delivery is an important method for both local therapy and systematic therapy. In order to overcome the water-soluble problem of hydrophobic drug delivery, Gou *et al.* developed a composite system formed by drug-loaded poly(ε-caprolactone)-co-poly(ethylene glycol) nanoparticles incorporated in a thermosensitive hydrogels for injectable local delivery [156, 157]. This system has a great advantage over the others for sustained release of several weeks for example [158]. The model drug used was honokiol.

Blanco *et al.* [159] tested release behavior of 5-Fluorouracil, an antimetabolic drug extensively used in cancer chemotherapy, from copolymeric poly(acrylamide-co-monomethyl) itaconate hydrogels, a temperature-sensitive hydrogel system which exhibited a higher degree of swelling, a factor that, a priori, assures high biocompatibility because it increases the similarity of the gels with living tissues, as stated above. So, from this viewpoint these hydrogels are better than implants. It was observed that the drug delivery was more favorable. This is reflected in an increment of the diffusion coefficient value.

Another hydrogel delivery system for implants was reported by Franssen *et al.* [160] who studied a delayed release of IgG, a model protein, from enzymatically-degrading crosslinked dextran hydrogels. A correlation was found between the degradation rate of the hydrogel and the observed delay time: in the absence of dextranase, IgG was released very slowly for a period of 250 days. For gels containing dextranase, the release profile was highly dependent on the concentration of enzyme present in the gel. Moreover, Hennink *et al.* [161] studied the release behavior of proteins (lysozyme, albumin, IgG) from non-degrading and enzymatically degrading dextran hydrogels. From non-degrading dextran hydrogels a decreasing of water content resulted in a decreasing release rate.

## Topical and Transdermal Hydrogel Systems

Skin has the lowest enzymatic activity of the various delivery routes and presents an alternative for getting a systemic bioavailability but, the barrier properties of the epithelium limits the permeation and accordingly limits the passive diffusion of proteins/peptides for being usually charged and hydrophilic. Chen and Chien [162] have characterized the transdermal iontophoretic permeation of luteinizing hormone releasing hormone (LHRH) from polyacrylamide hydrogels. Iontophoresis-facilitated transdermal delivery has several advantages over other skin permeation-enhancing techniques and holds promise for controlled delivery of a variety of compounds [163]. The on/off ratios of pulsed direct current showed a significant effect on transdermal iontophoretic permeation of LHRH, and demonstrated that the higher the on/off ratio (duty cycle), the greater the skin permeation of LHRH is. Several studies were undertaken on transdermal iontophoresis.

Fang *et al.* [164] have reported the transdermal iontophoresis of sodium nonivamide acetate (SNA) from two polymers as binary system formulations made of methylcellulose, polyvinylpyrrolidone, chitosan, hydroxypropyl cellulose, Hydroxypropyl methylcellulose and Plastoid®. It was found that fluxes of SNA from hydrogels were all suitable to fit using a zero-order equation indicating that a steady-state permeation rate of nonivamide acetate may be achieved and, thus, also the sustained release effect.

## Wound Healing Hydrogel Systems

Hydrogels can be used as a barrier and local drug delivery system for the local control of wound and vessel wall healing. Many studies have been undertaken in this perspective. Hubbell [165] investigated the use of hydrogel systems for an effectively healing of wound serving as barriers to cell attachment and migration and functioning as depots for the local release of therapeutic macromolecular drugs. A block copolymer was formed by polyethylene glycol and lactic acid oligomers [165, 166]. The release of proteins occurs differently depending on their molecular mass and on the polyethylene glycol. Small proteins were released by diffusion in the absence of degradation, whereas larger proteins were released by diffusion exclusively following degradation.

The cooling of wound from burn is an emergency measure which demonstrated a beneficial approach in clinical and experimental practices. Depending on the extension of lesion, hydrogel-based product can be used for immediate dressing for burns. Boles *et al.* [167] demonstrated the effectiveness of Burnshield®, tea tree oil-based hydrogel in burned tissue healing.

## Hydrogel-based Ocular Delivery Systems

Delivery of drugs to the anterior side of the eye is routinely done with eye drops, but this method results in low bioavailability and low patient compliance. In other to overcome lachrymal clearance, blinking and unproductive systemic absorption, contact lenses as medicated contact lenses could incorporate drug-grafted hydrogels for a smart delivery system. In this perspective, Santos *et al.* [168] reported the grafting of β-cyclodextrin to Poly(hydroxyethylmethacrylate) hydrogels which were prepared by copolymerization with glycidyl methacrylate (GMA) at various proportions for the release of diclofenac for the treatment of ocular inflammatory disorders. On the one hand, mechanical properties and the biocompatibility of the hydrogels were not altered and no biocompatibility problems were pointed out. On the other hand the β-cyclodextrin grafting improved diclofenac loading by 1300%, enhanced drug affinity 15-fold and prevented drug leakage during storage for soft contact lenses enabling accordingly its delivery for at least two weeks in the tear film.

## Gastrointestinal (GI) Hydrogel Systems

The main factors that hinder an efficient stomach absorption and treatment of local microorganisms, especially *Helicobacter pylori*, reducing luminal delivery of drugs are gastric emptying, gastric acidity and the epithelial mucus layer [169]. According to the authors, for better prolongation of the time contact between drugs and the stomach mucosa for *Helicobacter pylori*, treatment floating systems and mucoadhesive or bioadhesive systems are relevant.

He *et al.* [170] developed a system composed of two bonded hydrogels layers. One is a pH-sensitive hydrogel based on crosslinked poly(methyacrylic acid) (PMAA) that swells significantly when in contact with body fluids, while the other is a non-swelling layer based on poly(hydroxyethyl methacrylate) (PHEMA) where the latter layer can serve as a barrier for drug leakage in the GIT. This system is of paramount benefit for drugs that present high sensibility within the GIT environment.

Patel and Amiji [171] developed novel drug delivery systems with pH-sensitive swelling and drug release properties for localized antibiotic delivery in the stomach for the treatment of *Helicobacter pylori* infection in peptic ulcer disease. The hydrogel consisted of chitosan-co poly(ethylene oxide) semi-IPN for the delivery of amoxicillin and metronidazole as model drugs. The results showed that this system could be useful for localized delivery of antibiotics in the acidic environment of the gastric fluid.

Park *et al.* [172] reported a study on the behavior of superporous pH-responsive chitosan and glycol chitosan hydrogels for their application for gastric retention device. The hydrogels were swollen and deswollen reversibly, depending on the acidity of media. Superporous hydrogels kept equilibrium swelling ratios for a certain period of time in acidic environment. The weight loss of glycol chitosan hydrogels occurred after 12 h and the degradation kinetics investigated in this study will be another important aspect for applying the present materials to a gastric retention device.

Daily subcutaneous injections of insulin are a tedious treatment for patients with insulin-dependent diabetes. Among the many routes investigated for insulin administration, the oral route seems the most convenient for ease of administration in long-term treatments and because this peptide undergoes a hepatic first-pass effect which contributes to the inhibition of the hepatic glucose output. But the main problem is that after oral administration of non-encapsulated insulin, only less than 0.5% of the dose is absorbed due to the extremely slow rate of transport across the mucosal membrane [148]. Nakamura *et al.* developed an insulin sustained hydrogel delivery system. The hydrogel consists of poly(methacrylic acid-g-ethylene glycol) prepared by free-radical, solution photo-polymerization. In the *in vitro* release studies, no insulin was released from any of the gels in the acidic media (simulating stomach conditions) but rapid release occurred in the higher pH solution (pH = 7.4), indicating the potential for drug release in the upper small intestine.

Morimoto *et al.* [173] reported an indomethacin release profile from poly vinyl alcohol hydrogels. It was found in several animals that rectal administration of indomethacin hydrogels yielded high indomethacin plasma concentrations, without producing a sharp peak, and a sustained-release effect.

Gutowska *et al.* [174] reported a study on squeezing hydrogels for controlled oral drug delivery with acetaminophen and dyphylline, used as model drugs. They prepared and analyzed the release profiles of three hydrogels systems: a temperature sensitive hydrogel made of N-Isopropylacrylamide and acrylic acid (NiPAAm/AAc), a pH/temperature sensitive hydrogel made of N-Isopropylacrylamide, butylmethacrylate and diethylaminoethyl methacrylate (NiPAAm/BMA/DEAEMA) and a pH sensitive cationic hydrogel of N-Isopropylacrylamide and diethylaminoethyl methacrylate (NiPAAm/DEAEMA). The release profiles of these systems were determined including a delayed, a zero-order and an 'on-off' release profile. The NiPAAm/AAc hydrogel demonstrated a squeezing controlled 'on-off' release profile in response to a temperature change between 35°C and 40°C with the 'on' phase at higher temperature. Thus, it is a good candidate for the squeezing controlled 'on-off' delivery of antipyretic agents. The two others presented a delayed and a zero-order release profile in response to a pH change from pH 2.0 to 7.4. The systems with a delayed and minimal release profile at pH 2.0 and a zero-order release at pH 7.4 may be used for squeezing controlled delivery of drugs that decompose in the stomach or irritate the stomach and cause adverse effects in the upper GI tract.

Dorkoosh et al. [175] held an *in vitro* release study of three peptides drugs that are buserelin, octreotide and insulin from superporous hydrogel (SPH) and SPH composite (SPHC) which are able to achieve mechanical fixation of the dosage form at desirable sites of absorption in the intestine by controlled rapid swelling. These SPH and SPHC consisted of acrylic acid, acrylamide and N,N′-methylenebis acrylamide. In the release profiles from SPH a slightly quicker peptide release was found. The release characteristics of all peptides from the designed oral delivery systems demonstrated a time-controlled release profile controlled by the swelling rate and pore size of the two polymers.

Bowersock *et al.* [176] have used sodium alginate hydrogel microspheres to deliver various vaccines to several animal species by oral administration. Oral administration of vaccines generally requires large doses of antigen due to inefficient stimulation of gut-associated lymphoid tissue (GALT). Oral administration of vaccines offers several advantages: vaccine can be administered to a large number of individuals via the food or water with minimal restraint and labor and adverse immune reactions are less likely to occur compared to parenteral methods. It was observed that sodium alginate hydrogel can be used as adjuvant for vaccine delivery. The use of this system for oral delivery of vaccines presented several advantages: possibility to be used for live bacteria and viruses, protection of vaccines against degradation in the GI tract and provide means to bypass the stomach to deliver antigens to GALT in man as well as animals.

## Vaginal Hydrogel Systems

During sexual intercourse or other conditions that expose the vaginal vault to disease and infection, many pathogenic agents act in this part of the body. So, it is necessary the use of spermicidal or nonspermicidal microbicidal agents in local delivery systems. D'Cruz and Uckun [177] reported on the possibility to use gel-microemulsions as vaginal spermicides/contraceptive and intravaginal drug delivery vehicles in order to improve bioavailability of poorly soluble antimicrobial agents without causing significant vaginal damage. The main drawback is the destruction of the normal vaginal flora leading to an increased risk of opportunistic infections. In addition, irritation caused by these products yield mucosal erosion and local inflammation. To overcome all these drawbacks, the authors cited above developed a system that was found to be safe and did not cause local, systemic, or reproductive damage and warrant further *in vivo* evaluation in humans as intravaginal/rectal drug delivery vehicles.

## CONCLUSION

Advances in technology development of hydrogels have promoted a real revolution in the design and efficiency of these materials as controlled released systems for drug delivery and bioactive agents. The advantage of being able to shape the chemical, physical and biological features of these systems facilitates their application in different routes of administration. The choice of ingredients to compose a system is according to its functionality and destination. So, the properties of the active ingredient, the chemical nature of polymers and the preparation technique of the hydrogel are important factors for this determination. And in order to generate an entirely biomedical product, it is essential to ensure the safety and efficacy of the hydrogel as a controlled release system through assessments of physicochemical characterization in all stages, from the pre-formulation up to the final product.

## ACKNOWLEDGMENT

All the authors acknowledge the National Council for Scientific and Technological Development (CNPq), Brazilian Federal Agency for Support and Evaluation of Graduate Education (CAPES), Pará Research Foundation (FAPESPA) and Federal University of Pará (UFPA) by financial support to works performed.

## REFERENCES

[1] Alves, T., et al., Thermal Analysis Characterization of Paam-≪I≫ Co≪ /I≫-Mc Hydrogels. *Journal of Thermal Analysis And Calorimetry,* 2011: P. 1-8.

[2] Peppas, N.A., Hydrogels In Medicine And Pharmacy. 1987: Crc Pr.

[3] Peppas, N.A., et al., Hydrogels In Pharmaceutical Formulations.European Journal of Pharmaceutics And Biopharmaceutics : of*ficial Journal of Arbeitsgemeinschaft Fur Pharmazeutische Verfahrenstechnik E.V,* 2000. 50(1): P. 27-46.

[4] Saravanan, P., M. Padmanabha Raju, And S. Alam, A Study On Synthesis And Properties of Ag Nanoparticles Immobilized Polyacrylamide Hydrogel Composites. *Materials Chemistry And Physics*, 2007. 103(2-3): P. 278-282.

[5] Hamidi, M., A. Azadi, And P. Rafiei, Hydrogel Nanoparticles In Drug Delivery. *Advanced Drug Delivery Reviews*, 2008. 60(15): P. 1638-1649.

[6] Rosiak, J.M., et al., Radiation Formation of Hydrogels For Biomedical Purposes. Some Remarks And Comments. *Radiation Physics And Chemistry,* 1995. 46(2): P. 161-168.

[7] Peppas, N.A. And A.R. Khare, Preparation, Structure And Diffusional Behavior of Hydrogels In Controlled Release. *Advanced Drug Delivery Reviews.* 11(1-2): P. 1-35.

[8] Zheng, Y., et al., Peg-Based Hydrogel Synthesis Via The Photodimerization of Anthracene Groups. *Macromolecules,* 2002. 35(13): P. 5228-5234.

[9] Oviedo, I.R., et al., Design of A Physical And Nontoxic Crosslinked Poly(Vinyl Alcohol) Hydrogel. *International Journal of Polymeric Materials,* 2008. 57(12): P. 1095-1103.

[10] Deligkaris, K., et al., Hydrogel-Based Devices For Biomedical Applications.*Sensors And Actuators B: Chemical,* 2010. 147(2): P. 765-774.
[11] Moura, M.R.D., A.F. Rubira, And E.C. Muniz, Hidrogéis Semi-Ipn Baseados Em Rede De Alginato-Ca2+ Com Pnipaam Entrelaçado: Propriedades Hidrofílicas, Morfológicas E Mecânicas.*Polímeros,* 2008. 18: P. 132-137.
[12] Haider, M., Z. Megeed, And H. Ghandehari, Genetically Engineered Polymers: Status And Prospects For Controlled Release. *Journal of Controlled Release,* 2004. 95(1): P. 1-26.
[13] Chen, H., et al., Microemulsion-Based Hydrogel Formulation of Ibuprofen For Topical Delivery. *International Journal of Pharmaceutics,* 2006. 315(1-2): P. 52-58.
[14] Daugherty, A.L. And R.J. Mrsny, Formulation And Delivery Issues For Monoclonal Antibody Therapeutics. *Advanced Drug Delivery Reviews,* 2006. 58(5-6): P. 686-706.
[15] Chen, F.-M., et al., Novel Glycidyl Methacrylated Dextran (Dex-Gma)/Gelatin Hydrogel Scaffolds Containing Microspheres Loaded With Bone Morphogenetic Proteins: Formulation And Characteristics. *Journal of Controlled Release,* 2007. 118(1): P. 65-77.
[16] Vila, A., et al., Design of Biodegradable Particles For Protein Delivery. *Journal of Controlled Release,* 2002. 78(1-3): P. 15-24.
[17] Angelova, N. And D. Hunkeler, Rationalizing The Design of Polymeric Biomaterials.*Trends In Biotechnology, 1*999. 17(10): P. 409-21.
[18] Peppas, et al., Hydrogels In *Pharmaceutical Formulations.* Vol. 50. 2000, Amsterdam, Pays-Bas: Elsevier.
[19] Brant, A.J.C., Preparação E Caracterização De Hidrgéis A Partir De Misturas De Soluções De Quitosana E Poli(N-Vinil-2-Pirrolidona), In *Instituto De Química Da Universidade De São Paulo.* 2008, Universidade De São Paulo: São Paulo.
[20] Colombo, P., R. Bettini, And N.A. Peppas, Observation of Swelling Process And Diffusion Front Position During Swelling In Hydroxypropyl Methyl Cellulose (Hpmc) Matrices Containing A Soluble Drug. *Journal of Controlled Release,* 1999. 61(1-2): P. 83-91.
[21] Gabrielii, I. And P. Gatenholm, Preparation And Properties of Hydrogels Based On Hemicellulose. *Journal of Applied Polymer Science,* 1998. 69(8): P. 1661-1667.
[22] Kumar, et al., *Biodegradable Block Copolymers.* 2001, Kidlington, Royaume-Uni: Elsevier. 131.
[23] Faisant, N., J. Siepmann, And J.P. Benoit, Plga-Based Microparticles: Elucidation of Mechanisms And A New, Simple Mathematical Model Quantifying Drug Release. *European Journal of Pharmaceutical Sciences*, 2002. 15(4): P. 355-366.
[24] Rajeev A, J., The Manufacturing Techniques of Various Drug Loaded Biodegradable Poly(Lactide-Co-Glycolide) (Plga) Devices. *Biomaterials,* 2000. 21(23): P. 2475-2490.
[25] Park, C., et al., Effect of Hydrophilicity On The Biodegradability of Polyesteramides.*Journal of Applied Polymer Science,* 2003. 90(10): P. 2708-2714.
[26] Pavan, F.A., et al., Extração De Al(Iii), Cr(Iii) E Fe(Iii) De Meio Etanólico Usando O Xerogel Anilinapropilsílica.*Química Nova,* 2004. 27: P. 730-733.
[27] Jenquin, M.R. And J.W. Mcginity, Characterization of Acrylic Resin Matrix Films And Mechanisms of Drug-Polymer Interactions. *International Journal of Pharmaceutics,* 1994. 101(1-2): P. 23-34.

[28] Zhu, Y., *Properties of Polymeric Drug Delivery Systems Prepared By Hot-Melt Extrusion,* In Faculty of The Graduate School of The University of Texas. 2002, University of Texas: Austin.

[29] Sousa, R.G., et al., Dependence of Copolymer Composition, Swelling History, And Drug Concentration On The Loading of Diltiazem Hydrochloride (Dil.Hcl) Into Poly[(N-Isopropylacrylamide)-Co-(Methacrylic Acid)] Hydrogels And Its Release Behaviour From Hydrogel Slabs.*Journal of Controlled Release,* 2005. 102(3): P. 595-606.

[30] Aouada, F.A., et al., Correlação Entre Parâmetros Da Cinética De Intumescimento Com Características Estruturais E Hidrofílicas De Hidrogéis De Poliacrilamida E Metilcelulose. *Química Nova,* 2009. 32: P. 1482-1490.

[31] Klug, M., et al., Análise Das Isotermas De Adsorção De Cu(Ii), Cd(Ii), Ni(Ii) E Zn(Ii) Pela N-(3,4-Dihidroxibenzil) Quitosana Empregando O Método Da Regressão Não Linear. *Química Nova,* 1998. 21: P. 410-413.

[32] Oliveira, M.W.D., et al., Estudo Da Adsorção De Brometo De Etídeo Em Resina Xad-7.*Química Nova*, 2009. 32: P. 1134-1138.

[33] Akpa, O.M. And E.I. Unuabonah, Small-Sample Corrected Akaike Information Criterion: An Appropriate Statistical Tool For Ranking of Adsorption Isotherm Models.*Desalination, 2*011. 272(1-3): P. 20-26.

[34] Lopes, C.M., J.M.S. Lobo, And P. Costa, Formas Farmacêuticas De Liberação Modificada: Polímeros Hidrifílicos.*Revista Brasileira De Ciências Farmacêuticas,* 2005. 41: P. 143-154.

[35] Abdullah, M.A., L. Chiang, And M. Nadeem, Comparative Evaluation of Adsorption Kinetics And Isotherms of A Natural Product Removal By Amberlite Polymeric Adsorbents. *Chemical Engineering Journal,* 2009. 146(3): P. 370-376.

[36] Washington, C., Drug Release From Microdisperse Systems: A Critical Review. *International Journal of Pharmaceutics,* 1990. 58(1): P. 1-12.

[37] Canal, T. And N.A. Peppas, Correlation Between Mesh Size And Equilibrium Degree of Swelling of Polymeric Networks. *Journal of Biomedical Materials Research,* 1989. 23(10): P. 1183-1193.

[38] Rios, M., Polymers For Controlled Release: Formulation Follows Function.*Pharm. Technol.,* 2005. 6(29): P. 42-50.

[39] Alves, T.V.G., Obtenção E Caracterização De Hidrogéis De Poliacrilamida-Co-Metilcelulose Como Sistemas Carreadores De Cloridrato De Propranolol, In *Programa De Pós Graduação Em Ciências Faramacêuticas.* 2011, Universidade Federal Do Pará: Belém.

[40] Zhang, J.-T., et al., Preparation And Properties of Poly(N-Isopropylacrylamide)/Poly(N-Isopropylacrylamide) Interpenetrating Polymer Networks For Drug Delivery. *Journal of Polymer Science Part A: Polymer Chemistry,* 2004. 42(5): P. 1249-1254.

[41] Ratner Buddy, D. And S. Hoffman Allan, Synthetic Hydrogels For Biomedical Applications, In Hydrogels For Medical And Related Applications. 1976, *American Chemical Society.* P. 1-36.

[42] Micic, M., et al., Comparative Studies of Surface Topography And Mechanical Properties of A New, Photo-Switchable Peg-Based Hydrogel. *Colloids And Surfaces B: Biointerfaces,* 2003. 27(2-3): P. 147-158.

[43] Hoare, T.R. And D.S. Kohane, Hydrogels In Drug Delivery: Progress And Challenges. *Polymer,* 2008. 49(8): P. 1993-2007.
[44] Hennink, W.E. And C.F. Van Nostrum, Novel Crosslinking Methods To Design Hydrogels.*Advanced Drug Delivery Reviews,* 2002. 54(1): P. 13-36.
[45] Aouada, F., Síntese E Caracterização De Hidrogéis De Poliacrilamida E Metilcelulose Para Liberação Controlada De Pesticidas, In *Programa De Pós Graduação Em Química.* 2009, Universidade Federal De São Carlos: São Carlos.
[46] Aouada, F.A., et al., Physicochemical And Morphological Properties of Poly(Acrylamide) And Methylcellulose Hydrogels: Effects of Monomer, Crosslinker And Polysaccharide Compositions. *Polymer Engineering & Science,* 2009. 49(12): P. 2467-2474.
[47] Tomihata, K. And Y. Ikada, Crosslinking of Hyaluronic Acid With Glutaraldehyde. *Journal of Polymer Science Part A: Polymer Chemistry,* 1997. 35(16): P. 3553-3559.
[48] Park, J.S., J.W. Park, And E. Ruckenstein, Thermal And Dynamic Mechanical Analysis of Pva/Mc Blend Hydrogels. *Polymer,* 2001. 42(9): P. 4271-4280.
[49] Mandal, B.B., S. Kapoor, And S.C. Kundu, Silk Fibroin/Polyacrylamide Semi-Interpenetrating Network Hydrogels For Controlled Drug Release. *Biomaterials,* 2009. 30(14): P. 2826-2836.
[50] Hassan, C.M., J.E. Stewart, And N.A. Peppas, Diffusional Characteristics of Freeze/Thawed Poly(Vinyl Alcohol) Hydrogels: Applications To Protein Controlled Release From Multilaminate Devices. *European Journal of Pharmaceutics And Biopharmaceutics,* 2000. 49(2): P. 161-165.
[51] Martin, L., et al., The Release of Model Macromolecules May Be Controlled By The Hydrophobicity of Palmitoyl Glycol Chitosan Hydrogels. *Journal of Controlled Release,* 2002. 80(1-3): P. 87-100.
[52] Qiu, Y. And K. Park, Environment-Sensitive Hydrogels For Drug Delivery. *Advanced Drug Delivery Reviews,* 2001. 53(3): P. 321-339.
[53] Liu, W., et al., A Rapid Temperature-Responsive Sol–Gel Reversible Poly(N-Isopropylacrylamide)-G-Methylcellulose Copolymer Hydrogel. *Biomaterials,* 2004. 25(15): P. 3005-3012.
[54] Park, T.G. And A.S. Hoffman, Synthesis And Characterization of Ph- And/Or Temperature-Sensitive Hydrogels. *Journal of Applied Polymer Science,* 1992. 46(4): P. 659-671.
[55] De Moura, M.R., et al., Release of Bsa From Porous Matrices Constituted of Alginate–Ca2+ And Pnipaam-Interpenetrated Networks. *Materials Science And Engineering:* C, 2009. 29(8): P. 2319-2325.
[56] He, H., J. Guan, And J.L. Lee, An Oral Delivery Device Based On Self-Folding Hydrogels. *Journal of Controlled Release,* 2006. 110(2): P. 339-346.
[57] Gupta, R.B. And U.B. Kompella, *Nanoparticle Technology For Drug Delivery.* 2006: Taylor & Francis.
[58] Kim, J.-H. And T.R. Lee, Thermo- And Ph-Responsive Hydrogel-Coated Gold Nanoparticles. *Chemistry of Materials,* 2004. 16(19): P. 3647-3651.
[59] Na, K., et al., Self-Assembled Hydrogel Nanoparticles From Curdlan Derivatives: Characterization, Anti-Cancer Drug Release And Interaction With A Hepatoma Cell Line (Hepg2). *Journal of Controlled Release,* 2000. 69(2): P. 225-236.

[60] Vázquez, B., et al., Polymeric Hydrophilic Hydrogels With Flexible Hydrophobic Chains. Control of The Hydration And Interactions With Water Molecules. *Macromolecules,* 1997. 30(26): P. 8440-8446.

[61] Bajpai, A.K. And A. Giri, Water Sorption Behaviour of Highly Swelling (Carboxy Methylcellulose-G-Polyacrylamide) Hydrogels And Release of Potassium Nitrate As Agrochemical. *Carbohydrate Polymers,* 2003. 53(3): P. 271-279.

[62] Gad, S.C., Pharmaceutical Manufacturing Handbook: Production And Processes. 2008: Wiley-Interscience.

[63] Akiyoshi, K., et al., Microscopic Structure And Thermoresponsiveness of A Hydrogel Nanoparticle By Self-Assembly of A Hydrophobized Polysaccharide. *Macromolecules,* 1997. 30(4): P. 857-861.

[64] Falk, B., S. Garramone, And S. Shivkumar, Diffusion Coefficient of Paracetamol In A Chitosan Hydrogel. *Materials Letters,* 2004. 58(26): P. 3261-3265.

[65] Kabiri, K., H. Omidian, And M.J. Zohuriaan-Mehr, Novel Approach To Highly Porous Superabsorbent Hydrogels: Synergistic Effect of Porogens On Porosity And Swelling Rate. *Polymer International,* 2003. 52(7): P. 1158-1164.

[66] Xu, X.-D., et al., Fabrication of Fast Responsive, Thermosensitive Poly(N-Isopropylacrylamide) Hydrogels By Using Diethyl Ether As Precipitation Agent. *Journal of Biomedical Materials Research Part A,* 2008. 86a(4): P. 1023-1032.

[67] Wu, X.S., A.S. Hoffman, And P. Yager, Synthesis And Characterization of Thermally Reversible Macroporous Poly(N-Isopropylacrylamide) Hydrogels. *Journal of Polymer Science Part A: Polymer Chemistry,* 1992. 30(10): P. 2121-2129.

[68] Saalwächter, K. And D. Reichert, Polymer Applications of Nmr, In *Encyclopedia of Spectroscopy And Spectrometry (Second Edition*), L. Editor-In-Chief: John, Editor. 2010, Academic Press: Oxford. P. 2221-2236.

[69] Barbieri, R., et al., Investigation of Water Dynamic Behaviour In Poly(Hema) And Poly(Hema-Co-Dhpma) Hydrogels By Proton T2 Relaxation Time And Self-Diffusion Coefficient N.M.R. Measurements. *Polymer*, 1998. 39(5): P. 1059-1066.

[70] Bhalerao, V.S., et al., Thermoreversible Hydrogel Based On Radiation Induced Copolymerisation of Poly(N-Isopropyl Acrylamide) And Poly(Ethylene Oxide). *Polymer,* 1998. 39(11): P. 2255-2260.

[71] Huang, X., B.R. Nayak, And T.L. Lowe, Synthesis And Characterization of Novel Thermoresponsive-Co-Biodegradable Hydrogels Composed of N-Isopropylacrylamide, Poly(L-Lactic Acid), And Dextran. *Journal of Polymer Science Part A: Polymer Chemistry,* 2004. 42(20): P. 5054-5066.

[72] Ishizone, T. And M. Ito, Synthesis of Well-Defined Poly(N-Isopropylacrylamide) By The Anionic Polymerization of N-Methoxymethyl-N-Isopropylacrylamide. *Journal of Polymer Science Part A: Polymer Chemistry,* 2002. 40(23): P. 4328-4332.

[73] Dorkoosh, F.A., et al., Preparation And Nmr Characterization of Superporous Hydrogels (Sph) And Sph Composites. *Polymer,* 2000. 41(23): P. 8213-8220.

[74] Vervoort, L., et al., Inulin Hydrogels As Carriers For Colonic Drug Targeting: I. Synthesis And Characterization of Methacrylated Inulin And Hydrogel Formation. *Pharmaceutical Research,* 1997. 14(12): P. 1730-1737.

[75] Díez-Peña, E., et al., Advanced 1h Solid-State Nmr Spectroscopy On Hydrogels, 2. *Macromolecular Chemistry And Physics,* 2004. 205(4): P. 438-447.

[76] Elvira, C., et al., Starch-Based Biodegradable Hydrogels With Potential Biomedical Applications As Drug Delivery Systems. *Biomaterials,* 2002. 23(9): P. 1955-1966.
[77] Yasunaga, H., et al., Dynamic Behaviour of Water In Hydrogel Containing Hydrophobic Side Chains As Studied By Pulse 1h Nmr. *Journal of Molecular Structure,* 2002. 602-603(0): P. 399-404.
[78] Baumgartner, S., et al., Investigation of The State And Dynamics of Water In Hydrogels of Cellulose Ethers By≪ Sup≫ 1≪/Sup≫H Nmr Spectroscopy. *Aaps Pharmscitech,* 2002. 3(4): P. 86-93.
[79] Di Colo, G., et al., A New Hydrogel For The Extended And Complete Prednisolone Release In The Gi Tract. *International Journal of Pharmaceutics,* 2006. 310(1-2): P. 154-161.
[80] Capitani, D., et al., Nmr Study of A Novel Chitosan-Based Hydrogel. *Carbohydrate Polymers,* 2001. 45(3): P. 245-252.
[81] Suzuki, S., F. Horii, And H. Kurosu, Theoretical Investigations of The Γ-Gauche Effect On The 13c Chemical Shifts Produced By Oxygen Atoms At The Γ Position By Quantum Chemistry Calculations. *Journal of Molecular Structure,* 2009. 919(1-3): P. 290-294.
[82] Shi, S. And L. Liu, Microwave-Assisted Preparation of Temperature Sensitive Poly(N-Isopropylacrylamide) Hydrogels In Poly(Ethylene Oxide)-600. *Journal of Applied Polymer Science,* 2006. 102(5): P. 4177-4184.
[83] Seida, Y., K. Takeshita, And Y. Nakano, Phase Behavior of N-Isopropylacrylamide/Acrylic Acid Copolymer Hydrogels Prepared With Ultrasound. *Journal of Applied Polymer Science,* 2003. 90(9): P. 2449-2452.
[84] Kaneko, T., T.-A. Asoh, And M. Akashi, Ultrarapid Molecular Release From Poly(N-Isopropylacrylamide) Hydrogels Perforated Using Silica Nanoparticle Networks. *Macromolecular Chemistry And Physics,* 2005. 206(5): P. 566-574.
[85] Zhao, Q., et al., Effect of 1,4-Dioxane On Synthesis of Macroporous Poly(N-Isopropylacrylamide) Hydrogels. *Journal of Polymer Science Part A: Polymer Chemistry,* 2008. 46(19): P. 6594-6603.
[86] Callister, W.D., Materials Science And Engineering: An Introduction. 2007: John Wiley & Sons.
[87] Morimoto, N., et al., Thermo-Responsive Hydrogels With Nanodomains: Rapid Shrinking of A Nanogel-Crosslinking Hydrogel of Poly(N-Isopropyl Acrylamide). *Macromolecular Rapid Communications,* 2008. 29(8): P. 672-676.
[88] Zhang, J.-T. And K.D. Jandt, A Novel Approach To Prepare Porous Poly(N-Isopropylacrylamide) Hydrogel With Superfast Shrinking Kinetics. *Macromolecular Rapid Communications,* 2008. 29(7): P. 593-597.
[89] Liu, H., et al., Synthesis of Poly(N,N-Diethylacrylamide-Co-Acrylic Acid) Hydrogels With Fast Response Rate In Nacl Medium. *Journal of Applied Polymer Science,* 2008. 109(5): P. 3037-3043.
[90] Caykara, T., S. Küçüktepe, And E. Turan, Swelling Characteristics of Thermo-Sensitive Poly[(2-Diethylaminoethyl Methacrylate)-Co-(N,N-Dimethylacrylamide)] Porous Hydrogels. *Polymer International,* 2007. 56(4): P. 532-537.
[91] Zhang, X.-Z., C.-C. Chu, And R.-X. Zhuo, Using Hydrophobic Additive As Pore-Forming Agent To Prepare Macroporous Pnipaam Hydrogels.Journal of Polymer Science Part A: *Polymer Chemistry,* 2005. 43(22): P. 5490-5497.

[92] Ferreira, A.A.P. And H. Yamanaka, Microscopia De Força Atômica Aplicada Em Imunoensaios.*Química Nova,* 2006. 29: P. 137-142.

[93] González-Méijome, J.M., et al., Microscopic Observation of Unworn Siloxane–Hydrogel Soft Contact Lenses By Atomic Force Microscopy. *Journal of Biomedical Materials Research Part B: Applied Biomaterials,* 2006. 76b(2): P. 412-418.

[94] Zhang, X.-Z., et al., Preparation And Characterization of Fast Response Macroporous Poly(N-Isopropylacrylamide) Hydrogels. *Langmuir*, 2001. 17(20): P. 6094-6099.

[95] Huang, J., et al., Synthesis And In Situ Atomic Force Microscopy Characterization of Temperature-Responsive Hydrogels Based On Poly(2-(Dimethylamino)Ethyl Methacrylate) Prepared By Atom Transfer Radical Polymerization†. *Langmuir,* 2006. 23(1): P. 241-249.

[96] Ohya, S., S. Kidoaki, And T. Matsuda, Poly(N-Isopropylacrylamide) (Pnipam)-Grafted Gelatin Hydrogel Surfaces: Interrelationship Between Microscopic Structure And Mechanical Property of Surface Regions And Cell Adhesiveness. *Biomaterials,* 2005. 26(16): P. 3105-3111.

[97] Grobe, G.L., P.L. Valint, And D.M. Ammon, Surface Chemical Structure For Soft Contact Lenses As A Function of Polymer Processing.*Journal of Biomedical Materials Research,* 1996. 32(1): P. 45-54.

[98] Schaffazick, S.R., et al., Caracterização E Estabilidade Físico-Química De Sistemas Poliméricos Nanoparticulados Para Administração De Fármacos.*Química Nova,* 2003. 26: P. 726-737.

[99] Ricciardi, R., et al., X-Ray Diffraction Analysis of Poly(Vinyl Alcohol) Hydrogels, Obtained By Freezing And Thawing Techniques. *Macromolecules,* 2004. 37(5): P. 1921-1927.

[100] Qu, X., A. Wirsén, And A.C. Albertsson, Novel Ph-Sensitive Chitosan Hydrogels: Swelling Behavior And States of Water. Polymer, 2000. 41(12): P. 4589-4598.

[101] Li, J., et al., Self-Assembled Supramolecular Hydrogels Formed By Biodegradable Peo–Phb–Peo Triblock Copolymers And A-Cyclodextrin For Controlled Drug Delivery. *Biomaterials,* 2006. 27(22): P. 4132-4140.

[102] Dumitriu, S., *Polymeric Biomaterial*s. 2002: Marcel Dekker.

[103] Paul J, F., Molecular Theory of Rubber Elasticity. *Polymer,* 1979. 20(11): P. 1317-1320.

[104] Flory, P.J., Network Topology And The Theory of Rubber Elasticity. *British Polymer Journal,* 1985. 17(2): P. 96-102.

[105] Deam, R.T. And S.F. Edwards, The Theory of Rubber Elasticity. Philosophical Transactions of The Royal Society of London. *Series A, Mathematical And Physical Sciences,* 1976. 280(1296): P. 317-353.

[106] Kou, J., G. Amidon, And P. Lee, Ph-Dependent Swelling And Solute Diffusion Characteristics of Poly(Hydroxyethyl Methacrylate–Co–Methacrylie Acid) Hydrogels. *Pharmaceutical Research,* 1988. 5(9): P. 592-597.

[107] Siepmann, J., et al., Hpmc-Matrices For Controlled Drug Delivery: A New Model Combining Diffusion, Swelling, And Dissolution Mechanisms And Predicting The Release Kinetics. *Pharmaceutical Research,* 1999. 16(11): P. 1748-1756.

[108] Kim, S.W., Y.H. Bae, And T. Okano, Hydrogels: Swelling, Drug Loading, And Release. *Pharmaceutical Research,* 1992. 9(3): P. 283-290.

[109] Stammen, J.A., et al., Mechanical Properties of A Novel Pva Hydrogel In Shear And Unconfined Compression. *Biomaterials,* 2001. 22(8): P. 799-806.
[110] Haraguchi, K., et al., Compositional Effects On Mechanical Properties of Nanocomposite Hydrogels Composed of Poly(N,N-Dimethylacrylamide) And Clay. *Macromolecules,* 2003. 36(15): P. 5732-5741.
[111] Masaro, L. And X.X. Zhu, Physical Models of Diffusion For Polymer Solutions, Gels And Solids. *Progress In Polymer Science,* 1999. 24(5): P. 731-775.
[112] Kim, S.J., et al., Properties of Interpenetrating Polymer Network Hydrogels Composed of Poly(Vinyl Alcohol) And Poly(N-Isopropylacrylamide). *Journal of Applied Polymer Science,* 2003. 89(8): P. 2041-2045.
[113] Murali Mohan, Y., P.S. Keshava Murthy, And K. Mohana Raju, Preparation And Swelling Behavior of Macroporous Poly(Acrylamide-Co-Sodium Methacrylate) Superabsorbent Hydrogels. *Journal of Applied Polymer Science,* 2006. 101(5): P. 3202-3214.
[114] Crank, J. And G.S. Park, Diffusion In Polymers. 1968: Academic Press.
[115] Caykara, T., S. Küçüktepe, And E. Turan, Thermosensitive Poly[(2-(Diethylamino) Ethyl Methacrylate)-Co-(N,N-Dimethylacrylamide)] Cryogels Prepared By A Two-Step Polymerization Method. *Macromolecular Materials And Engineering,* 2006. 291(10): P. 1278-1286.
[116] Pastoriza, A., I.E. Pacios, And I.F. Piérola, Kinetics of Solvent Responsiveness In Poly(N,N-Dimethylacrylamide) Hydrogels of Different Morphology. *Polymer International,* 2005. 54(8): P. 1205-1211.
[117] Zhang, X.-Z. And C.-C. Chu, Synthesis And Properties of The Semi-Interpenetrating Polymer Network–Like, Thermosensitive Poly(N-Isopropylacrylamide) Hydrogel. *Journal of Applied Polymer Science,* 2003. 89(7): P. 1935-1941.
[118] Johnson, B.D., D.J. Beebe, And W.C. Crone, Effects of Swelling On The Mechanical Properties of A Ph-Sensitive Hydrogel For Use In Microfluidic Devices. *Materials Science And Engineering:* C, 2004. 24(4): P. 575-581.
[119] Jeon, O., et al., Mechanical Properties And Degradation Behaviors of Hyaluronic Acid Hydrogels Cross-Linked At Various Cross-Linking Densities. *Carbohydrate Polymers,* 2007. 70(3): P. 251-257.
[120] Serizawa, T., et al., Thermoresponsive Properties of Porous Poly(N-Isopropylacrylamide) Hydrogels Prepared In The Presence of Nanosized Silica Particles And Subsequent Acid Treatment. *Journal of Polymer Science Part A: Polymer Chemistry,* 2002. 40(23): P. 4228-4235.
[121] Gutowska, A., et al., Thermosensitive Interpenetrating Polymer Networks: Synthesis, Characterization, And Macromolecular Release. *Macromolecules,* 1994. 27(15): P. 4167-4175.
[122] Mounir, A., N. Darwish, And A. Shehata, Effect of Crosslinking Concentration On The Micelle Formation Inside The Polymer Gel Network And Mechanical Properties of Sensitive Hydrogel Polymer In Contact With Surfactant Solutions. *Journal of Applied Polymer Science,* 2004. 91(6): P. 3921-3926.
[123] Djonlagić, J. And Z.S. Petrović, Semi-Interpenetrating Polymer Networks Composed of Poly(N-Isopropyl Acrylamide) And Polyacrylamide Hydrogels. *Journal of Polymer Science Part B: Polymer Physics,* 2004. 42(21): P. 3987-3999.

[124] Haraguchi, K. And T. Takada, Characteristic Sliding Frictional Behavior On The Surface of Nanocomposite Hydrogels Consisting of Organic-Inorganic Network Structure. *Macromolecular Chemistry And Physics,* 2005. 206(15): P. 1530-1540.

[125] Zhang, J.-T., et al., Poly(N-Isopropylacrylamide) Nanoparticle-Incorporated Pnipaam Hydrogels With Fast Shrinking Kinetics. *Macromolecular Rapid Communications,* 2005. 26(16): P. 1346-1350.

[126] Zhang, X.-Z., D.-Q. Wu, And C.-C. Chu, Effect of The Crosslinking Level On The Properties of Temperature-Sensitive Poly(N-Isopropylacrylamide) Hydrogels. Journal of Polymer Science Part B: *Polymer Physics,* 2003. 41(6): P. 582-593.

[127] Cai, W. And R.B. Gupta, Fast-Responding Bulk Hydrogels With Microstructure. *Journal of Applied Polymer Science,* 2002. 83(1): P. 169-178.

[128] Kim, S.J., et al., Thermal Characteristics of Interpenetrating Polymer Networks Composed of Poly(Vinyl Alcohol) And Poly(N-Isopropylacrylamide). *Journal of Applied Polymer Science,* 2003. 90(3): P. 881-885.

[129] Aouada F A, C.B., Orts W J, Mattoso L H C, Physicochemical And Morphological Properties of Poly(Acrylamide) And Methylcellulose Hydrogels: Effects of Monomer, Crosslinker And Polysaccharide Compositions. *Polym Eng Sci.,* 2009. 49(12): P. 2467–74.

[130] Kulkarni R V, S.B., Electroresponsive Polyacrylamide-Graftedxanthan Hydrogels For Drug Delivery. *J Bioact Compat Polym.* , 2009. 24(4): P. 368–84.

[131] Sutar P, M.R., Pal K, Banthia A, Development of Ph Sensitive Polyacrylamide Grafted Pectin Hydrogel For Controlled Drug Delivery System. *J Mater Sci: Mater Med.* , 2008. 19(6): P. 2247–53.

[132] Nakamura K, M.R.J., Joseph J I, Peppas N A, Morishita M, Lowman A M, Oral Insulin Delivery Using P(Maa-G-Eg) Hydrogels: Effects of Network Morphology On Insulin Delivery Characteristics. Journal of *Controlled Release*, 2004. 95: P. 589– 599.

[133] G, E.A., Electroconductive Hydrogels: Synthesis, Characterization And Biomedical Applications. *Biomaterials* 2010. 31: P. 2701–2716.

[134] Moura M R, A.F.A., Guilherme M R, Radovanovic E, Rubira A F, Muniz E C, Thermo-Sensitive Ipn Hydrogels Composed of Pnipaam Gels Supported On Alginate-Ca2+ With Lcst Tailored Close To Human Body Temperature. *Polymer Testing* 2006. 25: P. 961–969.

[135] Hoffman, A.S., Hydrogels For Biomedical Applications. *Advanced Drug Delivery Reviews* 2002. 43: P. 3–12.

[136] Peppas N A, B.P., Leobandung W, Ichikawa H, Hydrogels In Pharmaceutical Formulations. *European Journal of Pharmaceutics And Biopharmaceutics* 2000. 50: P. 27-46.

[137] Rosiak J M, U.I.P., Pjewski L A, Yoshii F, Makuuchi K, Radiation Formation of Hydrogels For Biomedical Purposes. Some Remarks And Comments. *Radiat. Phys. Chem.* , 1995. 46(2): P. 161-168.

[138] Hamidi M, A.A., Rafiei P Hydrogel Nanoparticles In Drug Delivery. *Advanced Drug Delivery Reviews* 2008. 60: P. 1638–1649.

[139] Gutowska A , B.J.S., Kwon I C , Bae Y H , Cha Y, Kim S W, Squeezing Hydrogels For Controlled Oral Drug Delivery. *Journal of Controlled Release* 1997. 48: P. 141–148.

[140] Dorkoosh F A, V.J.C., Ambagts M H C, Tehrani M R, Borchard G, Junginger H E, Peroral Delivery Systems Based On Superporous Hydrogel Polymers: Release

Characteristics For The Peptide Drugs Buserelin, Octreotide And Insulin. *European Journal of Pharmaceutical Sciences* 2002. 15(433–439).

[141] Bowersock T L , H.H., Suckow M, Porter R E, Jackson R, Park H K, Park Oral Vaccination With Alginate Microsphere Systems. *Journal of Controlled Release* 1996. 39 P. 209-220.

[142] Wu J, W.W., Wang L Y, Su Z G, Ma G H , A Thermosensitive Hydrogel Based On Quaternized Chitosan And Poly(Ethylene Glycol) For Nasal Drug Delivery System. *Biomaterials,* 2007. 28: P. 2220–2232.

[143] Gou M L, L.X.Y., Dai M, Gong C Y, Wang X H, Xie Y, Deng H X, Chen L J , Zhao X, Qian Z Y, Wei Y Q, A Novel Injectable Local Hydrophobic Drug Delivery System: Biodegradable Nanoparticles In Thermo-Sensitive Hydrogel. *International Journal of Pharmaceutics* 2008. 359 P. 228–233.

[144] Leach J B, S.C.E., Characterization of Protein Release From Photocrosslinkable Hyaluronic Acid-Polyethylene Glycol Hydrogel Tissue Engineering Scaffolds. *Biomaterials,* 2005 26: P. 125–135.

[145] Bhattarai N , G.J., Zhang M Chitosan-Based Hydrogels For Controlled, Localized Drug Delivery. *Advanced Drug Delivery Reviews* 2010. 62: P. 83–99.

[146] Blanco M D , G.O., Trigo R M, Teijón J M, Katime I, 5fluorouracil Release Fkom Copolymeric Hydrogels of Itaconic Acid Monoester. *Biomaterials* 1996. 17 P. 1061-1067.

[147] Franssen O, V.O.P., Hennink W E, Delayed Release of A Model Protein From Enzymatically-Degrading Dextran Hydrogels. *Journal of Controlled Release* 1997. 44: P. 237–245.

[148] Hennin K W E, F.O., Wolthuis W N E D, Talsma H, Dextran Hydrogels For The Controlled Release of Proteins. *Journal of Controlled Release* 1997. 48: P. 107–114.

[149] Chen L H, C.Y.W., Transdermal Iontophoretic Permeation of Luteinizing Hormone Releasing Hormone: Characterization of Electric Parameters. *Journal of Controlled Release* 1996. 40: P. 187 - 198.

[150] K, S.J.B.A.S., Lontophoresis In Dermatology. *A Review. J. Am. Acad. Dermatol.* , 1986. 15 P. 671-684.

[151] Fang J Y, H.Y.B., Lin H H, Tsai Y H, Transdermal Iontophoresis of Sodium Nonivamide Acetate. Iv. Effect of Polymer Formulations. *International Journal of Pharmaceutics* 1998. 173 P. 127–140.

[152] A, H.J., Hydrogel Systems For Barriers And Local Drug Delivery In *The Control of Wound Healing. Journal of Controlled Release* 1996. 39: P. 305-313.

[153] Sawhney A S, P.C.P., Hubbell J A, Hydrogels Based On Photopolymerized Poly(Ethylene Glycol)-Co-Poly(0/-Hydroxy Acid) Diacrylate Macromers.Bioerodible, *Macromolecules* 1993. 26: P. 581-587.

[154] Boles M M, L.M.C., Flório F A, Orgaes S, Marques B P A, Gonella H A, Curativo Imediato Com Hidrogel À Base De Água, Óleo De Melaleuca Alternifolia E Emulsificantes Após Queimaduras. *Rev. Bras. Cir. Plást.,* 2008. 23(4): P. 328-31.

[155] Santos J F R, L.C.A., Silva M, Balsa L, Couceiro J, Labandeira J J T , Concheiro A, Soft Contact Lenses Functionalized With Pendant Cyclodextrins For Controlled Drug Delivery.*Biomaterials* () 2009. 30: P. 1348–1355.

[156] Bardonnet Pl, F.V., Pugh W J, Piffaretti J C, Falson F Gastroretentive Dosage Forms: Overview And Special Case of Helicobacter Pylori. *Journal of Controlled Release,* 2006. 111: P. 1 – 18.

[157] He H, G.J., Lee J L, An Oral Delivery Device Based On Self-Folding Hydrogels. *Journal of Controlled Release* 2006. 110: P. 339 – 346.

[158] Patel V R, A.M.M., Preparation And Characterization of Freezedried Chitosan-Poly(Ethylene Oxide) Hydrogels For Site-Specific Antibiotic Delivery In *The Stomach. Pharm. Res.,* 1996. 13: P. 588±593.

[159] Park, H., K. Park, And D. Kim, Preparation And Swelling Behavior of Chitosan-Based Superporous Hydrogels For Gastric Retention Application. *Journal of Biomedical Materials Research Part A,* 2006. 76a(1): P. 144-150.

[160] Morimoto, K., et al., Evaluation of Polyvinyl Alcohol Hydrogel As A Sustained-Release Vehicle For Rectal Administration of Indomethacin. *Pharmaceutical Research*, 1989. 6(4): P. 338-341.

[161] D'cruz O J, U.F.M., Gel-Microemulsions As Vaginal Spermicides And Intravaginal Drug Delivery Vehicles. *Contraception* 2001. 64: P. 113–123.

Hydrogels: Synthesis, Characterization and Applications ISBN: 978-1-61942-842-3
Editors: F. V. Câmara and L. J. Ferreira 

***Chapter 4***

# INCORPORATION OF HYPROMELLOSE IN FORMULATING NIACIN CONTROLLED RELEASE ORAL DOSAGE FORMS

***Monica C. Chuong***
Massachusetts College of Pharmacy and Health Sciences, Boston, MA, US

## ABSTRACT

This chapter describes and discusses (1) the disadvantage of formulating an acidic drug with hydrogel using traditional approach, (2) preformulation of niacin including drug particle characterization using Accusizer™ 780 A, (3) high performance liquid chromatographic and UV-Vis spectrophotometric assay developments for the active pharmaceutical ingredient, and the correlation of the two assays, (4) formulation procedures, (5) proposals of two pH stage *in vitro* drug release study to simulate fasting and fed conditions, (6) application of FDA similarity factor to determine the similarity of release profiles, (7) percentage of drug remained in the accelerated temperature testing, (8) thermal analysis using a differential scanning calorimeter to examine whether there were melting point shifting in drug and three important excipients contributing the inert core in the samples underwent accelerated stability study, and (9) challenges encountered in selecting excipients for making drug dough extrudable into pellets, but facing accelerated temperature test in the later stage to determine shelf life.

## 1. INTRODUCTION

Some hygrogels are thermosensitive and biodegradable, while some are stimuli-sensitive. The stimuli that they are responsive are changes in pH, glucose, electricity, inflammation, antigen, etc.[1]. Hypromellose, widely used in oral, ophthalmic, nasal and topical pharmaceutical formulations, is the nonproprietary name for cellulose hydroxypropyl methyl ether, while hydroxypropyl methylcellulose is a synonym. There are four-digit number appending to Hypromellose[2]. The first two digits and second two digits refer respectively to the approximate percentage contents of the methoxy ($OCH_3$) and hydroxypropoxy groups[2].

For example, Type 1828 (hypromellose 1828) contains 16.5-20% methoxy and 23-32% hydroxypropoxy groups, while Type 2208 contains 19-24% methoxy and 4-12% hydroxypropoxy groups[2, 3]. Hypromellose 2906 contains 19.0-24.0% methoxy and 4-12% hydroxypropoxy contents. And Hypromellose 2910 contains 28-30% methoxy and 7-12% hydroxypropoxy contents calculated on dried basis. In oral products, it is primarily used as a matrix for use in extended release tablet formulations. High viscosity grade of hypromellose at 10% to 80% w/w is generally used to retard the release of drugs from a tablet or capsule pellet matrix[2]. This chapter presents a commercial polygel product and two laboratory preparations using niacin (nicotinic acid) as the model drug and hypromellose as an excipient to formulate into oral dosage forms and propose *in vitro* release experiments which simulate fasting and fed states. Utilization of other excipients in order to control drug release rates was also recommended.

# 2. Polygel Controlled-Release Tablets

## 2.1. Formulation Ingredients and Clinical Issues

Polygel Controlled-Release (CR) Niacin 500 mg tablets contains hypromellose, hydrogenated vegetable oil, silicon dioxide, magnesium stearate, glyceryl behenate, and Red 40 as inactive ingredients for making into a CR delivery system[3]. The active ingredient, niacin (Vitamin $B_3$), is not only used to prevent and treat niacin deficiency (pellagra), but also improve cholesterol and lower triglycerides in the blood[4-6] under the supervision of doctors. It is generally used after non-drug treatments have not been fully successful at lowering cholesterol (e.g., diet change, increase in exercise, weight loss if overweight). Since the Polygel CR tablets by name are a slow release dosage form, the consumers are warned that taking niacin on an empty stomach increases side effects, such as flushing, upset stomach[7-9]. To prevent flushing, it is advised against co-administered with alcohol or hot drinks. Instead, taking a plain (non-enteric-coated, 325mg) aspirin or a nonsteroidal anti-inflammatory drug (e.g., Ibuprofen) 30 minutes before taking niacin may help prevent flushing[4]. Two questions were thus raised. First, how long can a patient keep stomach empty in order to cope with the administration of a CR dosage form. Second, in addition to the acidity comes from niacin, aspirin is also an acidic compound. For niacin (nicotinic acid), 1 gram dissolves in 60 mL of water[10] and the pH of standard aqueous solution was measured as 2.7. For aspirin (salicylic acid), 1 g dissolves in 100 mL water at $37^{o}C$[10] and pKa is 3.0[11]. Both are rather acidic. Administered simultaneously in a close time frame these two drugs may aggravate gastric discomfort, or even develop ulcer as anti-cholesteremia requires a longer term treatment strategy. Therefore, the mechanism of drug release has to be investigated involving simulating fast and fed studies.

## 2.2. Selection of the USP Dissolution Apparatus 1 versus Apparatus 2

The role of dissolution plays in the absorption of drug products is still far from being understood perfectly. Refinement of current dissolution methods in USP-NF[3] or development of new methods to yield information about the biological availability of a drug are missions for pharmaceutical scientists[12-13]. Prior to the study of this project, all apparatuses were calibrated according to the USP protocols[14]. The USP dissolution apparatus engaged was seven-vessel Distek Premiere 5100 (North Brunswick, NJ). A non-disintegrating calibrator, USP salicylic acid non-disintegrating calibrators (300 mg Lot Q0D200), was utilized to validate the USP dissolution apparatus as both the commercial reference tablet and the hydrophobic matrix candidate formulations of our lab belonged to the non-disintegrating system. The maximal absorption wavelength for salicylic acid was measured at 296 nm. The average amount of salicylic acid released from a USP non-disintegrating calibrator was determined at 24.7 ± 0.4 % (with the relative standard deviation at 1.62 %). The calibration results suggested that the dissolution apparatus would be used in this study met the required percentage range of release from 23 – 30 % mandated in the calibrator instruction sheet sent by the USP.

The reason for choosing the USP dissolution apparatus I (basket method) not apparatus 2 (paddle method) was related to the nature of hypromellose. When a polygel CR tablet was dropped into the dissolution medium, it remained in the bottom of the vessel and spun locally following the turbulence caused by the stirring rod of the compendium apparatus because to its initial density is greater than that of 0.1N HCl (commonly used simulated gastric medium). As water penetrated into the matrix, a certain ingredients of tablet such as hypromellose, which polymer chains formed a three-dimensional structure holding together by physical or chemical bonds would absorb water in the network and swell. Thus, the density of the dosage form became lowered. Resultantly, after being in the dissolution medium for two hours the tablet moved away from the initial location, that is, the bottom of the dissolution vessel. The tablet grew from a caplet shape into a round ball floating and rolling in the dissolution vessel actively after being in dissolution medium for three and half hours to four hours. Since the tablets are marketed as a CR product, a prolonged dissolution study would have to conduct in order to receive the entire drug release profile. If such a dosage form were allowed to remain in the vessel of a USP dissolution apparatus 2 (paddle method) instead of apparatus 1 (basket method), the tablet might be hit by the spinning paddle. This might distort the data of drug release due to the alternation of surface configuration. Surface area is a well know parameter for drug release. Thereby, the USP dissolution apparatus 1, basket method[3] was chosen for the study.

## 2.3. Medium Change Methods Affect Polygel Drug Release

Since Polygel CR tablet is an extended release dosage form, the drug release is expected to extend into the enteric region of the human gastrointestinal tract (Table 1). Therefore, medium change is unavoidable in such an *in vitro* drug release study.

Two medium change methods in the USP-NF[3] are available as guidelines for pharmaceutical research and development. They are Method A: Add Directly and Method B: Drain and Add. Are both methods equally good for studying polygel tablet products? Investigation on whether these two medium change methods provided the similar drug release was conducted as below.

**Table 1. Comparison of Different Parts of the Gastrointestinal Tract†**

| | Fasting State | | Fed State | |
|---|---|---|---|---|
| | pH | Residence Time | pH | Residence Time |
| Esophagus | 6.8 | ~ 30 sec | | |
| Stomach | 1-2 | 1-5 h | 2-5* | * |
| Duodenum | 6.1 (5-6.5) | > 5 min | 4.5-5.5 (1 h) | 1 h |
| Jejunum | 6.5 (6.0-7.0) | 1-2 h | 4.7 (2 h) | 2 h |
| Ileum | 6.5 | 2-3 h | 6.5 | * |
| Colon | 5.5-7.8 | 15-48 h | 8.0 | up to 72 h |

*Dependent on volume, pH and buffer capacity of the food.

† References 15- 17.

For Method A (Add Directly), a Polygel CR tablet was placed in the USP Apparatus 1 with 500 mL of deaerated 0.1 N HCl at the stirring rate of 50 rpm. After two hours, 500 mL of deaerated 0.0375 M tribasic sodium phosphate ($Na_3PO_4$) preheated to 37°C was added. The resulting buffer solution was adjusted to pH 6.8 by 2 M NaOH and deaerated to mimic the intestinal fluid. For method B (Drain and Add) the initial dissolution operating conditions (in the acidic stage) were the same as those for Method A. But at the end of two hours, the dissolution platform was lifted into the air in order to change medium according to the protocol[3]. The acidic medium was drained from the dissolution vessel and 1000 mL of preheated, deaerated pH 6.8 phosphate buffer was added. The dissolution apparatus platform was then lowered and stirring was resumed at 50 rpm to operate for another four hours. For comparison purpose, the sampling schedules were the same for both methods. The acidic stage samples were quantified by a spectrophotometer at 261 nm, while the buffer stage samples were assayed at 263 nm. These two wavelengths were determined by scanning a known concentration standard solution using the Peak Scan program of the UV/Vis spectrophotometer prior to the dissolution study.

## 2.4. Niacin Standard Curves in Water, 0.1 N HCl, pH 2.2 Hydrochloric Buffer and pH 6.8 Phosphate Buffer

The quantification method using high performance liquid chromatography (HPLC) is not only more time consuming than using UV/Vis spectrophotometer but also has the disadvantage of hazardous organic chemical waste produced by the mobile phase. The niacin standard curves for 0.1 N HCl, pH 2.2 hydrochloric buffer and pH 6.8 phosphate buffer were measured through HPLC and UV/Vis spectrophotometer. The absorbance wavelengths of niacin in 0.1 N HCl and in pH 6.8 phosphate buffer have already mentioned in the study of Method A and Method B. The optimal absorbance for pH 2.2 HCl buffer was scanned as 261 nm and for deionized water as 263 nm. After good correlations between UV absorbance and HPLC AUC ratio of niacin and its internal standard, Pyridoxine HCl (vitamin $B_6$), in three simulated physiologic media, this project continued the remaining quantifications by using UV. The established standard curves were then used to qualify the *in vitro* dissolution of Polygel CR tablets as well as the ER oral dosage candidates formulated in my laboratory.

The niacin released from the Polygel CR 500 mg tablets in acid stage during simulated fasted condition was biphasic: curved in the acid stage and linear in the buffer stage (n = 8, Figure 1c). The best-fit equation obtained for the release versus time profile in the acid stage is described as:

$$Y = 19.482\,X^3 - 88.033\,X^2 + 183.06\,X + 1.624$$

where $X$ is time in hr and $Y$ is the amount of niacin (mg) released into the medium, and the correlation coefficient, $r^2$, 0.999.

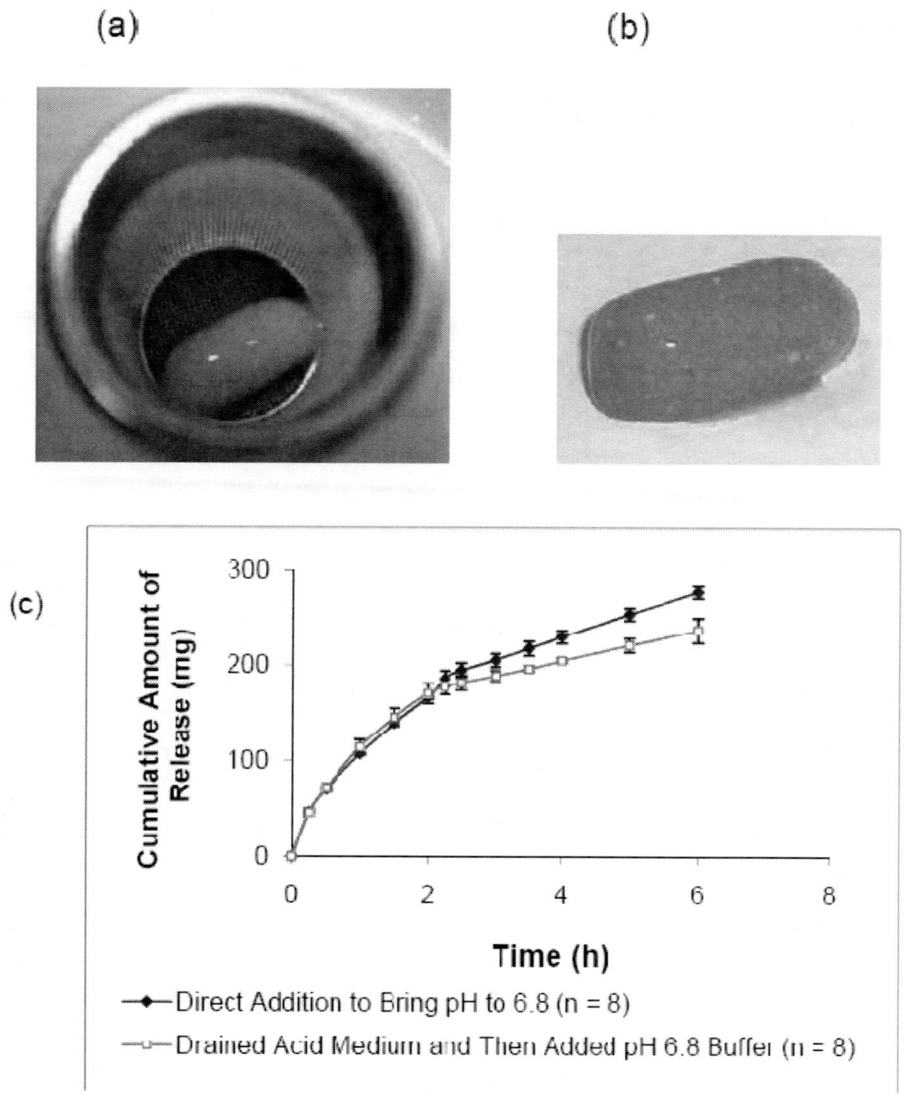


Figure 1. Niacin release from a Polygel 500 mg CR tablet using Method A vs. Method B for medium change. In Method B, the dissolution apparatus platform had to be lifted up into air to change medium (Figures a - b). The drug release rate in the simulated small intestinal medium (Phosphate Buffer, pH 6.8) was significantly reduced in Method B than in Method A as shown in the slopes of two profiles from two to six hours ($P < 0.001$, Figure c). Please refer to the text for the proposed rationale behind.

When Method A (Add Directly) was used to change dissolution medium from acidic pH to pH 6.8, the 4-h niacin release profile from a Polygel CR 500 mg tablet was linear, and the release rate during this buffer stage was 27.5 ± 1.5 mg/h. But when Method B (Drain and Add) was used to change medium, the release rate was significantly slowed. The computed average of the release rate was as low as 16.5 ± 5.8 mg/h ($n = 8$, $p < 0.001$, Chuong). Method A (Add Directly) is thus superior to Method B (Drain and Add) during the dissolution study for testing a Polygel CR tablet for two main reasons. First, after tablet exposure to an acidic medium for 2 h, the swollen three dimensional scaffold of the Polygel tablet may deswell due to the lack of medium support when the dissolution apparatus platform with the basket holding the study tablet is lifted into the air to change dissolution medium as required in Method B. The air might also block the channels and thus affected the diffusion of drug molecules after the platform was returned to the system (Figures 1a - 1c) and required time to reabsorb medium to open the porosity of the tablet matrix. Second, Method A simulates the physiological environment more closely as it is a continuous process for a tablet to travel in the medium from the stomach into the small intestine (Table 1). Hence, Method A is recommended for the dissolution study of a Polygel CR tablet.

## 2.4. Simulated Fasting Study

In the simulated fasted condition, Polygel 500 mg CR tablets were tested with the USP dissolution apparatus 1 as well as the USP disintegrating apparatus. For the dissolution study to simulate drug release in the fasting state, each tablet was immersed in 500 mL 0.1 N HCl in each vessel of a USP dissolution apparatus 1 and stirred at 50 rpm for 2 h to mimic the gastric content, pH and slow activity. Then 500 mL of 0.0375 M tribasic sodium phosphate ($Na_3PO_4$) was added to make into 1000 mL phosphate buffer pH 6.8 based on the USP Dissolution Medium Change Method A[3, 18-19]. The stirring continued at the same rate of 50 rpm for another 4 h (to simulate fasting intestinal content and pH). The test duration for each phase was chosen based on the gastrointestinal transit time compiled from literature[15-17] (Table 1). The excipients controlling the niacin release from the Polygel CR product (e.g., glycerol behenate, hypromellose) seemed to be pH-independent in this acidic pH region. In the beginning of the dissolution study, which was in the acidic medium, the niacin molecules present on and near the surface of the matrix tablets dissolved quickly. As time progressed, the release rate slowed down due to the increased diffusion distance. This may explain why the niacin release in acidic medium is a curve. But during the pH 6.8 phosphate buffer stage, the release profile became linear instead of continuously moving upward as a curve (Figure 2a). Although the reason is unknown, we speculate that this observation may be linked to the charge interactions among niacin ions, hydrogen ions, chloride ions, sodium ions, and mono- and bi-basic phosphate ions present in the medium with the tablet ingredients, making the release rate of niacin from the inner part of the tablet through gel material constant (Figure 2a).

For the disintegration study simulating the fasting state, the to-be-used USP disintegrating apparatus was calibrated to ensure that the USP-NF standard[3] had met. After that, 800 mL of preheated (39 ± 1°C) deaerated 0.1 N HCl was used as the test immersion fluid to mimic a tablet in an pre-prandial stomach. At the end of two hours, the apparatus was paused temporarily. The basket containing the Polygel CR tablet was quickly lifted and the test beaker was switched to another one containing 800 mL of deaerated pH 6.8 phosphate buffer preheated to 39 °C. This medium preparation was to mimic the intestinal fluid. No USP disk was applied onto a test Polygel CR tablet, because the experiment was designed to study the fasting state. The disintegration apparatus was resumed operation in this buffer medium for 4 h.

Sampling schedule were the same for both dissolution and disintegrating apparatuses. Acidic samples were assayed spectrophotometrically at 261 nm, and buffer stage samples at 263 nm based on the wavelength scans described in Session 2.3. The amounts of release at different sampling times were computed according to the established standard curves and a correction factor was made to adjust the medium volume used by each apparatus[18-19] to compile its release profile (Figure 2a).

## 2.5. Simulated Fed Study

In the simulated fed study a Polygel CR tablet was designed to study in a pH 2.2 HCl buffer for 6 h using both the USP dissolution apparatus and the USP disintegration apparatus as contrast[18-19]. In brief, the USP dissolution apparatus 1 was chosen for the *in vitro* dissolution study for the same reason mentioned in the simulated fast study. After a tablet was dropped into the mesh basket, the dissolution apparatus platform was lowered into vessels containing 1000 mL preheated pH 2.2 HCl buffer each (instead of being 500 mL of 0.1 N HCl in the fasting study). The operating condition to simulate the fed state was: (1) the basket speed spun at 100 rpm, not 50 rpm (to simulate active gastric digestion and agitation) for 6 h (prolonged gastric content empty time).

For using the USP disintegration apparatus to simulate the fed condition[18-19], the apparatus conditions differed from those of a simulated fasting study were: (1) 800 mL pH 2.2 HCl buffer was used (instead of 0.1 N HCl, which pH is approximate 1.0 to 1.4), and (2) a USP disk weighing 3.05 g was placed flat on the top of a test tablet to mimic food weight effect in maximal contact with a test tablet after a heavy meal (Figure 2).

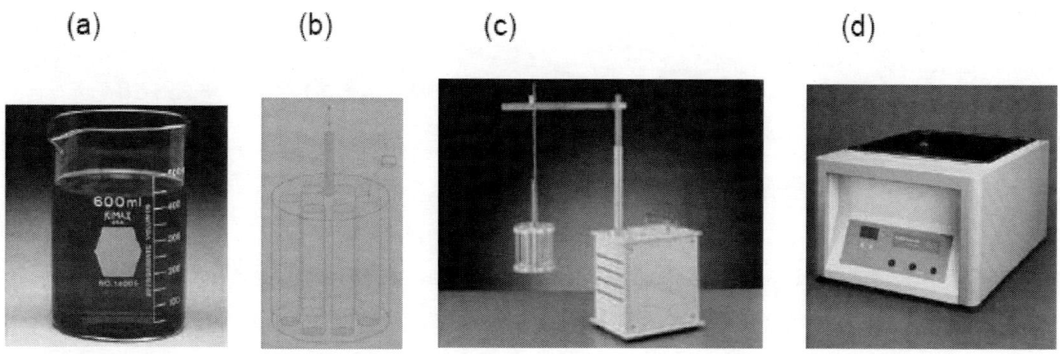


Figure 2. *In vitro* disintegration studies were performed with a USP disintegration apparatus in a water bath: (a) a 1-liter size glass beaker with 800 mL test medium, (b) placement of a tablet into a basket vessel and then an USP disk on top of the tablet, (c) the position of the basket-rack assembly in the USP disintegration apparatus, (d) the water bath to maintain the test medium at 39±1°C.

**Table 2. Niacin released from a Polygel CR tablet in 6 h using both the USP dissolution apparatus I and the USP disintegrating apparatus. In the disintegration apparatus a USP disk weighing 3.05 g was placed on top of the tablet to mimic food weight effect**

| | Disintegration Apparatus | Dissolution Apparatus |
|---|---|---|
| Fasting | 66.3 ± 2.1% (n = 6) | 55.3 ± 1.3% (n = 8) |
| Fed | 88.9 ± 5.9% (n = 6) | 47.5 ± 0.8% (n = 5) |

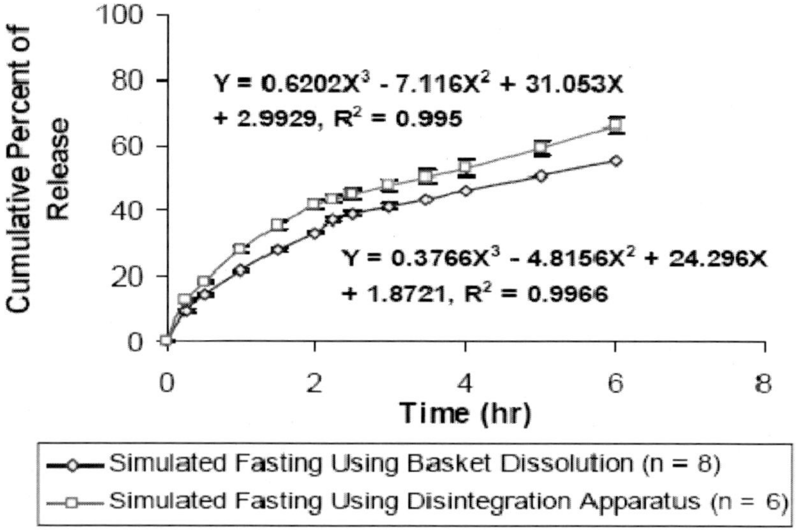


a

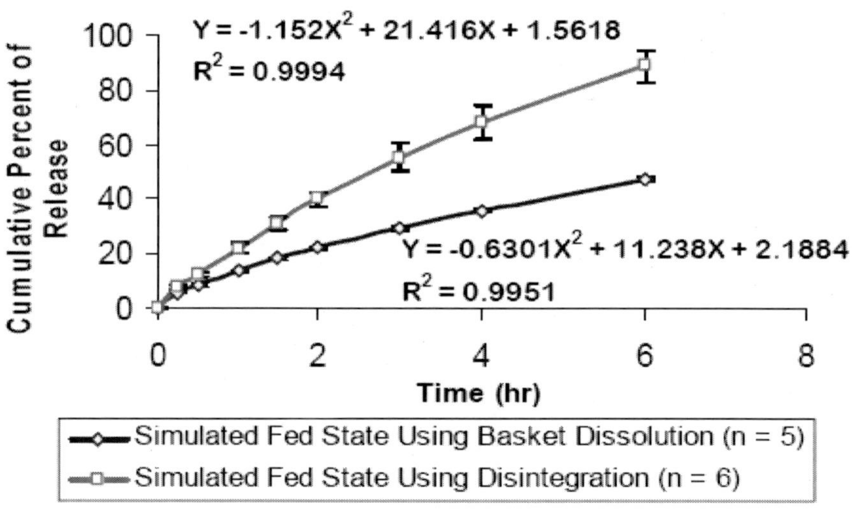


b

Figure 3. Niacin release from a Polygel CR tablet in fasting and fed conditions using the USP dissolution apparatus 1 and the USP disintegrating apparatus. There was no application difference in either apparatus in the fasting study (Figure a). But the release from the USP disintegration apparatus with a disk laid flat on top of the tablet to mimic food weight effect during the fed state was higher than that using the USP dissolution apparatus (Figure b). See text.

This project found that use of a disintegration apparatus or a dissolution apparatus did not seem to affect the cumulative percent niacin release from a Polygel CR tablet during simulated fasting condition at the end of 6-h experimental period (66.3 ± 2.1%, n = 6 versus 55.3 ± 1.3%, n = 8, Table 2 and Figure 3a). However, the cumulative percent of drug release at the end of 6 h was much higher in the simulated fed state using the USP disintegration apparatus with a 3.5 g disk on a tablet top to simulate food weight effect on a drug tablet (Figure 3b). The results were 88.9 ± 5.9%, n = 6 versus 47.5 ± 0.8%, n = 5 (Table 2).

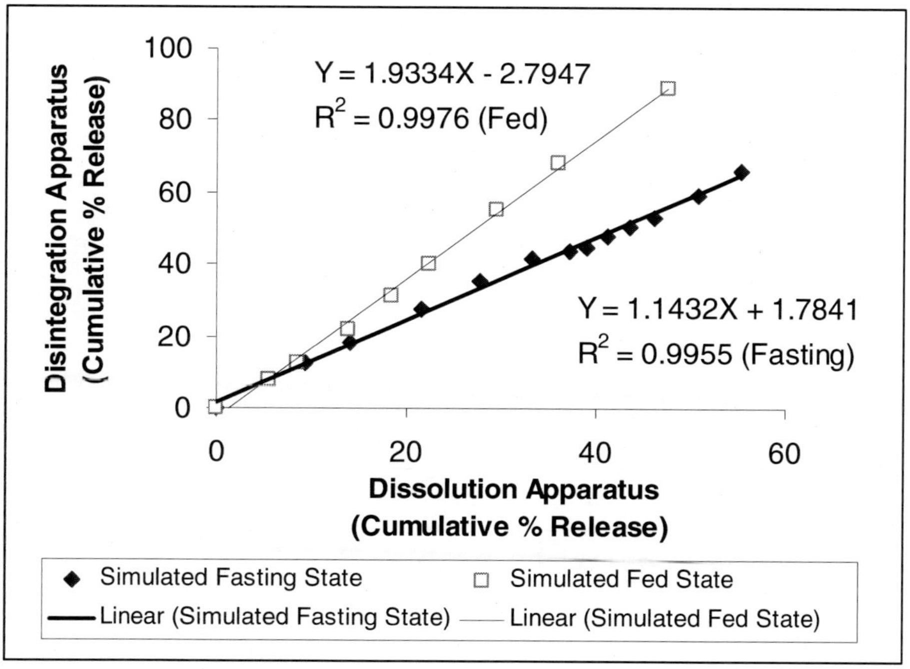


Figure 4. The USP Dissolution Apparatus 1 and the USP disintegration apparatus correlation when niacin release from a Polygel CR tablet was studied in the simulated fasting state (no disk added) and simulated fed state (with a 3.05-g USP disk on top of a test tablet).

When the cumulative percent of niacin release from a Polygel CR tablet in the simulated fasted and simulated fed states were plotted for USP dissolution apparatus 1 and the USP disintegration apparatus, the correlation coefficient, $R^2$, was 0.9976 for the simulated fasted study, while the $R^2$ value was 0.9955 for the simulated fed study (Figure 4). The slope of the simulated fed state trend line was 1.7 times that of the simulated fasted state trend line (1.933 versus 1.143, Figure 4) reflecting the former had a much faster drug release.

## 3. PREFORMULATION

Since the Polygel niacin CR tables were not able to release its drug completely in the aforementioned *in vitro* fasting study, but they released drug too rapidly during the *in vitro* fed study, development of an innovative ER niacin capsule with a near zero-order release profile in a two-stage *in vitro* dissolution study was desired. In order to achieve so, preformulation works, such as drug solubility and stability in different physiological media would have to be included to comprehend the physicochemical properties of the active pharmaceutical ingredient is essential in succeeding a formulation. The solubility of niacin as 1 g dissolved in 60 mL of water[10] was used to validate the purity of the niacin USP powder purchased for use in formulation and also to determine the solubility of niacin in the solvents of 0.1 N HCl, pH 2.2 HCl medium, and pH 6.8 phosphate buffer. Niacin solubilities were found as 0.1 N HCl > pH 6.8 phosphate buffer > pH 2.2 HCl buffer > water (Table 3). The stability of niacin in the former thee media (0.1 N HCl, pH 2.2 hydrochloric acid buffer, pH 6.8 phosphate buffer) was monitored up to 76 h, while niacin stability in 3% hydrogen peroxide was only monitored for 3 h due to the volatility and nature of rapid decomposition of hydrogen peroxide.

**Table 3. Solubility of Niacin USP powder in various simulated gastrointestinal media**

| Solvent | Solubility in 60 mL (Mean ± SD, n = 3) |
|---|---|
| Water | 0.97 g |
| 0.1N HCl | 1.80 ± 0.01 g |
| pH 2.2 HCl buffer | 1.05 ± 0.03 g |
| pH 6.8 phosphate buffer | 1.20 ± 0.01 g |

The Angle of Repose was measured as 50.12±1.93° (n = 4), which is defined as poor flow property and must be agitated or vibrated according to the scale of Flow Properties and Corresponding Angles of Repose listed in USP-NF[3]. The measurement of Compressibility Index and Hausner Ratio also indicated the same (Table 4).

**Table 4. Flow Properties and Corresponding Angles of Repose[3]**

| Flow Property | Angle of Repose (degrees) |
|---|---|
| Excellent | 25–30 |
| Good | 31–35 |
| Fair—aid not needed | 36–40 |
| Passable—may hang up | 41–45 |
| Poor—must agitate, vibrate | 46–55 |
| Very poor | 56–65 |
| Very, very poor | > 66 |

**Table 5. Scale of Flowability in USP-NF[3]**

| Compressibility Index (%) | Flow Character | Hausner Ratio |
|---|---|---|
| 10 | Excellent | 1.00–1.11 |
| 11–15 | Good | 1.12–1.18 |
| 16–20 | Fair | 1.19–1.25 |
| 21–25 | Passable | 1.26–1.34 |
| 26–31 | Poor | 1.35–1.45 |
| 32–37 | Very poor | 1.46–1.59 |
| > 38 | Very, very poor | >1.60 |

**Table 6. Determination of angle of repose, compressibility index and Hausner ratio of Niacin powder, USP grade**

| Flowability Parameters | Value and Prediction |
|---|---|
| Angle of Repose | 50.12±1.93 (n = 4)<br>Indicating poor flowability - must agitate, vibrate |
| Compressibility (Carr) Index | 41.02±2.29 (n = 3)<br>Indicating very, very poor flowability |
| Hausner Ratio | 1.70±0.07 (n = 3)<br>Indicating very, very poor flowability |

Niacin USP powder very rarely stands alone as a single particle when being examined under an optical microscope. The shape of this chemical taken under Raman Microscope 20X Lens (Figure 5 a-b) is regarded as irregularly shaped interlocking particles according characterization tabled by RJ Lantz[20]; is best described as aggregated based on the system defined by Randall[21]; and considered as column shape based on the USP/NF system[3]. Powder with this particle shape often shows poor flowability and causes bridging[21]. The high degree of aggregation will require aggressive vibration, tapping and gyratory motion of the sieves[22]. The particle flowability study such as angle of repose, compressibility index and Hausner ratio, agreed with the findings of optical microscope that the flowability of Niacin USP powder is very, very poor. All these information presented a problem in formulation that needed to be overcome in order to scale up later.

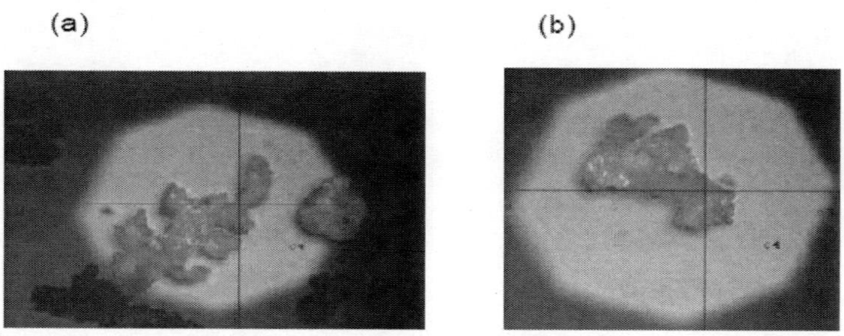

Figure 5. Niacin powder taken under Raman Microscope 20X Lens (a) and (b) suggesting as column, high degree aggregates and irregularly shaped interlocking particles (see text).

Niacin powder under Raman Microscope 20X (Figures 4 a-b) illustrating that niacin USP powder is irregularly shaped interlocking particles. This was corresponded to niacin being characterized as poor flowability based on the measurements of angle of repose (50.12±1.093, n = 4), compressibility index (41.02±2.29, n = 3) and Hausner ratio (1.70±0.07, n = 3, Table 6).

The particle size distribution[3,23] of niacin USP analyzed by Accusizer™ 780 A - PSS-NICOMP (Particle Sizing Systems, Santa Barbara, California) linked via a single computer to the Nicomp 380 and windows based software controller so that the main peak of a sample could be analyzed by Nicomp 380[24]. Analysis was performed on 60 mL saturated filtered niacin solution prepared by bidistilled water as tested medium. Sample time was set at 60 seconds, resolution 256 channels (0.5-500 microliters), flow rate 1 mL/sec. Assuming the tested Niacin USP was spherical in shape, the particle size distribution was reported in three different Y-axes: (a) number, (b) projection surface area, and (c) volume with respect to equivalent particle diameter as X-axis. The median particle size output by number particle size distribution of Accusizer™ was 7.13 μm (Figure 6 a-c).

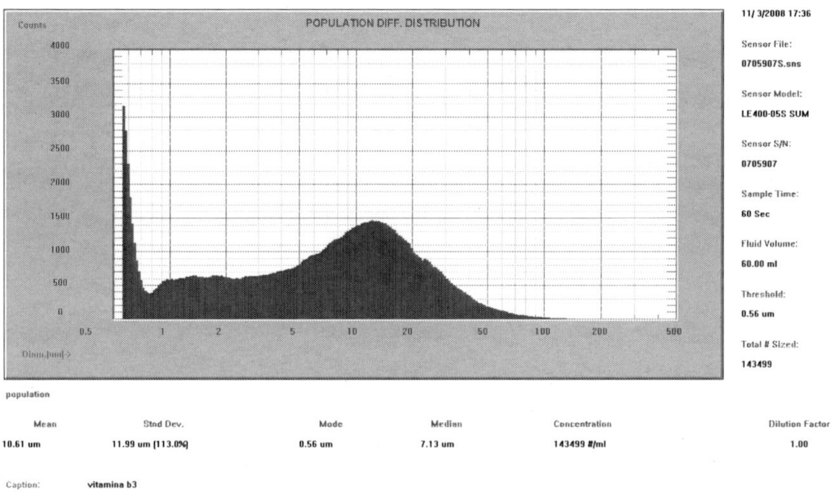


a

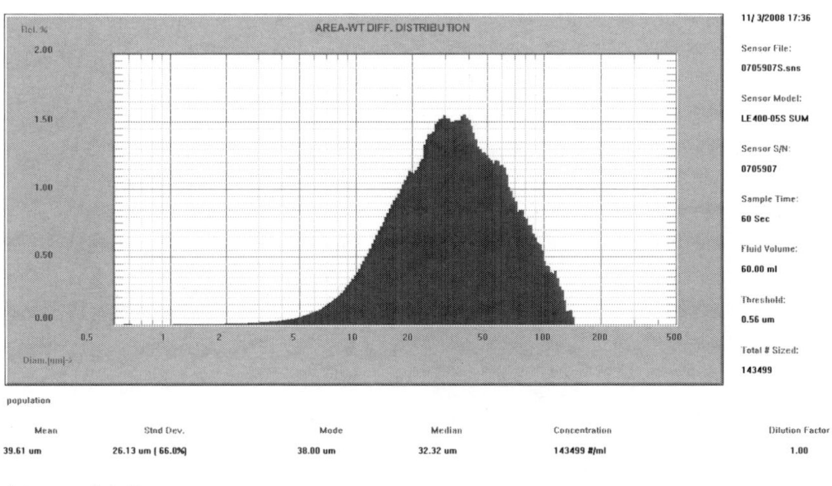


b

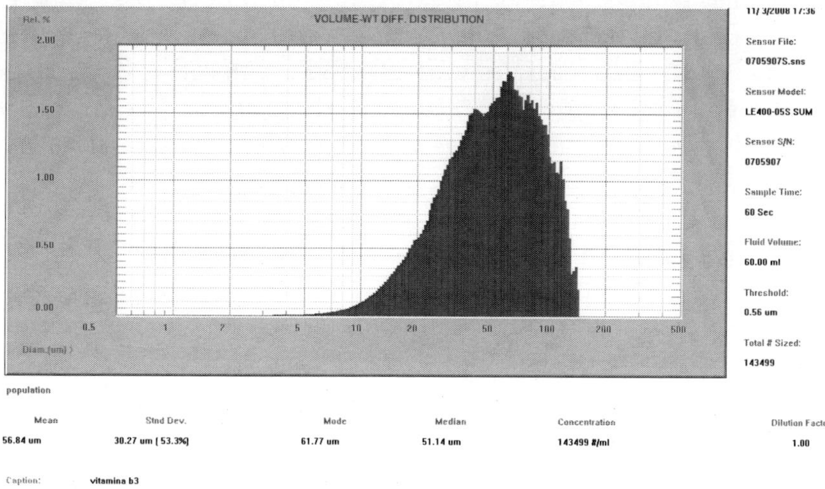


c

Figure 6. Accusizer™ 780 A output the particle size distribution of niacin powder USP by: (a) number, (b) projection surface area, and (c) volume, with respect to equivalent particle diameter (see text).

# 4. Formulation of Novel Niacin Extended-Release Capsules

The materials selected to embed niacin in a lipophilic matrix in order to control its release profile included white wax and Gelucire 43/01[25]. After being melted in a beaker on a heat plate, the other excipients were added and stirred. Before complete cooling, the ingredients were molded into a cohesive doughy mass and removed from the beaker. The mass was then further blended on a tile plate to ensure homogeneous dispersion of all ingredients. Once the mass was thoroughly kneaded, it was extruded though sieve size 18 to form granules, then manually spheronized to create round shaped pellets and allowed to cool to room temperature. Pellet flowability and size distribution were next characterized. Finally, the shaped pellets were once again sieved through size 16 (1.18 mm) and size 20 (0.85 mm) sieves[3] to ensure a relatively homogeneous size distribution. The pellets were then stored in a container overnight. In the next morning a calculated weight (based upon the amount of excipients present in the formulation candidate) that included 275 mg of niacin (110% of the 250 mg label claim) was filled into an appropriate size of hard gelatin capsule. The formulation of a novel dosage form proved to be a challenge. Since the goal of formulation was to slow the release profile of niacin in the acidic stage, hydrophilic excipients commonly used for capsule formulations, including lactose and starch, were immediately excluded from formulation testing. Eventually, Carbomer (Carbopol® 940) and Methocel® K4M CR was chosen as a controlled release polymer to help linearize the release profile of the candidate formulations. Both Gelucire® 43/01 and white wax (beeswax) were chosen for their highly lipophilic characteristics, an essential embedding substance to control the release of niacin (a highly water soluble compound) from the formulation. A surfactant in tiny amount was added to help with homogeneous blending of the hydrophilic and lipophilic ingredients together to form a workable doughy mass. After being extruded through sieve size #20 (0.85 mm), the needed pellets to contain 110% of 250 mg niacin was determined. In order to decrease variability in dissolution results, capsules were filled only with pellets sieved through sieves #20 and #18 (1 mm) in order to ensure homogeneity of particle size, because the release performance in an *in vitro* dissolution study is pellet specific surface dependent.

## 4.1. Pellet Characterization

After pellet size distribution was determined and before filling pellets into a hard gelatin capsule, a compressed size distribution of pellets was chosen to improve the consistency of particle surface area as the niacin release profile in simulated *in vitro* studies would be pellet surface area related. The optimal candidate formulation (6.78 g of pellets) was poured through a funnel to determine the angle of repose. The angle of repose for innovative formulation was determined to be 35.13°. This angle predicted good flow properties to the pellets according to the scale of Flow Properties and of Repose given by USP-NF[3]. This proved that the pellet formulation was able to overcome the poor flow properties of niacin powder USP and increased ease of capsule filling at the industrial level. The compressibility index was determined to be 12.9 while the Hausner ratio was computed to be 1.15 (Table 7). Combining these two values, the pellet flowability of innovative formulation was again determined to be good based on the table values given in USP-NF[3] (Table 5).

**Table 7. Pellet flowability of the innovated niacin ER pellet: angle of repose, compressibility index and Hausner ratio**

| Flowability Parameter | Tested Pellets 6.78 g (n = 1) |
|---|---|
| Tangent | 0.704 |
| Angle of Repose (Flow Property) | 35.13°(Good) |
| Compressibility (Carr) Index | 12.9 |
| Hausner Ratio | 1.15 |
| Flow Character | GOOD |

When sieved through 10 different sieve sizes of increasing diameter to determine the size distribution of the innovative pellets, it was found that the largest amount of pellets were from sieve #20 and sieve #18. The largest concentration of pellet size was thus determined to be between 0.850 and 1.0 mm. In order to decrease variability during dissolution studies, capsules for *in vitro* studies were filled only with pellets sieved through sieves #20 and #16 (0.85 mm to1.18 mm).

## 4.2. *In Vitro* Dissolution of the Innovative Niacin ER Pellets

As mentioned in the Methods section, the methods listed in the USP-NF[3] were used to compare niacin ER pellet-filled capsules with a commercial reference. To simulate a fasting state *in vivo* study, the first 2 h of the *in vitro* dissolution test were conducted in 750 mL 0.1

N HCl[3]. At the completion of acidic stage, 250 mL of 0.20 M tribasic sodium phosphate were added, and the pH was adjusted to 6.8 ± 0.5 with 2 N sodium hydroxide and/or 2 N HCl. The last 4 h of the *in vitro* study were conducted at pH 6.8 ± 0.5, simulating the *in vivo* path of the formulation through to the more basic small intestinal medium[15-17] (Table 1). The goal of the optimal formulation was to create a release profile that was both slower than the commercial reference and also more linear. Both formulations #36 and #37 were able to achieve a release profile slower than the commercial reference. At the end of 6 h, formulation #36 released an average of 182.5 ± 4.5 mg of niacin (n = 3), while formulation #37 released an average 223.9 ± 23.8 mg (n = 4). This was compared to the incomplete release of the commercial reference after 6 h averaging 259.4 ± 0.6 mg of release (n = 3, Figure 7). The commercial reference which loaded with 500 mg of niacin only had a 51.9% release. Our formulation candidates loaded with 250 mg of niacin. Formulation #36 demonstrated a 73% release, while #37 a 89.6% release. Both of the most optimal formulations of the project were thus superior to the commercial reference based upon the total release. Formulation #37 had the most efficient release of all.

However, a formulation candidate cannot be judged alone by the total amount of release at the end of a dissolution study. Achieving a linear release is also a goal of the optimal formulation candidate, allowing for a zero-order release factor that would allow dissolution of the dosage form to be consistent in possible future *in vivo* studies. Therefore, the linearity of the release profiles of formulations #36 and #37 were further compared to that of thc commercial reference. Both formulations #36 and #37 had slower release profiles than the commercial reference. However, based upon the linear regression coefficient ($R^2$ value) the linearity of each release profile from least to most, formulation #36 (0.84, n = 3) < commercial reference (0.86, n = 3) < formulation #37 (0.93, n = 4). When comparing #37 to #36, it was found that although the rate of release under *in vitro* gastric conditions (during the first 2 h in 0.1 N HCl) was approximately the same, formulation #37 actually sped up its release in *in* v*itro* small intestine pH medium.

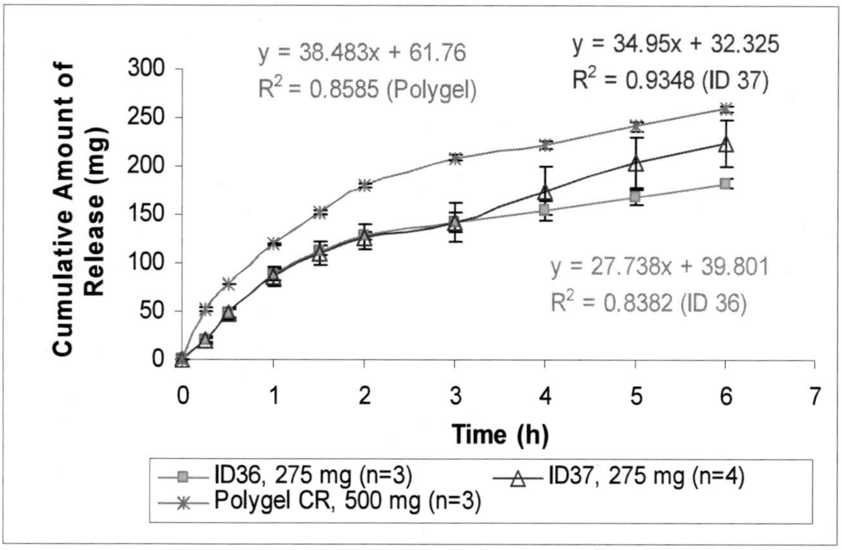


Figure 7. Comparative Dissolution of Polygel CR reference dosage form, innovative formulations ID 36 and ID 37 (see text).

## 4.3. Pharmaceutical Stability and Thermal Analysis

After formulation candidate #37 showed near zero order *in vitro* release pattern for six hours and most efficient in drug release (89.6%), more capsules were fabricated by filling pellets equivalent to 275 mg of niacin into a size 00 hard gelatin capsule. They were stored in a 40°C, 60% RH laboratory incubator to conduct accelerated stability study[26-27]. Heat stressed capsules were removed from the incubator according to a preset schedule to examine the content, *in vitro* release and heat flow behaviors. The obtained results were then compared to the data obtained from analyzing the capsules stored at room temperature for the same time period.

Capsules filled with the most optimal formulation pellets were stored in 40°C 60% RH to conduct an accelerated storage test. After being in an incubator for two weeks and four weeks respectively, three capsules each time were taken out and place in 750 mL 0.1 N HCl to conduct dissolution study with the basket stirring at 100 rpm for 2 h and then add 250 mL of 0.20 M tribasic sodium phosphate into 1000 mL of pH 6.8 phosphate buffer. The stirring rate of the apparatus was reduced to 50 rpm to continue dissolution study for another 4 h to examine the *in vitro* release pattern of the capsules.

### *4.3.1. Differential Scanning Calorimetry*

Differential scanning calorimetry (DSC, Seiko Instruments model SSC/220C) linked via a single computer to the EXSTAR software (Woodland CA) was conducted at different sampling times: niacin USP powder (drug reference) and formulation #37 pellets stored at 40°C for time zero (formulation time 0 reference) and 4 weeks (accelerated temperature study). The melting point of niacin, 236.6°C, was consistent among three DSC curves (Figures 8). The melting event at 43°, 67° and 180°C in the reported DSC temperature heat flow curves (Figures 8) matched the melting points of Gelucire 43/01, white wax and Methocel K4M CR as listed in the Handbook of Pharmaceutical Excipients[2]. Niacin was found stable after being in a 40°C incubator for 4 weeks.

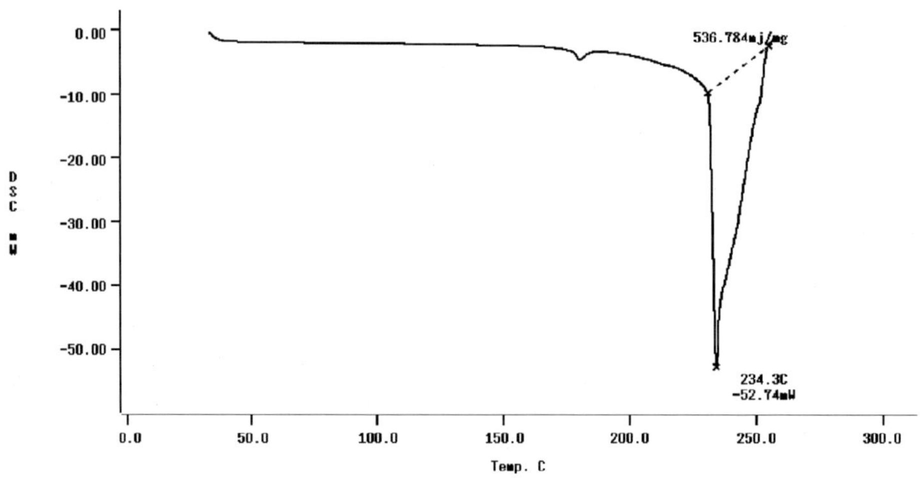

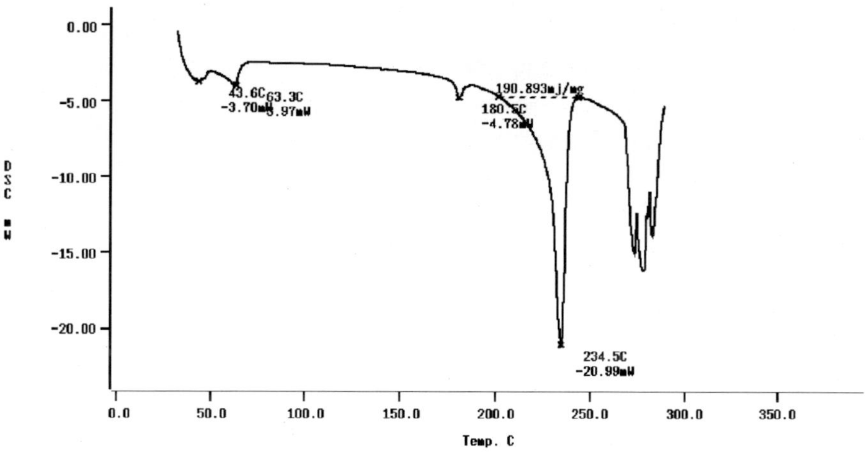


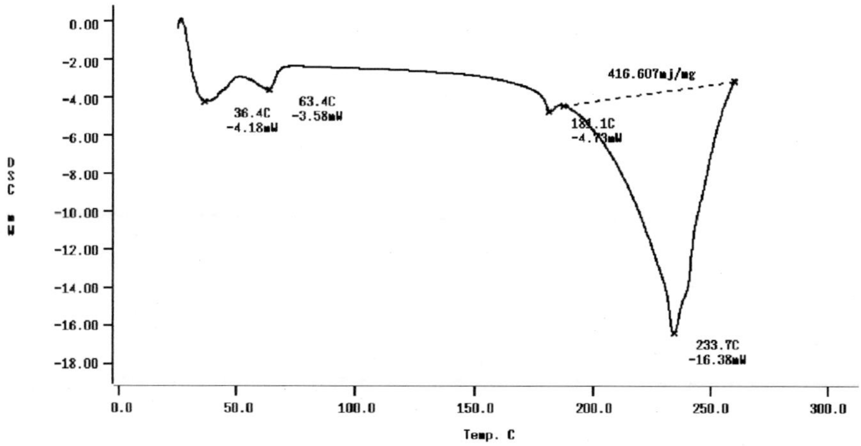


Figure 8. Differential scanning calorimetric curves of Niacin USP powder (figure a), formulation ID #37 time 0 pellets (figure b), 4-week pellets (figure c). The melting point of niacin, 236.6°C, was consistent among all DSC curves. The melting at 43°, 67° and 180°C in Figures b and c matched the melting points of Gelucire 43/01, white wax and hypromellose[2].

In order to verify stability of our proposed formulation strength, 250 mg, a content assay was expected to be over 90% of niacin present in a tested capsule. The capsule batch stored at room temperature up to 6 months were determined to be 107.4±0.2% (n = 3). Its relative standard deviation (RSD) was 0.21%. Average niacin retrieved from the capsule batch stored at 40°C for 6 months was determined to be 107.1±0.7% (n = 3) and the RSD was 0.69%.

The gastric release profile was the most important aspect due to the acidity of niacin determining the more acidic stomach to be its primary site of absorption and its impeded absorption due to ionization in the basic pH of the small intestine. The differences between formulations #36 and #37 were achieved by altering the amounts of the excipients Methocel and Carbomer. Methocel® K4M CR is a tablet binder and viscosity increasing agent. This agent was used in formulation #36, but not in formulation #37, which might be explained why niacin release in pH 6.8 phosphate buffer from formulation #36 was hindered. The amount of Carbomer in formulation #37 was almost three times that of formulation #36 (20 mg versus 7.5 mg). Carbomer (Carbopol® 940) is an aqueous gel and a pH responsive polymer. It becomes ionized and expels each other at approximately pH 7.0 into the largest volume. This character allowed niacin to diffuse through open pore channels in the dosage form and release into the dissolution medium. Since both the gastric release profiles of formulations #36 and #37 were similar, ultimately formulation #37 was chosen over #36 because its more efficient release profile in phosphate buffer stage dissolution, the simulated small intestine condition (Figure 7). Formulation #37 was thus the most optimal formulation candidate of all for the reasons of being efficient in niacin release from the lipid based pellets, higher in release linearity. Additional merits added to the preference of this formulation, (ID #37) are its shorter list of required excipients and an outer coating over the pellets is not required. All of these therapeutic and economical strengths endorsed the selection of formulation #37.

# 5. Formulation of Novel Niacin ER Tablets

## 5.1. Preparations of Innovative Niacin CR Tablets

Niacin formulations would be prepared with various excipient combinations and compressed with a laboratory scale Healthstar tablet press (Type JC-RT-16H, Jenn-Chiang Machinery Co., LTD, Taiwan) with automotive mode. The tablet compression tooling purchased to be used in this project was the D type of TSM standard[9]. TSM is the acronym for the Tablet Specification Manual published, revised and distributed by the American Pharmacist Association in Washington DC. The D type has a nominal barrel diameter of 1 in. (25.4 mm) and a die diameter of 1.500 in (38.10 mm)[26]. Because good granulation is important for compressing quality tablets, sufficient but not excessive amount of lubricant/glidant such as magnesium stearate or stearic acid would be incorporated into the formulation. The granules would then be ground and check flowability before and after sieved though two particular sieve sizes to assure when the technology is transferred to scale up in industry, homogeneous ingredient mixture may be flowed into the punch dies without segregation issue.

Proper choice of tableting tool is essential[26]. Therefore, other than fabrication of elegant products and reduction of manufacture loss, tablet shape would also have to be chosen carefully to enhance consumer compliance. A too long or too big tablet is hard to swallow. Therefore, whether selecting an oblong (also known as capsule shape) tooling set or a round tooling set for the ease of administration orally would be judged on the total weight of ingredients incorporated into compression into a single unit. Based on these reasons the pharmaceutical category and amount of an excipient used in the formulation would be carefully controlled. Three different sets of upper and lower punches and dies would be experimented: two round shapes and one oblong (Figure 9). The tablet being compressed by two different shapes: oblong tool set is described as capsule shape and convex cup. The configuration of the punches purchased to be used in this project is classified as the American TSM domed head with standard head flat and normal relief[26], which is suitable for all high-speed tablet pressers. The domed head configuration provides a smoother transition into the compression cycle of the table press, reducing stress and premature wear of the pressure roller. The capsule shape tool set contained Woodruff key[26].

## 5.2. Formulation Considerations

Forty-seven formulation candidates, not counting same composition with different compression pressures, were prepared with various excipient combinations and compressed drug mixture which had 325 mg of strength of niacin into hydrophobic core tablets (Table 8) with a Healthstar laboratory scale tablet press. The capsule shape (oblong) punch set had the length of 17.12 mm and the width of 7.59 mm which might compress tablet weight up to 900 mg into different thicknesses based on the amount of ingredient in each unit. The tablet strength of this innovative formulation was then reduced to 325 mg (Figure 9). The reason was that Polygel CR 500 mg tablets were not able to release its drug content completely at the end of *in vitro* dissolution study. If not absorbed in the small intestine, niacin drug molecules are believed to be less likely to enter circulation in the color because this region of GIT can only absorb nonpolar small molecules[29]. Niacin powder is a water soluble chemical[10] with pKa at 2.2 and log P at 1.0, which exists as ionic species in a basic pH medium.

The prepressure, depth and main pressure of the tablet press were adjusted to achieve a predetermined hardness[26]. The prepressure was set to remove air pockets present in the powder particles in the die. The main pressure was important to achieve the desired hardness which was not only important for drug release pattern, but also essential for package and shipment. The postcompressed hardness was confirmed by a hardness tester (Schleuniger-2E, Vector Co. Marion Iowa). Friability tests were also conducted with Erweka (D-33150 Heusenstamm Germany, Model TA 100) to make sure the candidate fulfilled the requirements set in the USP-NF[3], not losing more than 1% of tablet weight. Two potential candidates were further compressed into different hardness to compare drug release behaviors.

**Table 8. Ingredients and compositions of the innovative formulations**

| Ingredient Name | Function | 100 Tablets |
|---|---|---|
| Nicotinic Acid | Drug | 32.50 g |
| Carnauba Wax | Hydrophobic base | 12.57 g |
| Stearic Acid | Hydrophobic base/lubricant | 0.30 g |
| Carbomer 940 NF (Carbopol) | Release modifying agent | 0.61 g |
| Hydroxypropyl Methylcellulose (Methocel K4M CR) | Rate-controlling polymer | 3.03 g |
| Sodium Starch Glycolate (Explotab) | Disintegrant | 0.61 g |
| Povidone (Kollidon VA 64) | Binder | 0.87 g |

The innovative formulation was conducted by mixing niacin with carnauba wax and stearic acid and then heated on a heating plate until carnauba wax was melt (mp 82- 85.5°C)[2] and well blended with the other two ingredients. Immediately removed from heat, the rest of excipients of that particular formulation candidate were added and mixed (Table 8). Once cooled to room temperature, the solid mixture was transferred to a mortar and be grinded by a pestle to reduce particle size. The granules were then passed though sieve sizes 18. The granules larger than this size was returned to mortar for trituration and then sieved again.

After the particles were harvested, yield percentage was calculated to determine the amount required to have the desired niacin strength prior to compress into a core tablet. Prepressure, depth and main pressure were adjusted until the core tablet has less than 1% loss during friability test and owns the hardness of approximately 9.2 kiloponds (kp), 12.4 kp or 14.6 kp.

The Polygel CR tablets, a hydrogel and lipophilic matrix system, allows niacin quickly released into acidic media during the simulated fasting study as well as simulated fed study. But the drug release was much slower in phosphate buffer pH 6.8 (Figure 1c). The excipients controlling the drug release from the Polygel CR product (e.g. glycerol behenate, hypromellose) are pH independent. In the beginning of the dissolution study which was in the acidic medium, the niacin molecules present on and near the surface of the tablets dissolved into the medium quickly. As time progressed, the release rate slowed down due to the increased diffusion distance, which may explain why the release in acidic medium was a curve. But when the medium was adjusted into the pH 6.8 phosphate buffer, the release profile became a line with a slow rising slope instead of being continued to move upward in curvature. Although the reason is unknown, this observation is speculated to probably link with the charge interactions among tablet excipients, niacin ions, hydrogen ions, chloride ions, sodium ions, mono- and bi-basic phosphate ions present in the medium making the release of niacin from the inner part of the tablet through gel material at a more constant but slower rate.

## 5.3. Friability and Hardness Tests of Core Tablets

Friability and hardness tests will be conducted according to the guidelines of USP-NF[3], less than 1% weight loss in friability test. There is no specific requirement for tablet hardness, but 9 to15 kiloponds was chosen as the range for this project.

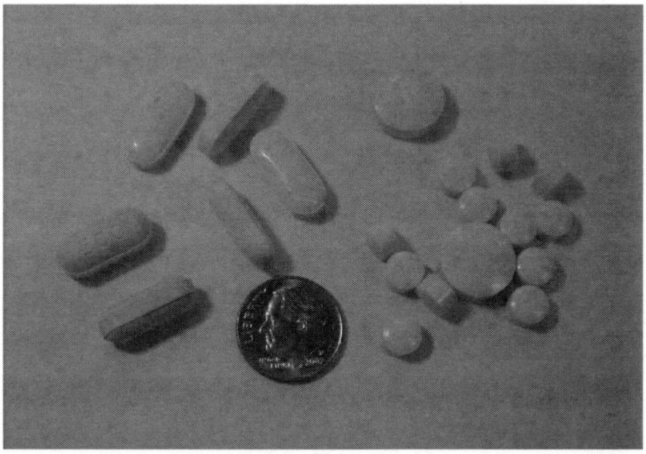

Figure 9. Reference tablet and products developed in this project from left to right: Polygel CR 500 mg tablets (pink, caplet shape), innovative formulation ID #45 of 325 mg (pale yellow caplet shape), innovative formulation ID #45 of 90 mg (small convex cup round shape, pale yellow) and innovative formulation ID #45 of 500 mg (large convex cup round shape, pale yellow). A US dime (ten cent coin) was included in the figure as the tablet size reference. Friability test conducted according to the guidance of USP-NF was found to fulfill the guideline for less than 1% weight lost. The tablet hardness of the innovative formulation candidates was 10-14 kiloponds.

# 6. *In Vitro* Dissolution and Disintegration Studies

The testing of the commercial references, Polygel CR 500 mg tables has been described thoroughly in Session 2. In brief, There are dissolution guidelines inside the niacin monograph, USP-NF[3] for immediate release niacin tablet, which stated to use 900 mL 0.1 N HCl with dissolution apparatus I, basket method, at the stirring rate of 100 rpm for 60 min. But there is no specific dissolution guideline for niacin CR tablet, because a CR dosage form generally does not finish release within one hour. Therefore, the dissolution guidelines in the General Chapter: Dissolution for the extended release dosage form in USP-NF[3] was adopted to be used for this project instead of the dissolution method listed in niacin monograph. The guidelines for the ER dosage form cited from the compendium was "The study tablet is placed in 750 mL 0.1 N HCl with basket stirring rate at 100 rpm for 2 h and then add 250 mL of 0.20 M tribasic sodium phosphate in the vessel and if necessary, with 2 N hydrochloric acid or 2 N sodium hydroxide to a pH of 6.8 ± 0.05." But the USP-NF[3] does not specify the test duration in pH 6.8 phosphate buffer as this guidelines are written as a general purpose for all extended release dosage forms. Therefore, the dissolution time in pH 6.8 in this project was chosen as 4 hour and stirring rate at 50 rpm. The retention time of a tablet in the small intestine in general is regarded as 3.5 to 4 h based on the literature information[15-17] constructed into Table 1. The passage of a tablet in this physiological region is not influenced very much by tablet size and presence of food. Samples were collected at 0.25, 0.5, 1, 1.5, 2, 2.5, 3, 4, 5 and 6 h. However, food effect is important for a CR solid dosage form in the stomach as this drug substance which pKa is 2.2. The environmental pH (pH 1.0-1.4 versus pH 2.5) affects the rate of drug release. Therefore, a single USP protocol cannot represent both the fasting state and fed state.

## 6.2. Our Proposed Fasting State Dissolution Method for Niacin CR Products

We proposed to conduct Polygel CR tablets experiment by using 500 mL of 0.1 N HCl in the USP dissolution apparatus, basket method operated at 50 rpm for 1.5 h, then adding 500 mL of 0.070M sodium phosphate tribasic, dodecahydrate ($Na_3PO_4 \cdot 12H_2O$) and if necessary, adjust with 2 N hydrochloric acid or 2 N sodium hydroxide to pH of 6.8 ± 0.05. The dissolution study continued at 50 rpm for another 4 h to simulate the fasting condition in the small intestine. Samples were collected every 30 min and assayed by a UV spectrophotometer (Hewlett Packard Model 8453) at wavelength of 261nm and 263 nm for acidic stage and pH 6.8 phosphate buffer stage, respectively.

## 6.3. Our Proposed Fed State Dissolution Method for Niacin CR Products

We proposed to conduct Polygel CR tablets by using 1000 mL of pH 2.2 HCl buffer to simulate food in the stomach in the USP dissolution apparatus, basket method operated at 100 rpm for 6 h to mimic the tablet residence time in the stomach after meal. After that, in order to simulated the fed condition in the small intestine, the apparatus speed reduced to 50 rpm and 15 mL of 0.289 M sodium phosphate tribasic, dodecahydrate ($Na_3PO_4 \cdot 12H_2O$) was added to bring pH up to 4.5 for 15 min to simulate duodenum pH. Then another 5 mL of 0.289 M sodium phosphate tribasic, dodecahydrate was added. Hydrochloric acid or sodium hydroxide of 2 N was used to adjust the pH to 6.8 ± 0.05. The experiment continued for another 3.75 h at the same stirring rate, 50 rpm. The collected samples were assayed by a UV spectrophotometer (Hewlett Packard Model 8453) at the wavelengths of 261 nm and 263 nm for pH 2.2 HCl buffer and pH 6.8 phosphate buffer.

## 6.4. Our Proposed Disintegration Methods of Niacin CR Products in Simulated Fasting and Fed States

Since the USP dissolution apparatus does not incorporate food weight and churning effect on drug release from a dosage form, the Polygel CR tablet release studied by using USP disintegration apparatus with a USP disk to mimic food weight effect[2]. The USP disintegration apparatus used in this project consisted of a basket-rack assembly, a 1000-mL beaker with immersion fluid maintained at 39±1°C by a water bath (Precision Scientific, Model 183, Chicago, IL) which is the known body core temperature. The test apparatus was calibrated first to ensure that the USP standard[3] had been met. That is, the basket raised and lowered in the 800 ml of deionized water as the immersion fluid at 29-32 cycles per minute[3]. The volume of the fluid in the vessel allowed at the highest point of the upward stroke the wire mesh remains at least 2.5 cm below the surface of the fluid and descended to not less than 2.5 cm from the bottom of the vessel on the downward stroke[3].

### *6.4.1. Simulated Fasting State*

Eight hundred mL of preheated (39±1°C) deaerated 0.1 N HCl was used as the test immersion fluid mimicking a tablet present in an empty stomach. No disk was applied on a test tablet. At the end of two hours, the apparatus was paused. The basket with a Polygel tablet was quickly switched to another beaker containing 800 mL of deaerated and preheated to 39°C phosphate buffer pH 6.8 to mimic the intestinal fluid. The apparatus was resumed for 4-h operation. Samples collected in the acidic phase were assayed by a spectrophotometer (Hewlett Packard Model 8453) at 261 nm wavelength and samples collected during the buffer stage were quantified at 263 nm.

### *6.4.2. Simulated Fed State*

Eight hundred mL preheated to 39±1°C deaerated of pH 2.2 HCl buffer was used as the test immersion fluid mimicking a tablet present in a fed stomach. The study was conducted for 6 h or until the tablet was totally disintegrated and no residue left on the mesh of the USP disintegration apparatus basket tube where the tablet was dropped onto in the beginning of the study.

## 7. FINDINGS OF *IN VITRO* DISSOLUTION AND DISINTEGRATION TESTS

The current USP dissolution method[3] does not reflect the drug release in either fed or fasting state, because the test period was two hours in acidic medium at the stirring rate of 100 rpm and 4 hours in pH 6.8 buffer. Therefore, the suggestion of new *in vitro* dissolution methods for simulated fasting and simulated fed states are indeed needed. The experimental results of all three methods are described as follows.

### 7.1. Release Profiles of Innovative Niacin CR Tablets Using the USP Dissolution Method

When the tablets of 325 mg ID #45 and #47 were subjected to the USP dissolution method (0.1 N HCl, 100 rpm for 2 h and then pH 6.8 phosphate buffer 50 rpm for 4 h), the release profiles were much slower than the Polygel 500 mg CR tablets throughout the entire test period (201.2±10.6 mg, 192.5±2.7 mg versus 256.6 mg at 6 h; Figure 10a). These innovative formulations may reduce the concern of undesired side effects such as cardiovascular dilation, intolerance and flushing due to the less amount of drug release in time course. Powder mixture of the same formulation was compressed into tablets by two main compression pressures, 9 kp and 14 kp, to study the difference in *in vitro* drug release. The batches of tablets that held the same formulation but had the tablet hardness of 9 kp released its active ingredient slightly faster in phosphate buffer pH 6.8 than the batches with hardness of 14 kp (218.8±4.6 mg versus 192.6±2.7 mg release at 6 h, n = 3). However, drug release from the tablets made of either hardness showed no remarkable difference in the acidic stage (Figure 10b).

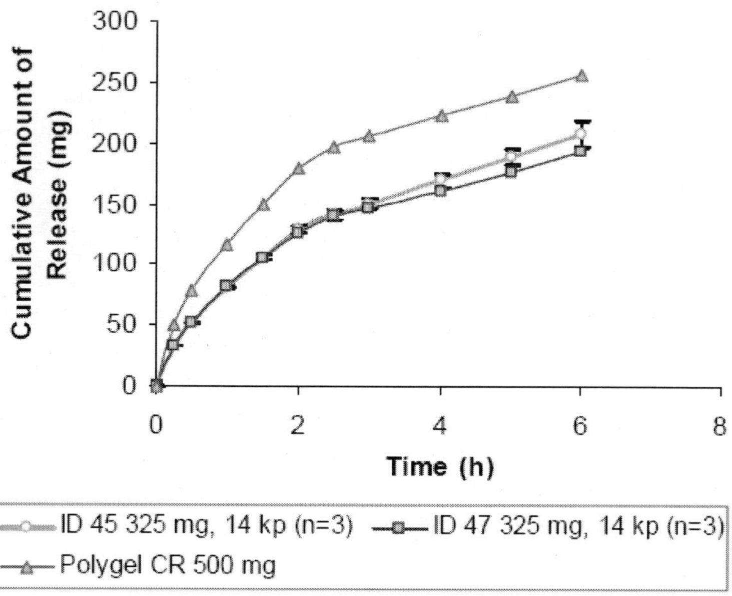


a

Figure 10 (Continues)

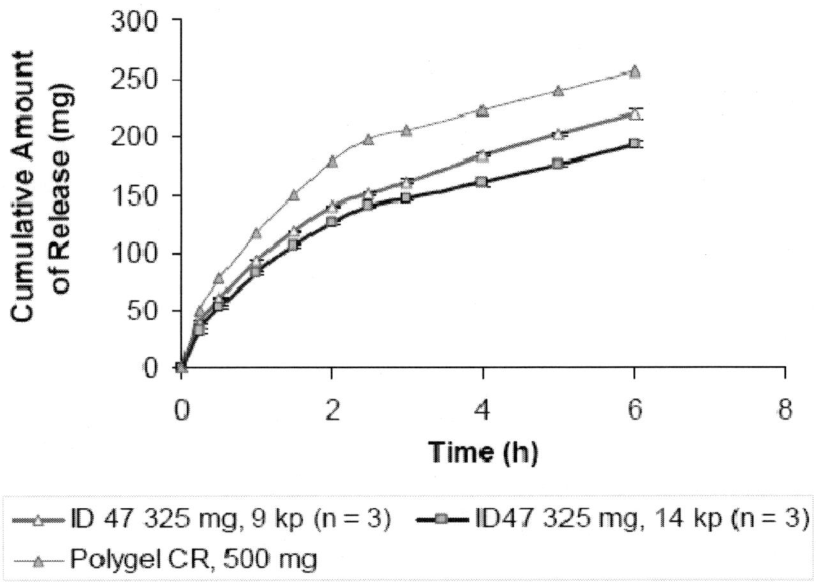


b

Figure 10. The *in vitro* dissolution profiles of two 325 mg innovative formulations and the reference tablet, Polygel CR 500 mg tablets. The releases from both formulations of ID #45 and #47 were much slower than that of Polygel CR tablet throughout the entire test period (Figure a). Tablets of 9 kp hardness released active ingredient slightly faster than those of 14 kp in phosphate buffer pH 6.8, but no difference in the acidic stage (Figure b).

## 7.2. The Release Profiles of Two Laboratory Innovative Niacin CR Tablets Using Our Proposed Simulated Fasting State and Simulated Fed State Dissolution Method

When the tablets of 325 mg ID #45 and #47 were subjected to the simulated fasting state dissolution method as we proposed (500 mL of 0.1 N HCl at 50 rpm for 1.5 h and then 1000 mL of pH 4.0 phosphate buffer at 50 rpm for 15 min prior to progress to pH 6.8 phosphate buffer at 50 rpm for another 3.75 h), the release profiles were slower than the Polygel CR 500 mg tablets throughout the entire test period (200.5±8.1 mg, 194.0±7.6 mg vs. 238.3±4.9 mg at 6 h; Figure 10a). No significant difference in release was observed between these two innovative formulations. Again, the innovative formulations may lessen the concern of the adverse effects of niacin, such as cardiovascular dilation, intolerance and flushing due to less amount of the drug loaded (325 mg) in the tablets of innovation. The release profile of ID #45 was more linear than that of ID #47 as well as that of the Polygel CR 500 mg tablets, judging from the $R^2$ values of linear regression in Figure 10a and Table 9.

When the tablets of 325 mg ID 45 and 47 were subjected to our proposed simulated fed state dissolution method (1000 mL of pH 2.2 HCl buffer at 100 rpm for 6 h and then 1000 mL of pH 4.0 phosphate buffer at 50 rpm for 15 min prior to progress to pH 6.8 phosphate buffer at 50 rpm for another 3.75 h), the release profiles were not different from that of Polygel CR 500 mg tablet throughout the entire test period (265.3±11.6 mg, 261.8.0±12.0 mg vs.

298.8.3±33.3 mg at 10 h; Figure 11b). There were no significant difference in release between ID 45 and ID 47. Again, Polygel CR tablet never finished releasing its 500 mg even the dissolution time was extended to 10 h (Figure 11b). Therefore, there is questionable need to dose a patient with such a high strength to risk the remaining 200 mg of niacin irritates the colon. The rationale is the absorption of ionized drug molecules is very poor. According to Henderson–Hasselbalch equation Niacin, a rather acidic drug, is ionized in the colonic medium which pH is greater than 7.0 in the most region. Giving 325 mg is surely enough.

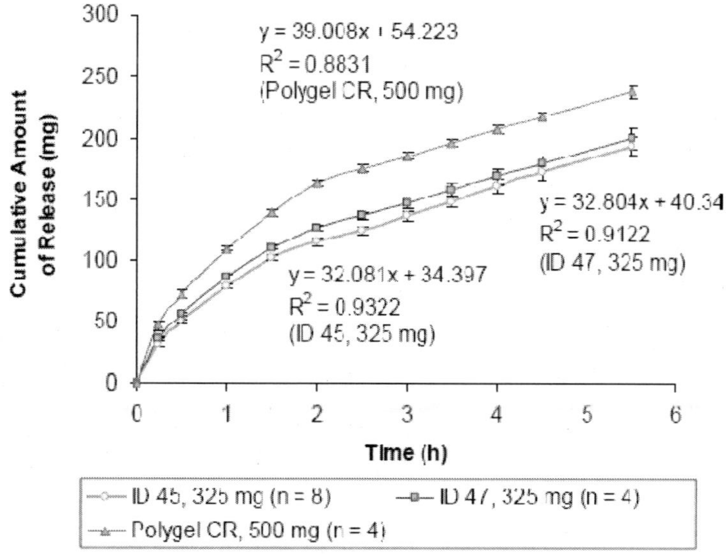


a

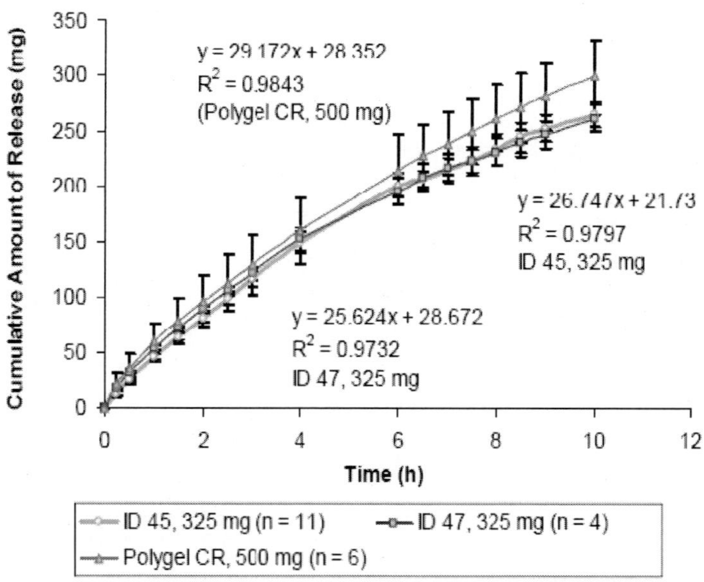


b

Figure 11. The release profiles of 325 mg innovative formulations, ID #45 and ID #47, using our proposed simulated fast state (Figure a) and simulated fed state (Figure b) dissolution methods. The Polygel CR 500 mg tablets served as the reference.

## 7.3. The Release Profiles of Innovative Niacin CR Tablets Using Our Proposed *In Vitro* Disintegration Method

The USP dissolution apparatus is important for formulation development and quality assurance and quality control monitor. However, this machinery either spinning a basket (USP Dissolution Apparatus 1) or spinning a paddle (USP Dissolution Apparatus 2) does not yield the best model for the fed condition. Formulation ID #45 was thus evaluated with our laboratory proposal. We modified the USP disintegration apparatus by placing an USP disk (Figure 2b) lay flat on top of the test tablet to mimic food weight effect[18].

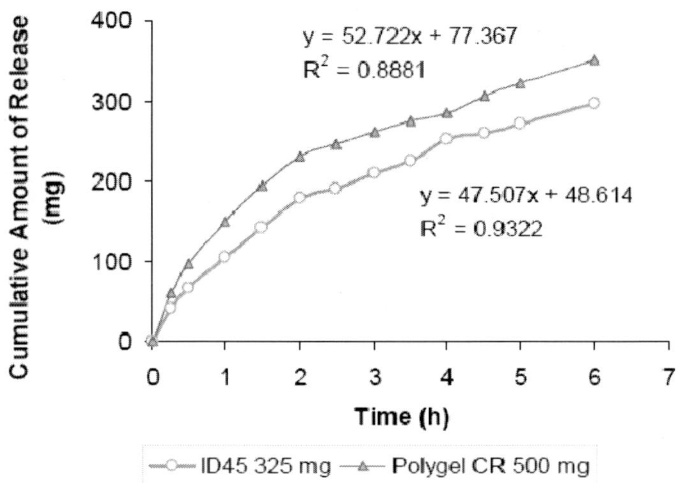


a

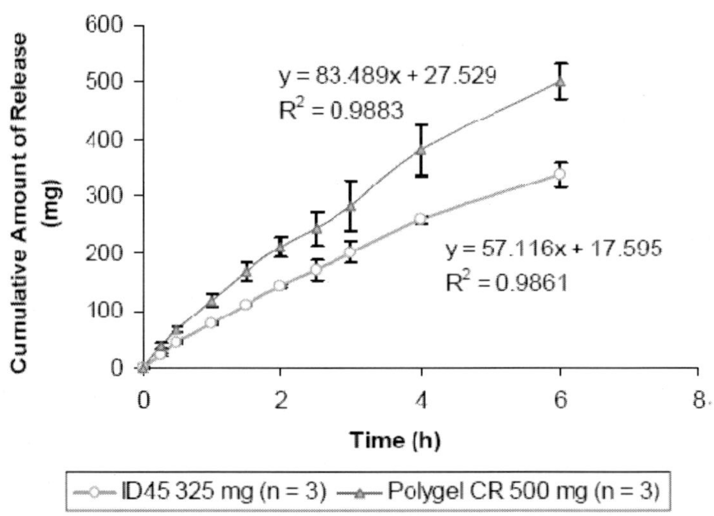


b

Figure 12. The release profiles of innovative niacin CR tablet ID #45 using our proposed simulated fasting state disintegration method (Figure a), and simulated fed state disintegration method with a 3.05-g USP disk lay flat on top of the test tablet to mimic food weight effect (Figure b).

In simulated fasting state Polygel CR 500 mg tablets released 56% in 6 h while the innovative formulation ID 45 released 87% in 6 h (Figure 12a). In simulated fed state the innovative ID 45 tablet released niacin completely in 6 h (103.8%), so did Polygel CR tablet 500 mg (100.1%) (Figure 12b). Since the two release profiles in simulated fasting state were similar to each other as seen in Figure 11a, the FDA similarity factor, $f_2$ value, was also computed. The FDA similarity factor revealed that these two disintegration profiles (Figure 12a) were similar, except the time point at 6 h, which $f_2$ value was calculated as 48.8.

**Table 9. The correlation coefficient, $R^2$ values, of the dissolution and disintegration release profiles to simulate the administration of the innovative niacin tablet (#45, 325 mg) in fasting and fed states to compare with those of the reference, Polygel niacin 500 mg CR tablets**

| | Dissolution | | Disintegration | |
|---|---|---|---|---|
| | Polygel CR Niacin Tablet | ID #45 | Polygel CR Niacin Tablet | ID #45 |
| Simulated Fasting | 0.8831 | 0.9322 | 0.8881 | 0.9322 |
| Simulated Fed | 0.9843 | 0.9797 | 0.9883 | 0.9861 |

## 8. Tablet Accelerated Stability Study and Drug Content Assay

Accelerated storage testing is invaluable in identifying stable formulation candidates[27, 28]. After the innovative formulation was achieved, that is, the candidate owned a near zero order release profile in all study methods: the USP method, our proposed dissolution methods for both fasting and fed states and disintegration method, multiple units were prepared, placed into a 50 mL non-pyrogenic polypropylene centrifuge tube (Corning Inc., Corning NY) and subjected into 40°C and 60°C incubators for stability study. At a predetermined time point, tablets were taken out from the stability chamber. One at a time, they were crushed and ground into a mortar with a pestle into fine powder and then transferred into 1000 mL of 0.1 N HCl to dissolve. UV absorbance of the filtered solution was assayed to determine the percent of drug remained in the tablet. The innovative formulation was considered stable and might be projected into a shelf life of two year being in a 60°C incubator for 3 months or 40°C for 6 months and still had the average of more than 90% label claim.

The niacin inside the innovative ID 45 tablets (325 mg) showed stable in both 40°C (Figure 13a) and 60°C up to 6 months (Figure 13b). Both still had approximately 105% of the active ingredient remained after 110% of 325 mg of niacin was loaded to make into each tablet 6 months ago. The mp of niacin is 236.6°C, which is much higher than the two temperatures, 40°C and 60°C, used to conduct the accelerated stability study.

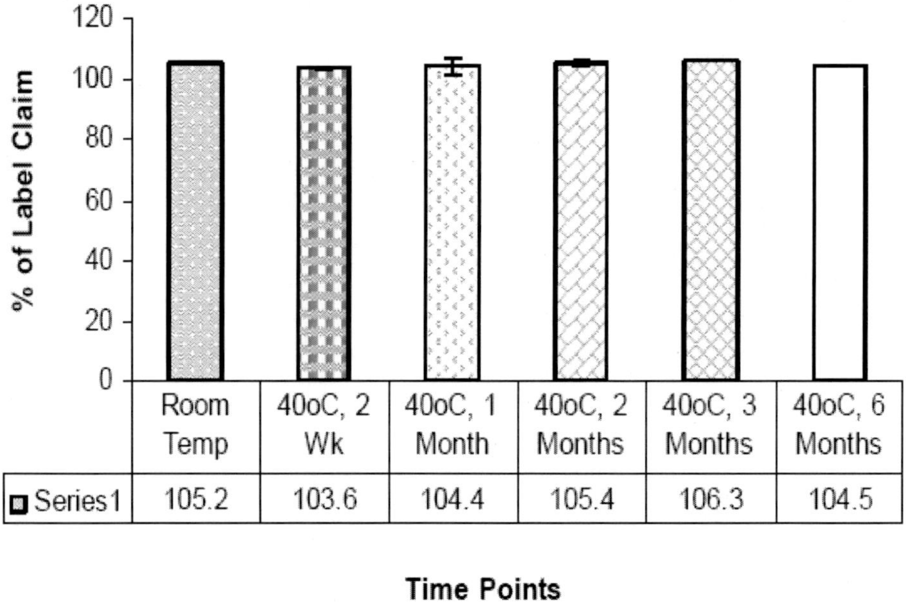


a

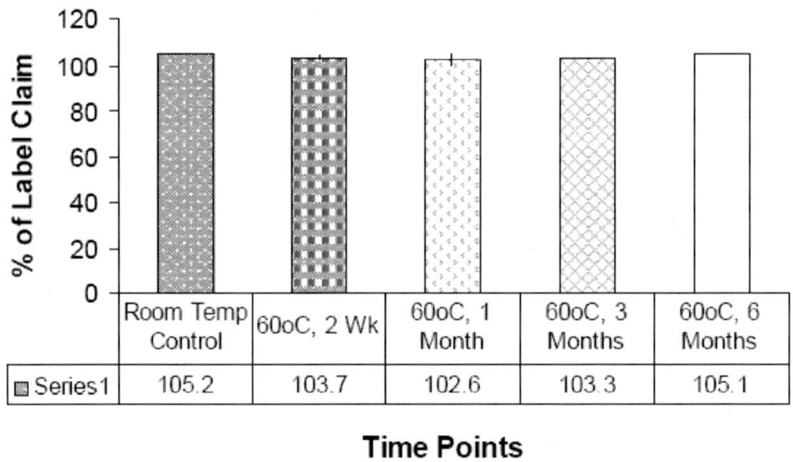


b

Figure 13. Stability study of innovative niacin 325 mg tablet (ID #45) in (a) 40°C, and (b) 60°C up to 6 months (n=3).

## SUMMARY

Two novel oral dosage forms of niacin were achieved: 250 mg ER capsules and 325 mg ER tablets. The optimal formulation, ID #37 (250 mg in strength) is in hope of fewer adverse effects include skin flushing and gastrointestinal distress, in percent of release, linearity of release, and slower release profile. The *in vitro* release study for this innovative CR 250 mg capsules which had been stored in 40°C up to 6 months may be projected as the shelf life in room temperature up to two years.

An innovative 325 mg CR tablet dosage form was also developed by melting lipophilic carnauba wax and minimal amount of stearic acid to embed hydrophilic niacin USP powder. After being removed from a heat plate, the preparation was further mixing with Carbopol® 940, Methocel® K4M CR, Explotab® and Kollidon® VA64. The granules were then transferred to a mortar to reduce in size and compressed into tablets containing 325 mg strength. The niacin release from such an innovative tablet was near zero-order in the simulated fasting and fed dissolution studies and disintegration experiments. Why was only 325 mg of drug loaded instead of being 500 mg? The human colon allows the permeation of small nonionic molecules only[16, 29]. Even if niacin may be released into the colon from an oral solid dosage form, it will become a charged species in the medium of this physiological region and is less likely be absorbed from the colonic lumen. Therefore, based on the data of *in vitro* dissolution study whether it was simulated fasting (6 h) or simulated fed state (10 h), there is no need to load the innovative CR tablet developed in this project with niacin in 500 mg strength. Technically it is achievable to design a formulation which may release the entire 500 mg before the tablet reaches the colon by reduction in the compression force of tableting or incorporation of a superdisintegrant into the formulation. The consequence for such attempts would be high amount of niacin released and dissolved in the gastric medium which leads to even more prominent side effects as the manufacturer of Polygel CR tablets has warned the consumers, such as gastrointestinal distress and skin flushing[6-9].

The choice of a particular excipient was considered up to the particle size due to the procedure of granulation involved. For example Avicel® PH 200 (which has a mean diameter of 180 micron) was used in the formulation work of this project instead of Avicel® PH 102 (90 micron) or Avicel® PH 101 (50 micron). Avicel® PH 200 is larger in its particle size and will have less chance to segregate out from carnauba wax-niacin granules during mixing and during tablet die filling. In addition, this chemical (Avicel) is known to have better tableting property due to its good flowability. The Hypromellose 2208 used in formulating ID #45 and ID #47 tablets were Methocel® K4M CR, which viscosity was 4000 centipoises (cps) when was prepared into 2% aqueous solution. Another two excipients used in this project to prepare some candidates were Precirol® ATO and Compritol® 888 ATO. These two compounds are novel materials with lubricant, glidant and putative binding action[30]. But the releases of these candidates in acidic pH were similar to that of Polygel CR niacin tablet. However, they may be of importance in other formulations.

The test results for angle of repose were reported to be very dependent upon the method used[3], such as funnel height, amount of powder test and particle size distribution and powder cohesivity. Despite the difficulties, the method of measuring angle of repose continues to be used in the pharmaceutical industry, and numerous examples appeared in the literature demonstrating its value in predicting manufacturing problems[3]. Compressibility index and Hausner ratio are generally two acceptable methods of flowability[3]. Never the less, deviations caused by measuring these two flowability parameters of our laboratory innovative niacin ER tablets were possibly present among the repeated measures: (1) the diameter of the cylinder used to measure, (2) the number of times the tablets were tapped to achieve the tapped density, (3) the mass of material used in the test, and (4) rotation of the sample during tapping. These variations may be in better control if a tapping machine with the preset tapping force and duration is applied when the product is in the industrial scale-up stage.

Literature suggests that niacin tablet be administered with food to maximize bioavailability and minimize gastrointestinal intolerance, because peripheral vasodilation is seen within 20 minutes after taking an immediate-release product and may last for up to 1 h, while the rate of onset of vasodilation is slower with sustained-release forms and may attenuate the severity of flushing, a side effect of large dose[6]. But this public advice is contrary to our finding. Our disintegration study using an USP disk indicated that simulated food weight significantly speeds up the release of niacin from a Polygel CR tablet (Figure 4). The possible explanation is that Polygel is a hydrogel which absorbs water. The swollen surface as water diffused in allows the active ingredient to diffuse out. But the simulated food weight caused by the weight of an USP disk pounding upon a tablet brushed away the fragile softened gel material abundant with water and thus continued to expose the tablet core to the medium for more water to diffuse in and more active ingredient, niacin, to diffuse out due to the shortened pathway. Thus, the entire amount of active ingredient in a commercial Polygel CR 500 mg tablet and the innovative niacin formulation ID #45 (325 mg) was found completely released within 6-h test period whether it was in simulated fasting study (2 h in gastric and 4 h in small intestine media) or simulated fed study (6 h in gastric medium) (Figure 12). This observation opposites from the advice[6] that Polygel CR 500 mg tablet be taken with food or pre-administered with an aspirin tablet because niacin belongs to the Biopharmaceutical Classification System (BCS) Class I compounds. Since a CR dosage form would mostly likely to encounter food, reduction of the strength of ID 45 niacin CR tablet from 500 mg to 325 mg is a strategy to prevent uncomfortable flushing and irritating effect.

Accelerated storage testing is invaluable in the development of stable formulations. It is important to be aware of a limitation of this technique, that is, the order of reaction in higher temperature is no different from that in room temperature. Light, oxygen and moisture content are also factors to take count of. The innovative formulation has been stable in both 40°C and 60°C up to six months (Figure 13). The results imply that not only niacin is stable in these two tested temperatures, but there was also no interaction between niacin and excipients or interaction between niacin and the package containers that we used to store the innovative tablets which was made of polypropylene.

The FDA similarity factor, $f_2$ value, was only used once in this project to compare a set of dissolution profiles (Figure 12a) because it was hard to judge by observation alone whether ID #45 was different from the Polygel CR tablets in the simulated fasting state disintegration study. As guided by the U.S. FDA, this similarity factor equation is used to compare the similarity of two release profiles between a test product and the reference based on the percentage of release at time points. A question raised here is that whether FDA similarity factor can be used to compare a test product having the strength of 325 mg with the reference tablet which containing 500 mg of drug. Or the strengths of the two products have to be the same to use the equation? Can the amounts of drug release at different time points be used since it is not logical to use the percentage of drug release when the two products to be compared contain different strengths? Hope the topics discussed in this book chapter lead to more research in the future.

## ACKNOWLEDGMENTS

Author would like to thank the Faculty Development Committee and the Summer Undergraduate Research Committee of Massachusetts College of Pharmacy and Health Sciences for funding this research project.

## REFERENCES

[1] Jeong SH, Fu Y, and Park K. Hydrogels for Oral Administration. Kwon YM, Kim SW. Thermosenstive Biodegradable Hydrogels for the Delivery of Therapeutic Agents. Kikuchi A and Okano T. Stimuli-Sensitive Hydrogels. In: Polymeric Drug Delivery Systems; Kwon GS, Ed.; Taylor & Francis Group, Boca Raton, FL, 2005, p. 196, 251, and 275.

[2] Rowe RC, Sheskey J, Quinn M.E. Handbook of Pharmaceutical Excipients. $6^{th}$ ed., Pharmaceutical Press, Grayslake, IL, 2009, p. 111, 327, 517, 535, 579, 587, 595, 772 and 779.

[3] Disintegration (Chapter 701), Dissolution (Chapter 711), Optical Microscopy (Chapter 776), Particle Size Distribution (Chapter 786), and Powder Flow (Chapter 1174), Buffer Solutions, Hypromellose Monograph and Niacin Monograph In: *United States Pharmacopeia and National Formulary USP*34–NF 29; The United States Pharmacopeial Convention, Inc., Rockville, MD, 2011; pp. 266-267, 278, 304, 308, 639, 813-4, 2799, 3096, 3650.

[4] Bamba R and Scanu AM "Niacin and Lipoprotein (a): Facts, Uncertainties, and Clinical Considerations." Am J Cardiol 2008; 101 (8A): 44B-47B.

[5] Dale MM, Moore PK, Rang HP, and Ritter JM. Pharmacology. Scotland: Elsevier Science Ltd, 2003; 309-313.

[6] Davidson M. Niacin Use and Cutaneous Flushing: Mechanisms and Strategies for Prevention. Am J Cardiol 2008; 101 (8A): 14B-19B.

[7] Marks JW. Niacin. MedicineNet, Inc. Website. http://www.medicinenet.com/niacin-oral/article.htm. Accessed Oct. 15, 2011.

[8] Sanborn KD. Physician Desk Reference, $62^{th}$ ed. SloNiacin Tablet, Niaspan ER Tablet, Montvale New Jersey: Thomson, 2008, pp. 1759, 1768, 3279.

[9] MicroMedex. "DRUGDEX® Evaluations – Niacin". New Hersey: Thomson Healthcare, 2008, <http://www.thomsonhc.com/hcs/librarian/ND_T/HCS/ND_CPR/SearchByCategory/ND_PR/Drugs/CS/532650/DUPLICATIONSHIELDSYNC/7BBE60/ND_PG/PRIH/ND_B/HCS/SBK/1/ND_P/Drugs/PFPUI/b8Y1pS2rM9KA5/PFActionId/hcs.common.RetrieveDocumentCommon/DocId/1278/ContentSetId/31/SearchTerm/niacin>.

[10] Budavari S. The Merck Index: An Encyclopedia of Chemicals, Drugs and Biologicals. New Jersey: Whitehouse Station: Merck & Co., Inc.; 2001, p. 6552.

[11] Lemke TL, Williams DA, Roche VF and Zito SW. Foye's Principles of Medicinal Chemistry. Philadelphia: Lippincott Williams & Wilkins, 2008. p. 1349.

[12] Kramer, J.; Grady, L. T.; Gajendran. J. Historical Development of Dissolution Testing. Gray, V. A. Compendial Testing Equipment: Calibration, Qualification, and Sources of

Error. Klein, S.; Stippler E.; Wunderlich, M.; Dressman, J. Development of Dissolution Tests on the Basis of Gastrointestinal Physiology. In *Pharmaceutical Dissolution Testing*, Dressman, J. J, Kramer, J., Eds.; Taylor & Francis Group: New Your, 2005; pp. 8, 43, 90-93, 215, and 335-6.

[13] Wong, G.; Collins, C. C. Dissolution Testing. In *Preformulation in Solid Dosage Form Development*; Adeyeye, M. C., Brittain, H. G., Eds.; Informa Healthcare USA Inc.: New York, 2008; pp 478–480.

[14] www.usp.org

[15] Gibson M. Pharmaceutical Preformulation and Formulation, Interpharm/CRC Baco Raton, FL, 2004, p. 108.

[16] Washington N, Washington, C.; Wilson, C. G. Physiological Pharmaceutics. NY: Taylor and Francis, 2001, p. 95, 98-101, 123-8, 149, and 157.

[17] Avdeef, A. Physicochemical profiling (solubility, permeability and charge state). *Curr. Topics Med. Chem.* 2001, *1* (4), 277–351.

[18] Chuong MC, Lee R, Saxena V and Palugan L. Examining Whether USP Disintegration Apparatus with a Disk Is Superior To USP Dissolution Apparatus in Predicting Drug Release During Simulated Fed State. Dissolution Technologies. 2009; 16(1):25-32.

[19] Chuong MC and Palugan L. Innovative Sustained and Modified Release Niacin Tablets: Formulation and Drug Release Patterns. Massachusetts College of Pharmacy and Health Sciences, Faculty Development Grant Report; 2008.

[20] Lantz RJ. Size reduction. In: Liberman HA, Schwartz JB, eds. Pharmaceutical Dosage Forms, 2nd ed. Vol. 2. NY: Marcel Dekker, Inc., 1990, p. 107-200.

[21] Randall CS. Chapter 6: Particle Size Distribution. In: Physical Characterization of Pharmaceutical Solids. New York: Marcel Dekker, Inc.; 1995. p. 158 and 170.

[22] Hoag SW, Lim HP. Particle and Powder Bed Properties. In: Pharmaceutical Dosage Forms: Tablets. Volume 1: Unit Operations and Mechanical Properties. NY: Informa Health care USA. Inc., 2008, p. 51-2, 59.

[23] Newman AW, Brittain HG. Chapter 5: Particle Morphology: Optical and Electron Microscopies. Randall CS. Chapter 6: Particle Size Distribution. In: Physical Characterization of Pharmaceutical Solids. NY: Marcel Dekker, Inc., 1995, pp. 130, 158 and 170.

[24] http://www.labrecyclers.com/product.html?InventoryID=1156

[25] http://www.gattefosse.com/node.php?articleid=176?

[26] Natoli D. Tooling for Pharmaceutical Processing. In: Augsburger LL and Hoag SW, eds. Pharmaceutical Dosage Forms: Tablets, Volume 3: Manufacture and Process Control. Edited by. Informa, New York, 2008, p. 2, 7, 12.

[27] Cha J, Ranweiler JS, Lane PA. Stability Study. In: Ahuja S, Scypinski S, eds. Handbook of Modern Pharmaceutical Analysis. San Diego: Academic Press, 2001: 445-483.

[28] Florence AT, Attwood D. Drug Stability. In: Physicochemical Principles of Pharmacy. Pharmaceutical Press, London, 2006, p. 127-8.

[29] Wilson CG. Colonic Drug Delivery. In: Rathbone MJ, Hadgraft J, Roberts MS, Lane ME eds. Modified-Release Drug Delivery Technology, Volumes 1, Informa Healthcare, New York, 2008, p. 288-9.

[30] http://www.gattefosse.com/document-center/. Accessed November 27, 2011.

Hydrogels: Synthesis, Characterization and Applications
Editors: F. V. Câmara and L. J. Ferreira
ISBN: 978-1-61942-842-3


*Chapter 5*

# PROTEIN-BASED HYDROGEL FROM FISH INDUSTRIALIZATION WASTES

***Carlos Prentice*[*] and Vilasia Guimarães Martins***
Federal University of Rio Grande, School of Chemistry and Food,
Rio Grande RS, Brazil

## ABSTRACT

Chemical modification of proteins has been largely used with the subject of expose functional groups "hidden" before into the protein chains, determinate functional groups of enzymes and also help in understanding of drug action. Animal proteins are rich in lysine, which is an essential amino acid with many functional properties, to the opposite of vegetal proteins, such as soybean. The fish proteins tend to be very big, with an inherent capacity of water uptake. For the application in hydrogel, the physical properties of the fish polymers are more superior to vegetable polymers. Hydrogel is a light and wet class material, whose properties depend on the polymer net constructed and the water content. Proteins permeability of different molecular weight in a biodegradable hydrogel has been researched. Many synthetic polymers have limited structural and functional properties, while the natural polymers are unique, with various functional and structural properties. The mechanical properties and swelling properties of hydrogel made by using natural polymers can be improved by chemical or physical modification of the functional groups. Modification does not alter the biodegradable and biocompatible characteristics of the proteins. Therefore, if suitable functionality can be imparted, the natural polymers could replace some potentially toxic synthetic polymers, which are unsuitable for many biological applications. The hydrogel with high water retention capacity has a great number of attributes that make them attractive in several applications. The basic property of uptake water has been suggesting the use of absorbent materials in many applications, such as towel paper, surgery sponge, meat tray, bath rug and medical wound. One of the greatest challenges that confront the science nowadays is the development of a new generation of biomaterials for human organism repair and also to combat environmental problems relate with pollution.

[*] Email: dqmprent@furg.br.

# INTRODUCTION

For the synthesis of a protein hydrogel, there is a long way to travel, which is composed to many steps. First of all, it need the protein, preferably a isolated or hydrolyzed protein, after it suffer a chemical modification for expose functional groups that were inaccessible before, the next step is the crosslinking that will form a net and then it has a protein hydrogel with superabsorbent properties. The process will be described in more details along the chapter. Because of their hydrophilic character and potential for biocompatibility, hydrogels have been of great interest to biomaterial scientists for many years (Crescenzi et al., 2000). According to Yanas et al. (1989), in the 1980s, was demonstrated the successful application of natural polymer hydrogels as artificial burn dressings. After these, hydrogels have become especially attractive to tissue engineers as matrices for regenerating a wide variety of tissues and organs (Hubbell et al., 1998; Kim et al., 1999).

# PROTEIN STRUCTURE

The proteins are complex macromolecules, composed of amino acids, and necessary for the chemical processes that happen in the alive organisms. They are the basic representatives of the life, so much that your name flows of the Greek word “proteios”, that means “in first place”. In the animals, the proteins correspond about 80% of the weight of the dehydrated muscle, about 70% of the skin and 90% of the dried blood. Proteins are complex nitrogen substances with high molecular weight, which originate from amino acids from chemical or enzymatic hydrolysis. The proteins are also called polypeptide, because the amino acids that compose are them united for peptide bonds (a peptide bond is the union of the group amino (-$NH_2$) of an amino acid with the group carboxyl (-COOH) of another amino acid, through the formation of an amide, as display the Figure 1. The bond C-N, in a peptide, is special, because this is 10% more “short” than a union normal C-N, and the connection angle is also different from the expected for a carbon $sp^2$. This because the peptide bond, actually, presents a resonance structure, tends a fort character of double bond. Through these connections, the amino acids form long chains; most of the proteins have more than 200 amino acids. All the amino acids have, in common, a linked carbon to a group amino and a group carboxyl (Fennema, 1993).

A chemical hydrolysis is realized by breaking peptide bonds with acid or alkaline. The importance of proteins in the production of superabsorbent hydrogel is the amount of lysine that these proteins have.

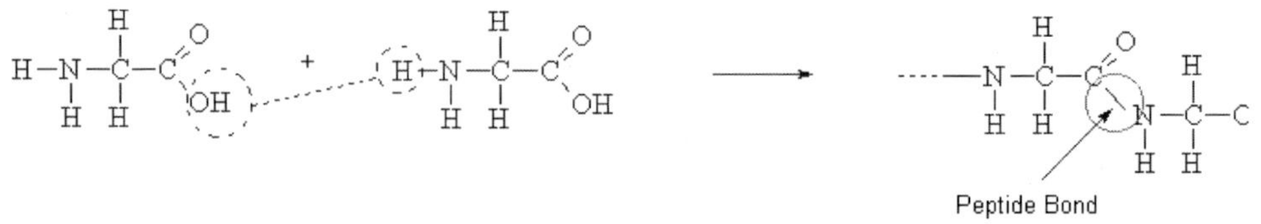


Figure 1. Formation of peptide bond.

Animal proteins are rich in lysine, which is an essential amino acid with many functional properties, unlike vegetal proteins, such as soybean. The fish proteins tend to be very large, with an inherent capacity of water uptake. Rathna and Damodaran (2002) reported that for the application in hydrogel, the physical properties of the fish polymers are higher than vegetable polymers. The lysyl residues (–$NH_2$) of proteins can be modified with carboxylic groups by reacting with ethylenediamine tetraacetic dianhydride (EDTAD) (Rathna e Damodaran, 2001).

## ENZIMATIC PROCESS

The biochemical hydrolysis for the production of protein hydrolysates is accomplished by the use of enzymes that hydrolyzed the peptides bonds. This can already be realized through proteolitic enzymes presents in the material, for example, viscera or muscles of the fish (endogenous proteins), or for the addition of enzymes of other sources (Kristinsson and Rasco, 2000).

Enzymes are essential in the biochemical catalysis, because they accelerate the chemical reactions among the organic representatives and the interior of the cells that it would delay a long time to complete in another way. The enzymes are used to accomplish functions wanted in the processing and analysis and to facilitate the conversion of by-products in products of high quality (Richardson and Hyslop, 1984). The enzymes get to execute this activity because the active site of an enzyme is highly specific for certain substrate. The enzymes catalyze only a reaction and specific function, by the formation of a complex with a substrate which will modify (Kristinsson and Rasco, 2000).

The enzymatic hydrolysis of the proteins is a complex process due to the many peptide bonds and your specific accessibility for enzymatic reactions (Linder et al., 1995). The specificity of the enzymes is not the only factor that affects the profile of the peptides in the final product. Environmental factors, such as temperature and pH also are extremely important, because they affect extensively the kinetics of the enzymatic reactions, and the effect of these factors is different for each enzyme. Usually, there is an optimum combination of pH and temperature, where the enzyme is more reactivates. Temperatures and extreme pH disable the enzymes by the denaturation of the same ones (Kristinsson and Rasco, 2000).

In the enzymatic hydrolysis, the capacity of the proteases breaking the peptide bonds is dependent of the physical interactions between the substrate and the enzyme in the environment aqueous present during the hydrolysis. The great portion of hydrophobic amino acids resides inside of hydrophobic area of the peptide chain of the raw materials, and it is for this reason that the access of the proteases in those areas can be limited (Chothia, 1974). Therefore, the fractions with insoluble peptides can present high proportion of hydrophobic amino acids when compared with the protein hydrolyzed (Liaset and Espe, 2008).

Proteolysis means enzymatic hydrolysis of peptide bonds in proteins. In that reaction, for each peptide bond cleaves by the enzyme, release a mol of carboxyl group and a mol of amino group. When it is allowed the reaction to be completed, the final product is the mixture of all the constituent amino acids of the protein. The incomplete proteolysis results in the liberation of a mixture of polypeptides coming from the original protein. The functional properties of protein hydrolysate depend on the degree of hydrolysis and of the physical-chemical properties, that is, size, solubility, etc., of the hydrolyzed polypeptides. The degree of hydrolysis is defined as the fraction of peptides bonds cleaves (Damodaran et al., 2010).

The degree of hydrolysis is monitored using the method of the pH-Stat. The beginning to that method is that, when the peptide bond is hydrolyzed newly, the carboxyl group formed-recently ionize entirely in a pH>7 and release an ion $H^+$. As result, the pH of the protein solution decreases progressively with the time of hydrolysis. In the pH among 7-8, the number of mols ions $H^+$ released is equivalent to the number of mols of the hydrolyzed peptide bonds. In the method of the pH-Stat, the pH of the protein solution is maintained in constant pH by the titulation with NaOH. The number of mols of NaOH consumed during the proteolysis is equivalent to the number of mols of break peptides bonds.

Several proteases can be used in the preparation of protein hydrolysate. Some of those proteases are enzymes of specific sites (Table 1). Because of your specificity, the types of fragments polypeptides released in the hydrolyzed can be different among the proteases. Alcalase originating from *Bacillus licheniformis* is to main commercial enzyme used in the production of the protein hydrolysates (Damodaran et al., 2010).

The microbial enzymes, when compared with enzymes from animals and plants, it offer several advantages, including a wide variety of catalytic activities available, and great stability in different pHs and temperatures. The resulting protein hydrolysate will present peculiar properties in agreement with the new generated peptides (Guerard et al., 2002).

**Table 1.Specificity of some proteases for hydrolysis**

| Protease | Type | Specificity |
|---|---|---|
| Elastase | Endoprotease | Ala-aa; Gly-aa |
| Bromelain | Endoprotease | Ala-aa; Tyr-aa |
| Trypsin | Endoprotease | Lys-aa; Arg-aa |
| Chymotrypsin | Endoprotease | Phe-aa; Trp-aa; Tyr-aa |
| Pepsin | Endoprotease | Leu-aa; Phe-aa |
| Protease V-8 | Endoprotease | Asp-aa; Glu-aa |
| Thermolysine | Endoprotease | aa-Phe; aa-Leu |
| Alcalase | Endoprotease | Unspecific |
| Papain | Endoprotease | Lys-aa; Arg-aa; Phe-aa; Gly-aa |
| Prolil endopeptidase | Endoprotease | Pro-aa |
| Subtilisin | Endoprotease | Unspecific |

Source: Damodaran et al. (2010).

## CHEMICAL PROCESS

In agreement with authors as Cheng et al. (1998), Kristinsson and Rasco (2000), Slizyte et al. (2005) and Freitas (2011), the protein isolate is basically an extraction process of alkaline or acid solubilization and precipitation on isoelectric point, in order to obtain a product free from interference. The protein isolate can be manufactured from a variety of raw materials, such as soybean, milk, bean, or fish.

Protein isolate is usually obtained by chemical solubilization, acid or alkaline, or for enzymatic road. These can be used as functional ingredients in a wide and always growing application strip in several foods and in other products.

To accomplish the isolate, chemical and biological methods are used thoroughly; the chemical hydrolysis is more commonly used in practice industrial, however the biological processes that use addition of enzymes they are more promising when it is wanted products with high functionality and nutritional value. There are many potential techniques for the extraction of the protein from animal tissue. These include the use of water and organic solvents, conventional processes of roast, application of high pressures, drying and extraction from hot oil (Pigott, 1982). The functional properties of protein isolate are influence by production method used.

In general, the hydrolysis is an efficient process for the protein solubilization. The production of soluble materials that constitute the final product of the hydrolysis depends on several factors, such as, chemical reagents, enzyme type, substrate, pH, temperature, incubation time and enzyme concentration (Adler-Nissen, 1986).

## ISOLATION OF PROTEIN

In 1999, was realized the largest innovation considering the protein isolate from muscles of raw materials from low commercial value. Hultin and Kelleher (1999) patented the process of acid solubilization as a way of improving the productivity and the stability of protein isolate. Few years later, a similar process, but based on the alkaline solubilization it was patented. Three main advantages related with the acid and alkaline technologies need to be enhanced.

The first is that the muscle doesn't need to be removed mechanically from bones and skins to be processed. The raw material or homogenized could be submitted directly to the protein solubilization, acid or alkaline, since all the contaminate materials have been removed. Another advantage is that the sarcoplasmic proteins is also recovered, increasing the amount of protein in the isolate. Last, the neutral lipids and the lipids membranes can be removed during the process, reducing the risk of lipid oxidation during the storage. The solubilization process, acid and alkaline, use the beginning that the solubility of the material containing protein when homogenized with water it is affected by the pH of the mixture. In extremely acid or alkaline conditions, repulse forces act on the myofibrils and cytoskeletal proteins, happening interactions with water, and occurring the solubilization (Nolsoe and Undeland, 2008).

An important characteristic of the processes for solubilization acid or alkaline is that when the protein muscle is submitted to extreme values of pH, the proteins are unfolded partially. This unfolding partial induce to substantial changes in the structural and conformational proteins parts, which lead to different properties when they are recovered (Kristinsson and Hultin, 2003). Kristinsson (2001) evaluated conformational, structural and functional changes of two proteins, hemoglobin and myosin, during exhibition the acid and alkaline pHs, as well as subsequent recovery of the proteins in the isoelectric point. The pH changes were crucial for the hemoglobin related to lipid oxidation. Changes in the myosin, on the other hand, influenced largely in the functionality of the protein isolate, such as, capacity of retention of water, gelatinization and emulsification.

## Protein Content

The protein production obtained during the process by pH change (known as pH shifting process) is determinate for three main factors, the solubility of the proteins in extreme conditions of pH, the size of the insoluble sediment formed during the centrifugation process and the solubility of the proteins in the pH selected for the precipitation (Nolsoe and Undeland, 2008).

Using processes of protein solubilization, acid and alkaline, Undeland et al. (2002) found 74% and 68%, respectively, of protein recovered from the white muscle of herring (*Clupea harengus*). The low production presented by the alkaline process it was attributed to the great amount of sediment formed in the first centrifugation.

In similar comparison among the processes with pH, acid and alkaline, Kristinsson and Ingadottir (2006) evaluated the protein production from tilapia (*Oreochromis niloticus*). The authors reported that found among 56 and 61% of protein for the acid process and 61 to 68% for the alkaline process.

Martins et al. (2009) evaluated the protein production from Whitemouth croaker (*Micropogonias furnieri*) wastes. These authors used acid and alkaline solubilization, they obtained 86.94% and 72.34% of recovered protein, respectively.

It is evident that most of the species presents different answer for the acid and alkaline solubilization, when the two methods are compared. It is observed that the acid process usually possess certain advantages, it usually presents larger protein production, while the alkaline process possess other, as larger removal of lipids, color clearer, larger force of the gel, among others. Then, the selection of the method will depend on the subsequent application of the isolate.

# Chemical Modification of Proteins

The primary structure of the proteins can be modified chemically to improve it functional properties and also to study the relation between structure and function of the proteins (Feeney et al., 1982).

The oldest and simple way to modify proteins is to treat the same ones with a determined reagent which will react with the functional groups exposed in the surface of the proteins. For instance, acetic anhydride, $CH_3COOCOCH_3$ reacts with the amino groups of the lateral chain of the lysine resulting in $–NH_2COCH_3$. An important effect of this reaction is that the positive charge of the lysine is lost. These reagents are frequently not selective. A challenge is to produce a reagent that only reacts with a position of the protein (Doig and Stemberg, 1995).

Natural polymers, such as, dextrane (Hovgaard and Brondsted, 1995), pectins (Munjeri et al., 1997), gum guar (Gliko-Kabir et al., 1998) and inulin (Vervoort et al., 1997), they have been modified chemically to assist characteristics of the textile industry. After modification, these polysaccharides are chemically or physically linked to reduce it solubility in aqueous systems (Reis et al., 2006).

Chemical modification of proteins has been largely used to expose functional groups previously hidden into the protein chains, determining functional groups of enzymes and also helping to understand the action of drugs (Rathna and Gunasekaran, 2004; Léonard et al.,

2005; Zhang and Ping, 2006; Kurniawar et al., 2007; Solanki et al., 2008). Many synthetic polymers have limited structural and functional properties, while the natural polymers are unique, with various functional and structural properties (Amiva and Tanaka, 1987). The mechanical and swelling properties of hydrogel by using natural polymers can be improved by chemical or physical modification of the functional groups (Rathna and Damodaran, 2002; Lu and Chen, 1999). Modification does not alter the biodegradable and biocompatible characteristics of the proteins. Therefore, if suitable functionality can be imparted, the natural polymers could replace some potentially toxic synthetic polymers, which are unsuitable for many biological applications.

## Chemical Modification: Procedures

The chemical modification of proteins is based on the differences of the individual reactivate of the lateral chains of the amino acids. These include the imidazole of the histidine, the indol of the tryptophan, the p-hydroxyphenyl of the tyrosine, the thioether of the methionine, the thiol groups of the cysteine, the connection dissulphidic of the cysteine, the carboxylic groups of the aspartic and glutamic acids, the terminal carboxylic groups of the amino acids and the groups amino of the lysine and terminal amino groups of other amino acids. These lateral chains react in different proportions, not only with different reagents, but also under different conditions and locations of specific environments of a particular protein (Means and Feeney, 1998).

Heating up proteins in pH acid or alkaline, it is possible to modify the lateral chains of the residues of amino acids. Like this, for instance, the heating of the glutenin in acid environment get desamidation of 30% of the residues of glutamine and asparagine, what improves the solubility and the properties of surface of the protein. These beneficial effects are resulted of the conformational changes produced by the decrease of the number of hydrogen bridges and the increment of the repulses electrostatics (Fennema, 1993).

The main classes of reactions used to modify the lateral chains of the proteins chemically are acetylation, alkylation, oxidations and reductions. A certain lateral chain can react with several classes of reagents and a determined reagent can also make it with different lateral chains resulting in the introduction of identical lateral groups in several amino acids. During the acetylation, the acetyl reactive groups react with the amine, hydroxyl, phenol, imidazole and thiol groups (Fennema, 1993).

A determined type of lateral chain doesn't have to carry out in all proteins the same reactivity front a certain reagent, because its reactivity is function of the closer amino acids of the protein chain and its conformation. The introduction of the ionized groups, usually, improves the capacity of absorption water of the protein and its stability front to the heat and increases the stability with relation to the precipitation for calcium ions. Such effects have been observed in the fish proteins, soy, gluten, etc.

The degree of hydrophobicity from modified protein depends on the amino acid type or fatty acid and the proportion of susceptible protein groups to changes to form derived. For instance, the acetylation improves the emulsification properties of milk proteins, however it impair the absorption of water in the proteins of the soybeans (Fennema, 1993).

The modification can cause unfolding of proteins due to the repulses electrostatics between added carboxyl groups and native adjacent carboxyl groups, producing more interaction protein-water and less interaction protein-protein (Ponnampalam et al., 1988).

The anhydrides, acetic and succinic are the acetyls of larger use in proteins for human consumption. Those reagents are considered GRAS (Generally Recognized as Safe) for the Food and Drug Administration - FDA of the United States. The acetylation have been applied in vegetable proteins of several species and parts of the plants, such as, leafs (Franzen and Kinsella, 1976), peanut (Beuchat, 1977), sunflower (Canella et al., 1979), *Vicia faba* (Schmandke et al., 1981), cotton (Choi et al., 1982), pea (Johnson, 1982), oat (Ma, 1984), winged bean (Narayana and Rao, 1984), corn germ (Messinger et al., 1987), oat protein concentrate (Goulet et al., 1987), canola (Paulson and Tung, 1988), huarango seeds protein isolate (Silva et al., 1997), among others.

The chemical modification for acetylation is the more used because of it specificity with free amino groups, mainly the lysine. The esterification of the free amino groups with a neutral acetyl group results in the reduction of the positive charge of the isolate under acid conditions. This loss of the groups with positive charge results in a possible change in the isoelectric point for smaller pH values, it does with the solubility increase in the pH of 4.5 up to 7. Besides, the loss of these groups still results in a decrease of the number of water molecules for protein molecules bonds (Kuntz, 1971) and it also reduces the ionic attractions among close molecules that are partly responsible for the stabilization of the protein gels.

Considine (2002) reported that for the application in hydrogels, the physical properties of the fish polymers are very superior the ones of the vegetable proteins, because animal proteins are rich in lysine, an essential amino acid and that can be easily modified chemically. The proteins of the fish tend to be very large, with an inherent capacity of water retention. The solubility of a protein in water is connected with the nature of its lateral groups. Polar groups, such as, OH, CO, $NH_2$, interact with the water, stabilizing the hydrogen bridges among the chains. Charged groups, such as, $NH_3^+$ or $COO^-$, also influence in the hydrophylicity of the chains. The amount of charged groups or not, changed with the pH (Iannace and Nicolais, 1997).

Hwang and Damodaran (1996) verified that through the chemical modification of the residues of the lysine with tetracarboxylic dianhydride, it is possible to introduce a great number of carboxylic groups inside of the protein molecule. Theoretically, for each residue of modified lysine, three carboxylic groups were incorporate in the protein molecule. Those carboxyl groups, together with the extensive unfolding of the protein molecule through intramolecular repulses electrostatics, it provided a polyanionic character for the protein with numerous connection sites for the water. Polyanionic proteins with agents of crosslinking can produce a hydrogel with superabsorbent properties.

In research accomplished by Hwang and Damodaran (1996) with soybean protein isolate and several tetracarboxylics dianhydride, such as, benzenetetracarboxylic dianhydride, diethylenetriaminepentaacetic dianhydride, cyclobutane tetracarboxylic dianhydride and ethylenediaminetetraacetic dianhydride (EDTAD), it was verified that EDTAD is the best reagent to be used in the chemical modification of the lysine residues, because it possess a high rate of reaction with lysine residues, it doesn't possess toxicant potential and it has a smaller cost when compared to the other reagents. The structure of EDTAD is presented in the Figure 2. EDTAD is a bifuncional reagent. Consequently, in aqueous solution the reaction of EDTAD with lysine residues can follow two different ways, as shown in the Figure 3.

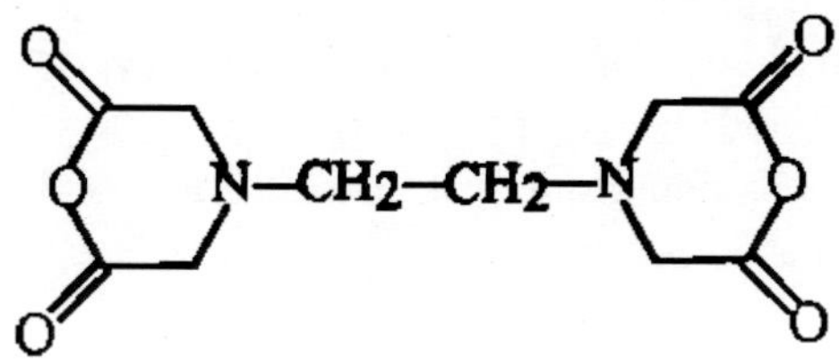


Figure 2. Molecular structure of tetraacetic ethilenediamino dianhydride.

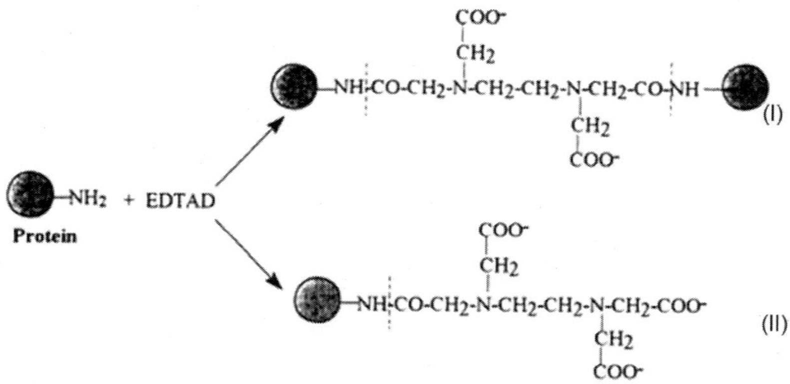


Figure 3. Reactions of lysine residues with EDTAD (Hwang and Damodaran (1996).

In the reaction I, a molecule of EDTAD reacts simultaneously with two lysine residues from two protein molecules or two polypeptides chains of an oligomeric protein to form a bond. When the reaction proceeds to the reaction I, the result is the incorporation of only a carboxyl group for lysine residue. Besides, if the reaction I occur among subunits of the protein molecule, the intramolecular crosslinking can impact the water absorption of the modified molecule. In the reaction II, a molecule of EDTAD reacts with a lysine residue and a molecule of water. In that reaction, three carboxyl groups for each lysine residue are incorporate in the protein and crosslinking are not formed. This great increase in the anionic charge of the modified protein net can help in the unfolding of the protein structure. How doesn't form crosslinking, the swelling of the modified protein is not impacted (Hwang and Damodaran, 1996).

## Crosslinking Agents

The polymers with crosslinking are obtained through conditions homogeneous or heterogeneous by the addition of crosslinking agent bi or polifunctional, such as, epichlorhydrin (Chiou et al., 2004; Delval et al., 2000), ethylene diglycol ether (Chiou and Li, 2003; Mi et al., 2002), glutaraldehyde (Rathna and Gunasekaran, 2004; Chiou and Li, 2003; Arrascue et al., 2003; Rathna and Damodaran, 2001; Hwang and Damodaran, 1996), benzoquinone (Mcfee et al., 2001), phosphorus oxychloride (Kim and Lim, 1999), carboxylic acids (Seidel et al., 2001), anhydride resin (Girek et al., 2000) or isocyanates (Lee et al., 2002; Mocanu et al., 2001).

Study accomplished with hydrogels produced from gelatin showed that the reaction between the glutaraldehyde and the gelatin involved only the lysine groups of the proteins chains (Chatterji, 1989). The lysine group can also suffer typical reactions of primary amino forming products, as, amides (reaction with carboxylic acids, esters or anhydrides), urea (reaction with isocyanates), imine (reaction with aldehyde or ketones), salts (reaction with salts) and urethane (Iannace and Nicolais, 1997).

## HYDROGEL

Hydrogels are solid polymeric structures, which contain a significant fraction of water (regularly above 90%). The three dimensional polymeric structure in the hydrogel usually formed by crosslinking, not only for physical interactions, such as, Van der Waals or hydrogen bridges, but also for covalent bonds formed by reagents of crosslinking or $\gamma$ -irradiation (Kunioka and Choi, 1998).

The polymeric gels possess interesting characteristics, because it has a great capacity of water uptake; however they are insoluble in the same. When the gel absorbs water, it swelling and increase your volume considerably and the gel maintain the form and the elasticity (Peppas, 1985). These peculiarities are consequences of different factors, such as:

a) the capacity of water uptake is due the presence of hydrophilic groups, such as, -OH, -COOH, $-CONH_2$ or $-SO_3H$;
b) the insolubility in water of the hydrogel is originated by the existence of a three-dimensional net in the structure;
c) the soft appearance and elastic consistence is due to the high content of water and the low density of polymer crosslinking.

Initially, the hydrogels were prepared polymerizing only one hydrophilic monomer, however subsequent studies demonstrated that when using two or more monomers in different proportions it allowed to vary some characteristics of these materials, what resulted in hydrogels with better mechanical, optical and water uptake properties (Pal et al., 2007).

Most of the hydrogels is made of polyacrylate, synthetic substance with high capacity of water absorption. A gram of polyacrylate absorbs around 400 g of water and above 35 g of saline solution. Polyacrylate molecules possess carboxylic groups linked. These groups contain negative charges, then it attract molecules of water that are polar. More than seven molecules of water can link to each carboxylic group. The polyacrylic powder absorbs water, it expands and becomes a gel (Considine, 2002).

The polymers that form hydrogels are preferably of light crosslinking to have substantial insolubility in water. The crosslinking can be accomplished by the irradiation for covalents, ionic, Van der Waals bonds or interactions of hydrogen bridges (Reeves et al., 2002).

Considine (2002) in the attempt of finding a biodegradable hydrogel that it could substitute the polyacrylic in disposable diapers, made a study with production of a hydrogel based on soybean proteins, obtaining as result that it absorbed 300 g of water and above 22 g of saline solution for gel gram. The manufacturers of disposable diapers, however, want a more absorbent product than the soybean hydrogel, such as the polyacrylic that absorbs 400 g of water and 35 g of saline solution.

Many disadvantages accompany the synthetic hydrogel (not biodegradable), it can be overcome by the use of derived hydrogels of natural sources of polymers. The most critical solicitations for any type of produced hydrogel from biopolymers are that, those gels must have the capacity to absorb a great amount of water relative to it mass and the material must be resist to the dissolution (Pal et al., 2007).

However, the conventional protein hydrogels thermally induced don't return to original volume after its have been dehydrated. This decrease in the swelling capacity is related with the increase of hydrogen bonds, as well as the electrostatics and hydrophobics interactions which occur in the protein dehydration. The loss of the swelling capacity of protein hydrogels thermally induced limits the industrial applicability (Hwang and Damodaran, 1996).

Hwang and Damodaran (1997) described a protein hydrogel which is superabsorbent, reversible to the swelling, biodegradable, and capable of cations linking. The protein hydrogel described was produced through the treatment of the protein with an acetyl agent and a crosslinking agent (glutaraldehyde), however the researchers pointed out that a residual of the crosslinking agent can remain in the gel and it does that the gel went less desirable for some applications.

## HYDROGEL TYPES

Hydrogels are hydrophilic polymers that can retain a significant amount of water while it maintains a different three-dimensional structure. In the sixties it was proposed that the hydrogels could be used in biomaterials. Based on the nature of the crosslinking force, the hydrogels can be characterized as chemists (covalents) or physical (Park et al., 1993). The structure and the properties of the primary chains, as well as the crosslinking density, contribute to the global properties of the chemical hydrogels. Physical gels are nets that set for the molecular arrangements and/or secondary molecular interactions. These interactions can be undone by the changes in the environment, such as, temperature, pH, ionic forces, presence of specific solutes and stress, consequently, the formation of physical hydrogels can be reversible. The formation of the physical hydrogels depends on the spontaneous organization and of the specific association of molecules through a number of interactions non-covalents (Rajapopal and Schneider, 2004; Zhang, 2002).

Hybrid hydrogels are usually defined as hydrogels systems which the components are formed by a minimum of two different classes of molecules (for instance, synthetic polymers and biological macromolecules) linked for both covalents and non-covalents bonds (Vandermeulen and Klok, 2004; Kopecek et al., 2001). According to Xu and Kopecek (2007) the peptides conjugation and the synthetic polymers, can lead to new materials with superior properties to those of the individual components.

Synthetic polymers have been elaborated with crosslinking by oligopeptides sequences (West and Hubbell, 1999; Subr et al., 1990), with proteins, such as, bovine albumin (Park, 1988), oligodeoxiribonucleotides (Nagahara and Matsuda, 1996) and polysaccharides (De Jong et al., 2000).

Biodegradable polymers can be consumed by microorganisms and they are reduced to composed simple, such as, carbon dioxide, water and ammonia (Hideki, 1992). However, based hydrogels in natural polymers are less hydrophylic and provide low mechanics force, what is unwanted in synthetic polymers. The natural polymers are unique; they possess several functional groups, which can develop the wanted properties through chemical modifications of based hydrogels in natural polymers (Rathna and Gunasekaran, 2004).

A hydrophilic gel, such as the polyacrylamide, contracts during the dehydration realized by an entropy process until the complete collapse or if contains rigid particles, it contracts the volume until the percolation point (Eichler et al., 1998).

A hydrophobic gel, such as, the poly-n-isopropiacrylamide (PNIPA), exhibits a separation phase due to the temperature or concentration change (Shibayama and Tanaka, 1993; Schild, 1992). This process causes in the gel a spontaneous contraction and collapse, while it forms the two phases (Panyukov and Rabin, 1996). One of the phases of the matrix dense is "rich in polymers" where the mechanical properties increase significantly (Shibayama et al., 1994).

## HYDROGEL FORMATION

Figure 4 shows the formation of chemical and physical hydrogels. When a gel begins to absorb water, firstly the hydrophilic groups are moisturized by the molecules of water that enter in the matrix. After the polar groups being moisturized, the three-dimensional net absorbs water leading to the exhibition of the hydrophobic groups, which also interact with the water molecules. After the polar and hydrophobic groups have interacted with the water molecules, the hydrogels retain more additional water, due to the osmotic forces of the chains of the polymeric net. That additional retention is promoted by the covalents or physics bonds, what takes a force of elastic recoil of the gel (Pal et al., 2007). In that additional swelling it is assumed that the water fills out spaces among the chains of the polymeric net.

Diffusion through ionically crosslinked polymer is mainly influenced by ionic interactions between polyionic chains of polymer and the crosslinking density set during the formation of the network. An increase in crosslinking density decreases swelling and thus diffusion by improving the stability of the network (Pal et al., 2007).

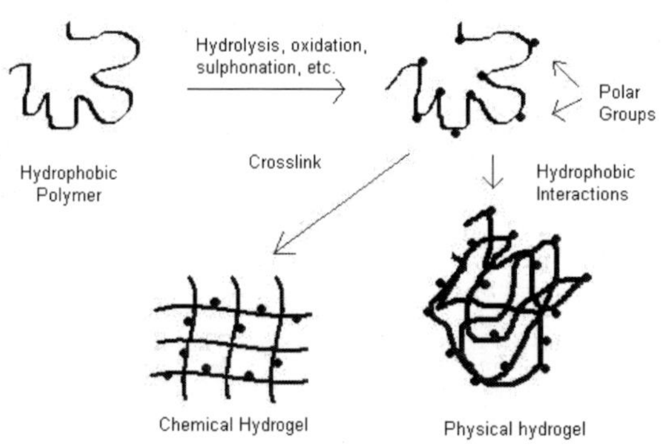


Figure 4. Schematic representation of methods for formation of hydrogels by chemical modification of hydrophobic polymers representative (adapted of Pal et al., 2007).

## BIOPOLYMERS-BASED HYDROGEL

There is an explosion of interests in hydrogels derived of biological molecules, since its components are similar with those that belong to the cells. Polysaccharides have been used in hydrogels formation, including agar (Liu et al., 2004), dextrane, chitosan and alginate (Zhang et al., 2005; Donati et al., 2005; Cascone et al., 2001), and also proteins (Damodaran, 2004; Martins et al, 2011a).

### Hydrogel Based on Natural Proteins

Hydrogel from natural proteins, such as collagen and gelatin, has been elaborated. Collagen is a structural protein with a triple helix in your secondary structure. In high temperatures or content of salt, the macromolecules of the collagen change the conformation for the semi crystalline form and lengthen the triples helixes for amorphous and spirals compact, resulting in the contraction of the collagen hydrogels (Oplatka and Yonath, 1968). The route that have been used to synthesize hydrogels and shown schematically in Figure 5.

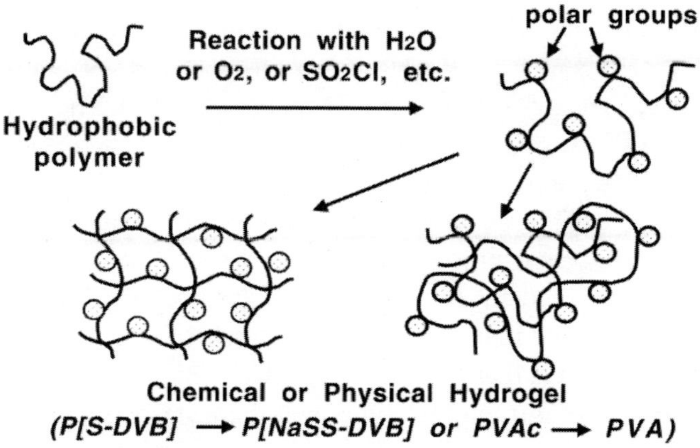


Figure 5. Synthesis of hydrogels by chemical modification of polymers (Hoffman, 2006).

### Hydrogel Based on Synthetic Polypeptides

The form of the hydrogels frequently based on proteins reflects in the macromolecules biological structure. Hydrogel based on elastine (Urry, 1999) were designated and synthesized with a repeated sequence of pentapeptides (VPGVG)m(VPGXG)n where X can be any one of the 20 amino acids. The disadvantage of those hydrogels is that contract when the temperature is above 25ºC. In low temperatures, the protein chains continue extended, because of the pentagonal water network around. When the temperature increases, the pentagons of water lose the structure allowing the protein chains to bend in compact structures (Xu and Kopecek, 2007).

## Hydrogel from Fish Protein Wastes

Recovery and alteration of fish protein present in co-products and its application, industrial products has been a promising and exciting alternative for the industry. In addition, developing processes to recover and applying the wastes of fish processing has become more feasible economically than discarding them (Kristinsson and Rasco, 2000; Martins et al., 2011a).

From the whole fish, approximately 50% are considered waste after processing and they are not applied as food (Martins et al., 2011b). The world population has grown and so has fish caught nowadays. It is highly likely that the estimative of sustainability has exceeded, which suggests figures of approximately 100 millions/year and, obviously, this increase must lead to a more intelligent and cautious use of waterway resources.

According to Damodaran (2004) and Martins et al., (2011a), wastes generated from fish processing operations contains 27–49% protein on a dry weight basis. Besides a variety of chemical modifications employed to modify the functional properties of fish protein (FP) concentrate, researchers have investigated a nonfood application for FP to be used as a superabsorbent hydrogel.

Following the introduction of a large number of hydrophilic groups into FP by modification with EDTAD, the proteins were crosslinked by sulfhydryl-disulfide interchange reaction between the endogenous sulfhydryl groups (–SH) and –S–S– bonds to produce a super absorbent polymer network (Hwang and Damodaran, 1997). The swelling capacity of a 76% EDTAD-modified FP was reported to be 540 g/g at 214 g. It depended on pH and ionic strength of the swelling media, similar to that observed for EDTAD-modified SPI hydrogels (Hwang and Damodaran, 1997; Hwang and Damodaran, 1997; Damodaran, 2004)

When glutaraldehyde was employed as a crosslinker, the super absorbent polymer swelling ability was diminished to 150–200 g/g, whereas the gel rigidity was enhanced. Therefore, these super absorbent polymers will be preferable to use for water absorbing under pressure in real applications, e.g. diapers (Zohuriaan-Mehr et al, 2009).

The swelling capacity of the protein hydrogels may be manipulated by changing the reaction variables, e.g. the extent of chemical modification, the degree of crosslinking, and the protein concentration at the time of crosslinking. However, it has been observed that the rate of swelling of these hydrogels was slower than that of synthetic super absorbent polymers. This is principally attributed to the presence of a significant amount of folded a-helix and b-sheet structures in protein monomers even after denaturation at basic pH and modification with the dianhydride (Rathna and Damodaran, 2002).

## Treatment with Ethanol

For improving the swelling rate, it is essential to minimize the folded secondary structure content of the polypeptide and increase the aperiodic (or random-coil) structure content. This was partially accomplished by treating the crosslinked hydrogel (without drying) with an organic solvent, such as ethanol (Rathna and Damodaran, 2001; Rathna and Damodaran, 2002; Damodaran, 2004; Lennartsson et al., 2011).

According to Rathna and Damodaran (2001, 2004), an ethanol treatment caused an extraordinary enhancement in the water uptake capacity. This improvement has been attributed to the ethanol-induced conformational reorganization in protein chains and hence the rate and extent of relaxation of the macromolecular network as water diffuses into the network (Damodaran, 2004).

Denaturation of the 80% EDTAD-modified FP hydrogels by treatment with absolute ethanol increased the equilibrium swelling of the gel from 200 to 425 g/g. The gel uptake of saline (NaCl 0.1 M) also enhanced from 24 g/g for the control gel to 35 g/g for the ethanol-treated gel (Rathna and Damodaran, 2001). Additional advantages have been claimed for the ethanol treatment, e.g. the gel dehydration which eliminates the need for drying the gel, extraction of any residual unreacted glyceraldehydes and low molecular weight off-odor compounds from the fish protein-based super absorbent polymer and thereby eliminating the unpleasant fishy odor (Rathna and Damodaran, 2001; Zohuriaan-Mehr et al., 2009).

Protein hydrogels with superabsorbent properties produced with crossed links when treated with ethanol, had their water swelling properties enhanced, as for the acid protein modified, it obtained an increase of 209% and alkaline of 172%, when compared with modified in proportion of 0.5:1 (EDTAD:protein, p/p) without adding ethanol (Martins et al., 2011a). As well as raising the capacity of water retention of hydrogel, the treatment with ethanol offers other advantages, such as, dehydration of gel which does not require dryness for a long period, odorous extraction compounds of low molecular weight, which improves its acceptance by the consumer.

## FISH HYDROGEL APPLICATIONS

Fish protein hydrogel has a great technological and economical importance for the wide field of applications, it can be used in absorbent materials, contact lenses, paper towel, surgical sponges, meat trays, rugs of bathrooms, diapers, prostheses of fabrics, hemodialysis membranes, support for catalysts, deposits of water, nutritious for plants and cultivations (Martins et al., 2011a; Shih et al., 2001), material for immobilization of enzymes (Park and Hoffman, 1992), processes of ionic change, remediation of polluted soils with heavy metals (Qin, 1993; Mitani et al., 1992) and support biodegradable for herbicide and pesticide (Weinhold et al., 1993; Schreiber et al., 1993).

Lim et al. (1980) demonstrated the application of calcium alginate microcapsules for cell encapsulation. Later in the 1980s, Yannas et al. (1989) used natural polymers such as collagen and shark cartilage into hydrogels for use as artificial burn dressings. Hydrogels based on both natural and synthetic polymers are being used for encapsulation of cells (Gin et al., 1990; Hsu et al., 2000) and in tissue engineering as matrices for repairing and regenerating a wide variety of tissues and organs (Woerly, 1997; Suggs and Mikes, 1999).

The applications of larger relevance at the present time are in the biomedicine and in the production of absorbent materials. In the biomedicine it is requested that the material that composes the hydrogel has good compatibility with the tissues which it will be in contact, cannot be altered front to degenerative processes and possesses appropriate mechanical properties for the use. The presence of water is beneficial for the biocompatibility of the hydrogels, however it causes a decrease in mechanical properties, for that reason are requested hydrogels that possess great absorption capacity, however that maintain good mechanical properties (Pal et al., 2007).

## Conclusion

In general, the synthesis of a protein-based hydrogel involves essentially three steps: first, dissociation and partial denaturation of protein; second, modification of lysyl residues with ethylenediaminetetraacetic dianhydride, or with any other tetracarboxylic dianhydrides, to cause unfolding and to enhance the polyanionic character of the protein; and third, cross-linking of the modified protein with a bifunctional reagent to obtain an insoluble polymer network capable of absorbing a large amount of water.

Results corroborate the possibility to produce hydrogel, a biopolymer with great capacity of water uptake from fish protein isolates wastes. This is a hopeful alternative to increase the productivity in the fish plants, because at the same time, as the residue is being treated, another product is made.

## References

Adler-Nissen, J. A review of food hydrolysis specific areas. In: Enzymic hydrolysis of food proteins. Elsevier Applied Science Publishers. Copenhagen, *Denmark*, 57-109, 1986.

Amiya, T.; Tanaka, T. Phase transitions in crosslinked gels of natural polymers. *Macromolecules*, 20, 1162, 1987.

Arrascue, M. L.; Garcia, H. M.; Horna, O.; Guibal, E. Gold sorption on chitosan derivates. *Hydrometallurgy*, 71, 191-200, 2003.

Beuchat, L. R. Functional and eletrophoretic characteristics of succinylated peanut flour protein. *Journal of Agricultural and Food Chemistry*, 25, 258-261, 1977.

Buchholz, F.L. In: Superabsorbent Polymers: Science and Technology, ACS Symposium Series 573, American Chemical Society, Washington, DC, 1994.

Canella, M.; Castriotta, G.; Bernardi, A. Functional and physicochemical properties of succinylated and acetylated sunflower protein. *Lebensmittel-Wissenschaft and Technologie*, 12, 95-101, 1979.

Cascone M. G.; Barbani, N.; Cristallini, C.; Giusti, P.; Ciardelli, G.; Lazzeri, L. Bioartificial polymeric materials based on polysaccharides. *Journal of Biomaterials Science*, Polymer Edition, 12, 267-281, 2001.

Chatterji, P. R. Gelatin with hydrophilic hydrophobic grafts and glutaraldehyde crosslinks. *Journal of Applied Polymer Science*, 37, 2203-2212, 1989.

Cheng, G. X.; Liu, J.; Zhao, R. Z.; Yao, K. D. Studies on dynamic behavior of water cross Anked chitosan hydrogel. *Journal of Applied Polymer Science*, 67, 983-988, 1998.

Chiou, M. S.; Ho, P. Y.; Li, H. Y. Adsorption of anionic dyes in acid solutions using chemically cross-linked chitosan beads. *Dyes Pigments*, 60, 69-84, 2004.

Chiou, M. S.; Li, H, Y. Adsorption behaviour of reactive dye in aqueous solution on chemical cross-linked chitosan beads. *Chemosphere*, 50, 1095-1105, 2003.

Choi, Y. R.; Lusas, E. W., Rhee, K. C. Effects of acylation of defatted cottonseed flour with various acid anhydrids on protein extractability and functional properties of resulting protein isolates. *Journal of Food Science*, 47, 1713-1716, 1982.

Chothia, C. Hydrophobic bonding and acessible surface-area in proteins. *Nature*, 248, 338-339, 1974.

Considine, E. A fish solution to landfills. *Wisconsin Engineer Magazine*, 1-3, 2002.

Crescenzi, V.; M. Dentini, M.; De Nooy, A. E. J. Novel synthetic routes to hydrogels. *ACS Polymer Preprints*, 41 (1), 718. 2000.

Damodaran, S. Carboxyl-modified superabsorbent protein hydrogels. European Patent. EP 1263883B1. 2004.

Damodaran, S.; Parkin, K. L.; Fennema, O R. Química de Alimentos de Fennema. 4ª Edição. Editora Artmed. Porto Alegre, 2010.

De Jong, S. J.; De Smedt, S. C.; Wahls, M. W. C.; Demeester, J.; Kettenes-van den Bosch, J. J.; Hennink, W. E. Novel self-assembled hydrogels by stereocomplex formation in aqueous solution of enantiomeric lactic acid oligomers grafted to destran. *Macromolecules*, 33, 3680-3686, 2000.

Delval, F.; Vebrel, J.; Pont, P.; Morcellet, M.; Janus, L.; Crini, G. Sorption properties toward aromatic compounds of insoluble crosslinked polymer containing starch derivates. *Polymer Recycling*, 5, 137-143, 2000.

Desantis, G.; Jones, B. Chemical modification of enzymes for enhanced functionality. *Current Opinion in Biotechnology*, 10, 324-330, 1999.

Doig, A. J.; Sternberg, M. J. Side-chain conformational entropy in protein folding. *Protein Science*, 4, 247-251, 1995.

Donati, I.; Holtan, S.; Morch, Y. A.; Borgogna, M.; Dentini, M.; Skjak-Braek, G. New hypothesis on the role of alternating sequences in calcium-alginate gels. *Biomacromolecules*, 6, 1031-1040, 2005.

Du Toit, L. C.; Pillay, V.; Danckwerts, M. P. Application of synergism and variation in ionic compatibilities within a hydrophilic polymeric sodium starch glycolate-kappa-carrageenan combination: textural profiling of the suspension behavior. *Journal of Bioactive and Compatible Polymers*, 21, 107-122, 2006.

Eichler, S.; Ramon, O.; Ladyzhinski, I.; Cohen, Y.; Mizrahi, S. Collapse processes on shrinkage of hydrophilic gels during dehydration. *Food Research International*, 30, 719-726, 1998.

Feeney, R. E.; Yamasaki, R. B.; Geochegan, K. F. Chemical modification of proteins: An Review, in Modifications of Proteins. *American Chemistry Society*, 3-55, 1982.

Fennema, O. R. Química de los alimentos. Editora Acribia. Zaragoza, Espanha, 1993.

Franzen, K. L.; Kinsella, J. E. Functional properties of succinylated and acetylated leaf protein. *Journal of Agricultural and Food Chemistry*, 24, 788-795, 1976.

Freitas, I. R. Otimização do processo de obtenção de isolados protéicos provenientes de pescado de baixo valor comercial. M. Sc. Dissertation, FURG, Rio Grande. 2011.

George, M.; Abraham, T. E. pH sensitive alginate-guar gum hydrogel for the controlled delivery of protein drugs. International *Journal of Pharmaceutics*, 335, 123-129, 2007.

Girek, T.; Shin, D. H.; Lim, S. T. Polymerization of β-cyclodextrin with maleic anhydride and structural characterization of the polymers. *Carbohydrate Polymers*, 42, 59-63, 2000.

Gliko-Kabir, I.; Yagen, B.; Penhasi, A.; Rubinstein, A. Low Swelling, Crosslinked Guar and Its Potential Use as Colon-Specific Drug Carrier. *PHARMACEUTICAL RESEARCH*, 15, 1019-1025, 1998.

Goulet, G.; Ponnampalam, R.; Amiot, J.; Roy, A.; Brisson, G. J. Nutritional value of acylated oat protein concentrates. *Journal of Agricultural Food and Chemistry*, 35, 589-592, 1987.

Guerard, F.; Guimas, L.; Binet, A. Production of tuna waste hydrolysates by a commercial neutral protease preparation. *Journal of Molecular Catalysis B: Enzymatic*, 20, 489-498, 2002.

Hall, R. J.; Trinder, N.; Givens, D. I. Observations on the use of 2,4,6-trinitrobenzenesulphonic acid for the determinarion of available lysine in animal protein concentrates. *Analyst*, 98, 673-686, 1973.

Hideki, O. Handbook of polymer degradation. Marcel Dekker, New York, 1992.

Hoffman, A. S. Hydrogels for Medical Applications. *Annals of the New York Academy of Sciences*, 949, 62-73, 2006.

Hovgaard, L.; Brondsted, H. Dextran hydrogels for colon-especific drug delivery. *Journal of Controlled Release*, 36, 159-166, 1995.

Hubbell, J. A. Synthetic biodegradable polymers for tissue engineering and drug delivery. *Current Opinion in Solid State Material Science*, 3, 246–251. 1998.

Hultin, H. O.; Kelleher, S. D. Process for isolating a protein composition from a muscle source and protein composition. U. S. Patent. US6005073, 1999.

Hwang, D. C.; Damodaran, S. Chemical modification strategies for synthesis of protein-based hydrogel. *Journal of Agricultural and Food Chemistry*, 44, 751-758, 1996.

Hwang, D. C.; Damodaran, S. Synthesis and properties of fish protein-based hydrogel. *Journal of the American Oil Chemical Society*, 74, 1165-1171, 1997.

Iannace, S.; Nicolais, L.; Water sorption of glycol-modified cross-linked gelatin-based hydrogels. *Journal of Materials Science*, 32, 1405-1408, 1997.

Johnson, E. A. Functional properties of acylated pea protein isolates. Dissertação de Mestrado. Washington State University, 1982.

Kim, B. S.; Lim, S. T. Removal of heavy metal íons from water by cross-linked carboxymethyl corn starch. *Carbohydrate Polymers*, 39, 217-223, 1999.

Kim, B. S., Nikolovski, J.; Bonadio, J.; Mooney, D. J. Cyclic mechanical strain regulates the development of engineered smooth muscle tissue. *Nature Biotechnology*, 17, 979–983. 1999.

Kopeček J, Tang A, Wang C, Stewart R. De novo design of biomedical polymers: hybrids from synthetic macromolecules and genetically engineered protein domains. *Macromolecular Symposia*, 174, 31-42, 2001.

Kristinsson, H. G. Conformational and functional changes of hemoglobin and myosin induced by pH: functional role in fish quality. Ph.D. thesis. University of Massachusetts, Amherst, MA, USA, 2001.

Kristinsson, H.; Hultin, H. O. Changes in conformation and subunit assembly of cod myosin at low and high pH and after subsequent refolding. *Journal of Agricultural and Food Chemistry*, 51, 7187-7196, 2003.

Kristinsson, H.; Ingadottir, B. Recovery and properties of muscle proteins extracted from tilapia (*Oreochromis niloticus*) light muscle by pH shift processing. *Journal of Food Science*, 71, E132-E141, 2006.

Kristinsson, H. G.; Rasco, B. A. Fish protein hydrolysates: production, biochemical, and functional properties. *Critical Reviews in Food Science and Nutrition*, 40, 43-81, 2000.

Kunioka, M.; Choi, H. J. Hydrolytic degradation and mechanical properties of hydrogels prepared from microbial poly(amino acids). *Polymer Degradation and Stability*, 59, 33-37, 1998.

Kuntz, I. D. Hydration of macromolecules. III. Hydration of polipeptides. *Journal of the American Chemical Society*, 93, 514-526, 1971.

Kurniawan, L.; Qio, G. G.; Zhang, X. Q. Chemical modification of wheat protein-based natural polymers: grafting and cross-linking reactions with poly(ethylene oxide) diglycidyl ether and ethyl diamine. *Biomacromolecules*, 8, 2909-2915, 2007.

Lee, K. P.; Choi, S. H.; Ryu, E. N.; Park, J. H.; Kim, Y. Preparation and characterization of cyclodextrin polymer and its high-performance liquid-chromatography stationary phase. *Analytical Science*, 18, 31-34, 2002.

Lennartsson, P.R.; Niklasson, C.; Taherzadeh, M. J. A pilot study on lignocelluloses to ethanol and fish feed using NMMO pretreatment and cultivation with zygomycetes in an air-lift reactor. *Bioresource Technology*, 102 (6), 4425-4432. 2011.

Léonard, R.; Lhernould, S.; Carlué, M.; Fleurat, P.; Maftah, A.; Costa, G. Biochemical characterization of silene alba alpha4-fucosyltransferases and Lewis products. *Glycoconjugate Journal*, 22, 71-78, 2005.

Liaset, B.; Espe, M. Nutritional composition of soluble and insoluble fractions obtained by enzymatic hydrolysis of fish-raw materials. *Process Biochemistry*, 43, 42-48, 2008.

Lim, F.; Sun, A. M. Microencapsulated islets as bioartificial pancreas. *Science*, 210, 908-910, 1980.

Linder, M.; Fanni, J.; Parmentier, M.; Sergent, M.; Phan-Than-Luu, R. Protein recovery from veal bones by enzymatic hydrolysis. *Journal of Food Science*, 60, 949-952, 1995.

Lu, G. H.; Chen, T. C. Application of egg white and plasma powders as muscle food binding agents. *Journal of Food Engineering*, 42, 147-151, 1999.

Ma, C-Y. Functional properties of acylated oat proteins. *Journal of Food Science*, 49, 1128-1131, 1984.

Mackie, I. M. Fish protein hydrolysates. *Process biochemistry*, 17, 26-28, 1982.

Martins, V. G.; Costa, J. A. V.; Prentice, C. H. Hidrolisado protéico de pescado obtido por via química e enzimática a partir de corvina (*Micropogonias furnieri*). *Química Nova*, 32, 61-66, 2009.

Martins, V. G.; Costa, J. A. V.; Damodaran, S.; Prentice, C. Chemical modification and structural analysis of protein isolates to produce hydrogel using Whitemouth croaker (*Micropogonias furnieri*) wastes. *Applied Biochemistry and Biotechnology*, 165, 279-289, 2011.

Martins, V. G.; Costa, J. A. V.; Silveira, S. T.; Brandelli, A.; Prentice, C. Protein and amino acid solubilization by using *Bacillus cereus*, *Bacillus velesensis*, and *Chryseobacterium* sp. from chemical extraction of fish protein residue. *Food Bioprocess and Technology*, 4, 116-123, 2011.

Matsushima, A.; Kodera, Y.; Hiroto, M.; Nishimura, H.; Inada, Y. Bioconjugates of proteins and polyethylene glycol: potent tools in biotechnological processes. *Journal of Molecular Catalysis B: Enzymatic*, 2, 1-17, 1996.

McAfee, B. J.; Gould, W. D.; Nadeau, J. C.; Da Costa, A. C. A. Biosorption of metal ions using chitosan, chitin and biomass of *Rhizopus oryzae. Separation Science and Technology*, 36, 3207-3222, 2001.

Means, G. E.; Feeney, R. E. Chemical modifications of proteins: a review. *Journal of Food Biochemistry*, 22, 399-425, 1998.

Messinger, J. K.; Rupnow, J. H.; Zeece, M. G.; Anderson, R. L. Effect of partial proteolysis and succinylation on functionality of corn germ protein isolate. *Journal of Food Science*, 52, 1620-1624, 1987.

Mi, F. L.; Shyu, S. S.; Chen, C. T.; Lai, J. Y. Adsorption of indomethacin onto chemically modified chitosan beads. *Polymer*, 43, 757-765, 2002.

Mitani, T.; Moriyama, A.; Ishii, H. Heavy metal uptake by swollen chitosan beads. Bioscience, *Biotehnology and Biochemistry*, 56, 985-990, 1992.

Mocanu, G.; Vizitiu, D.; Carpov, A. Cyclodextrin polymers. *Journal of Bioactive and Compatible Polymers*, 16, 315-342, 2001.

Munjeri, O.; Collett, J. H.; Fell, J. T. Hydrogel beads based on amidated pectins for colon-specific drug delivery: the role of chitosan in modifying drug release. *Journal of Controlled Release*, 46, 273-278, 1997.

Nagahara, S.; Matsuda, T. Hydrogel formation via hybridization of oligonucleotides derivatized in water-soluble vinyl. *Polymer Gels and Networks*, 4, 111-127, 1996.

Narayana, K.; Rao, N. Effect of acetylation and succinylation on the functional properties of winged bean *(Prosopis carpus tetragonolobus) flour. Journal of Food Science*, 49, 547-550, 1984.

Nolsoe, H.; Undeland, I. The acid and alkaline solubilization process for the isolation of muscle proteins: state of the art. *Food and Bioprocess Technology*, 2, 1-27, 2008.

Oplatka, A.; Yonath, J. Mechanochemical melting of collagen fibers. II. Diffusion-controlled contractions. *Biopolymers*, 6, 1147-1158, 1968.

Pal, K.; Banthia, A. K.; Majumdar, D. K. Hydrogels for biomedical applications: a short review. *Journal of Material Science: Materials in Medicine*, 2007.

Panyukov, S.; Rabin, Y. Volume transitions, phase separation, and anisotropic surface phases in charged gels. *Macromolecules*, 29, 8530-8537, 1996.

Park, K.; Shalaby, W. S. W.; Park, H. Biodegradable hydrogels for drug delivery. Technomic Publishing Co., Inc., Lancaster, PA, 1993.

Park, K. Enzyme-digestible swelling hydrogels as platforms for long-term oral drug delivery: synthesis and characterization. *Biomaterials,* 9, 435-441, 1988.

Park, T. G.; Hoffman, A. S. Synthesis and characterization of pH-and/or temperature sensitive hydrogels. *Journal of Applied Polymer and Science*, 46, 659-664, 1992.

Paulson, A. T.; Tung, M. A. Emulsification properties of succinylated canola protein isolate. *Journal of Food Science*, 53, 817-820, 1988.

Peppas, N. Hydrogels in Medicine. Boca Raton, FLA.: CRC, Press, 1985.

Peppas, N. A.; Mikes, A. G. Hydrogels in Medicine and Pharmacy. Vol. 1, CRC Press: Boca Raton, FL, 1986.

Pigott, G. M. Enzyme modifications of fishery byproducts, in: Chemistry and Biochemistry of Marine Food Products. AVI Publishing Company, Westport, CT, 1982.

Ponnampalam, R.; Goulet, G.; Amiot, J.; Chamberland, B.; Brisson, G. Some functional properties of acetylated and succinylated oat protein concentrates and a blend of succinylated of oat protein and whey protein concentrates. *Food Chemistry*, 29, 109-118, 1988.

Pourjavadi, A.; Kurdtabar, M.; Mahdavinia, G. R.; Hosseinzadeh, H. Synthesis and super-swelling behavior of a novel protein-based superabsorbent hydrogel. *Polymer Bulletin*, 57, 813-824, 2006.

Qin, V. The chelating properties of chitosan fibers. *Journal of Applied Polymer Science*, 49, 727-731, 1993.

Rajapopal, K.; Schneider, J. P. Self-assembling peptides and proteins for nanotechnological applications. *Current Opinion in Structural Biology*, 14, 480-486, 2004.

Rathna, G. V. N.; Damodaran, S. Effect of nonprotein polymers on water-uptake properties of fish protein-based hydrogel. *Journal of Applied Polymer Science*, 85, 45-51, 2002.

Rathna, G. V. N.; Damodaran, S. Swelling behavior of protein-based superabsorbent hydrogels treated with ethanol. *Journal of Applied Polymer Science*, 81, 2190-2196, 2001.

Rathna, G. V. N.; Gunasekaran, S. Functionally-modified egg white albumen hydrogels. *Polymer International*, 53, 1994-2000, 2004.

Reeves, W. G.; Damay, E. C.; Hamilton, W. L.; Hansen, P. A.; Lindon, J. N.; Sorebo, H. A. Process for preparing superabsorbent-containing composites. *US Patent*, US6376011, 2002.

Reis, A. V.; Guilherme, M. R.; Cavalcanti, O. A.; Rubira, A. F.; Muniz, E. C. Synthesis and characterization of pH-responsive hydrogels based on chemically modified Arabic gum polysaccharide. *Polymer*, 47, 2023-2029, 2006.

Richardson, T.; Hyslop, D. B. Enzymes in Food Chemistry, 2nd ed. Fennema, O. R. Ed., Marcel Dekker Inc., New York, 1984.

Sadeghi, M.; Hosseinzadeh, H. Synthesis of starch–poly(sodium acrylate-co-acrylamide) superabsorbent hydrogel with salt and pH-responsiveness properties as a drug delivery system. *Journal of Bioactive and Compatible Polymers*, 23, 381-404, 2008.

Schild, H. G. Poly(n-isopropylacrylamide): experiment, theory and application. *Progress in Polymer Science*, 17, 163-249, 1992.

Schmandke, N.; Maune, R.; Scuhmann, S.; Schultz, M. Contribution to the characterization of acetylated protein fractions of *Vicia faba* in view of their functional properties. *Die Nahrung*, 25, 99-109, 1981.

Schreiber, M. M.; Hickman, M. V.; Vail, G. D. Starch-encapsulated atrazine: efficacy and transport. *Journal of Environment Quality*, 22, 443-453, 1993.

Scopes, R. K. Measurement of protein by spectrophotometry at 205 nm. *Analytical Biochemistry*, 59, 277-282, 1974.

Seidel, C.; Kulicke, W.M.; Heb, C.; Hartamann, B.; Lechner, M; D.; Lazik, W. Influence of the cross-linking agent on the gel structure of starch derivates. *Starch/Stärke*, 53, 305-310, 2001.

Shibayama, M.; Tanaka, T. Volume phase transition and related phenomena of polymer gels. In K. Dusek (Ed.). *Advances in polymer science*, 109, 1-62, 1993.

Shibayama, M.; Morimoto, M.; Nomura, S. Phase separation induced mechanical transition of poly(n-isopropylacrylamide) water isochore gels. *Macromolecules*, 27, 5060-5066, 1994.

Shih, I. L.; Van, Y. T. The production of poly-(γ-glutamic acid) from microorganisms and its various applications. Bioresource *Technology*, 79, 207-225, 2001.

Silva, J. B.; Bora, P. S.; Neto, V. Q. Caracterização de propriedades funcionais do isolado protéico de sementes de algaroba (*Prosopis juliflora* (SW) D. C.) modificado por acetilação. *Ciência e Tecnologia de Alimentos*, 17, 263-269, 1997.

Slizyte, R.; Dauksas, E.; Falch, E.; Storro, I.; Rustad, T. Characteristics of protein fractions generated from hydrolysed cod (*Gadus morhua*) by-products. *Process Biochemistry*, 40, 2021-2033, 2005.

Solanki, K.; Shah, S.; Gupta, M. N. Chemical modification of alpha-chymotrypsin for obtaining high transesterification activity in low water organic media. *Biocatalysis and Biotransformation*, 26, 258-265, 2008.

Šubr, V.; Duncan, R.; Kopeček, J. Release of macromolecules and daunomycin from hydrophilic gels containing enzymatically degradable bonds. *Journal of Biomaterials Science Polymers*, 1, 261-278, 1990.

Suggs, L. J.; Mikos, A. G. Synthetic biodegradable polymers for medical applications. *Cell Transplantation*, 8, 345-350, 1999.

Thiansilakul, Y.; Benjakul, S.; Shahidi, F. Composition, functional properties and antioxidative activity of protein hydrolysates prepared from round scad (*Decapterus maruadsi*). *Food Chemistry*, 103, 1385-1394, 2007.

Tsai, H-S.; Wang, Y-Z. Properties of hydrophilic chitosan network membranes by introducing binary crosslink agents. *Polymer Bulletin*, 60, 103-113, 2008.

Undeland, I.; Kelleher, S. D.; Hultin, H. O. Recovery of functional proteins from herring (*Clupea harengus*) light muscle by an acid or alkaline solubilization process. *Journal of Agricultural and Food Chemistry*, 50, 7371-7379, 2002.

Urry, D. W. Elastic molecular machines in metabolism and soft-tissue restoration. *Trends in Biotechnology*, 17, 249-257, 1999.

Vandermeulen, G. W.; Klok, H. A. Peptide/protein hybrid materials: enhanced control of structure and improved performance through conjugation of biological and synthetic polymers. *Macromolecular Bioscience*, 4, 383-398, 2004.

Vervoort, L.; Van den Mooter, G.; Augustijns, P.; Busson, R.; Toppet, S.; Kinget, R. Inulin Hydrogels as Carriers for Colonic Drug Targeting: I. Synthesis and Characterization of Methacrylated Inulin and Hydrogel Formation. *Pharmaceutical Research*, 14, 1730-1737, 1997.

Weinhold, B. J.; Sadeghi, A. M.; Gish, T. J. Organic chemicals in the environment: effect of starch encapsulation and temperature on volatilization of atrazine and alachlor. *Journal of Environment Quality*, 22, 162-166, 1993.

West, J. L.; Hubbell, J. A. Polymeric biomaterials with degradation sites for proteases involved in cell migration. *Macromolecules*, 32, 241-244, 1999.

Woerly, S. Porous hydrogels for neural tissue engineering. *Porous Material for Tissue Engineering*, 250, 53-68, 1997.

Xu, C.; Kopecek, J. Self-assembling hydrogels. *Polymer Bulletin*, 58, 53-63, 2007.

Yannas, I. V.; Lee, E.; Orgill, D. P.; Skrabut, E. M.; Murphy, G. F. Synthesis and characterization of a model extracellular matrix that induces partial regeneration of adult mammalian skin. Proceedings of the National Academy of Sciences of the United States of America, 86, 933-937, 1989.

Zhang, J.; Yuan, K.; Wang, Y. P.; Shen-Tang Zhang, S. T.; Zhang, J. Preparation and pH responsive behavior of poly(vinyl alcohol) – chitosan- poly(acrylic acid) full-IPN hydrogels. *Journal of Bioactive and Compatible Polymers*, 22, 207-218, 2007.

Zhang, R.; Tang, M.; Bowyer, A.; Eisenthal, R.; Hubble, J. A novel pH and ionic-strength sensitive carboxy methyl dextran hydrogel. *Biomaterials*, 26, 4677-4683, 2005.

Zhang, S. Emerging biological materials through molecular self-assembly. *Biotechnology Advances*, 20, 321-339, 2002.

Zhang, S-P.; Ping, W. Chemical modification of enzymes – An important tool to enhance the performance of biocatalysts. Shengwu Jiagong Guocheng. 4, 4-8, 2006.

Zohuriaan-Mehra, M. J.; Pourjavadib, A.; Salimib, H.; Kurdtabarb, M. Protein- and homo poly(amino acid)-based hydrogels with super-swelling properties. *Polymers Advanced Technologies*, 20, 655–671, 2009.

Hydrogels: Synthesis, Characterization and Applications
Editors: F. V. Câmara and L. J. Ferreira
ISBN: 978-1-61942-842-3


*Chapter 6*

# ENCAPSULATION IN POLYSACCHARIDE-BASED HYDROGELS: THE ALGINATE-CHITOSAN "DUO"

***Massimiliano Borgogna*, Barbara Bellich and Attilio Cesàro***
Department of Life Sciences,
University of Trieste Via L. Giorgieri, 1 – I-34127 Trieste (Italy)

## ABSTRACT

In nature polysaccharide-based 3D-hydrogel structures, such as cell walls, biofilms, and extracellular matrices, play a fundamental role in providing mechanical support and protection to biological entities, also controlling the diffusion of metabolites and potentially harmful molecules. Hence, natural saccharidic 3D-structures have always been regarded both as important models and as excellent materials for human applications (food, pharmaceutical, and biomedical).

Alginate and chitosan are two families of polysaccharides widely studied and employed as natural materials for bio-oriented applications, and they are exploited as immobilization and protective structures for sensible materials (from drug molecules and protein to cells).

Moreover, polysaccharide-based hydrogels are characterized by the possibility to engineer the polymer-based structure, thus controlling the resulting biological, mechanical, and transport properties. The encapsulation in polysaccharide matrices represents since ever the best example of protection. A variety of methods to encapsulate has been developed, by exploiting different materials and physico-chemical principles. In this paper, some relevant examples of micro- and nano-encapsulation based on alginate and chitosan in several bio-oriented technological fields are reviewed.

* Corresponding author:E-mail: mborgogna@units.it

# 1. INTRODUCTION

Polysaccharides are biopolymers composed of sugar repeating units linked by glycosidic bonds, which form linear and branched structures. They are characterized by high compositional and molecular weight variability, and by high branching possibility, that results in the formation of homo- and hetero-structures with very distinct physico-chemical properties.

Polysaccharides are abundant and obtained from renewable sources (such as plants, algae, bacteria); they present a variety not easily reproducible by synthetic routes; their production is generally easier and cheaper than for synthetic polymers. The number of polysaccharides investigated for technological applications of hydrogels is extremely large [1]. Polysaccharides are also employed as derivatives, to tailor the final properties of interest; moreover, biotechnology can enable the in vitro production of high levels of polysaccharides from micro-organisms [2-4].

Polysaccharide-based 3D-structures are widespread in nature, in form of cell walls, biofilms, and extracellular matrices, and provide mechanical support and protection to biological entities, also controlling the diffusion of metabolites and potentially harmful molecules. Natural saccharidic 3D-structures represent both important models and excellent materials for human applications.

Hence, polysaccharide-based matrices are versatile scaffolds characterized by the possibility to engineer the polymer-based structure, controlling mechanical and transport properties, and reproducing biocompatible microenvironments.

The polysaccharides most commonly used for the preparation of both micro- and nano-particles include agarose, alginate, carrageenan, chitosan, dextran, gum arabic, heparin, pullulan, and starch [1]. They are effective colloidal stabilizers, which increase the biocompatibility, and confer chemical functionality towards nanostructures, from carbon nanotubes, to colloidal silver and gold particles, quantum dots and iron oxide magnetic nanoparticles [5].

Biomaterial hydrogels are defined as “three-dimensional, hydrophilic, polymeric networks capable of imbibing large amounts of water or biological fluids” [6]. Hydrogels are chemically or physically cross-linked: in the former case by covalent bonds, in the latter by reversible molecular entanglements (such as ionic and hydrophobic interactions).

Due to their high water content, hydrogels present soft consistency, and are very similar to natural living tissues, overcoming the performances of the other classes of biomaterials. Generally, hydrogels are prepared for technological applications either from natural or synthetic polymers, and also from a combination of the two types. Several polymeric materials are employed as biomaterials, each one with specific properties which influence the hydrogel design parameters. Polysaccharide hydrogels are exploited as immobilization matrices and protective structures (as scaffolds and capsules) for sensible materials, such as living cells and active compounds. The food, pharmaceutical and biomedical applications of hydrogels based on two representative polysaccharides, alginate and chitosan, will be reviewed.

# 2. ENCAPSULATION IN POLYSACCHARIDE HYDROGELS

## 2.1. Alginate

Alginate is a linear polysaccharide produced by brown algae and bacteria, with variable composition and sequence, which depend from the specific origin. The recent progress in the regulation of alginate biosynthesis in bacteria, combined with the capability to easily engineer the bacterial strains, may enable the large scale biotechnological production of alginates with controlled features [7].

In molecular terms, it consists of two sugar residues 1→4 linked (D-mannuronic acid, M and L-guluronic acid, G), which form homopolymeric block structures along the chains (M-blocks and G-blocks) interspaced by alternate MG sequences [8]. The structure of the residues is schematically illustrated in Figure 1. The interaction between G-blocks and divalent cations results in the formation of interchain crosslinks, following the “egg-box” model proposed by Rees and coworkers [9, 10], and leads to the formation of the hydrogel 3D-network.

The relation between the cooperative sequence of guluronan residues and the “egg-box” structure has been pointed out since the initial works on the elucidation of alginate binding properties [11-13]. This model has been central for the phenomenological description of alginate gelation, and recently both a refinement and an extension of the original model shed some more light on the structural features of the “dimeric forms” of guluronan sequences involved in the chelation with calcium ions and responsible for the gel strength [14-16]. Several features of co-monomer fraction and sequence in the ion-binding properties of alginate towards divalent ions have been described, including the study of the thermodynamic aspects of the “egg-box” model and of the mechanism of hydrogel formation [17-20]. The concluding remarks on the gelation features of alginate attribute to calcium ions the right size and charge to fit into the distorted egg-box structure, including a non-negligible effect of the secondary junctions due to the alternating sequences of MG-type [21].

The biocompatibility of the alginate enables several applications, which will be reviewed throughout the paper, ranging from the food industry to the health technologies (development of drug delivery systems, wound healing dressing, cell immobilization matrices,…) [8].

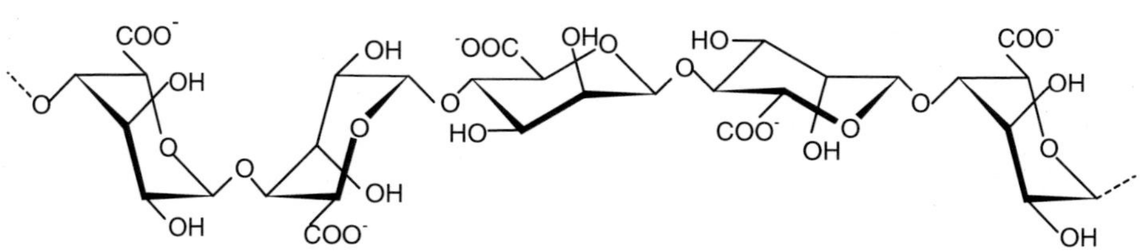


Figure 1. Chemical structure of alginate: G-G-M-M-G residues.

## 2.2. Chitosan

Chitosans are linear binary hetero-polysaccharides derived from the de-acetylation of the chitin (extracted from crustacean and insect exoskeleton or from some algae and fungi). The production of chitosans with controlled properties by means of micro-organisms (fungi) fermentation is of increasing interest [22, 23]. Chitosans are composed of 1→4 linked units (N-acetyl-D-glucosamine, A and D-glucosamine, D) in varying composition, sequence, and molecular weight. A- and D-units are randomly distributed. The structure of the residues is schematically illustrated in Figure 2.

The presence of a variable number of free amino groups and substitutions is responsible for the tunable interaction with anionic and polyanionic systems [24]. Chitosan conformational properties are mainly ascribed to the two-fold cellulose-like helix [25]. These features largely explain the intrinsic conformational stiffness as measured by several hydrodynamic and scattering experiments, sometime affected in the past by uncertainties due to the microgel formation for samples at high degree of acetylation (DA) [26]. Recent experimental results converge toward assigning to random acetyl substitution the relevant key-factor in controlling the changes in the chain stiffness, with a significant flexibility achieved at moderate to medium DA (<50%). This result is at least in part due to the common "co-monomer" effect, but must contain other contributing factors, such as short-range nearest-neighbor interactions [27] and long range excluded volume effect [28]. More recently, enhancement of chain flexibility and solubility properties has been reached by controlled chemical-etching of the glucosidic ring with periodate oxidation [29], as well as by introducing several types of pendant groups at the amino functionality [30]. Chitosan bio-oriented applications are related to its biodegradability and biocompatibility, and to some intrinsic unique features. It is used as antimicrobial agent [31]; it is exploited as bioadhesive pharmaceutical material, since it strongly interacts with epithelial cells and the overlying mucus layer, prolonging the contact time between the drug formulation and the target sites [32, 33]; it is explored as nucleic acid (DNA, siRNA) vector [34]; it is studied and applied in tissue engineering and regenerative medicine as scaffold material and as wound healing system [32, 35].

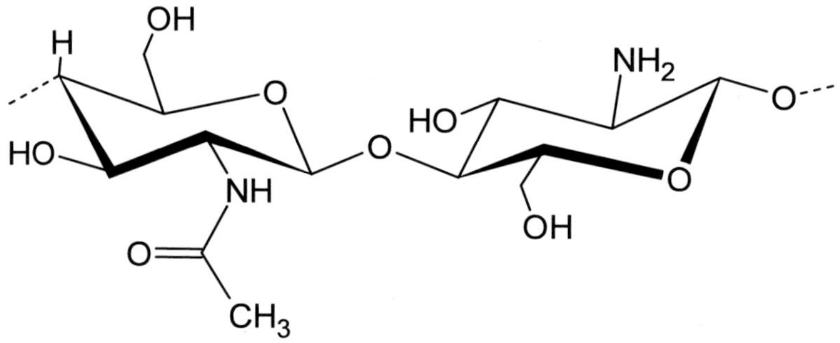


Figure 2. Chemical structure of chitosan: A-D residues.

## 2.3. Microencapsulation in Polysaccharide Hydrogels

The encapsulation in polymeric hydrogels is applied to a wide range of materials of biological interest, from molecules to cells, of bacterial, yeast, plant and animal origin. A variety of encapsulation methods has been developed, based on different materials and physico-chemical principles. The best choice is based on the final requirements, such as mean particle size, physical/chemical properties, payload application, desired release mechanisms, industrial manufacturing scale and process cost [36].

Among all the existing methods, here the attention will be primarily given to the microencapsulation by ionotropic gelation, that exploits the capability of some polymers to form hydrogels in the presence of proper multivalent counterions: divalent cations (in particular calcium ions) for alginate [37], and triphosphate (TPP) for chitosan [38]. The polymeric matrix controls the exchanges between the internal and the external parts, thus enabling the protection of the biological structures immobilized. Moreover, the matrix structure ensures the exchange of molecules, thus permitting the metabolic activity of the cells and the (controlled) release of the entrapped drugs, when used as drug depot. Figure 3 schematizes the concept of the encapsulation of big biological entities (cells) and/or small ones (such as macromolecular drugs). The encapsulation in hydrogels protects from external hazards of both biological and chemical nature (immunological response, enzymatic degradation, extreme pH,…).

In the literature definition [39, 40], gel particle technology mainly includes three different methods to prepare microspheres by ionotropic gelation (as schematically represented in Figure 4): a) dropping the polymer solution into a gelling ion solution, b) using a water in oil emulsification technique and c) exploiting the complexation of oppositely charged polyelectrolytes by mixing. Such classification merged with the years into the other approaches that use, indeed, several different techniques and procedures.

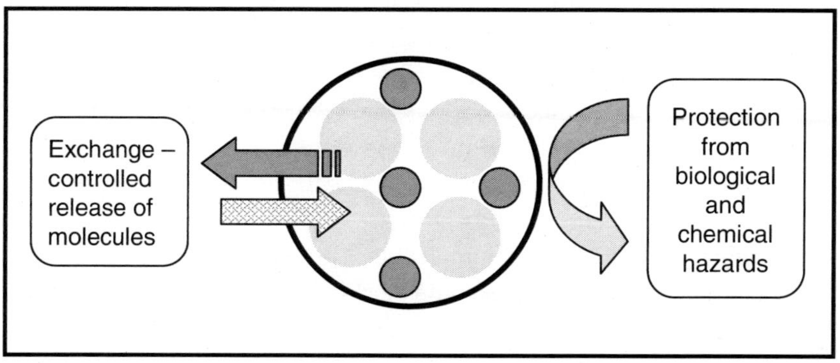


Figure 3. Schematic representation of the encapsulation of biological entities (cells, active compounds).

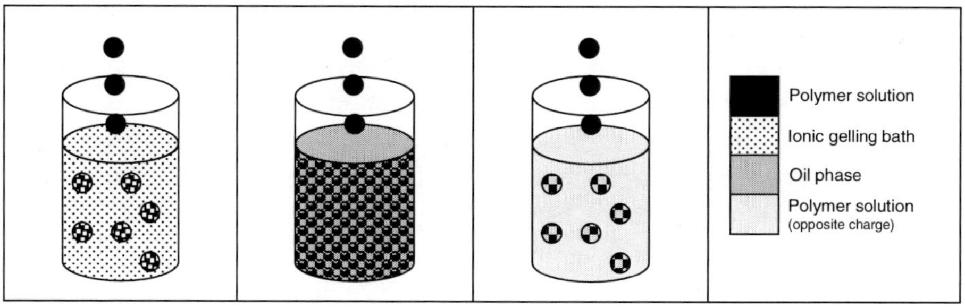


Figure 4. Schematic representation of the methods described in the literature for gel particle technology (redrawn from Borgogna et al. [41]).

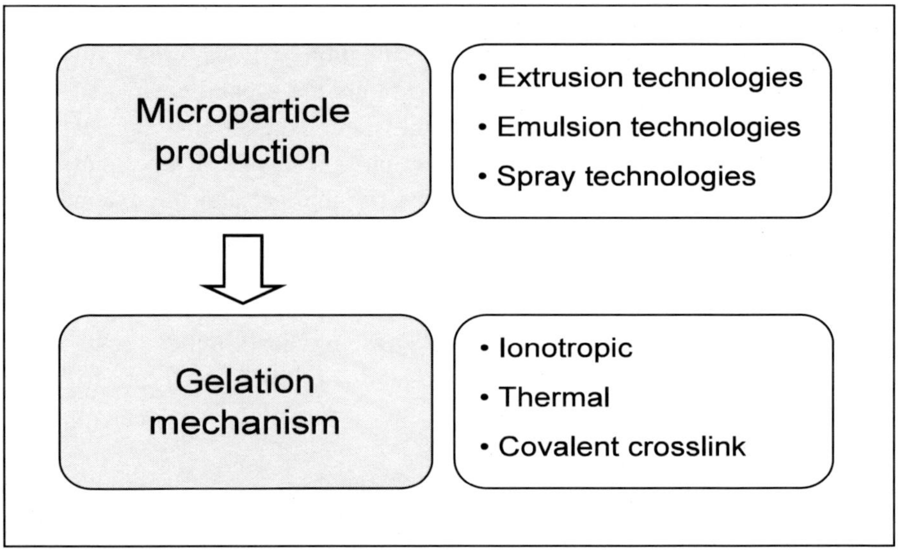


Figure 5. Graphical classification of the procedures and techniques to encapsulate in polysaccharide hydrogels.

A graphical classification of the procedures and techniques to encapsulate in polysaccharide hydrogels is proposed in Figure 5. The microparticle productions is regarded as the preliminary step for the physical generation of spherical microdomains, independently of the final stage of preparation. Then, the appropriate gelation mechanism is exploited to stabilize the preformed microdomains.

In the following classification, based on a revisitation of some recent publications [36, 40, 42, 43], the methods to obtain microscale droplets/particles and those to transform them in hydrogels will be separately described [41]. The focus will be on the encapsulation in hydrogels of biological entities (cells, biomolecules, drugs,…), taking into account that the literature nomenclature for the procedures could vary in different application fields [44].

#### *2.3.1. Microparticle Production*

The first step in the preparation of hydrogel microcapsules is to obtain micro-droplets to be then gelled in different ways. Three procedure families will be presented: extrusion, emulsion and spray-technologies.

The "extrusion" technique is widely exploited and can be performed by using a syringe either manually or with a perfusor with a needle. As a general principle, the polymer solution is extruded through an orifice (e.g., needle) and dripped into a hardening (gelling) bath. The size of the droplets is responsible of the final dimensions of the gel particles, and can be controlled through several system parameters: the diameter of the orifice, the flow rate, the viscosity of the solution, the distance between the hardening solution and the orifice, the polymer concentration and the temperature [36]. The reduction of the droplet (and gel particle) size is a fundamental technical problem, which has been handled by developing several devices based on different physical mechanisms. The most used are the coaxial laminar air-flow, the air atomization, the electrostatic potential, the vibrating nozzle, the jet cutting and the spinning disk atomization [45-47]. The resulting particle size is not the only factor to be considered, since other aspects are relevant, such as reproducibility, size tunability and distribution, time and yield of production. It is worth to emphasize that the extrusion methods include some of the most "cell friendly" techniques that can be applied to animal cell encapsulation. The other techniques described in the following part find application mainly in microbial cell (probiotics), drug and other bioactive compounds encapsulation.

The emulsification is the process of dispersing a liquid in a second one that is immiscible [44]. The resulting emulsion can be defined basically as a system of two phases (dispersed and dispersing) in which droplets of one liquid are dispersed in another one: when the emulsion is made of oil droplets in an aqueous phase it is called "oil-in-water" (O/W), while the opposite system is called "water-in-oil" (W/O). The stability of an emulsion can be regulated by the addition of a surfactant. Gelation of O/W systems can produce hydrogel sponges [48]. By working on the composition of the two phases, the type of polymer and surfactant, this technique has been employed also for the production of nanometric droplets and particles. Advances in this technology have been obtained also by the application of microfluidic principles [49, 50].

Among the spray-technologies, spray-drying is based on the formation of dried particulate starting from a fluid material (a solution, an emulsion or a suspension), by the atomization in heated gas (air or nitrogen), and the fast removal of the solvent (water). The powdered particles are then separated from the drying air. This continuous process is influenced by the temperatures of the system (feed, air inlet and air outlet temperatures), the viscosity and concentration of the feed solution, the morphology of the polymer. Hydrogel particles can be thereafter produced by re-hydrating the powders under gelling conditions. Other methods, such as spray-cooling and spray-chilling, are based on a different principle and in the absence of solvent: the dispersion of a bioactive product in a polymeric matrix is cooled (or chilled) to allow solidification and immobilization [36, 44, 51]. Finally, the spray-coating technology is often reported in association to particle production. It is an efficient coating technique, based on devices such as the fluidized bed, to be used with virtually any polymer. The core particles are suspended by an air flow and then coated with the polymer solution sprayed from different directions (top, bottom and tangential spray). This technique can be used on pre-formed microcapsules after the gelation step [36, 40].

### *2.3.2. Gelation Methods*

The step that follows the preparation of micro-droplets of the desired dimension is the gelation, which can occur in different ways, through physical and chemical mechanisms. Physical gelation is mainly done by ionotropic and thermal gelation.

The ionotropic gelation is in turn carried out by two main techniques, diffusion and internal setting. The first one is based on the introduction (mainly a dripping) of the polymer solution into an ionic solution: gelation occurs when the ions diffuse into the polymer solution droplets. The second one is based on the addition of the gelling ions as inactive form (such as calcium carbonate for calcium alginate gels). The gelation is triggered by changes in some system properties, generally pH changes [36]. The ionotropic gelation is suitable for alginate and chitosan, and also for materials of different origin (for example carrageenan and pectins). Microencapsulation in alginate hydrogels is probably the most exploited [52-55]. Also polyelectrolyte complexation (PEC), between polymers of opposite charges, is associated to the hydrogel formation, either as method to form the gel structure, or as technique to improve the quality (mechanical strength and permeability barrier) of hydrogel beads. In the latter case, the complexation can be concurrent with the ion-based gelation, or subsequent (coating step). PEC formation between chitosan and alginate (or other biopolymers) has been since long time exploited mainly for pharmaceutical applications for the production of micro- and nano-particles based systems [56-60].

Thermal gelation is applied to polymers that present a conformational order-disorder transition with the temperature, such as agarose, gelatin, maltodextrin and several synthetic polymers, such as polymethylmethacrylate, polystyrene, poly(N-isopropyl-acrylamide), etc. Concerning cell encapsulation, the agarose-based hydrogels have been widely studied for different biological applications [61, 62].

Chemical gelation is based on the covalent cross-linking of the polymer chains, which results in the formation of the 3D-matrix. Several examples have been reported in literature, the oldest one is the reticulation of dextrans to prepare gel beads for size-exclusion chromatography. The formation of chitosan microcapsules was initially based on the glutaraldehyde cross-linking. However, its use is avoided in cell and human applications, because the toxicity of such reactant. Covalent cross-link can be obtained also by enzymatic routes, among which the safe cross-linker genipine has been employed for the preparation of hydrogels and beads of chitosan [1]. There is also a long list of covalent cross-linking reactions induced by the exposure to UV or visible light (photo-crosslinked hydrogels) [63].

In conclusion, several strategies of combinations of physical and chemical hydrogel formation techniques have been proposed, leading to "in tandem" mechanisms, in order to tailor and enhance the properties of the microcapsules produced [64-66].

# 3. Applications

## 3.1. Food Technology

The food industry represents the oldest application field of polysaccharide hydrogels, that are naturally responsible of the food texture, and are exploited for a variety of applications: as thickeners, stabilizers or restructuring agents, for example in juices and chocolate. A lot of GRAS biopolymers are compatible with many food components and are already used in the food industry (such as alginate, pectins, dextran, starch, cellulose and derivatives). Often these polymers are modified to optimize the compatibility with the food matrices [44, 67].

Alginate is employed as an additive in both USA and Europe, and is used to control the texture and the structural properties of the foodstuffs [68]. In addition to the most traditional applications, alginate is used for the preparation of restructured food, characterized by uniform size and consistency, such as onions, meat, seafood, fruit, and vegetables [68]. From the physico-chemical point of view, the use of pectin as gelling agent parallels the application of alginate. Chitosan has been approved for dietary applications in some countries (such as Japan, Italy and Finland) [69].

In general, the use of microencapsulation with biopolymers has gained great importance in the development of food enriched with bioactive components and nutraceuticals, for the protection against inactivation or degradation, and also for the controlled delivery of the compounds. Microencapsulation is used to stabilize reactive, sensitive, or volatile ingredients, and is the source of totally new ingredients with exceptional properties, thanks to the tailored controlled-release properties: targeted and controlled release enhances the effectiveness of food additives, ensuring the optimal dosage, and improves the cost effectiveness [70]. Microencapsulation is defined as "the technology of packaging solid, liquid and gaseous materials in small capsules that release their contents at controlled rates over prolonged periods of time" [40]. The food industrial production requires more and more the addition of functional ingredients, able to control flavour, colour, texture or preservation properties. The general definition of bioactive component refers to lipids, vitamins, peptides, fatty acids, antioxidants, minerals and also living cells (in the case of probiotics). One of the more recent developments in alginate use in the food technology has been in the encapsulation of probiotics [68]. Nutraceuticals are ingredients with potential health benefits [40, 44, 71]. The addition of nutraceutical substances to food matrices is pursued to produce physiological benefits or reduce the risk of given diseases. However, only a reduced amount of molecules remains available after the oral administration, due to several factors, such as low permeability or solubility within the gut, and instability under food processing conditions (temperature, oxygen, light) or in the gastro-intestinal tract (pH, enzymes, presence of other nutrients) [71]. Many encapsulation procedures have been developed specifically for food industry: each technology suits to different systems, to meet the physico-chemical and molecular requirements of a specific bioactive component. The influences from the environment are fundamental: in addition to the degradation during the transit through the gastrointestinal tract, the possible damage during the storage in the final product vehicle is relevant; moreover, many food ingredients may interfere with the activity of the bioactive component. Thus, the encapsulation procedure should protect the bioactive component during the entire period of processing, storage, and transport, without interfering with the final texture and taste. Finally, a system might be required which releases the bioactive component in a specific site of the gastrointestinal tract [44].

## 3.2. Health Applications

The attention for the application of polysaccharide hydrogels in drug delivery systems increased dramatically in the past decades [4, 33, 58, 72-74]. Pharmaceutical applications of hydrogels are based on their capability to act as drug delivery vehicles. Drugs and bioactive compounds are incorporated into the matrices and can be released according to various release profiles depending on the hydrogel properties [75]. It is generally accepted that the drug release from the matrix follows two main mechanisms, that is diffusion of the protein through the pores of the polymer network and degradation of the polymer network [39]. In addition, water diffusion and swelling through the hydrogel are some of the major factors affecting drug release rate [76]. Thus, drug release from a hydrogel is governed by several parameters: pore volume fraction; pore sizes; extent of interconnections; size and physico-chemical nature of the drug molecule, and in general type and strength of interactions between drugs and polymeric chains [77]. Some peculiar properties of hydrogels are shown in physiological conditions: for example, they can have a swelling dependent on the external environment, behaving as physiologically-responsive hydrogels. Some hydrogels (smart and stimuli responsive) show significant changes in the swelling ratio, and thus in release capability, in relation to pH, ionic strength, temperature or electromagnetic radiation [6]. Moreover, hydrogels are used as muco-adhesive drug carriers able to interact with the mucosal surface (for example in the gastrointestinal tract or the respiratory epithelium) prolonging their residence time at the site of delivery. Hydrogel-based systems have been exploited for protection and delivery of low-molecular weight drugs and macromolecular payloads, such as peptide and protein drugs (insulin, melatonin, heparin, haemoglobin, parathyroid hormone, calcitonin,...) [78-80], nucleic acids (DNA, siRNA) [81, 82] and antigens (from pathogens responsible of influenza, pertussis, diphtheria, tetanus,...) [73, 79, 83, 84]. Also vaccines have been encapsulated (attenuated or inactivated pathogens) in both medical and veterinary fields, combining the cell immobilization to a drug delivery system [41]. Such technology is gaining great attention, due to the possibility to protect the vaccine during the administration (also for the oral route) and hopefully to control the targeting and delivery rate [85, 86].

Hydrogels, thanks to their peculiarities (biocompatibility, physical properties, flexibility in the synthesis, variety of constituents), are the most relevant materials for tissue engineering and regenerative medicine approaches [87]. Tissue engineering has been described as "an exciting and revolutionary strategy to treat patients who need a new organ or tissue" thanks to the possibility to supply man-made organs or tissues [88]. Hydrogels are currently used with several applications: as scaffolds, as tissue barriers and bioadhesives, as drug depots, as delivery systems for bioactive agents (that enhance the natural reparative process), as matrices to encapsulate and deliver cells [87].

The traditional approaches to tissue or organ failure have been based mainly on the replacement of the tissue with a synthetic implant (if possible) or in the transplantation of the organ (if available). Also extracorporeal treatments, based on passive membrane exchange or experimental biohybrid systems of encapsulated cells, are used (see the scheme in Figure 5). Thus tissue engineering represents a great potential for the regeneration of tissues and organs. Among the several strategies, the most appealing is based on the combination of cells (generally isolated from a small tissue biopsy from the patient) and polymer scaffolds [77]. These 3D-scaffolds are used as substrates for supporting and guiding tissue regeneration in several *in vitro* and *in vivo* systems. Tissue-specific cells are collected *in vitro* and subsequently incorporated into the scaffold that mimics the extracellular matrix (ECM). ECM is a very intriguing structure, made of protein and sugar-based macromolecules: it is a substrate for cell adhesion and migration and provides the physical support for cells, tissue and organ growth. ECM enables also the diffusion of nutrients and metabolites and regulates, thanks to the presence of growth factors, cellular proliferation and functionality. The preparation of ECM-mimicking biomaterials with defined shape and complex porous architecture represents still a technical challenge. The scaffold design is necessary to regulate various aspects of the tissue-formation process; the spatial and temporal control of biochemical and physical cell properties has been pursued by incorporating molecules involved in cellular processes [89].

Hydrogels designed as 3D-scaffolds should present pores with the ideal size to accommodate living cells; moreover, they can be designed to dissolve, releasing active molecules [77, 88]. Nowadays a variety of tissues has been engineered, with systems at the stage of clinical trial or application, including artery, bladder, skin, cartilage, bone, ligament, and tendon [35, 36, 88]. Among other known systems, the hydrogel capsules represent a very peculiar type of scaffold.

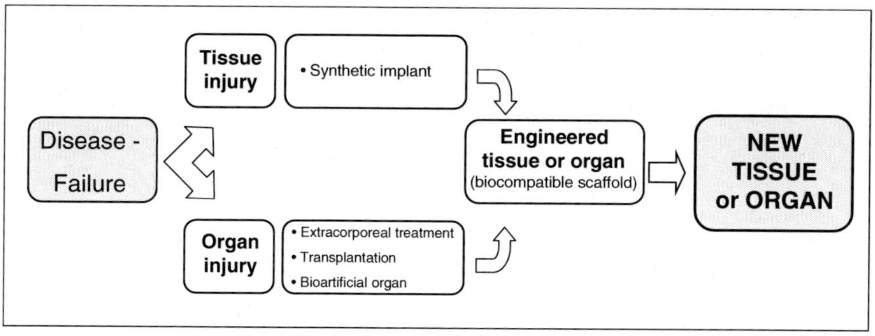


Figure 6. Schematization of the evolution of the treatment of organ and tissue injury and diseases. Redrawn from Hoffmann [77].

Although cell immobilization and encapsulation systems have a broad range of applications, the most studied and described is the transplantation of microencapsulated cells, proposed as a therapy for the treatment of a wide variety of diseases. This technology is based on the principle that foreign cells (xenogeneic or allogeneic) are physically segregated and protected from the host immune system by an artificial membrane [37, 90, 91].

Thus hydrogel microcapsules can provide the immunoisolation, without interfering with the diffusion of oxygen, nutrients, and metabolic products [87]. Some relevant examples are the treatment of cardiovascular diseases, kidney and liver failure, and diabetes mellitus (by the microencapsulation of islets of Langerhans) [36]. Various cell types including primary cells, stem cells or bioengineered cells have been investigated for the treatment of several diseases, with clinical application [36, 75, 92-95]. Considering the field of stem cell biology [96], biomimetic constructs are used to prepare model systems and to develop carriers for stem cells, facilitating their *in vivo* use in cell therapy, tissue engineering and regenerative medicine. Examples are the application to diabetes therapy (with the use of stem cells both for the replacement of damaged islets, and as cell source for islet-cell transplantation [97]), and the application to cartilage regenerative medicine [98].

## 4. Nanoscale Carriers: Nanocapsules and Nanoparticles

Bio-based polymers present important applications also at the nanoscale level. Biopolymeric nanocapsules, nanogels and nanoparticles have been developed since the early nineties, in parallel to the liposome and the other lipid-based nanosystems, and have the logical application to the encapsulation of biological entities of smaller dimensions (mainly drugs, therapeutic proteins, nucleic acids) [99-102]. Drug delivery systems based on polysaccharide nanoparticles present several advantages: the capability to penetrate cells and tissues; the possibility to improve the bioavailability of drugs, reducing toxic side effects; the ability to control release properties due to the biodegradability and the stimuli (pH, ion and temperature) sensibility of materials [58]. The rationalization of the assembly mechanisms and the capability to tailor the properties (size, charge, and loading capability) to desirable levels are essential goals to advance biodegradable polysaccharidic nanoparticles as efficient drug delivery vehicles. Nowadays the results obtained by the researchers in this field has generated a huge number of patents [103, 104] and a remarkable scientific production: for historical and technical overviews of these systems, the reader is referred to some exhaustive recent reviews [50, 105-109].

Nanoscale systems of various origin are presently developed mainly for pharmaceutical and biomedical applications, without involving all other nanotechnological products for non-human uses. There are social and ethical issues related to the use of nanotechnology in the food and drug sector that must be considered, since the potential risks of nanomaterials to human health and to the environment are still partially unknown [110, 111]. Thus, world-wide research is strongly promoted by several agencies, and is going on also in the food sector, in the production of functional food, but also in packaging and food-sensors. In this respect, the use of biocompatible and biodegradable materials (in particular protein, lipids and polysaccharides) represents a great opportunity for all nanotechnological applications [112, 113].

Advanced nanotechnological applications of polysaccharides include their use as coating material as agents for increased biocompatibility of less safe materials. As a typical example, being effective colloidal stabilizers, polysaccharides confer new chemical functionalities to encapsulated inorganic nanosystems. Thus, several polysaccharide coatings have been tested with carbon nanotubes, colloidal silver and gold particles, quantum dots and iron oxide magnetic nanoparticles. Some of these systems are already on the market [5].

## 5. Biopolymer Engineering for "Biosmarterials"

The biomaterial research is aimed at the production of "bioSMARTerials" (smart biomaterials), designed with molecular precision, which present the required properties for advanced technological applications. This is pursued by controlling the material characteristics at the nanoscale, in particular at the natural building blocks level [114, 115]. The specific material requirements include some biological activity, in addition to the mechanical stiffness or toughness and the controlled swelling and degradation. Most commonly, the additional biological properties include binding or release of bioactive molecules, control of cell attachment, and direct interface with cells/tissues through defined molecular pathways, in order to guide the biological responses. The possibility to engineer biological activity into biomaterials improves the performances in the traditional applications, and increases the number of new potential uses [116]. For polysaccharide-based biomaterials, the molecular up-grade and control can be achieved by engineering the polysaccharide molecules and gels by means of chemical, physico-chemical and enzymatic strategies [117].

Chemical modifications can be obtained by using traditional post-polymerization reactions, thus by exploiting the functional groups already present on the polysaccharide molecules (in particular amine, carboxyl and hydroxyl groups) in order to introduce selected "decorations" for specific applications, such as those widely exploited for chitosan [32]. An enlightening example of this approach is the introduction of bioactive molecules of saccharidic or peptidic origin to enhance cooperatively the interaction with cells and tissues, through the so-called cluster effect [118]. This strategy leads to the production of polymers conjugated with molecules and other functional groups, able to interact with cellular receptors and tissue structures (i.e. mucosal barriers). Relevant examples are the introduction of saccharidic (galactose) residues on alginate and chitosan for cell (hepatocytes, chondrocyte,...) embedding, to develop biomedical devices for tissue engineering [119-122]. Other examples are the synthesis of *N*-trimethyl chitosan for specific drug delivery applications [123], the functionalization of biopolymers with the Arg–Gly–Asp (RGD) peptide sequence or other similar sequences to enhance cell adhesion [7, 124-126], and the synthesis of amphiphilic alginate for drug delivery systems [7].

Thiolated polymers (thiomers) also represent an important class of functional polymers obtained by chemical modification, characterized by high mucoadhesive properties. These materials (obtained by the modification of chitosan, hyaluronan or other polymers) are capable of forming covalent bonds (disulfide bonds) with the mucus gel layer of the mucosa, in particular with the cysteine-rich sub-domains of mucus glycoproteins [127]. Chemical modifications have been also employed to bind chemical structures to be used to crosslink the polysaccharide chains to create new polymer networks or to reinforce the preexisting one [128-130].

Any chemical modifications of a polymer obviously influence its interacting features as well as its physico-chemical properties. Indeed, local change in a single monomer can even affect the conformation of the whole chain, resulting thus in both structural alterations, and in modified capability to interact with other polymer chains and molecules. For example, when alginate is modified by introducing residues of chemical or biological interest, the new polymer architecture may interfere with the calcium binding capability and thus with the gel formation process, leading to weak hydrogels, if the modifications are made on the G-units [131]. For this reason, a strict control of the functionalization conditions and of the substitution degree is needed, controlling both the amount of the pendant molecules or modification introduced and the selectivity of the reaction [132].

Physico-chemical strategy is mainly represented by the association of different biopolymers in order to design novel biomaterials, exploiting the new properties exhibited with respect of the parent polymer system. Examples of combined materials are the matrices obtained by blending alginate with hyaluronic acid, collagen and chitosan for cell immobilization in tissue engineering applications [133-137]. This approach requires a detailed study of the single components and of the conditions to allow the reciprocal compatibility, coupled with an extensive investigation on the synergistic interactions within the composite materials, which all determine both mechanical and biological properties [138, 139]. Another possible approach which exploits the physico-chemical modulation of the biomaterials has been described for alginate encapsulation and is based on the use of different crosslink cations as an alternative for calcium, such as barium, which has the capability to form stronger gels [140].

Finally, enzymatic strategies represent useful tools for selective biopolymer modifications, by exploiting the specificity of different enzyme classes for particular functional groups or polymer regions. Among the several examples of these modifications, it suffices to mention the alginate engineering by using mannuronan C5-epimerases [141], the enzymatic grafting of chitosan [32], and the use of transglutaminases (TGase) for hyaluronic acid (HA) and/or gelatin crosslink [142-144].

Undoubtedly, the richness of biopolymer structures, of modification strategies and of the nanogel assembly of the derived polymers thereof offers an almost unlimited combination properties that need only to be developed for advanced application.

## Conclusion

Although briefly, several features that make the polysaccharides alginate and chitosan worth of investigation have been described. The intention of assessing the performance of these biopolymers in the construction of hydrogel nanostructures has been made evident for the applications in food, biomedicals, and pharmaceuticals. The main reasons for the attention to polysaccharides are certainly their versatility and complete biocompatibility and biodegradability. It has been also stressed that additional performance, although often not fully explored, can be easily achieved by using a long list of potential derivatives, in which pendant moieties are introduced to modify the solution and gel properties.

In the block-type chart presentation, several points provide hints about how the final properties of the products may be more efficiently altered and how future developments can take full advantage of these biosystems. Still several issues of hydrogel structure remain unclarified or are not easily manipulated untill the full control of their chain structure is reached.

## Acknowledgments

The Authors acknowledge their colleagues and coworkers Sergio Paoletti, Ivan Donati, Gianni Sava and Moreno Cocchietto, for fruitful discussion and long standing collaboration. The encapsulation work has been carried out in the frame of projects "Oral vaccine carrier for fish farming of Friuli Venezia Giulia" and NanoBioPharmaceutics (FP6 EU Project no. 026723-2) with support of the University of Trieste. MB is recipient of a grant from MIUR (Rome).

## References

[1] Coviello T, Matricardi P, Marianecci C, Alhaique F. Polysaccharide hydrogels for modified release formulations. *Journal of Controlled Release* 2007;119:5-24.

[2] D'Ayala G, Malinconico M, Laurienzo P. Marine Derived Polysaccharides for Biomedical Applications: Chemical Modification Approaches. *Molecules* 2008;13:2069-106.

[3] Rinaudo M. Main properties and current applications of some polysaccharides as biomaterials. *Polymer International* 2008;57:397-430.

[4] Laurienzo P. Marine polysaccharides in pharmaceutical applications: an overview. *Mar. Drugs* 2010;8:2435-65.

[5] Dias AMGC, Hussain A, Marcos AS, Roque ACA. A biotechnological perspective on the application of iron oxide magnetic colloids modified with polysaccharides. *Biotechnology advances* 2011;29:142-55.

[6] Peppas NA, Bures P, Leobandung W, Ichikawa H. Hydrogels in pharmaceutical formulations. *European Journal of Pharmaceutics and Biopharmaceutics* 2000;50:27-46.

[7] Lee KY, Mooney DJ. Alginate: Properties and biomedical applications. *Progress in Polymer Science* 2012;37:106-26.

[8] Draget KI, Smidsrød O, Skjåk-Bræk G. Alginates from Algae. In: Steinbüchel A, Rhee SK, editors. *Polysaccharides and Polyamides in the Food Industry Properties, Production, and Patents*. Weinheim: WILEY-VCH; 2005. p. 1.

[9] Grant GT, Morris ER, Rees DA, Smith PJC, Thom D. Biological interactions between polysaccharides and divalent cations: The egg-box model. *FEBS Letters* 1973;32:195-8.

[10] Morris ER, Rees DA, Thom D, Boyd J. Chiroptical and stoichiometric evidence of a specific, primary dimerisation process in alginate gelation. *Carbohydrate Research* 1978;66:145-54.

[11] Smidsrød O, Haug A. Dependence upon gel-sol state of ionexchange properties of alginates. *Acta Chem. Scand.* 1972;26:2063-74.

[12] Smidsrød O. *Molecular basis for some physical properties of alginates in the gel state Faraday Discussions of the Chemical Society* 1974;57:263-74.

[13] Kohn R. Ion binding on polyuronates - alginate and pectin. *Pure and Applied Chemistry* 1975;42:371-97.

[14] Braccini I, Pérez S. Molecular Basis of $Ca^{2+}$-Induced Gelation in Alginates and Pectins: The Egg-Box Model Revisited. *Biomacromolecules* 2001;2:1089-96.

[15] Li L, Fang Y, Vreeker R, Appelqvist I, Mendes E. Reexamining the Egg-Box Model in Calcium−Alginate Gels with X-ray Diffraction. *Biomacromolecules* 2007;8:464-8.

[16] Sikorski P, Mo F, Skjåk-Bræk G, Stokke BT. Evidence for Egg-Box-Compatible Interactions in Calcium−Alginate Gels from Fiber X-ray Diffraction. *Biomacromolecules* 2007;8:2098-103.

[17] Donati I, Cesàro A, Paoletti S. Specific Interactions versus Counterion Condensation. 1. Nongelling Ions/Polyuronate Systems. *Biomacromolecules* 2005;7:281-7.

[18] Donati I, Benegas JC, Cesàro A, Paoletti S. Specific Interactions versus Counterion Condensation. 2. Theoretical Treatment within the Counterion Condensation Theory. *Biomacromolecules* 2006;7:1587-96.

[19] Donati I, Benegas JC, Paoletti S. Polyelectrolyte Study of the Calcium-Induced Chain Association of Pectate. *Biomacromolecules* 2006;7:3439-47.

[20] Donati I, Paoletti S. Material properties of alginates. In: Rehm BHA, editor. Alginates: *Biology and Applications.* Berlin Heidelberg: Springer-Verlag; 2009. p. 1.

[21] Donati I, Holtan S, Mørch YA, Borgogna M, Dentini M, Skjåk-Bræk G. New Hypothesis on the Role of Alternating Sequences in Calcium−Alginate Gels. *Biomacromolecules* 2005;6:1031-40.

[22] Niederhofer A, Müller BW. A method for direct preparation of chitosan with low molecular weight from fungi. *European Journal of Pharmaceutics and Biopharmaceutics* 2004;57:101-5.

[23] Wang W, Du Y, Qiu Y, Wang X, Hu Y, Yang J, et al. A new green technology for direct production of low molecular weight chitosan. *Carbohydrate Polymers* 2008;74:127-32.

[24] Vårum KM, Smidsrød O. Structure-Property Relationship in Chitosans. In: Dumitriu S, editor. *Polysaccharides Structural Diversity and Functional Versatility.* New York: Marcel Dekker; 2004. p. 625.

[25] Ogawa K, Yui T, Okuyama K. Three D structures of chitosan. *International Journal of Biological Macromolecules* 2004;34:1-8.

[26] Signini R, Desbrières J, Campana Filho SP. On the stiffness of chitosan hydrochloride in acid-free aqueous solutions. *Carbohydrate Polymers* 2000;43:351-7.
[27] Skovstrup S, Hansen SG, Skrydstrup T, Schiøtt B. Conformational Flexibility of Chitosan: A Molecular Modeling Study. *Biomacromolecules* 2010;11:3196-207.
[28] Lamarque G, Lucas J-M, Viton C, Domard A. Physicochemical Behavior of Homogeneous Series of Acetylated Chitosans in Aqueous Solution: Role of Various Structural Parameters. *Biomacromolecules* 2004;6:131-42.
[29] Christensen BE, Vold IMN, Vårum KM. Chain stiffness and extension of chitosans and periodate oxidised chitosans studied by size-exclusion chromatography combined with light scattering and viscosity detectors. *Carbohydrate Polymers* 2008;74:559-65.
[30] Ravi Kumar MNV, Muzzarelli RAA, Muzzarelli C, Sashiwa H, Domb AJ. Chitosan Chemistry and Pharmaceutical Perspectives. *Chemical Reviews* 2004;104:6017-84.
[31] Rabea EI, Badawy MET, Stevens CV, Smagghe G, Steurbaut W. Chitosan as Antimicrobial Agent: Applications and Mode of Action. *Biomacromolecules* 2003;4:1457-65.
[32] Alves NM, Mano JF. Chitosan derivatives obtained by chemical modifications for biomedical and environmental applications. *International Journal of Biological Macromolecules* 2008;43:401-14.
[33] Bonferoni MC, Sandri G, Rossi S, Ferrari F, Caramella C. Chitosan and its salts for mucosal and transmucosal delivery. *Expert Opinion on Drug Delivery* 2009;6:923-39.
[34] Mao S, Sun W, Kissel T. Chitosan-based formulations for delivery of DNA and siRNA. *Advanced Drug Delivery Reviews* 2010;62:12-27.
[35] Costa-Pinto AR, Reis RL, Neves MN. Scaffolds Based Bone Tissue Engineering: *The Role of Chitosan*. Tissue Engineering Part B: Reviews 2011;17:331-47.
[36] Brun-Graeppi AKAS, Richard C, Bessodes M, Scherman D, Merten O-W. Cell microcarriers and microcapsules of stimuli-responsive polymers. *Journal of Controlled Release* 2011;149:209-24.
[37] Smidsrød O, Skjåk-Bræk G. Alginate as immobilization matrix for cells. *Trends in Biotechnology* 1990;8:71-8.
[38] Janes KA, Calvo P, Alonso MJ. Polysaccharide colloidal particles as delivery systems for macromolecules. *Adv. Drug. Deliv. Rev.* 2001;47:83-97.
[39] Gombotz WR, Wee S. Protein release from alginate matrices. *Advanced Drug Delivery Reviews* 1998;31:267-85.
[40] Champagne CP, Fustier P. Microencapsulation for the improved delivery of bioactive compounds into foods. *Current opinion in biotechnology* 2007;18:184-90.
[41] Borgogna M, Bellich B, Cesàro A. Marine Polysaccharides in Microencapsulation and Application to Aquaculture: "From Sea to Sea". *Marine Drugs* 2011;9:2572-604.
[42] Gouin S. Microencapsulation. *Trends in Food Science and Technology* 2004;15:330-47.
[43] Gharsallaoui A, Roudaut G, Chambin O, Voilley A, Saurel R. Applications of spray-drying in microencapsulation of food ingredients: An overview. *Food Research International* 2007;40:1107-21.
[44] de Vos P, Faas MM, Spasojevic M, Sikkema J. Encapsulation for preservation of functionality and targeted delivery of bioactive food components. *International Dairy Journal* 2010;20:292-302.

[45] Cui JH, Goh JS, Park SY, Kim PH, Lee BJ. Preparation and physical characterization of alginate microparticles using air atomization method. *Drug Development and Industrial Pharmacy* 2001;27:309-19.
[46] Teunou E, Poncelet D. Rotary disc atomisation for microencapsulation applications-prediction of the particle trajectories. *Journal of Food Engineering* 2005;71:345-53.
[47] Prüsse U, Bilancetti L, Bučko M, Bugarski B, Bukowski J, Gemeiner P, et al. Comparison of different technologies for alginate beads production. *Chemical Papers* 2008;62:364-74.
[48] Didier M, Hamed L, Shanthi V S. A method to target bioactive compounds to specific regions of the gastrointestinal tract: *double gavage using polysaccharide hydrogels.* 2009.
[49] Pinto Reis C, Neufeld RJ, Ribeiro AJ, Veiga F. Nanoencapsulation I. Methods for preparation of drug-loaded polymeric nanoparticles. Nanomedicine: *Nanotechnology, Biology, and Medicine* 2006;2:8-21.
[50] Anton N, Benoit JP, Saulnier P. Design and production of nanoparticles formulated from nano-emulsion templates-a review. *J. Control Release* 2008;128:185-99.
[51] Barbosa-Cánovas GV, Ortega-Rivas E, Juliano P, Yan H. Food powders: *physical properties, processing, and functionality*. New York: Kluwer Academic/Plenum Publishers; 2005.
[52] Tønnesen HH, Karlsen J. Alginate in drug delivery systems. *Drug Development and Industrial Pharmacy* 2002;28:621-30.
[53] Zimmermann H, Shirley SG, Zimmermann U. Alginate-based encapsulation of cells: Past, present, and future. *Current Diabetes Reports* 2007;7:314-20.
[54] de Vos P, Bučko M, Gemeiner P, Navrátil M, Švitel J, Faas M, et al. Multiscale requirements for bioencapsulation in medicine and biotechnology. *Biomaterials* 2009;30:2559-70.
[55] Lee KY, Mooney DJ. Alginate: *Properties and biomedical applications. Progress in Polymer Science* 2011;doi:10.1016/j.progpolymsci.2011.06.003.
[56] Gåserød O, Smidsrød O, Skjåk-Bræk G. Microcapsules of alginate-chitosan - I. A quantitative study of the interaction between alginate and chitosan. *Biomaterials* 1998;19:1815-25.
[57] Ribeiro AJ, Silva C, Ferreira D, Veiga F. Chitosan-reinforced alginate microspheres obtained through the emulsification/internal gelation technique. European journal of pharmaceutical sciences: *official journal of the European Federation for Pharmaceutical Sciences* 2005;25:31-40.
[58] Liu Z, Jiao Y, Wang Y, Zhou C, Zhang Z. Polysaccharides-based nanoparticles as drug delivery systems. *Adv. Drug. Deliv. Rev.* 2008;60:1650-62.
[59] Hamman JH. Chitosan Based Polyelectrolyte Complexes as Potential *Carrier Materials in Drug Delivery Systems. Mar Drugs* 2010;8:1305-22.
[60] Patil JS, Kamalapur MV, Marapur SC, Kadam DV. Ionotropic gelation and polyelectrolyte complexation: The novel techniques to design hydrogel particulate sustained, modulated drug delivery system: A review. *Digest Journal of Nanomaterials and Biostructures* 2010;5:241-8.
[61] Batorsky A, Liao J, Lund AW, Plopper GE, Stegemann JP. Encapsulation of adult human mesenchymal stem cells within collagen-agarose microenvironments. *Biotechnology and bioengineering* 2005;92:492-500.

[62] Kumachev A, Greener J, Tumarkin E, Eiser E, Zandstra PW, Kumacheva E. High-throughput generation of hydrogel microbeads with varying elasticity for cell encapsulation. *Biomaterials* 2011;32:1477-83.

[63] Jeon O, Bouhadir KH, Mansour JM, Alsberg E. Photocrosslinked alginate hydrogels with tunable biodegradation rates and mechanical properties. *Biomaterials* 2009;30:2724-34.

[64] Cellesi F, Tirelli N. A new process for cell microencapsulation and other biomaterial applications: Thermal gelation and chemical cross-linking in 'tandem'. *Journal of Materials Science*: Materials in Medicine 2005;16:559-65.

[65] Shen F, Li AA, Cornelius RM, Cirone P, Childs RF, Brash JL, et al. Biological properties of photocrosslinked alginate microcapsules. *Journal of Biomedical Materials Research* - Part B Applied Biomaterials 2005;75:425-34.

[66] Rokstad AM, Donati I, Borgogna M, Oberholzer J, Strand BL, Espevik T, et al. Cell-compatible covalently reinforced beads obtained from a chemoenzymatically engineered alginate. *Biomaterials* 2006;27:4726-37.

[67] Wandrey C, Bartkowiak A, Harding SE. Materials for Encapsulation. In: Zuidam NJ, Nedović V, editors. *Encapsulation Technologies for Active Food Ingredients and Food Processing*: Springer; 2009.

[68] Brownlee IA, Seal CJ, Wilcox M, Dettmar PW, Pearson JP. Application of Alginates in Food. In: Rehm BHA, editor. Alginates: *Biology and Applications*. Berlin Heidelberg: Springer-Verlag; 2009. p. 211.

[69] Kean T, Thanou M. Biodegradation, biodistribution and toxicity of chitosan. *Advanced Drug Delivery Reviews* 2010;62:3-11.

[70] Desai KGH, Park HJ. Recent developments in microencapsulation of food ingredients. *Drying Technology* 2005;23:1361-94.

[71] Chen L, Remondetto GE, Subirade M. Food protein-based materials as nutraceutical delivery systems. *Trends in Food Science and Technology* 2006;17:272-83.

[72] Klein S. Polysaccharides in Oral Drug Delivery — Recent Applications and Future Perspectives. In: Edgar K, editor. Polysaccharide Materials: *Performance by Design*. Washington, DC: American Chemical Society; 2009. p. 13-30.

[73] Amidi M, Mastrobattista E, Jiskoot W, Hennink WE. Chitosan-based delivery systems for protein therapeutics and antigens. *Adv. Drug. Deliv. Rev.* 2010;62:59-82.

[74] Bhattarai N, Gunn J, Zhang M. Chitosan-based hydrogels for controlled, localized drug delivery. *Adv. Drug. Deliv. Rev.* 2010;62:83-99.

[75] Van Vlierberghe S, Dubruel P, Schacht E. Biopolymer-based hydrogels as scaffolds for tissue engineering applications: a review. *Biomacromolecules* 2011;12:1387-408.

[76] Faroongsarng D, Sukonrat P. Thermal behavior of water in the selected starch- and cellulose-based polymeric hydrogels. *International Journal of Pharmaceutics* 2008;352:152-8.

[77] Hoffman AS. Hydrogels for biomedical applications. *Adv. Drug. Deliv. Rev.* 2002;54:3-12.

[78] He P, Davis SS, Illum L. Chitosan microspheres prepared by spray drying. *Int. J. Pharm.* 1999;187:53-65.

[79] Illum L, Jabbal-Gill I, Hinchcliffe M, Fisher AN, Davis SS. Chitosan as a novel nasal delivery system for vaccines. Adv Drug Deliv Rev 2001;51:81-96.

[80] George M, Abraham TE. Polyionic hydrocolloids for the intestinal delivery of protein drugs: alginate and chitosan--a review. *J. Control Release* 2006;114:1-14.
[81] Borchard G. Chitosans for gene delivery. *Adv. Drug. Deliv. Rev.* 2001;52:145-50.
[82] Mao S, Sun W, Kissel T. Chitosan-based formulations for delivery of DNA and siRNA *Adv. Drug. Deliv. Rev.* 2010;62:12-27.
[83] Tafaghodi M, Sajadi Tabassi SA, Jaafari MR. Induction of systemic and mucosal immune responses by intranasal administration of alginate microspheres encapsulated with tetanus toxoid and CpG-ODN. *Int. J. Pharm.* 2006;319:37-43.
[84] Wen Z-S, Xu Y-L, Zou X-T, Xu Z-R. Chitosan Nanoparticles Act as an Adjuvant to Promote both Th1 and Th2 Immune Responses Induced by Ovalbumin in Mice. *Mar. Drugs* 2011;9:1038-55.
[85] Año G, Esquisabel A, Pastor M, Talavera A, Cedré B, Fernández S, et al. A new oral vaccine candidate based on the microencapsulation by spray-drying of inactivated Vibrio cholerae. *Vaccine* 2011;29:5758-64.
[86] Plant KP, Lapatra SE. Advances in fish vaccine delivery. *Developmental and comparative immunology* 2011; in press DOI: 10.1016/j.dci.2011.03.007.
[87] Slaughter BV, Khurshid SS, Fisher OZ, Khademhosseini A, Peppas NA. Hydrogels in regenerative medicine. *Advanced Materials* 2009;21:3307-29.
[88] Lee KY, Mooney DJ. Hydrogels for tissue engineering. *Chemical Reviews* 2001;101:1869-79.
[89] Luo Y, Engelmayr G, Auguste DT, Ferreira L, Karp JM, Saigal R, et al. Three Dimensional Scaffolds. In: Lanza R, Langer R, Vacanti JP, editors. *Principles of Tissue Engineering*: Elsevier Academic Press; 2007. p. 359.
[90] Dulieu C, Poncelet D, Neufeld RJ. Encapsulation and immobilization techniques. In: Kühtreiber WM, Lanza RP, Chick WL, editors. Cell *Encapsulation Technology and Therapeutics*. Boston: Birkhäuser 1999.
[91] de Vos P, Faas MM, Strand B, Calafiore R. Alginate-based microcapsules for immunoisolation of pancreatic islets. *Biomaterials* 2006;27:5603-17.
[92] Murua A, Portero A, Orive G, Hernandez RM, de Castro M, Pedraz JL. Cell microencapsulation technology: towards clinical application. *Journal of Controlled Release* 2008;132:76-83.
[93] Slaughter BV, Khurshid SS, Fisher OZ, Khademhosseini A, Peppas NA. Hydrogels in regenerative medicine. *Adv. Mater.* 2009;21:3307-29.
[94] Hernandez RM, Orive G, Murua A, Pedraz JL. Microcapsules and microcarriers for in situ cell delivery. *Advanced Drug Delivery Reviews* 2010;62:711-30.
[95] Hunt NC, Grover LM. Cell encapsulation using biopolymer gels for regenerative medicine. *Biotechnology letters* 2010;32:733-42.
[96] Lutolf MP, Gilbert PM, Blau HM. Designing materials to direct stem-cell fate. *Nature* 2009;462:433-41.
[97] Gangaram-Panday ST, Faas MM, de Vos P. Towards stem-cell therapy in the endocrine pancreas. *Trends in molecular medicine* 2007;13:164-73.
[98] Ghidoni I, Chlapanidas T, Bucco M, Crovato F, Marazzi M, Vigo D, et al. Alginate cell encapsulation: new advances in reproduction and cartilage regenerative medicine. *Cytotechnology* 2008;58:49-56.

[99] Rajaonarivony M, Vauthier C, Couarraze G, Puisieux F, Couvreur P. Development of a new drug carrier made from alginate. *Journal of Pharmaceutical Sciences* 1993;82:912-7.
[100] Skiba M, Morvan C, Duchene D, Puisieux F, Wouessidjewe D. Evaluation of gastrointestinal behaviour in the rat of amphiphilic β-cyclodextrin nanocapsules, loaded with indomethacin. *Int. J. Pharm.* 1995;126:275-9.
[101] Calvo P, Remuñan-López C, Vila-Jato JL, Alonso MJ. Chitosan and Chitosan/Ethylene Oxide-Propylene Oxide Block Copolymer Nanoparticles as Novel Carriers for Proteins and Vaccines. *Pharmaceutical Research* 1997;14:1431-6.
[102] Riley T, Govender T, Stolnik S, Xiong CD, Garnett MC, Illum L, et al. Colloidal stability and drug incorporation aspects of micellar-like PLA–PEG nanoparticles. Colloids and Surfaces B: *Biointerfaces* 1999;16:147-59.
[103] Bhavna MA, Sanjula B, Javed A. Patents on Nanoparticulate Drug Delivery Systems - A Review. *Recent Patents on Drug Delivery and Formulation* 2008;2:83-9.
[104] Wong TW. Chitosan and Its Use in Design of Insulin Delivery System. *Recent Patents on Drug Delivery and Formulation* 2009;3:8-25.
[105] Bawarski WE, Chidlowsky E, Bharali DJ, Mousa SA. Emerging nanopharmaceuticals. Nanomedicine: Nanotechnology, *Biology, and Medicine* 2008;4:273-82.
[106] Millotti G, Bernkop-Schnürch A. *Nano- and Microparticles in Oral Delivery of Macromolecular Drugs*. In: Bernkop-Schnürch A, editor.: Springer New York; 2009. p. 153.
[107] Mishra B, Patel BB, Tiwari S. Colloidal nanocarriers: a review on formulation technology, types and applications toward targeted drug delivery. Nanomedicine : *nanotechnology, biology, and medicine* 2010;6:9-24.
[108] Mora-Huertas CE, Fessi H, Elaissari A. Polymer-based nanocapsules for drug delivery. *Int. J. Pharm.* 2010;385:113-42.
[109] Phillips MA, Gran ML, Peppas NA. Targeted Nanodelivery of Drugs and Diagnostics. *Nano today* 2010;5:143-59.
[110] Grainger DW. Nanotoxicity assessment: all small talk? *Adv. Drug. Deliv. Rev.* 2009;61:419-21.
[111] Sozer N, Kokini JL. Nanotechnology and its applications in the food sector. *Trends in Biotechnology* 2009;27:82-9.
[112] Weiss J, Takhistov P, McClements DJ. Functional Materials in Food Nanotechnology. *Journal of Food Science* 2006;71:R107-R16.
[113] Peters R, Dam Gt, Bouwmeester H, Helsper H, Allmaier G, Kammer Fv, et al. Identification and characterization of organic nanoparticles in food. *TrAC Trends in Analytical Chemistry* 2011;30:100-12.
[114] Liu Tsang V, Bhatia SN. Three-dimensional tissue fabrication. *Advanced Drug Delivery Reviews* 2004;56:1635-47.
[115] Peppas NA, Webster JL. *Nanomaterials handbook*: CRC/Taylor and Francis; 2006.
[116] Huebsch N, Mooney DJ. Inspiration and application in the evolution of biomaterials. *Nature* 2009;462:426-32.
[117] Hoare TR, Kohane DS. Hydrogels in drug delivery: Progress and challenges. *Polymer* 2008;49:1993-2007.
[118] Chabre YM, Roy R. Design and creativity in synthesis of multivalent neoglycoconjugates. *Adv. Carbohydr. Chem. Biochem.* 2010;63:165-393.

[119] Yang J, Goto M, Ise H, Cho C-S, Akaike T. Galactosylated alginate as a scaffold for hepatocytes entrapment. *Biomaterials* 2002;23:471-9.

[120] Park I-K, Yang J, Jeong H-J, Bom H-S, Harada I, Akaike T, et al. Galactosylated chitosan as a synthetic extracellular matrix for hepatocytes attachment. *Biomaterials* 2003;24:2331-7.

[121] Donati I, Stredanska S, Silvestrini G, Vetere A, Marcon P, Marsich E, et al. The aggregation of pig articular chondrocyte and synthesis of extracellular matrix by a lactose-modified chitosan. *Biomaterials* 2005;26:987-98.

[122] Cho CS, Seo SJ, Park IK, Kim SH, Kim TH, Hoshiba T, et al. Galactose-carrying polymers as extracellular matrices for liver tissue engineering. *Biomaterials* 2006;27:576-85.

[123] Amidi M, Romeijn SG, Borchard G, Junginger HE, Hennink WE, Jiskoot W. Preparation and characterization of protein-loaded N-trimethyl chitosan nanoparticles as nasal delivery system. *Journal of Controlled Release* 2006;111:107-16.

[124] Rowley JA, Madlambayan G, Mooney DJ. Alginate hydrogels as synthetic extracellular matrix materials. *Biomaterials* 1999;20:45-53.

[125] Rowley JA, Mooney DJ. Alginate type and RGD density control myoblast phenotype. *Journal of Biomedical Materials Research* 2002;60:217-23.

[126] Hersel U, Dahmen C, Kessler H. RGD modified polymers: biomaterials for stimulated cell adhesion and beyond. *Biomaterials* 2003;24:4385-415.

[127] Bernkop-Schnürch A. Thiomers: A new generation of mucoadhesive polymers. *Advanced Drug Delivery Reviews* 2005;57:1569-82.

[128] Smeds KA, Pfister-Serres A, Miki D, Dastgheib K, Inoue M, Hatchell DL, et al. Photocrosslinkable polysaccharides for in situ hydrogel formation. *Journal of Biomedical Materials Research* 2001;55:254-5.

[129] Rokstad AM, Donati I, Borgogna M, Oberholzer J, Strand BL, Espevik T, et al. Cell-compatible covalently reinforced beads obtained from a chemoenzymatically engineered alginate. *Biomaterials* 2006;27:4726-37.

[130] Bhattarai N, Gunn J, Zhang M. Chitosan-based hydrogels for controlled, localized drug delivery. *Advanced Drug Delivery Reviews* 2010;62:83-99.

[131] Donati I, Vetere A, Gamini A, Coslovi A, Campa C, Paoletti S. Galactose-Substituted Alginate: Preliminary Characterization and Study of Gelling Properties. *Biomacromolecules* 2003;4:624-31.

[132] Donati I, Draget KI, Borgogna M, Paoletti S. Tailor-Made Alginate Bearing Galactose Moieties on Mannuronic Residues: Selective Modification Achieved by a Chemoenzymatic Strategy. *Biomacromolecules* 2004;6:88-98.

[133] Oerther S, Payan E, Lapicque F, Presle N, Hubert P, Muller S, et al. Hyaluronate-alginate combination for the preparation of new biomaterials: *investigation of the behaviour in aqueous solutions*. Biochimica et Biophysica Acta (BBA) - General Subjects 1999;1426:185-94.

[134] Gerard C, Catuogno C, Amargier-Huin C, Grossin L, Hubert P, Gillet P, et al. The effect of alginate, hyaluronate and hyaluronate derivatives biomaterials on synthesis of non-articular chondrocyte extracellular matrix. *Journal of Materials Science: Materials in Medicine* 2005;16:541-51.

[135] Li Z, Zhang M. Chitosan–alginate as scaffolding material for cartilage tissue engineering. *Journal of Biomedical Materials Research* Part A 2005;75A:485-93.

[136] Hahn MS, Teply BA, Stevens MM, Zeitels SM, Langer R. Collagen composite hydrogels for vocal fold lamina propria restoration. *Biomaterials* 2006;27:1104-9.
[137] Marsich E, Borgogna M, Donati I, Mozetic P, Strand BL, Salvador SG, et al. Alginate/lactose-modified chitosan hydrogels: A bioactive biomaterial for chondrocyte encapsulation. *Journal of Biomedical Materials Research* Part A 2008;84A:364-76.
[138] Donati I, Borgogna M, Turello E, Cesàro A, Paoletti S. Tuning Supramolecular Structuring at the Nanoscale Level: Nonstoichiometric Soluble Complexes in Dilute Mixed Solutions of Alginate and Lactose-Modified Chitosan (Chitlac). *Biomacromolecules* 2007;8:1471-9.
[139] Donati I, Haug IJ, Scarpa T, Borgogna M, Draget KI, Skjåk-Bræk G, et al. Synergistic Effects in Semidilute Mixed Solutions of Alginate and Lactose-Modified Chitosan (Chitlac). *Biomacromolecules* 2007;8:957-62.
[140] Mørch ÝA, Donati I, Strand BL. Effect of Ca2+, Ba2+, and Sr2+ on Alginate Microbeads. *Biomacromolecules* 2006;7:1471-80.
[141] Mørch ÝA, Donati I, Strand BL, Skjåk-Bræk G. Molecular Engineering as an Approach to Design New Functional Properties of Alginate. *Biomacromolecules* 2007;8:2809-14.
[142] Crescenzi V, Francescangeli A, Taglienti A. New Gelatin-Based Hydrogels via Enzymatic Networking. *Biomacromolecules* 2002;3:1384-91.
[143] Broderick EP, O'Halloran DM, Rochev YA, Griffin M, Collighan RJ, Pandit AS. Enzymatic stabilization of gelatin-based scaffolds. Journal of Biomedical Materials Research Part B: *Applied Biomaterials* 2005;72B:37-42.
[144] Bertoni F, Barbani N, Giusti P, Ciardelli G. Transglutaminase Reactivity with Gelatine: Perspective Applications in Tissue Engineering. *Biotechnology letters* 2006;28:697-702.

Hydrogels: Synthesis, Characterization and Applications
Editors: F. V. Câmara and L. J. Ferreira
ISBN: 978-1-61942-842-3


*Chapter 7*

# SILK FIBROIN HYDROGELS: A REVIEW UNDER BIOMATERIALS PERSPECTIVE

***Mariana Agostini de Moraes and Marisa Masumi Beppu****
School of Chemical Engineering,
University of Campinas, Brazil

## ABSTRACT

Silk fibroin (SF) is a protein that is spun into fibers by arthropods such as silkworms, spiders, scorpions, mites and flies. The process to obtain SF solution consists on fiber dissolution in concentrated saline solutions, followed by dialysis in water. SF aqueous solution (after dialysis) is thermodynamically metastable and may undergo sol–gel transition due to the lack of ions available to promote its solvation. The hydrogel formation occurs because SF chains tend to aggregate, passing from an amorphous conformation (random coil) to a more stable structure (β-sheet). The formation of β-sheets stabilizes the hydrogel and is irreversible under physiological conditions. SF hydrogels can be obtained by changing the protein concentration, pH or temperature, and by the addition of salts, metal ions and other components. This review describes the several methods available for SF hydrogel production and relates these methods with the final hydrogel properties and applications, focusing on biomaterials field.

## INTRODUCTION

Hydrogels are three-dimensional macromolecular networks, capable to absorb large quantities of water without losing their structural integrity. Because of their biocompatibility, ability to mimic body tissues and extracellular matrices (ECM), stimuli-responsive nature, high water content and good diffusive properties with respect to oxygen, nutrients and waste products, hydrogels are extensively used in biomedical research. They find uses as contact

* E-mail address: beppu@feq.unicamp.br

lenses, artificial tendons, artificial skins, maxillofacial reconstruction or vocal cord substitution [1]. Also, hydrogels can be used for enzyme or cell immobilization and drugs can be incorporated into hydrogels matrices, acting as drug delivery vehicles [2,3]. In tissue engineering (development of biological substitutes where cells are seeded in a 3D structure to generate a new tissue) hydrogels find uses as scaffolds (3D matrix) that support cell growth and degrade at the same rate of the new tissue formation. Also, physical, chemical and biological properties of hydrogels can be modulated to meet specific mechanical and functional requirements of the host tissues [4].

Natural polymers have the ability to stimulate tissue growth because they are composed of molecules presented on ECM and, therefore, they are polymers that naturally interact with cells. ECM plays an important role in the division, adhesion, motility, differentiation and migration of cells, thus natural polymers are used as starting points for the reconstruction of damaged tissues [5].

Hydrogels formed from natural polymers are eco-friendly and biodegradable materials, that usually does not present toxic residues and are biocompatible with a variety of cell lines and tissues. Among the natural polymers, silk fibroin (SF) appears as a potential material for hydrogel formation and applications in the biomaterials field.

SF is a protein fiber spun by silkworms that is very used in the textile industry, but also has applications in the biomaterials field. SF has been used as sutures for centuries and nowadays several research groups investigate SF as a promising material for biomedical and biotechnological applications.

SF has some advantages over other natural polymers for hydrogel formation, because its chains are capable to be physically crosslinked without any chemical modifications. SF molecules self-assemble naturally over time into stable crystalline β-sheet fibrils. Also, the mechanical properties and the degradation rate of SF hydrogels can be fine-tuned by adjusting SF hydrogel processing conditions and crystallinity.

SF is versatile and can be processed in vitro into different material formats such as films [6,7], powders [8], microspheres [9,10], nanofibers [11,12] and hydrogels [13,14]. The amazing properties that have attracted much interest on SF include good biocompatibility, good oxygen and water vapor permeability, biodegradability and minimal inflammatory reaction. Moreover, SF can be sterilized and prepared in different morphologies, and has high mechanical strength, high thermal stability and microbial resistance [15-17]. Additionally, SF exhibits slow degradation *in vitro* and *in vivo*, which is a good characteristic when looking for applications in tissue engineering, because SF materials are capable to maintain their mechanical integrity during the new tissue formation [18-20]. The use of SF as biomaterial is approved by Food and Drug Administration (FDA) as non absorbable sutures, and for soft tissue repair. SF scaffolds have been commercialized by Serica Technologies Inc., subsidiary of Allergan Inc. (www.allergan.com), for plastic and reconstructive surgery, for ligament and tendon repair, and for repair of connective tissues.

This chapter seeks to provide an overview of the production of silk fibroin hydrogels and the processing methods will be correlated to the hydrogel properties and applications in biomaterials field.

# SILK FIBROIN: GENERAL ASPECTS

Silkworm silk is composed by two proteins: fibroin and sericin. Sericin is a glue-like protein responsible to hold together the fibroin fibers and form the cocoon case that protects the growing worm [15]. SF molecules consists of a heavy chain of 350 kDa and a light chain of 26 kDa joined by disulfide bonds [21]. SF presents a repetitive amino acid sequence, with at least 16 amino acids, with predominance of alanine (A), glycine (G), serine (S) and in a lower content, tyrosine (T), which gives rise to very regular conformations at the primary level [17]. The mole fraction of glycine, alanine, serine and tyrosine combined is 90% [22]. The highly repetitive primary sequence GAGAGS forms antiparallel β-sheet crystalline regions. SF is composed of large internal hydrophobic blocks and small hydrophilic internal blocks with large hydrophilic chain end block [23,24].

SF has two major conformational states of the secondary structure: Silk I and silk II. Silk II is the solid hydrophobic SF that can be found in the spun silk fiber, characterized by a β-sheet structure, which is thermodynamically stable with strong hydrogen bonds and Van der Waals forces. Silk I is characterized by random coil and alpha helical structures; it is the water-soluble and metastable form of SF, stored in the silkworm gland [25,26]. The formation of insoluble crystallized structures of SF can be induced by exposure to heat [27], humidity [28,29], organic solvents [6,30], physical stretching [23] or shear stress [31].

SF aqueous solution is secreted in the posterior section of the silkworm gland and stored until reach a concentration of 30% (w/v) in the middle section, where sericin is also secreted. Then, fibroin and sericin pass to the anterior section without mixing and are spun in air to form the cocoon [21]. It is also thought that a combination of shear forces, active dehydration, changes in pH, and potassium and copper ions are involved in silk fiber synthesis, but mechanisms that transform water-soluble silk into insoluble fiber are still not completely understood [21,25].

SF fibers are purified from the sericin coating by a degguming process, in which the silkworm cocoons are immersed in boiling water or alkaline solution. In the degguming process, about 25% of the mass of the cocoon is removed due to the removal of sericin, and changes in the mechanical properties are also observed. Allergic reactions were observed in native silk and associated with the presence of sericin. When sericin is removed from SF, no inflammatory or allergic reactions are observed [15]. Additionally, SF degradation products proved to be not toxic to neuron cells [32,33].

SF fibers are stable at physiological temperatures, flexible and are insoluble in water and in the majority of organic solvents. SF dissolves in concentrated acid solutions and in concentrated aqueous, organic and aqueous-organic solutions of salts. The limited dissolution of SF is due to the presence of a large amount of intra- and intermolecular hydrogen bonds and its high crystallinity and specific physicochemical properties [22,34]. Nowadays, the most used solvents of SF fibers are LiBr and a ternary solvent made from $CaCl_2$, ethanol or methanol, and water. These solvents contain a high concentration of salts and high ionic strength, and they break down the intermolecular hydrogen bonds that stabilize the β-sheet structure of the fibers.

SF salt solution is salt supersaturated and can be stored for several months without undergoing gelation due to the strong ionic strength that promotes SF fibers solvation. However, to produce SF-based materials, it is necessary to remove the salts of the solvent by dialysis, obtaining the SF aqueous solution. During dialysis, the salt ions diffuse from SF solution to the dialysis water and the ionic strength of the solvent decreases; SF solution becomes metastable due to the lack of ions available to promote its solvation and may undergo a sol–gel transition through self-assembly into physically crosslinked β-sheet crystals, forming a network of SF fibrils. Any perturbation in the metastable system (SF aqueous solution) can result in instant phase separation and hydrogel formation.

All processing steps shown in Figure 1 will influence the final properties of SF-based materials. For example, the solvent used on dissolution process, as well as the time and the temperature used on dissolution will affect the crystallinity and the conformation of SF molecules, which will influence the material properties.

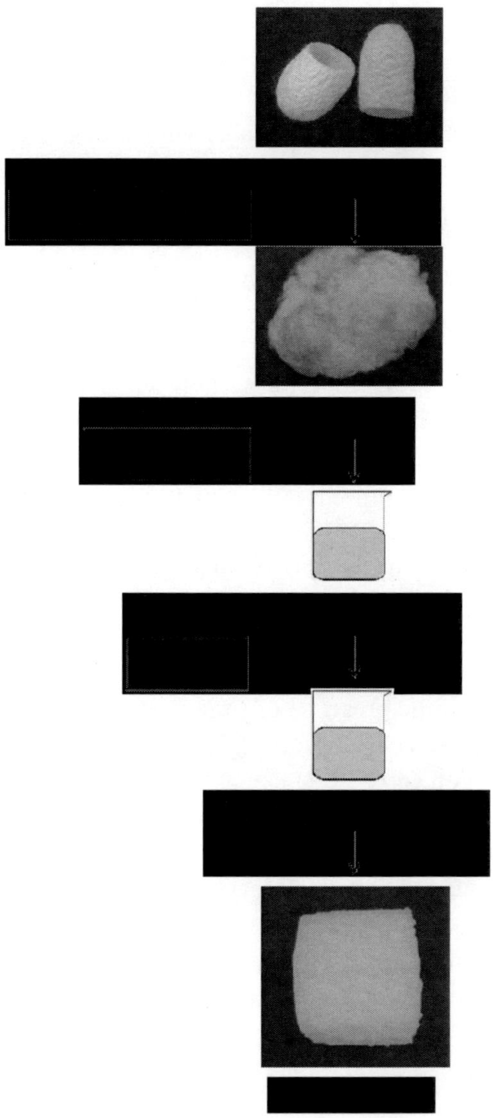

Figure 1. Schematic representation of processing steps for preparation of SF hydrogel.

The sol-gel transition occurs because SF chains, initially in amorphous conformation (random coil), become a more stable structure (β-sheet). Due to intra- and intermolecular hydrophobic interactions and hydrogen bonds, β-sheets aggregate and form molecular network structures, called hydrogels [35]. The formation of β-sheets stabilizes the hydrogel and it is irreversible under physiological conditions. This hydrogel can only be degraded by enzymatic or oxidative processes [14,36]. This transition occurs naturally in SF aqueous solution because stabilized β-sheet is the most thermodynamically favorable condition, and is only dependent on gelation kinetic. The sol-gel transition and the hydrogel properties are affected by several factors, such as protein concentration, temperature, pH and metal ions [14,36] and gelation can be induced by some recently developed methods such as sonication [3], vortexing [37] and electricity [38].

Protein concentration is an important factor that should be considered for SF hydrogel formation. At concentrations lower than 3 wt %, more than 10 days can be necessary for hydrogel formation, but increasing SF concentration up to 20 wt %, increases the molecules interactions and reduces gelation time to hours [14,36]. In the same manner, increasing temperature accelerates sol-gel transition by enhancing macromolecular interactions and inducing physical crosslinks. SF gelation is also dependent on the solution pH. At low pH values, SF molecules repulsion decreases and interactions among chains are facilitated, resulting in β-sheet formation through hydrophobic interactions and intermolecular hydrogen bonds. Like most proteins, near the isoelectric point (pI = 3.8), aggregation of SF molecules occurs and SF sol-gel transition is accelerated [14]. In all those cases, concurrently with hydrogel formation, a conformational change from random coil to β-sheet structure is observed, which characterizes the formation of a stable solid-like structure.

SF conformation transition from silk I (random coil and α-helix) to silk II (β-sheet) can also be induced by the addition of organic solvents, generally methanol or ethanol. These alcohols are hydrophilic and miscible in water. When in contact with SF, they dehydrate SF hydrophobic domains and allow chain-chain interactions, resulting in a β-sheet stable structure. Also, increasing methanol or ethanol concentration result in high thermal stability, high mechanical resistance, high crystallinity and high content of β-sheets due to more rapid chain rearrangements [39,40].

SF presents high mechanical resistance, thermal stability, and can be sterilized by common sterilization methods, which represent a major advantage when compared to other natural polymers like collagen, chitosan, alginate, among others. These properties are associated with SF crystallinity. The more crystalline (higher content of β-sheets) SF-based materials are, the more stable are their structures, and therefore better mechanical properties are achieved. They also become more resistant to high temperatures and are degraded at a slowly rate.

Due to the versatility of processing SF and to the possibility of its preparation in different forms, SF-based materials have distinguished properties. SF degradation, for example, depends directly on SF processing conditions, SF conformation and SF-material design or morphology. SF fibers (in native state) are crystalline, have high content of β-sheets and are classified, according to US Pharmacopeia, as non-degradable because they keep their tensile strength after 60 days of implantation [15]; however, studies demonstrated that SF fibers are proteolytically degraded *in vitro* [18] and *in vivo* [15] within longer periods (from 2 months to 2 years), depending on the degradation conditions, implantation site and animal model.

Nevertheless, when processing SF fibers to obtain new designed materials, β-sheets are broken and re-built, and degradation rate is modified, depending on β-sheet content and surface design (contact area, structure stability, among others). It was found that aqueous-derived electrospun SF fibers degraded *in vitro* within 24 days, while complete *in vivo* degradation was achieved after 8 weeks of implantation [19]. *In vivo* degradation of SF salt-leached scaffolds obtained from all-aqueous process and from organic process with hexafluoroisopropanol (HFIP) demonstrated that aqueous scaffolds degraded up to 6 months, while HFIP scaffolds degraded in more than 1 year of intramuscularly implantation in mice [20]. SF hydrogels obtained by sonication were found to degrade very slowly when implanted into dorsal subcutaneous pockets on Swiss nude mice and after 12 weeks of implantation the material was still clearly visible [41].

SF versatility is a powerful tool for material design. There is the possibility to fine-tune SF-based materials in order to obtain desired properties, including crystallinity, molecular conformation, mechanical and thermal resistance and degradation rate.

## Silk Fibroin Hydrogels

Hydrogels exhibit viscoelastic properties and, usually, a solid-like mechanical behavior with high compliance and high elastic-strain, even though consisting of more than 30% of liquid [42]. Gelation occurs by crosslinking of polymer chains into networks, by chemical reagents, called crosslinkers (chemically crosslinked hydrogels), or by physical stimulants, such as pH, temperature, concentration of polymer, among others (physically crosslinked hydrogels). Physically crosslinked hydrogels represent the most promising type of hydrogels for biomaterial applications because active molecules can be entrapped in the hydrogel, like proteins or even cells, without any toxic effect. On the other hand, chemically crosslinked hydrogels are usually crosslinked by UV radiation or X- and γ-rays, and toxic residues can affect the application of those hydrogels in the biomaterials field [42]. Yet, looking for applications in tissue engineering, the use of chemically crosslinked hydrogels is restrained because radiation destroys the cells and reduces their viability. By using chemically crosslinking methods, the cells can only be added on the top of the gel, after gelation is completed and exposure to radiation is finished.

SF solution naturally undergoes gelation by physical crosslinking. The environmental parameters that influence SF gelation, as well as the recently developed methods used to accelerate SF gelation will be discussed in this chapter.

### Parameters That Affect Fibroin Sol-Gel Transition

One of the first studies on SF gelation was accomplished by Ayub et al. [43], who observed accelerated gelation when SF solution pH was near its isoelectric point (pI = 3.8). A conformational transition from random coil to β-sheet was observed during gelation and it was suggested that hydrogen bonds between SF molecules are the responsible for hydrogel formation, with a network structure observed by scanning electron microscopy. Subsequently, the authors studied the influence of different dehydration levels on the conformation of SF hydrogels and found that SF hydrogels structural stability is independent of the water content [44]. Furthermore, increasing SF concentration decreased gelation time and a hydrogel with higher strength and thermal stability was obtained due to closer contact between SF molecules.

In order to mimic SF spinning from silkworm, the stability of SF aqueous solution was studied to determine the ideal storage conditions of SF aqueous solution [45]. It was found that SF gelation is accompanied with conformational transitions from random coil (solution) to β-sheet (gel) and that flow stability of SF aqueous solution is decreased by increasing temperature and increasing SF concentration. Hence, the authors recommended storage of diluted SF solution under refrigeration (low temperature) and the solution should be concentrated just before spinning.

Gelation of SF aqueous solutions in a pH lower than the isoelectric point (pI = 3.8) was studied by Nagarkar and co-authors [46]. They verified that at pH = 2, SF gelation proceeds in two steps: in the first gelation occurs immediately after lowering the pH, forming a weak gel; and the second occurs when self-assembling of stable SF structure takes place, and it is a function of gelation kinetics. SF gelation seems to be a nucleation and growth process, in which proteins aggregate and form stable β-sheet structures at the molecular level. Also, shear modulus and turbidity of SF hydrogels increase with SF concentration increase, but they are independent on temperature and pH, that ranged from 5 to 70ºC and 2 to 4, respectively [47].

Miyaguchi and Hu [48] evaluated the influence of the solvent of SF fibers on the SF solution properties and, consequently, on the gelation behavior. Calcium chloride/water and calcium chloride/water/ethanol were used to dissolve SF fibers, followed by dialysis. Results indicate that calcium chloride/water/ethanol is a better solvent for SF (even though 100% of protein solubility can be achieved by increasing salt concentration and heating time when calcium chloride/water was used as solvent); however, gelation time is reduced from 6 to 1 day when using calcium chloride/water as solvent. Moreover, hydrogels obtained when using calcium chloride/water as solvent showed better resistance to compression. Probably, reduced gelation time on calcium chloride/water systems is due to the reduced protein solubility provided by this solvent, and then, SF molecules are prone to self-assembly faster, forming the hydrogel.

Environmental factors that affect SF sol-gel transitions were studied by Kim et al. [14] in order to mimic SF processing in silkworm glands. Concentrated SF aqueous solutions were obtained by osmotic stress by dialysis of SF solution in poly(ethylene glycol) (PEG). This process allows the preparation of concentrated SF solutions (21 wt % after 9 hours of dialysis). SF concentrations higher than 21 wt % resulted in hydrogel formation in the dialysis cassette. Gelation time was highly decreased by decreasing pH and increasing temperature and SF concentration. Addition of poly(ethylene oxide) (PEO) significantly reduced gelation time at concentrations up to 5%; PEO concentrations higher than 5% did not change gelation kinetics. Increasing SF concentration and temperature produced a network with smaller pores and higher compressive strengths. $Ca^{2+}$ ions accelerated SF self-assembly, while $K^{+}$ ions did not change the gelation behavior. However, the subsequent study of the same authors found opposite results: gelation took place in the presence of NaCl, while hydrogel was not formed in the presence of $CaCl_2$ [49].

NaCl particles were added in SF aqueous solution to act as porogens [49]. By this method, SF hydrogels were prepared at *ca.* 24 hours, had porosity greater than 90%, with interconnected pores, good compressive strength and were degraded by protease within 21 days. Increasing SF concentration resulted in rapid hydrogel formation on the surface of SF aqueous solution when NaCl particles were added, thus those hydrogels could not be evaluated once the objective of the study was to obtain uniform pore size in the whole hydrogel, induced by NaCl particles. To better understand the effect of salts on SF gelation, the authors added a drop of 8 wt % SF aqueous solution on concentrated salt solutions (3M) of NaCl, KCl, $CaCl_2$ and $MgCl_2$. SF instantly undergoes gelation when in contact to NaCl and KCl solutions, while gelation was not observed for $CaCl_2$ and $MgCl_2$, attributed to the high charge density of $Ca^{2+}$ and $Mg^{2+}$ that bind adjacent water molecules more strongly. Addition of NaCl particles on SF solution enhances SF chain-chain interactions due to dehydration of hydrophobic SF domains, inducing the formation of stable β-sheet structures.

The sol–gel transition of SF was also studied by Matsumoto et al. [36], who reported that gelation is dependent on the protein concentration, temperature and pH. The authors improved gelation time from 15 to 8 days at room temperature by increasing chain interactions due to increased SF concentration in the hydrogel from 0.6 wt% to 7.2 wt%, respectively. The concentration-dependency was less pronounced when temperature was increased to 37°C, at which sol-gel transition can occur for a solution of 0.6 wt% SF in less than 8 days. By analyzing the secondary structure of SF during sol-gel transition the authors verified that at the initial states of gelation (until 20% of gelation) weaker interactions occur, and at this point gelation can be reversible; however, as gelation proceeds, strong β-sheets are formed and gelation is irreversible. pH plays an even more pronounced influence on SF gelation, and gelation occurs faster at acidic conditions. Protonation of carboxyl groups seams to play dominant role on gelation kinetics.

The authors suggest that at acidic pH the carboxyl groups of SF are neutralized, resulting in reduced hydrophilicity and charge repulsion, allowing hydrophobic interactions occurrence and quicker sol-gel transition. In the uncharged states, carboxyl groups promote hydrophobicity, while amino groups promote hydrophilicity. Also, at a neutral pH, the gelation kinetics is faster than at basic pH because at neutral pH, electrostatic interactions are more pronounced than hydrophobic interactions.

SF aqueous solution was mixed with ethanol in several ratios in order to induce gelation and the state of water on the hydrogels was analyzed [35]. Increasing ethanol content in the mixture decreased gelation time. A minimum of 2.5 wt% of SF concentration was necessary; otherwise gelation time began to increase. At SF concentration of 3.2 wt% and 50% of ethanol (in volume) sol-gel transition occurred within 5 seconds. Higher SF concentration resulted in higher β-sheet content, higher compressive modulus, higher cell viability and lower water content in the hydrogel. It was hypothesized that cells and cell-adhesion proteins in cellular matrix preferentially adhere on a surface containing more bound water.

A study of SF sol-gel transition during dialysis was performed by Nogueira et al. [13]. In this study, the SF solution was not removed from the dialysis bags and the influence of residual calcium concentration and temperature of dialysis on gelation time were evaluated. During dialysis, the salt ions diffuse from SF solution into the dialysis water and, as a result, the ionic strength decreases allowing more interactions among SF molecules. When the dialysis temperature increases, the time required for SF sol-gel transition decreases and a higher residual calcium concentration in the hydrogel is observed. When the dialysis temperature was 25°C the sol-gel transition occurred within 3 days. Additionally, those hydrogels were obtained with pure SF solution, without the addition of other components, and were not toxic to CHO cells as evaluated by in vitro cytotoxicity tests, indicating a promising material for use in the biomaterial field. Nogueira et al. [50] also observed that SF hydrogels obtained directly from dialysis can induce the deposition of calcium phosphates, which indicates the potential of use SF hydrogels for bone regeneration.

Looking for bone regeneration, Motta et al. [51] developed SF hydrogels from aqueous solutions at 4°C and room temperature, where glycerol was added at a ratio of 30%, in volume. Human osteoblasts-like cells were seeded in the hydrogels, in order to evaluate the possibility of application of these hydrogels as scaffolds for bone regeneration. Both hydrogels presented no cytotoxic effect, although osteoblast proliferation was higher on the hydrogel containing glycerol. On the other hand, the SF hydrogel produced at 4°C favored osteoblasts activity and differentiation, as indicated by alkaline phosphatase activity, type I collagen production and transforming growth factor-β1 level. Additionally, glycerol was found to accelerate the gelation rate of SF aqueous solution [51,52]. Moreover, the potential of SF hydrogels to heal bone defects in rabbit distal femurs was evaluated by Fini et al. [53]. The hydrogels were prepared by adding citric acid onto SF aqueous solution (pH < pI) and heated to 50°C overnight. No inflammatory effect or stimulation of TGF β1 production were observed by *in vitro* tests with MG63 osteoblasts. *In vivo*, SF hydrogel performance was comparable to the control (poly(D,L lactide-glycolide) copolymer) in terms of healing the critical size bone defect, but greater than the control in terms of bone remodeling and maturation. No inflammatory reaction was observed. Some degradation of SF hydrogel occurred, but the material was still present after 12 weeks of implantation. These results emphasize the potential of SF hydrogels to be used as bone replacements.

Generally, SF-based materials are obtained directly from silkworm cocoons or by silk fibers supplied by silk industry. All SF processing, as degguming to remove sericin, drying, dissolution (SF salt solution) and dialysis (SF aqueous solution) is done in the laboratory. Silk proteins can also be bought directly from the industry: silk proteins with high (76 kDa) and low (18 kDa) molecular weight, supplied by Aotesi Biochemical Co., mainly containing serine, aspartic acid and glutamine, were used to prepare silk hydrogels for drug delivery [54]. Silk proteins were added in citrate-phosphate buffer (pH 4) and heated at 60°C for 4 h. Then, at room temperature, the studied drugs (buprenorphine, nalbuphine, its prodrug, fluorescein isothiocynate and fluorescein isothiocynate-labeled dextran) were incorporated in the mixture, and refrigerated, in order to obtain the hydrogels. Increasing silk protein concentration in the hydrogel resulted in slower drug release, indicating that the drug release is diffusion-controlled through the matrix. Results showed that drug release from silk protein hydrogels was comparable to that from Pluronic F-127 and slower than chitosan hydrogels, which demonstrates that silk protein hydrogels are promising materials to be used in controlled drug delivery. Also, hydrogels can be prepared from SF peptides by adjusting the pH of the peptides solution. These hydrogels were adequate for in vitro culture of chondrocytes and may be used as scaffolds for cartilage tissue engineering [55].

## Methods to Accelerate Fibroin Hydrogel Formation

Understanding the sol–gel transition and the environmental factors that can influence this phenomenon is important to define the thermodynamic path to be followed to achieve the final product properties that are desired. However, for practical applications SF gelation time is too long to allow cells addition during hydrogel preparation. When looking for cell-based applications, gelation must occur in a short period of time and preferentially under physiological conditions; otherwise, cells will not support the severe conditions. The high temperatures and acidic pH found to accelerate SF gelation are not adequate conditions for cells survivor. That is the reason why several research groups all over the world try to improve SF gelation conditions and hydrogel properties. Recently developed methods, namely sonication, vortexing and electrogelation, hold promise on preparation of SF hydrogels with reduced gelation time that can be applied on cells encapsulation, drug release, among other biomedical applications. These gelation methods will be further discussed below.

Ultrasonication was proposed as a method to induce SF sol-gel transition and mimic the shear forces experienced at the anterior division of the silkworm gland without modification in the physicochemical parameters [3]. Ultrasonication accelerated the sol-gel transition of SF by altering the hydrophobic hydration and inducing rapid physical crosslinks via β-sheet crystal formation. Gelation occurs from 5 min to 100 h, depending on sonication parameters (such as power and time) and SF solution concentration (more concentrated solutions resulted in faster gelation). The authors also studied the influence of pH and salt concentration on SF gelation and verified that low pH and low concentrations of $K^+$ accelerate the gelation kinetics. This method also allows the addition of cells after the sonication, but prior to gelation. Mesenchymal stem cells were used in 4 wt% SF hydrogel and cultured for 21 days, retaining viability and proliferating in culture conditions, indicating that sonication can be used for cell encapsulation.

Additionally, chondrocytes behavior was evaluated in SF hydrogels obtained by ultrasonication and by salt leaching methods, aiming cartilage tissue engineering applications [56]. Besides, two different extraction methods of SF from silkworm cocoons were evaluated: typical extraction with sodium carbonate and extraction with sodium carbonate and detergent (Triton X-100), which is supposed to maintain the integrity of SF [57]. Ultrasonicated SF hydrogel, extracted with sodium carbonate and detergent, had mechanical properties, biochemical content and tissue formation similar to agarose constructs, considered as a model for chondrocytes culture [56].

Sonication was also used to prepare SF hydrogels containing human fibroblasts for soft tissue augmentation to be applied in periodontal and maxillofacial therapies [41]. Unconfined compression tests revealed that SF hydrogels (seeded or not with fibroblasts) had mechanical properties much greater than seeded collagen gels. Also, SF hydrogel degraded slowly, while collagen gels completely degraded after 2 weeks of subcutaneous implantation. *In vitro* culture demonstrates good cell survival and proliferation and *in vivo* subcutaneous implantation showed good implant stability and revascularization through the construct. Some inflammatory reaction was observed and disappeared after 12 weeks of implantation, proving that SF hydrogels are also valuable materials for engineering soft tissues.

Complementary, the authors used ultrasonication to entrap hyaluronic acid with SF during the sol-gel transition [42]. The possibility to physically encapsulate hyaluronic acid onto SF hydrogels without the need of chemical crosslinking is highly desired because residual components are avoided. Those hydrogels combine the mechanical integrity and slow degradation of SF with the control of water interactions and biological signaling of hyaluronic acid, useful for biomedical applications.

Looking for sustained therapeutic delivery, monoclonal antibodies were added in SF solution after sonication but prior to gelation, which happened within 2 hours [2]. The influence of lyophilization as an additional step after gelation was evaluated in terms of monoclonal antibody release. Lyophilization improved long-term release profile with sustained antibody release over 38 days, while SF hydrogel rapidly release the antibody in 10 days, proving to be a better option for local delivery of monoclonal antibody and a suitable choice for manufacturing and storage of the therapeutic protein and the matrix. Also, increasing SF concentration resulted in higher β-sheet density, higher hydration resistance and slower sustained release of the monoclonal antibody.

Vortexing SF aqueous solution was also found as a method to accelerate sol-gel transition [37]. It is known that SF aqueous solution is metastable and that water is a poor solvent for SF, so shear flow (or vortexing) increase the concentration fluctuations in SF aqueous solution and lead to SF self-assembly into clusters. Gelation time is decreased by increasing vortex time, postvortex assembly temperature and protein concentration; gelation time can be fine-tuned from hours to minutes by varying these parameters and a stable hydrogel with β-sheet structure is obtained. Additionally, in a similar manner of sonication, constructs can be prepared by adding cells in SF immediately before gelation, but after vortexing, and vortex-induced SF hydrogels can be used as injectable cell delivery scaffolds.

A recently developed method to produce SF gel is based on the application of electricity onto SF aqueous solution by exposure to direct current, forming the so-called electrogel or e-gel [58]. These SF gels are highly viscous and have interesting adhesive properties that can be controlled based on the hydration state: more hydrated gels have better adhesive properties than gels with lower water content. The gel is formed within seconds of voltage (25 $V_{DC}$) application on the positive electrode and retains the gel state after the electric field is turned off. More interestingly, reversing the electrode polarity, the gel return to the solution state and is formed on the new positive electrode. The same reversible sol-gel transition is observed when heating the gel up to 60°C, it returns to the solution state, and cooling at room temperature the gel is formed again. Fourier transformed infrared spectroscopy (FTIR) revealed that e-gels are formed by increasing α-helix content, but no increasing on β-sheets were observed when compared to SF aqueous solution [58]. It was found that SF amorphous micellar structures with intermicellar entanglement crosslinks are presented on SF e-gels. E-gel formation is mainly attributed to the increase in proton concentration in the vicinity of the positive electrode due to the applied electric field [38]. Further studies revealed that e-gel formation occurs due to aggregation of SF nanoparticles into SF nano and microspheres that are dispersed into solution when reversing sol-gel transition (solution state). Furthermore, increasing the duration of electric field application and SF concentration resulted in more stable gels that are difficult to return to the solution state [59]. Electrogelation appears as a novel method to create SF reversible gels, biomimetic adhesive systems and drug delivery vehicles.

## Combining Other Components in Fibroin Hydrogels

SF molecules can also be combined with other natural or synthetic polymers in order to conjugate their properties and prepare a final material with unique and tunable properties.

Blend films of gelatin (denatured collagen) and SF were prepared and hydrated to form hydrogels [60-62]. It was found that the helix-coil transition of gelatin was influenced by SF concentration and SF secondary conformation (e.g. β-sheets or random coils). Gelatin presents thermally reversible conformational transition and, at temperatures below 30ºC is in the gel state, while above 30ºC helix to coil transition occurs, and gelatin returns into the solution state. When SF was present in hydrogel, treated (β-sheet) or not (random coil) by methanol, the hydrogel exhibited extended solid-like behavior and the gel-sol transition for methanol-treated hydrogels occurred at temperatures above body temperature. These characteristics enabled these hydrogels to be used as thermally-triggered-release encapsulant. Gelatin/SF methanol treated hydrogels swells more at 37ºC than at 20ºC (fully gel state), but gelatin molecules are slowly released in aqueous solution at 37ºC, and SF β crystals are the only responsible for the stability of the hydrogels at higher temperatures [63].

Interpenetrating polymer network (IPN) hydrogels were prepared using gelatin methacrylate and SF [4]. UV light was used to photocrosslink gelatin methacrylate, while immersion in methanol was used to induce β-sheet transition onto SF. It was found that the immersion in methanol is crucial to achieve a crystalline material with the desired mechanical properties. Incorporation of SF in several ratios increased the compressive modulus and reduced the degradation rate of gelatin methacrylate. However, *in vitro* test with NIH-3T3 fibroblasts demonstrated that a maximum cell spreading and proliferation is achieved when SF content in the IPNs is 0.5 wt%. Higher SF content resulted in decreased cell proliferation.

Semi-IPN hydrogels were prepared from chitosan and SF by crosslinking chitosan with glutaraldehyde [64]. This material is a potential candidate to be used as artificial muscle because it could swell and shrink in a reversible way by immersion in buffer solutions with different pHs or $AlCl_3$ aqueous solutions with different concentrations.

Semi-IPN hydrogels of SF and polyacrylamide crosslinked with N,N'-Methylenebisacrylamide were investigated as drug delivery vehicles [65]. Even though no treatment was done to induce a conformation transition of SF, FTIR revealed β-sheet formation in these semi-IPN hydrogels. Increasing polyacrylamide content in the semi-IPN hydrogels resulted in increased mechanical strength, decreased swelling and decreased degradation rate, due to the high crosslinking density attributed to polyacrylamide. *In vitro* cytotoxicity test using AH927 feline embryo fibroblast cells showed good cell viability for all hydrogels. *In vitro* release of two model compounds (trypan blue and FITC-inulin) revealed sustained release up to 50 days and slower release profile for hydrogels with increased polyacrylamide content (higher crosslink density).

Poloxamer 407, a polymeric surfactant with fast sol-gel transition, was added in SF aqueous solution in order to accelerate the gelation kinetics [66]. Gelation time was significantly reduced by increasing temperature and concentrations of SF and poloxamer, at pH 7. Interestingly, SF/poloxamer sol-gel transition is reversible, even though SF β-sheets were observed in FTIR spectra.

SF hydrogels were also used to find out the mechanisms of individual components of nacre in controlling crystal growth [67]. A SF-like protein hydrogel is one of the components of nacre organic matrix, together with β-chitin and acidic proteins. SF hydrogels were prepared by heating SF solution at 60°C. Crystallization experiments were conducted in the presence of SF hydrogel and n16N-β-chitin and demonstrated that the presence of SF hydrogel changes the microenvironment for mineralization. Flat deposits of valerite and amorphous calcium carbonate were observed when SF hydrogel was present, while poly crystalline aragonite was formed in the absence of SF hydrogel, indicating that the substrates for nucleation affect the mineralization.

Genetically engineered silk-elastin-like polymer can also be used to prepare hydrogels with properties that are not sensitive to pH, temperature or ionic strength [68]. Gelation time and polymer concentration influenced the final water content of the hydrogel. Due to these properties, the hydrogels maintained their characteristics after gelation at physiological conditions and have potential uses in *in situ* gelation implants for drug delivery. Additionally, these hydrogels can be used to control the release theophylline, vitamin B12 and cytochrome c [69] and more complex molecules like DNA, for cancer gene therapy [70].

## Freeze-Dried and Salt-Leached Fibroin Scaffolds

SF hydrogels can be used to prepare porous scaffolds in dry state, also called sponges. Usually, SF hydrogels are obtained directly from SF aqueous solution and then they are freeze-dried [40,49,71] or freeze-thawed [72], to obtain a porous spongy material. The resulting SF-material needs to be immersed in organic solvents, to induce β-sheet conformation, and consists of a porous matrix where cells can be seeded.

SF sponges were prepared by adding water-miscible solvents, such as methanol, ethanol, propanol, butanol, DMSO and ethyleneglycol [72] onto SF aqueous solution, and the SF sponges obtained by freeze-thawing. Those materials were autoclavable, had adequate mechanical properties for MC3T3 proliferation and are adequate scaffolds for cartilage regeneration.

Chitosan/SF sponges were obtained by blending SF and chitosan solutions and adding genipin, a natural crosslinking agent [73]. The chitosan/SF hydrogels could be obtained from 12 to 20 min, observed by rheological measurements, depending on the hydrogel composition.

Chitosan reacted more quickly with genipin than SF and faster gelation was observed on hydrogels with high chitosan content. The hydrogels were frozen and freeze-dried, to obtain the spongy material with stable and ordered structures, that was non toxic to mouse fibroblast-like cell line, L929. These chitosan/SF sponges are potential materials for use in cartilage tissue engineering, since good adhesion, proliferation and matrix production were shown for ATDC5 chondrocyte-like cells.

SF/collagen hydrogels [74] were prepared by dissolving collagen gel into SF solution and crosslinking with 1-ethyl-3-(3-dimethylaminoprophy) carbodiimide hydrochloride (EDC), with a final pH equal to 4-5. The hydrogel was freeze-dried and studied in the dry state, showing great mechanical strength (storage moduli = 10 kPa) and thermal stability above 80°C. Vascular smooth muscle cells had better growth in the scaffolds crosslinked with EDC, indicating that the crosslinking did not jeopardize such application.

Another method to obtain SF-based scaffolds is called salt-leaching and consists on the addition of salts (NaCl) onto SF solution. In this method, SF fibers are dissolved in LiBr, the solution is dialyzed, lyophilized and re-dissolved in hexafluoroisopropanol (HFIP). Then, NaCl crystals with a predetermined size are added in the SF/HFIP solution, HFIP is evaporated forming a block (or gel) of SF. Subsequently, the block is immersed in organic solvents in order to stabilize SF β-sheets and NaCl is extracted in water [40]. By this method, SF scaffolds are prepared with controlled pore size, due to the NaCl crystals. It was found that those scaffolds are suitable materials for regeneration of hard tissues, such as bone [75-82] and cartilage [76,83,84], with several *in vitro* and *in vivo* studies, using different types of cells and different culture conditions. SF salt-leached scaffolds represent a good matrix for formation of cartilaginous tissue when seeded with human mesenchymal stem cells (hMSC) [76]; and also had suitable mechanical properties and better response to human bone marrow-derived mesenchymal stem cells (MSCs) than collagen scaffolds [84]. Moreover, good regeneration of bone tissue was achieved when implanting SF matrices seeded with hMSC cells and cultured on a bioreactor in critical cranial defects in mice [80]. Using the same *in vitro* procedure, bone marrow stromal cells (hBMSCs) are able to form bone-like tissue resembling to human trabecular bone when seeded on SF hydrogels prepared from salt leaching method [81]. Also, it was proved that HFIP salt-leached scaffolds showed good *in vivo* degradation profile, resisting beyond one year of implantation [20]. However, the lack of vascularization is often a limitation to 3D tissue cultures to cells migration and diffusion of nutrients.

Makaya et al. [85] studied two different salt-leaching methods used to prepare SF scaffolds for cartilage tissue engineering. The first one was based on formation of SF scaffolds from SF aqueous solution and addition of salt particles; while the second one proposed preparation of SF scaffolds from SF aqueous solution lyophilization and re-dissolution on HFIP and addition of sucrose particles to act as porogens. Although SF-salt scaffolds showed better mechanical properties, SF-sucrose scaffolds had slow degradation rate and better cartilage matrix deposition when seeded with rabbit ear chondrocytes, representing a better candidate than SF-salt scaffolds for cartilage regeneration.

## FINAL REMARKS

SF-based materials can be customized by changing SF processing methods. SF is a versatile natural polymer and it is possible to obtain SF-based hydrogels with different properties and different applications, by varying the method of SF gelation, protein concentration, pH, temperature and also by adding other components that can interact or not with SF molecules. In almost all studies reviewed in this chapter, SF-based hydrogels were excellent matrices for regeneration of several hard and soft tissues *in vitro* and, in some cases, SF-based hydrogels showed better performance than the controls.

Literature indicates that inflammatory or toxic effects to cells can be avoided by degguming silk (removing sericin) [15]. In addition, SF hydrogels can be used as matrices for tissue regeneration or local delivery of drugs. Some of these applications were studied *in vivo* using animal models producing promising results.

## REFERENCES


[1] B. D. Ratner, A. S. Hoffman, F. J. Schoen, and J. E. Lemons, in *Biomaterials science: an introduction to materials in medicine*, 2.ed. ed., edited by B. D. Ratner, A. S. Hoffman, F. J. Schoen, and J. E. Lemons (Elsevier Academic Press, San Diego, 2004), p. 1.

[2] N. Guziewicz, A. Best, B. Perez-Ramirez, and D. L. Kaplan, *Biomaterials* 32, 2642 (2011).

[3] X. Q. Wang, J. A. Kluge, G. G. Leisk, and D. L. Kaplan, *Biomaterials* 29, 1054 (2008).

[4] W. Xiao, J. He, J. W. Nichol, L. Wang, C. B. Hutson, B. Wang, Y. Du, H. Fan, and A. Khademhosseini, *Acta Biomaterialia* 7, 2384 (2011).

[5] J. Velema and D. Kaplan, in *Tissue Engineering I: Scaffold Systems for Tissue Engineering*; *Vol. 102* (2006), p. 187.

[6] G. M. Nogueira, A. C. D. Rodas, C. A. P. Leite, C. Giles, O. Z. Higa, B. Polakiewicz, and M. M. Beppu, *Bioresource Technology* 101, 8446 (2010).

[7] C. L. Du, J. Jin, Y. C. Li, X. D. Kong, K. M. Wei, and J. M. Yao, *Materials Science and Engineering C-Biomimetic and Supramolecular Systems* 29, 62 (2009).

[8] R. Rajkhowa, L. J. Wang, J. Kanwar, and X. G. Wang, *Powder Technology* 191, 155 (2009).

[9] Y. Srisuwan, P. Srihanam, and Y. Baimark, *Journal of Macromolecular Science Part a-Pure and Applied Chemistry* 46, 521 (2009).

[10] X. Q. Wang, E. Wenk, A. Matsumoto, L. Meinel, C. M. Li, and D. L. Kaplan, *Journal of Controlled Release* 117, 360 (2007).

[11] L. Jeong, I. S. Yeo, H. N. Kim, Y. I. Yoon, D. H. Jang, S. Y. Jung, B. M. Min, and W. H. Park, *International Journal of Biological Macromolecules* 44, 222 (2009).

[12] Z. X. Cai, X. M. Mo, K. H. Zhang, L. P. Fan, A. L. Yin, C. L. He, and H. S. Wang, *International Journal of Molecular Sciences* 11, 3529 (2010).

[13] G. M. Nogueira, M. A. de Moraes, A. C. D. Rodas, O. Z. Higa, and M. M. Beppu, *Materials Science and Engineering C-Materials for Biological Applications* 31, 997 (2011).

[14] U. J. Kim, J. Y. Park, C. M. Li, H. J. Jin, R. Valluzzi, and D. L. Kaplan, *Biomacromolecules* 5, 786 (2004).

[15] G. H. Altman, F. Diaz, C. Jakuba, T. Calabro, R. L. Horan, J. S. Chen, H. Lu, J. Richmond, and D. L. Kaplan, *Biomaterials* 24, 401 (2003).

[16] I. C. Um, H. Y. Kweon, Y. H. Park, and S. Hudson, *International Journal of Biological Macromolecules* 29, 91 (2001).

[17] A. C. MacIntosh, V. R. Kearns, A. Crawford, and P. V. Hatton, *Journal of Tissue Engineering and Regenerative Medicine* 2, 71 (2008).

[18] R. L. Horan, K. Antle, A. L. Collette, Y. Z. Huang, J. Huang, J. E. Moreau, V. Volloch, D. L. Kaplan, and G. H. Altman, *Biomaterials* 26, 3385 (2005).

[19] J. Zhou, C. Cao, X. Ma, L. Hu, L. Chen, and C. Wang, *Polymer Degradation and Stability* 95, 1679 (2010).

[20] Y. Wang, D. D. Rudym, A. Walsh, L. Abrahamsen, H. J. Kim, H. S. Kim, C. Kirker-Head, and D. L. Kaplan, *Biomaterials* 29, 3415 (2008).

[21] A. E. Terry, D. P. Knight, D. Porter, and F. Vollrath, *Biomacromolecules* 5, 768 (2004).

[22] E. S. Sashina, A. M. Bochek, N. P. Novoselov, and D. A. Kirichenko, *Russian Journal of Applied Chemistry* 79, 869 (2006).
[23] H. J. Jin and D. L. Kaplan, *Nature* 424, 1057 (2003).
[24] C. Z. Zhou, F. Confalonieri, N. Medina, Y. Zivanovic, C. Esnault, T. Yang, M. Jacquet, J. Janin, M. Duguet, R. Perasso, and Z. G. Li, *Nucleic Acids Research* 28, 2413 (2000).
[25] O. Hakimi, D. P. Knight, F. Vollrath, and P. Vadgama, *Composites Part B-Engineering* 38, 324 (2007).
[26] C. Vepari and D. L. Kaplan, *Progress in Polymer Science* 32, 991 (2007).
[27] S. Putthanarat, S. Zarkoob, J. Magoshi, J. A. Chen, R. K. Eby, M. Stone, and W. W. Adams, *Polymer 43*, 3405 (2002).
[28] Y. Kawahara, K. Furukawa, and T. Yamamoto, *Macromolecular Materials and Engineering* 291, 458 (2006).
[29] T. Hino, M. Tanimoto, and S. Shimabayashi, *Journal of Colloid and Interface Science 266,* 68 (2003).
[30] X. Chen, Z. Z. Shao, N. S. Marinkovic, L. M. Miller, P. Zhou, and M. R. Chance, *Biophysical Chemistry* 89, 25 (2001).
[31] G. M. Nogueira, R. F. Weska, W. C. Vieira, B. Polakiewicz, A. C. D. Rodas, O. Z. Higa, and M. M. Beppu, *Journal of Applied Polymer Science* 114, 617 (2009).
[32] K. Numata, P. Cebe, and D. L. Kaplan, *Biomaterials* 31, 2926 (2010).
[33] K. Numata and D. L. Kaplan, *Macromolecular Bioscience* 11, 60 (2011).
[34] X. Chen, D. P. Knight, Z. Z. Shao, and F. Vollrath, *Polymer 42,* 9969 (2001).
[35] K. Numata, T. Katashima, and T. Sakai, *Biomacromolecules* 12, 2137 (2011).
[36] A. Matsumoto, J. Chen, A. L. Collette, U. J. Kim, G. H. Altman, P. Cebe, and D. L. Kaplan, *Journal of Physical Chemistry B 110*, 21630 (2006).
[37] T. Yucel, P. Cebe, and D. L. Kaplan, *Biophysical Journal 97*, 2044 (2009).
[38] T. Yucel, N. Kojic, G. G. Leisk, T. J. Lo, and D. L. Kaplan, *Journal of Structural Biology 170,* 406 (2010).
[39] J. Nam and Y. H. Park, *Journal of Applied Polymer Science 81*, 3008 (2001).
[40] R. Nazarov, H. J. Jin, and D. L. Kaplan, *Biomacromolecules* 5, 718 (2004).
[41] O. Etienne, A. Schneider, J. A. Kluge, C. Bellemin-Laponnaz, C. Polidori, G. G. Leisk, D. L. Kaplan, J. A. Garlick, and C. Egles, *Journal of Periodontology* 80, 1852 (2009).
[42] X. A. Hu, Q. A. Lu, L. Sun, P. Cebe, X. Q. Wang, X. H. Zhang, and D. L. Kaplan, *Biomacromolecules* 11, 3178 (2010).
[43] Z. H. Ayub, M. Arai, and K. Hirabayashi, *Bioscience Biotechnology and Biochemistry* 57, 1910 (1993).
[44] Z. H. Ayub, M. Arai, and K. Hirabayashi, *Polymer* 35, 2197 (1994).
[45] H. Wang, Y. P. Zhang, H. L. Shao, and X. C. Hu, *International Journal of Biological Macromolecules* 36, 66 (2005).
[46] S. Nagarkar, A. Patil, A. Lele, S. Bhat, J. Bellare, and R. A. Mashelkar, *Industrial and Engineering Chemistry Research* 48, 8014 (2009).
[47] S. Nagarkar, T. Nicolai, C. Chassenieux, and A. Lele, *Physical Chemistry Chemical Physics* 12, 3834 (2010).
[48] Y. Miyaguchi and J. N. Hu, *Food Science and Technology Research* 11, 37 (2005).
[49] U. J. Kim, J. Park, H. J. Kim, M. Wada, and D. L. Kaplan, *Biomaterials* 26, 2775 (2005).

[50] G. M. Nogueira, C. G. Aimoli, R. F. Weska, L. S. Nascimento, and M. M. Beppu, (2008), p. 503.

[51] A. Motta, C. Migliaresi, F. Faccioni, P. Torricelli, M. Fini, and R. Giardino, in *Fibroin hydrogels for biomedical applications: preparation, characterization and in vitro cell culture studies*, 2004, p. 851.

[52] T. Hanawa, A. Watanabe, T. Tsuchiya, R. Ikoma, M. Hidaka, and M. Sugihara, Chemical and Pharmaceutical Bulletin 43, 284 (1995).

[53] M. Fini, A. Motta, P. Torricelli, G. Glavaresi, N. N. Aldini, M. Tschon, R. Giardino, and C. Migliaresi, *Biomaterials* 26, 3527 (2005).

[54] H. Y. Fang, J. P. Chen, Y. L. Leu, and H. Y. Wang, *Chemical and Pharmaceutical Bulletin* 54, 156 (2006).

[55] Y.-q. Jiang, M. Yao, D.-m. Dong, Y.-s. Wang, G.-j. Wei, C.-w. Zhou, and H.-y. Tian, *Micro and Nano Letters* 6, 125 (2011).

[56] P.-H. G. Chao, S. Yodmuang, X. Wang, L. Sun, D. L. Kaplan, and G. Vunjak-Novakovic, *Journal of Biomedical Materials Research Part B-Applied Biomaterials* 95B, 84 (2010).

[57] H. Yamada, H. Nakao, Y. Takasu, and K. Tsubouchi, *Materials Science and Engineering C-Biomimetic and Supramolecular Systems* 14, 41 (2001).

[58] G. G. Leisk, T. J. Lo, T. Yucel, Q. Lu, and D. L. Kaplan, *Advanced Materials* 22, 711 (2010).

[59] Q. Lu, Y. Huang, M. Li, B. Zuo, S. Lu, J. Wang, H. Zhu, and D. L. Kaplan, *Acta Biomaterialia* 7, 2394 (2011).

[60] E. S. Gil, R. J. Spontak, and S. M. Hudson, *Macromolecular Bioscience* 5, 702 (2005).

[61] E. S. Gil, D. J. Frankowski, M. K. Bowman, A. O. Gozen, S. M. Hudson, and R. J. Spontak, *Biomacromolecules* 7, 728 (2006).

[62] E. S. Gil, D. J. Frankowski, M. K. Bowman, A. O. Gozen, S. M. Hudson, and R. J. Spontak, *Biomacromolecules* 7, 728 (2006).

[63] E. S. Gil, D. J. Frankowski, R. J. Spontak, and S. M. Hudson, *Biomacromolecules* 6, 3079 (2005).

[64] X. Chen, W. J. Li, W. Zhong, Y. H. Lu, and T. Y. Yu, *Journal of Applied Polymer Science* 65, 2257 (1997).

[65] B. B. Mandal, S. Kapoor, and S. C. Kundu, *Biomaterials* 30, 2826 (2009).

[66] G. D. Kang, J. H. Nahm, J. S. Park, J. Y. Moon, C. S. Cho, and J. H. Yeo, *Macromolecular Rapid Communications* 21, 788 (2000).

[67] E. C. Keene, J. S. Evans, and L. A. Estroff, *Crystal Growth and Design* 10, 5169 (2010).

[68] A. A. Dinerman, J. Cappello, H. Ghandehari, and S. W. Hoag, *Biomaterials* 23, 4203 (2002).

[69] A. A. Dinerman, J. Cappello, H. Ghandehari, and S. W. Hoag, *Journal of Controlled Release* 82, 277 (2002).

[70] Z. Megeed, M. Haider, D. Q. Li, B. W. O'Malley, J. Cappello, and H. Ghandehari, *Journal of Controlled Release* 94, 433 (2004).

[71] Y. J. Ren, Z. Y. Zhou, B. F. Liu, Q. Y. Xu, and F. Z. Cui, International *Journal of Biological Macromolecules* 44, 372 (2009).

[72] Y. Tamada, *Biomacromolecules* 6, 3100 (2005).

[73] S. S. Silva, A. Motta, M. T. Rodrigues, A. F. M. Pinheiro, M. E. Gomes, J. F. Mano, R. L. Reis, and C. Migliaresi, *Biomacromolecules* 9, 2764 (2008).

[74] Q. Lv, K. Hu, Q. Feng, and F. Cui, *Journal of Biomedical Materials Research Part A 84A*, 198 (2008).

[75] L. Meinel, V. Karageorgiou, R. Fajardo, B. Snyder, V. Shinde-Patil, L. Zichner, D. Kaplan, R. Langer, and G. Vunjak-Novakovic, *Annals of Biomedical Engineering* 32, 112 (2004).

[76] L. Meinel, S. Hofmann, V. Karageorgiou, L. Zichner, R. Langer, D. Kaplan, and G. Vunjak-Novakovic, *Biotechnology and Bioengineering* 88, 379 (2004).

[77] L. Meinel, O. Betz, R. Fajardo, S. Hofmann, A. Nazarian, E. Cory, M. Hilbe, J. McCool, R. Langer, G. Vunjak-Novakovic, H. P. Merkle, B. Rechenberg, D. L. Kaplan, and C. Kirker-Head, *Bone* 39, 922 (2006).

[78] H. J. Kim, U. J. Kim, G. Vunjak-Novakovic, B. H. Min, and D. L. Kaplan, Biomaterials 26, 4442 (2005).

[79] S. Hofmann, H. Hagenmuller, A. M. Koch, R. Muller, G. Vunjak-Novakovic, D. L. Kaplan, H. P. Merkle, and L. Meinel, *Biomaterials* 28, 1152 (2007).

[80] L. Meinel, R. Fajardo, S. Hofmann, R. Langer, J. Chen, B. Snyder, G. Vunjak-Novakovic, and D. Kaplan, *Bone* 37, 688 (2005).

[81] D. Marolt, A. Augst, L. E. Freed, C. Vepari, R. Fajardo, N. Patel, M. Gray, M. Farley, D. Kaplan, and G. Vunjak-Novakovic, *Biomaterials* 27, 6138 (2006).

[82] J. Zhao, Z. Y. Zhang, S. Y. Wang, X. J. Sun, X. L. Zhang, J. Chen, D. L. Kaplan, and X. Q. Jiang, *Bone* 45, 517 (2009).

[83] Y. Z. Wang, U. J. Kim, D. J. Blasioli, H. J. Kim, and D. L. Kaplan, *Biomaterials* 26, 7082 (2005).

[84] S. Hofmann, S. Knecht, R. Langer, D. L. Kaplan, G. Vunjak-Novakovic, H. P. Merkle, and L. Meinel, *Tissue Engineering* 12, 2729 (2006).

[85] K. Makaya, S. Terada, K. Ohgo, and T. Asakura, *Journal of Bioscience and Bioengineering* 108, 68 (2009).

Hydrogels: Synthesis, Characterization and Applications ISBN: 978-1-61942-842-3
Editors: F. V. Câmara and L. J. Ferreira 

*Chapter 8*

# PVP-BASED HYDROGELS: SYNTHESIS, PROPERTIES AND APPLICATIONS

***Niladri Roy and Nabanita Saha****
Centre of Polymer Systems, Tomas Bata University in Zlin,
Zlin, Czech Republic

## ABSTRACT

Hydrogels are three dimensionally crosslinked polymeric network which can retain a huge amount of water but do not dissolve in water. Polyvinylpyrrolidone (PVP) is a synthetic polymer with good biocompatibility and transparency and with the action of different stimuli (radiation, heat, pressure, chemicals, etc.) undergoes crosslinking, and can form hydrogels. Due to their tissue compatibility and tissue like consistency, PVP hydrogels are very promising for different biomedical applications. However, because of inferior swelling capacity and poor mechanical property, the use of pure PVP hydrogels is limited. To overcome this problem, PVP is blended with different polysaccharides or other polymers according to requirement and from the standpoint of applications. This review article mainly focuses on different kinds of PVP based hydrogels; their modes of synthesis, properties and a range of applications of these blend hydrogels.

**Keywords:** Biomaterial, crosslinking, drug delivery, hydrogel, PVP, radiation, wound dressings.

## INTRODUCTION

Polyvinylpyrrolidone (PVP) is a biocompatible synthetic polymer, and for many years has been applied as a biomaterial or additive to drug compositions, e.g. as a blood plasma

* Corresponding author: Centre of Polymer Systems, Tomas Bata University in Zlin, Nam T.G. Masaryka 5555, 76001, Zlin, Czech Republic, E-mail ID: nabanitas@yahoo.com/nabanita@ft.utb.cz, Phone No: 00-420-57603 8013/8156, Fax No: 00-420-57603 1444

expander and as vitreous humor substitute [1]. Under action of different stimuli (radiation, heat, pressure, chemicals etc) PVP undergoes crosslinking and lead to the formation of PVP hydrogel. Hydrogels are extremely hydrated polymer gels with macromolecular three dimensional networks that swell but don't dissolve in water. They are physically or chemically cross-linked polymers and the ability of hydrogels to absorb water arises from hydrophilic functional groups attached to the polymeric backbone, while their resistance to dissolution arises from cross links between network chains [2-10]. PVP hydrogel has excellent transparency and biocompatibility.

Till date many findings in research regarding the synthesis and characterization of PVP hydrogels have been reported. However, it is also reported that because of inferior mechanical properties and low swelling capacity, the application of PVP hydrogel is not widespread [4, 11]. There are many detailed account of occurrence about PVP and polysaccharide (chitosan [1, 12-17], carboxymethyl cellulose [4-7, 11, 18], carrageenan [19-22], agar [4-7, 23-25], sodium alginate [26], etc) blended hydrogels, because when PVP is blended with polysaccharides, the properties of hydrogels improve. The hydrogels are generally synthesized by incorporating crosslinking in the polymer matrix. Crosslinked PVP based hydrogels mainly belong to the category of physical hydrogels, and they are synthesized by using $\gamma$-irradiation (most common method to prepare hydrogels) [23, 27-47], electron beam radiation [48-49], UV-photo crosslinking [50-60], freeze-thawing [46, 61], moist heat treatment [4-7], etc. They have diversified ways of application, mainly in biomedical field such as: human lens substitution, artificial cartilage, dermal wound healing, semi occlusive wound dressings, implants, controlled drug release, remineralization of enamel on teeth, and so on [23-25, 62-111]. In biomedical application, PVP based hydrogels possess the following important properties i.e. easy replacement, transparency, flexibility, barrier against bacteria, good adhesion, easy handling, oxygen permeability, control of drug dosage, absorption, and prevention of loss of body fluids, etc. According to the application point of view, the properties and structures of the PVP based hydrogels can be tailored, as there are different examples of semi interpenetrating networks (semi-IPN), grafted structure, electrospun mats, microspheres, nano-composites etc [24, 112-119]. There are many commercialized hydrogels (prepared with hydrophilic polymers) are available in the market throughout the world with different trade names. VIACELL s.r.o is one of the largest suppliers of hydrogels in Czech Republic. They manufacture hydrogels for normal wounds, burns and cuts as well as hydrogels to cure different ulcers [120]. Nu-Gel wound dressing is a contribution from Johnson and Johnson Consumer Products. They provide an optimal moist healing environment for partial thickness pressure ulcers, vascular ulcers, diabetic ulcers, and first and second degree burns. Nu-Gel cools and soothes as an occlusive-hydrogel dressing [121]. Kikgel Company from Poland supplies hydrogel dressing called Neoheal. This dressing material is a new, unconventional mean, facilitating the treatment and accelerating healing of different kinds of wounds, especially: burn wounds, ulcerations, bedsores, all kinds of skin damages in which humid medium is favorable. Hydrogel dressings Neoheal are an aqueous composition of natural and synthetic polymers such as: polyvinylopyrrolidone, polyethylene glycol and agar, cross linked by a beam of electrons. Due to radiation processing the product is fully sterile [122]. The Bhabha Atomic Research Centre (BARC) has developed a process for large-scale production of hydrogels based on cross-linking of PVA by radiation (national and international patent filed) instead of PVP. All ingredients used in the process are low cost, biocompatible and are locally available or can be readily imported. They have a cooling effect on the burn wound and thus reduce the severity of the pain. It further provides moist environment, soften any slough, provides sterile cover and regulates the oxygen supply to the wound site to enable faster healing. The hydrogel has the property of adhering firmly yet gently to the healthy surface. It does not adhere to wet wound surface. This results in painless dressing. The narcotic tissues in the wound stick to the inner surface of the hydrogel and come out with the dressing. Being transparent, the progress of the wound healing can be observed without removing the dressing. These dressings are available in various sizes [123]. This paper is focused about the review of PVP based blend hydrogels.

# PVP Based Blend Hydrogels

## PVA-PVP blend hydrogel

To increase the mechanical properties of the hydrogels, PVP hydrogel dressings were normally prepared with agar as a second component. But it had some disadvantages, because the presence of agar may facilitate the penetration of microorganisms into the hydrogel, especially in highly humid environments in tropical regions. It has been reported that incorporation of polyethylene glycol (PEG) could improve the barrier against bacteria. Razzak et al. reported about PVA and PVP blend hydrogel, where instead of using agar as a second material PVA was used, and they were crosslinked using gamma radiation [73, 75]. Irradiation is a very suitable method for hydrogel synthesis. Radiation process shows many advantages such as easy process control, possibility of joining hydrogel formation and sterilization in one technological step, no requirement of adding any initiators and crosslinkers which may be harmful and difficult to remove [124-125].

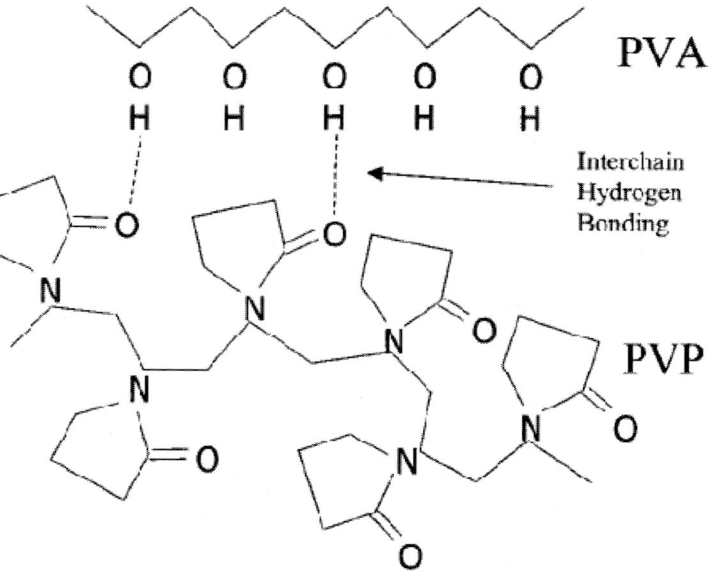


Figure 1. Interchain hydrogen bonding within a PVA/PVP blend occurs between carbonyl groups on PVP and hydroxyl groups on PVA. Reprinted with permission from Ref. [77].Copyright 2003, John Wiley and Sons

PVA/PVP interactions occur through interchain hydrogen bonding between the carbonyl group of PVP and hydroxyl group of PVA (Figure 1) resulting in physical crosslinking of these two polymers [77, 126].

## PVP-Chitosan Blend Hydrogel

Chitosan is a partially deacetylated product of chitin, containing D-glucosamine, which is a natural polymer present in exoskeleton of crustaceans and cell walls of fungi. Chitosan has been reported to be bioabsorbable, biocompatible and biodegradable with numerous applications in the biomedical, pharmaceutical, agriculture, packaging, and food biotechnology fields. By combining chitosan with the synthetic polymer PVP, a range of blends with a variety of properties and forms has been synthesized. They show blood compatibility [12-17].

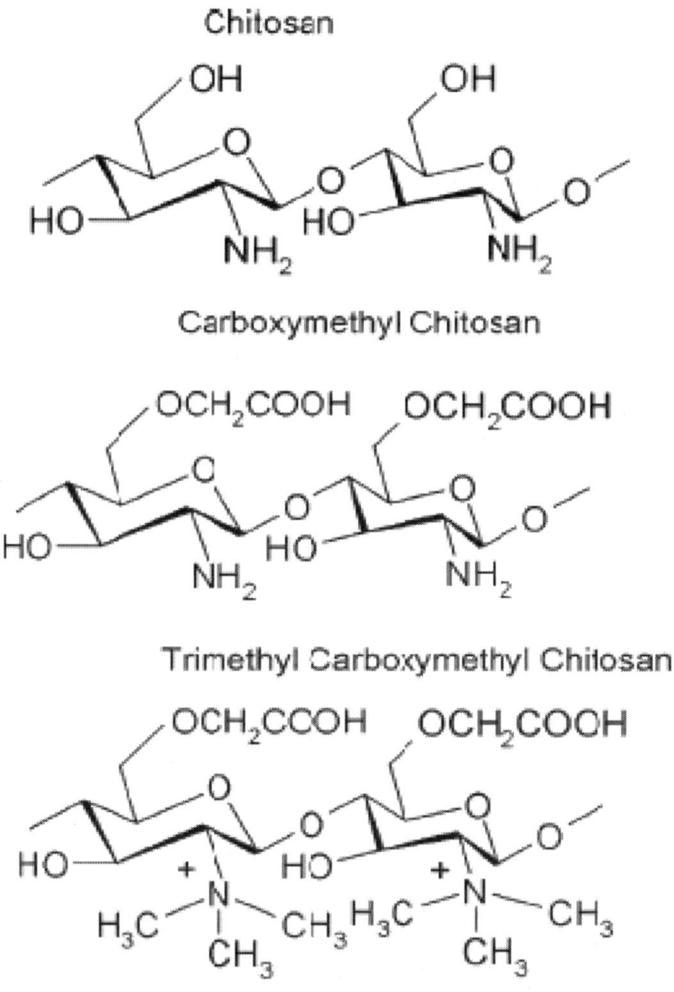


Figure 2. Chemical structures of chitosan, carboxymethyl chitosan and trimethyl carboxymethyl chitosan.. Reprinted with permission from Ref. [17].Copyright 2011, John Wiley and Sons

## PVP- Carboxymethyl Cellulose (CMC) Blend Hydrogels

Cellulose is a chain-shaped polymer which contains β-1,4-D-glucosidic linkages without any branches, is the most abundant natural polysaccharide. Carboxymethyl cellulose (CMC) is its ether derivate in which original H atoms of cellulose hydroxyl groups are replaced by carboxymethyl substituent, $-CH_2COOH$. Due to its good water-solubility, biocompatibility, high abundance and low price it has been widely used in many fields [4-7, 11 18].

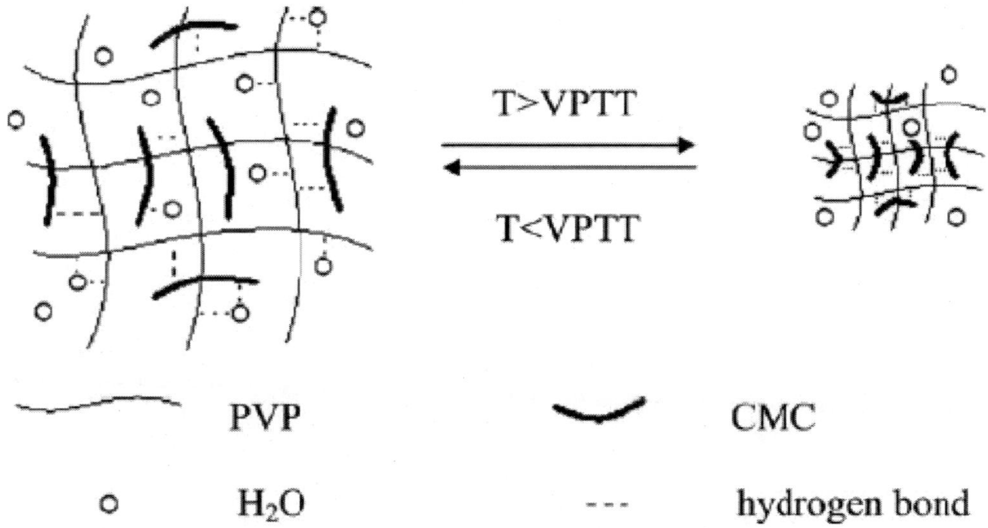


Figure 3. Schematic illustration of PVP/CMC semi-IPN hydrogels structure before and after phase separation. Reprinted with permission from Ref. [18].Copyright 2010, John Wiley and Sons

## PVP- Carrageenan Blend Hydrogels

Carrageenan is one of the most abundant polysaccharides, which exists as matrix materials, in numerous species of seaweeds. They are water soluble, sulfated polysaccharides extracted from different species of marine red algae of the class *Rhodophyceae*. Kappa-carrageenan and iota-carrageenan are considered as more important types of carrageenans as they are capable of forming thermoreversible gels. They have the backbone structure consisted of alternating α-1, 4 and β-1, 3-linked galactose residues. Several articles have been published regarding the synthesis of PVP-carrageenan hydrogels by radiation crosslinking [19-22]. Figure 4 shows a proposed mechanism of κ-Carrageenan grafted PVP.

Except these very common PVP-polysaccharide blend hydrogels, many different PVP blended hydrogels have been reported such as PVP-agar [4-7, 23-25], PVP-sodium alginate [26, 113], PVP-gelatin [127], PVP-polyethylene oxide (PEO) [128], PVP-I [70-71], PVP-polyacrylic acid (PAA) [72, 112, [129], PVP-2-hydroxyethyl methacrylate (HEMA) [114], and so on.

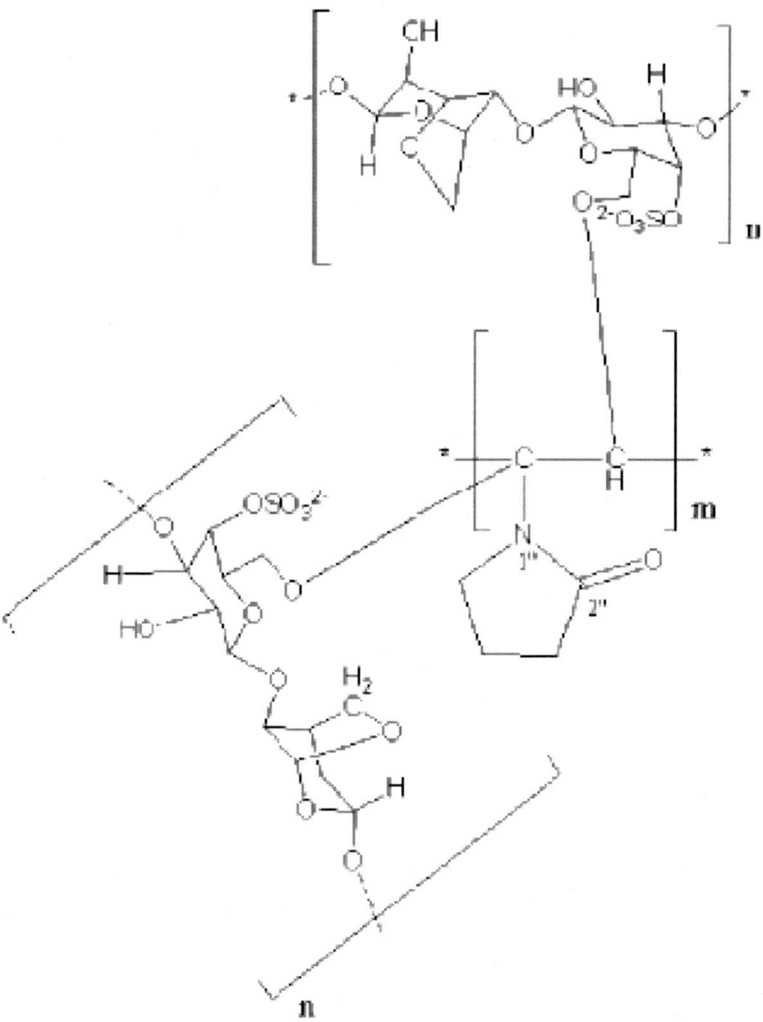

Figure 4. κ-Carrageenan – *graft* – PVP. Reprinted with permission from Ref. [22].Copyright 2006, John Wiley and Sons

## METHODS OF SYNTHESIS OF PVP-BASED HYDROGELS

### Radiation

i. γ- and electron beam radiation: In early 1950s, the pioneers of the radiation chemistry of polymers began some experiments with radiation crosslinking of hydrophilic polymers. The noticeable interest in the application of radiation techniques to obtain hydrogels for biomedical purposes began in the late sixties as a result of the papers and patents invented by Japanese and American scientists, headed by Kaetsu in Japan and Hoffman in USA. Hydrogels can be obtained by radiation technique in several ways, including irradiation of pure polymers, monomers or solution of polymers and/ or monomers in bulk, solution or emulsion.

The first method, i.e. irradiation of hydrophilic polymer in dry state has some drawbacks. It may require special sample preparation (like pressing or melting), and some difficulties may be encountered in obtaining homogeneous macroscopic gels. It requires usually much higher doses to obtain a gel as compared to irradiation in solution, and, furthermore, it may be difficult to remove the oxygen fully, a promoter of unwanted side reactions. More frequently the method of monomer irradiation is applied. In this technique, polymerization takes place in the first stage, followed by crosslinking of the formed chains. This way is possibly the most convenient when the chosen monomer is easily available but its polymer is not. Among various methods applied for the production of hydrogels, the radiation technique has many advantages, as a simple, efficient, clean and environment-friendly process. Application of radiation for the formation of hydrogels for biomedical use offers a unique possibility to combine the formation and sterilization of the product in a single technological step. This allows simplifying the technology and reducing production costs. Hydrogels based on PVP to be used as wound burn dressing were invented by Rossiak et al. Radiation production of hydrogel wound dressings started first in Poland at the Technical University of Lodz in 1992. PVP based dressings obtained by e-beam irradiation, were commercialized under the registered trademarks KIK-gel and AQUA-gel [30, 41, 44]. Baccaro and coworkers in 1995, investigated about the mechanical properties of the PVP hydrogels undergoing radiation. It was found that radiation dose increase has no effect on the modulus of elasticity but produces a negative effect on the break-strain and positive initial effect on the peak-load that develops into a negative effect above a dose value of 30kGy. It refers that peak-load depends strongly on the PVP concentration, the break-strain is inversely dependant on the radiation dose and the modulus of elasticity depends mainly on the PVP concentration [27]. New aspects of the synthesis and characterization of PVP and persulfate containing PVP-hydrogel irradiated by $\gamma$-ray has been investigate and discussed by Kaplan Can, et al. in their series of published articles. The gelation doses of persulfate containing polymer solutions were calculated by Charlesby – Pinner equation. It was found that the gelation dose values were shifted to higher values by increasing persulfate concentration in solution [28, 32, 34, 36, 37, 39]. Till now it has been found that using of gamma radiation is the most widespread technique to prepare different PVP-based hydrogels in different ways including synthesis of temperature sensitive hydrogel blends, tissue engineering scaffolds, for the grafting purpose, incorporation of electrical properties in hydrogel blends and so on [29, 31, 33, 35, 38, 40, 42, 43, 45, 47]. Electron beam radiation is also an effective mean of hydrogel crosslinking [48-49]. Dorit Meinhold and coworkers (2003) presented a novel approach to prepare PVP hydrogel coatings on poly (ethylene terephthalate) (PET) surfaces for the surface modification of materials in a wide variety of shapes like foils, fibers, tubes, or films. A schematic diagram is shown in Figure 5 [49].

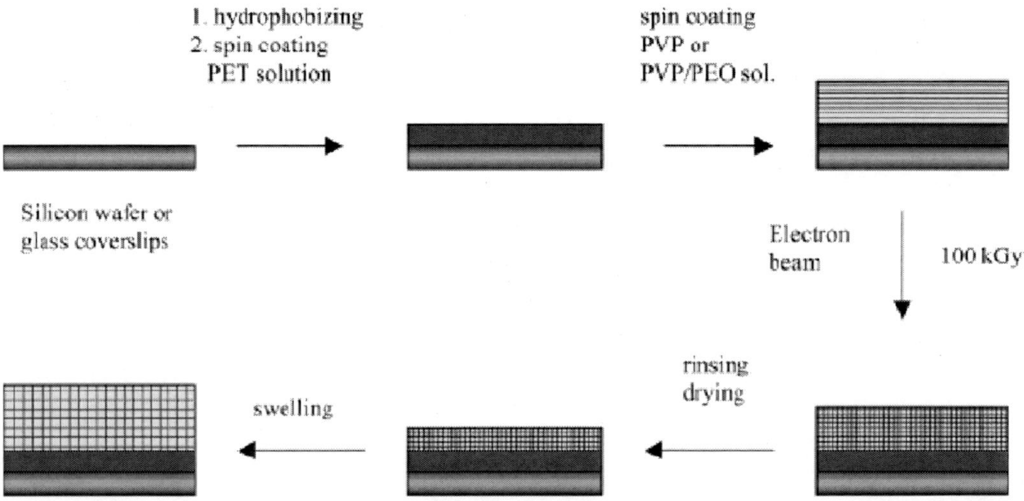


Figure 5. Scheme of the preparation of a PVP Hydrogel Layer on a PET Surface by Electron Beam Cross-Linking. Reprinted with permission from Ref. [49].Copyright 2004, American Chemical Society Publications

ii. Ultraviolet (UV) radiation may be considered as an alternative and effective method for the crosslinking of PVP hydrogels, because typical high energy radiation sources, e.g. γ-radiation, are very expensive and not easily available. Moreover, UV radiation involves very short time for reliable gel formation. This hydrogel preparation procedure appears very promising for the advantage that guest molecules or particles, such as cells or metal nanoparticles, can be easily incorporated in the gel by dissolving or suspending them in the polymer solution to be irradiated. By this simple procedure it is possible to include in the hydrogel formulation other components which could tune the hydrogel functional properties to the specific applications. Catalani and co-workers presented a study related to PVP crosslinking using UV-C radiation (λ=254 nm) from a low pressure arc mercury lamp. The process can be accelerated by the addition of hydrogen peroxide to the polymer aqueous mixture or by Photo-Fenton reaction. The peroxide addition accelerates the crosslinking kinetics through the hydroxyl radical (·OH) generation during its photo-homolysis. Hydroxyl radical reacts with polymer macromolecules giving rise to macroradicals. If macroradicals located on different polymer chains are favorably positioned, they may undergo recombination leading to crosslinking. Photo-Fenton reactions produce hydroxyl radicals through the absorption of the UVA – visible radiation (290 – 400 nm) using simple components like ferric aqueous solution in presence of hydrogen peroxide. UV crosslinked PVP hydrogels showed a strong gel mechanical behavior with viscoelastic moduli values similar to those of biological gels [50-59, 130]. Schemes of hydroxyl radical production are shown in Figure 6.

a) $H_2O_2 + h\nu_{254nm} \rightarrow 2HO\cdot$

b)

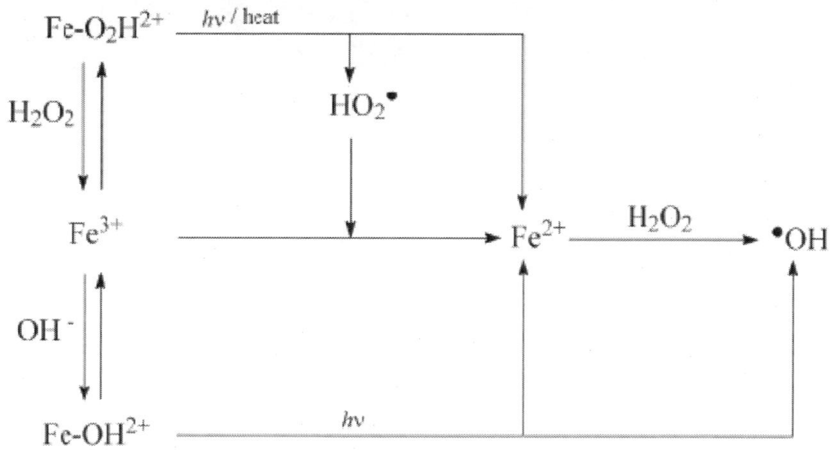


Figure 6. Scheme of production of hydroxyl radical from the (a) photo-homolysis of peroxide, and (b) photo-Fenton reactions Reprinted with permission from Ref. [51].Copyright 2004, Elsevier.

## Freezing-Thawing or Freezing-Thawing in Combination with Irradiation

Some researchers have reported about another approach for PVP based hydrogel preparation, repeated freezing-thawing or freezing-thawing in combination with γ-irradiation. For the freezing thawing, the polymer solution is kept under very cold condition, called freezing, (can vary from -20ºC to -70 ºC) for a period of time as required depending on the composition of hydrogel mixture and then temperature of the mixture is raised to the room temperature for a duration (thawing). The cycle is repeated several times [46, 61, 74].

## Moist Heat Treatment

A novel simple and cost effective method to prepare PVP based hydrogel has been developed in our lab. The technique of hydrogel preparation is designated as "moist heat treatment [4-7]. The hydrogels were developed by applying physical stimulation technique which is an operation friendly technique. The polymer solution was prepared in sealed bottles. Then, the polymer solution was treated in an autoclave sterilizer which provides moist heat condition (107 kPa, 120°C, and 20 minutes), where all the materials were sterilized and the cross-linking phenomena occurred in a single step. Afterwards, the solution was casted into moulds and allowed to cool at room temperature (20-22°C) under an aseptic environment to achieve the desired hydrogel. During cooling a 3- dimensional network of polymer is developed within the hydrogel and final product of hydrogel is achieved as shown in Figure 7 and 8.

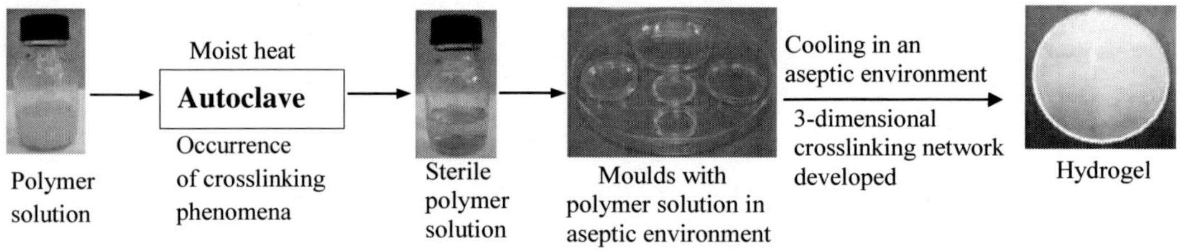


Figure 7. Schematic diagram of hydrogel preparation using moist heat treatment.

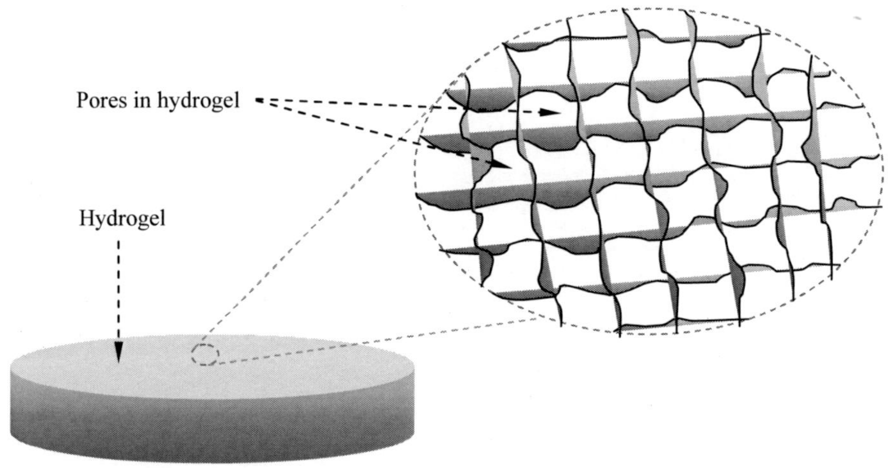


Figure 8. Schematic view of 3dimensionally crosslinked porous structure in hydrogel.

# PROPERTIES OF THE PVP-BASED HYDROGELS

## PVA-PVP Blend Hydrogels

The mechanisms of radiolysis and network formation (gelation) in mixed solution of PVA and PVP are quite complicated. Investigation of the dynamic rheological behavior of these hydrogels before and after gelation can provide useful information for elucidating these mechanisms, as upon irradiation, the formation of covalent bonds between adjacent polymer chains significantly affects the viscous and elastic properties of the matrix. Zainuddin et al. in their investigation represented the effect of varying the absorbed dose on the gelation behavior of PVA and PVP solutions. The dynamic rheological behavior of gamma-irradiated 12.8 wt% PVA, 12.8 wt% PVP, and a blend of 8 wt% PVA and 4.8 wt% PVP aqueous solutions have been studied pre- and post-gelation. The viscoelastic behavior observed by rheometry at each dose provides insight not only into the gelation process, but also into the effective network development and the final gel properties. Such a methodology allows us to elaborate the structure-function relationships between gel structure and strength and the radiation dose for these systems [131].

Jonathan and his coworkers showed the feasibility of PVA-PVP blend hydrogel for endoscopic replacement of the nucleus pulposus of a lumbar intervertebral disc. The hydrogels were processed with the use of three freeze/thaw cycles to induce crystallinity. Actually PVA/PVP hydrogels belong to the category of shape memory gels due to their elastic nature, so that they may exhibit predictable, reproducible changes in geometry from the hydrated to dehydrated states. This property of PVA/PVP gel was used in order to develop a procedure to insert the implant into the nuclear cavity endoscopically, implanting the dehydrated material and rehydrating to a desired shape and size *in situ* [76].

Due to their high water content, hydrogels possess excellent biocompatibility and a degree of flexibility similar to natural tissues, which minimizes potential irritation to surrounding membranes and tissues. By controlling the crosslink density, several parameters, such as pore size and elasticity, can be adjusted fitting the requirements of the tissue to be regenerated. Recent strong interest in the development of novel hydrogels and hydrogel composites can be attributed to their unique combination of properties, including biocompatibility, permeability, hydrophilicity, and low coefficient of friction. Zainuddin et al. continued their investigation on PVA/PVP blend hydrogels crosslinked via gamma radiation, and focused on their water diffusion and concentration distribution properties. They mainly extended the studies of Razzak and coworkers. They examined different polymer compositions and doses to identify the optimal condition. Aqueous solution of PVA and PVP containing ~80 wt% water and a mole fraction of PVP of 0.19 irradiated to 25 kGy were found to yield gels that have optimal bioadhesion and good mechanical properties, they were soft and robust and they evoked no significant body tissue response. These hydrogel dressings were capable of absorbing more water than usual and therapeutic drugs could be incorporated into the dressing and, when they are applied the drugs can be released slowly from the dressing into the wound environment [132-133]. In spite of radiation crosslinking, PVA-PVP hydrogels can be crosslinked chemically. As an example, crosslinked PVA-PVP hydrogels were prepared in aqueous media using $K_2S_2O_8$. In this case it was observed that the gel fraction increases with increasing potassium peroxodisulphate concentration. The swelling rate in water was faster in the beginning and the amount of absorbed water as immersion water uptake of greater than 1400% was obtained. The maximum of swelling was achieved at pH=7 because of the neutrality of the hydrogels and the maximum extent of swelling were followed at 25°C. However the rate of swelling decreased at higher temperature, thus the swelling behavior of PVA-PVP hydrogels depends on pH and temperature [134]. PVA-PVP hydrogels are ideal candidates for cartilage replacement or as artificial bearing surfaces, because they show better mechanical and network stability when compared with pure PVA hydrogels [78, 79]. Biomaterials developed for the purpose of cartilage replacement should be able to bear the high loads experience in human joints with minimal or no wear and a low coefficient of friction for millions of cycles. Jayanth K. Katta and coworkers have reported the friction/wear test of PVA-PVP hydrogels (prepared using solvent evaporation technique) using a Co-Cr pin articulation (as shown in Figure 9). The experiment was designed to determine the effects of load, lubricant and polymer content on the friction and wear characteristics of PVA-PVP hydrogels [78]. On the other hand Ruyin Ma, et al. reported about the friction properties of the PVA-PVP blend hydrogels (prepared by repeating freezing and thawing method) for the use as an artificial cartilage by using a self made ball-on-disc tribometer [79]. Instead of cartilage replacement PVA-PVP hydrogel can serve for human lens substitution [80].

PVA-PVP hydrogels are found to form a thermodynamically miscible pair and show biodegradability. El-Mohody and Ghanem investigated the biodegradation of PVA-PVP hydrogels by burial method in two types of soils (clay and sandy soils). The pure PVA was the most biodegradable, while the resistance to biodegradation increased with enhanced PVP ratio in the PVA-PVP hydrogel. The highest degradation rate was found to be achieved using clay soil [135].

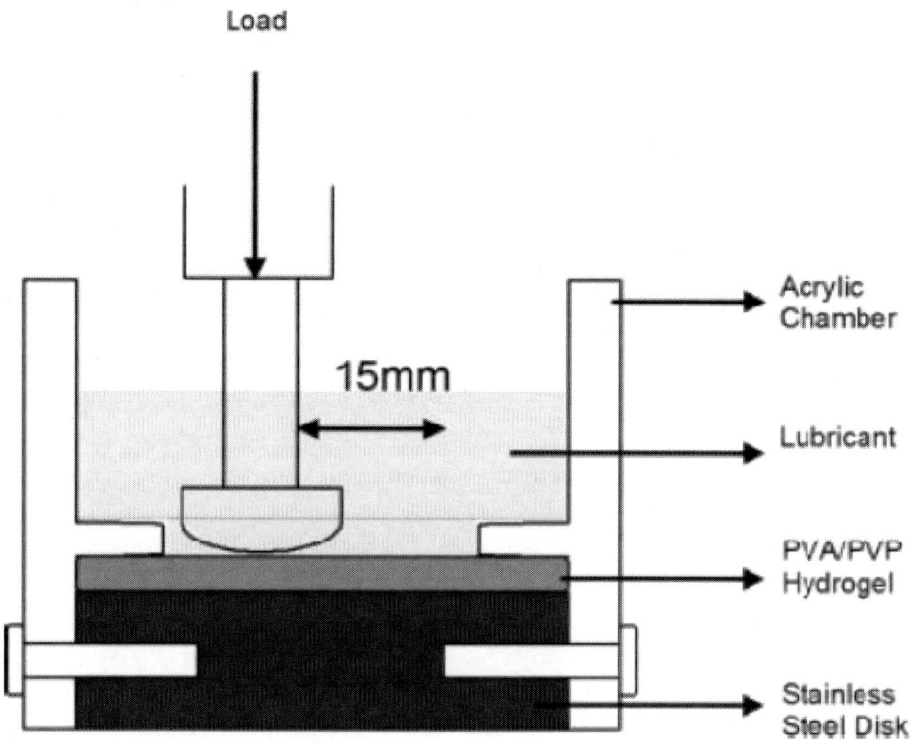


Figure 9. Two-dimensional schematic of a Co–Cr pin articulating against a PVA/PVP hydrogel sample. Reprinted with permission from Ref. [78].Copyright 2007, John Wiley and Sons

## PVP-Chitosan Blend Hydrogels

It has already been reported that incorporation of polysaccharides such as chitosan into the PVP/water system obviously increased the gel strength and equilibrium degree of swelling of PVP hydrogel. Comparing with chitosan, carboxymethyl chitosan is an amphoteric polyelectrolyte, which has good hydrophilicity and good miscibility with PVP in aqueous solution. In comparison to PVP hydrogels, these PVP-chitosan blend hydrogels can provide better properties like pH sensitivity, good swelling, good mechanical properties, improved surface property, potential adsorptive capacity for protein, etc. Investigation on PVP-chitosan hydrogels also demonstrated their suitability as a potential candidate for islet immunoisolation, and showed dual properties by supporting the growth of epithelial cells and selectively inhibiting fibroblast growth [12-17]. Chitosan contains hemostatic properties along with many other biological properties, including bacteriostatic and fungistatic properties that are particularly useful for wound treatment. Nho and Park reported about the hydrogels from a mixture of PVA, PVP and chitosan were made by the combination of freezing-thawing and γ-radiation. Swelling percent was greatly increased as the composition of chitosan in PVA/PVP–chitosan increased. Wounds of 1-cm diameter that formed on the back skin of the rat were covered with hydrogel samples (1.5 x 1.5 x 0.3 cm) and vaseline gauze. At the designated postoperative day, macroscopic observation of wound status was made. The Vaseline gauze dried quickly and stuck to the wound of rat. The PVA/PVP–chitosan hydrogel dressing stopped the bleeding from the wound and had the better curing effect than vaseline gauze [74] (see Figure 10).

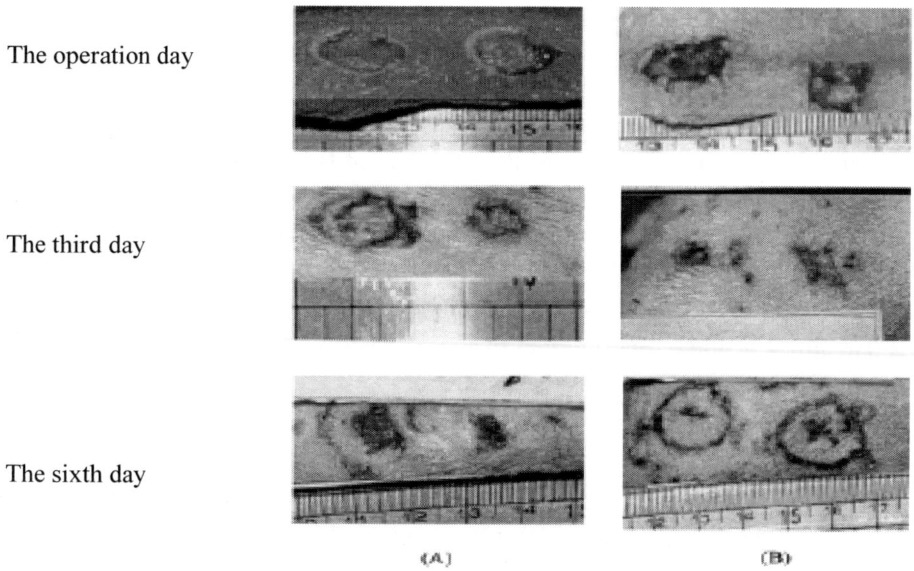


Figure 10. Healing process of wound using (A) no dressing (left)/gauze (right), (B) PVA/PVP–chitosan hydrogel (left)/gauze (right). Reprinted with permission from Ref. [74].Copyright 2002, John Wiley and Sons

## PVP-CMC Blend Hydrogels

Crosslinked CMC hydrogels possess many excellent properties for utilization of biomaterials, such as high water uptake, biodegradability, and innoxious quality. However they have very poor mechanical strength, especially in swollen states, which limits their applications. PVP-CMC hydrogel crosslinked by gamma radiation shows better mechanical strength than that of pure CMC hydrogel; better flexibility and swelling behavior than that of pure PVP hydrogel, and similar moisture retention capability to that of a commercial hydrogel wound dressing [11]. Due to the pH sensitive character of CMC, combination of CMC with PVP will produce dual-stimuli vehicles that respond to localized conditions of pH and temperature in the human body [18]. In our lab we have developed PVP-CMC based novel wound dressing hydrogels using a simple technique called "moist heat treatment". This method is very efficient, low cost, does not need extra laboratory set up, and moreover, crosslinking and sterilization of the hydrogels involve a single step like radiation crosslinking, but much safer as the technique does not deal with any kind of radiation [4-7, 11, 18].

made of synthetic polymers because they exert intrinsic bioactivity including certain cellular behaviour.

## Wound Healing

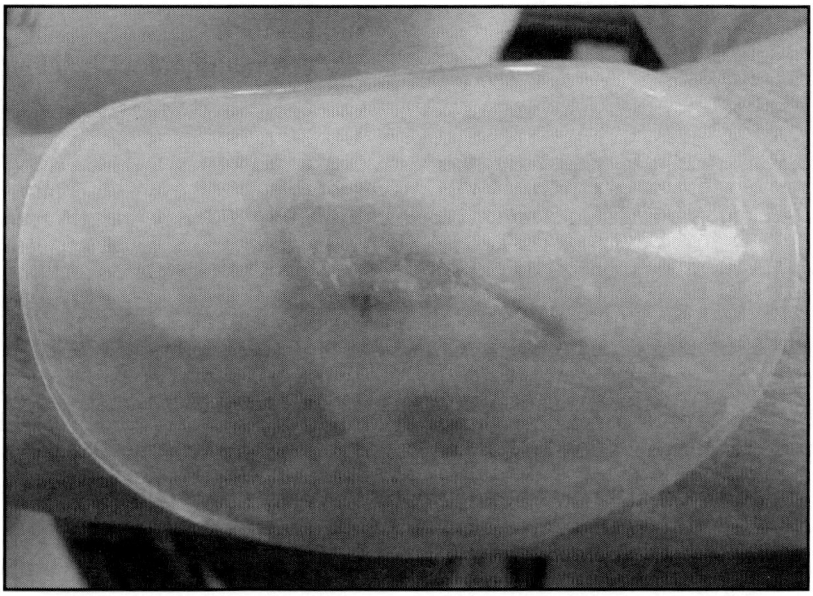

Figure 13. View of PVP-CMC hydrogel prepared by "moist heat treatment".

Wound healing or wound dressing is one of the major implementations of the hydrogels. There are two kinds of dressings: dry type and wet type. It has been reported that healing with a wet environment is faster than that with a dry environment [62, 66]. Hydrogel dressings in various forms have been used in wound management for over 20 years. Hydrogels for burn wound dressings were first invented by Rosiak et al. in 1989 based on PVP, PEG and agar [23]. Hydrogel wound dressings have many interesting properties like: immediate pain control effect, easy replacement, transparency, barrier against bacteria, good adhesion, easy handling, oxygen permeability, control of drug dosage, absorption and prevention of loss of body fluids, etc [64, 137]. The high moisture content maintains a desirable moist interface which facilitates cell migration and prevents dressing adherence. Hydrogel dressings are available in several forms, based upon a variety of different polymers. Some polymers are cross-linked to impart a degree of structural stability to the final product, which often takes the form of a thin sheet used for application to relatively shallow surface wounds. PVP hydrogel wound dressings are normally prepared in presence of agar as a second component to enhance the mechanical properties of the hydrogels [23, 25, 64, 137]. PVA hydrogel also has excellent transparency, biocompatibility and smoothness, but one of the major problems in the application of PVA hydrogels is their relatively poor mechanical strength, and the hydrogels become harder with the evaporation of water. The hard gels may damage the wound. This problem is reduced to a significant extent when PVA and PVP are blended together. The properties of the PVP based wound dressings can be monitored according to the composition, mainly upon the different polysaccharides blended with it [62, 69]. PVP-iodine hydrogels are very effective in wound therapy [65, 67]. In our work we have developed the PVP and carboxymethyl cellulose (CMC) based hydrogel by using moist heat treatment, and the hydrogel is designated as PVP-CMC hydrogel. The view of the PVP-CMC hydrogel has been shown on Figure 13.

## Drug Delivery

The concept of drug delivery or controlled drug release has emerged from the need for effective management of diseases. These include localized delivery of the drug to a particular part of the body. A variety of drug delivery systems have been developed till now and generally the drug release depends on the pH [100]. Figure 14 shows a model of drug release from swollen hydrogel. Controlled release systems have been developed over a range of pH-domains in the body, such as for the periodontal, oral, gastric and intestinal applications. Because of the soft tissue biocompatibility, hydrogels are of special interest in controlled release applications. It is easier to disperse the drug in the hydrogel matrix, and the high degree of control is achieved by physical and chemical properties of the polymer network.

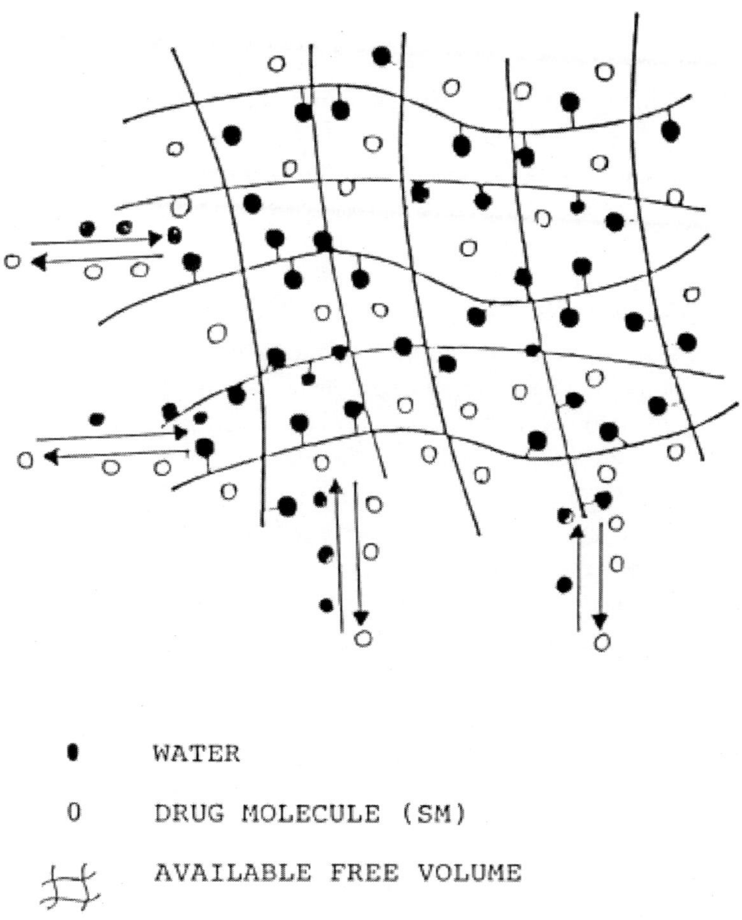


Figure 14. Model depicting release of drug from swollen hydrogel. Reprinted with permission from Ref. [86].Copyright 2001, John Wiley and Sons

Hydrogels have three dimensionally crosslinked mesh like structure, which provides a matrix for the entrapment of drugs. For example, pH-sensitive chitosan – PVP, semi interpenetrating based controlled released antibiotic delivery system that is well suited for use in a gastric environment [85]. Hydrogels may be impregnated with biologically active agents, such as antibiotics, enzymes, contraceptives, drug antagonists, anticoagulants, and anticancer drugs and may serve as systems for the controlled release of the agents absorbed over a prolonged time period at a specific site of the body. Normally there are two methods for the loading of hydrogels as drug carriers. In one method, the hydrogel monomer is mixed with the drug, an initiator, and a crosslinker and allowed to polymerize, trapping the drug within the matrix. In the second approach, a preformed hydrogel is allowed to swell to equilibrium in a suitable drug solution and then the drug-loaded hydrogel is dried and the device is obtained [86]. Generally, if the crosslinkings are based on physical bonds, such as hydrogen, ionic, or van der Waal's bonds, the responses of the hydrogels to external stimuli are often reversible [82].

Polymeric microsphere, beads, polymeric micelles etc have been shown to be effective in enhancing drug targeting specificity, lowering systematic drug toxicity, improving treatment absorption rates, and providing protection for pharmaceuticals against biochemical degradation [26, 90, 95, 96]. Sodium alginate beads can be prepared for drug release purpose crosslinked by calcium chloride solution due to the ionotropic bonding between carboxyl groups of sodium alginate and calcium ions according to the well known "egg-box model". Ionic crosslinking is an effective way for crosslinking of the hydrogels and generally carried out by dispersion of the polymer/drug solution into an aqueous gelation medium. It is a simple and mild process, and the biological activity of drugs can be well retained [26, 90, 95]. Hydrogel nonoparticles are also serving as very promising device for drug delivery. They are of great interest for drug encapsulation and delivery or as embolotherapic agents [81, 87, 98]. Oral delivery of drugs can be significantly improved by using nanoparticles as carriers. S.K. Sahoo and coworkers reported about the preparation of nanoparticles of up to 50 nm diameter which are co-polymers of biocompatible materials made from vinylpyrrolidone and acrylic acid monomers crosslinked with NN'methylene bis acrylamide and which were prepared in reverse micelles for precisely controlling the particle size. FITC-dextran was used as a marker compound which was entrapped in these nanoparticles. These smart hydrogel polymers were very sensitive to pH and temperature effects on the release of the entrapped marker compound [81]. Ferrogels are also in use for drug release purpose. Ferrogels are chemically crosslinked polymer network swollen by a ferrofluid. It's a new material with magnetic nanocrystals, embedded in a flexible polymer network, that provide high magneto-elasticity. This property makes the material applicable as a drug carrier [88, 89]. Jie Chen et al. reported about magnetic targeted drug controlled release hydrogel microspheres which were prepared by a radiation technique. Ferric oxide granules of size about 50 nm were used as the core for magnetic field. The PVP ferrogels, i.e. the ferromagnetic nanoparticles in hydrogel microsphere were obtained by irradiating an emulsion of PVP/ferromagnetic granule with cobalt 60 $\gamma$ – ray [88]. One of the major applications of PVP based hydrogels is transdermal drug delivery [83, 84]. A mathematical simulation was presented by A. L. Iordannskii which describes the in vitro drug delivery kinetics from hydrophilic adhesive water-soluble poly- N-vinylpyrrolidone (PVP)-polyethylene glycol (PEG) matrices of transdermal therapeutic systems (TTS) across skin-imitating hydrophobic Carbosil membranes. Propranolol is employed as the test drug [84]. Ibrahim A. Alsarra et al. represented the study related to different mucoadhesive polymeric hydrogels for nasal delivery of scyclovir. Gels containing

PVP were prepared with radiation crosslinking. They also evaluated the gels containing chitosan and carbopol [93, 99]. To determine the drug release capability of the hydrogels many model drugs such as methylene blue, ibuprofenate, salicylic acid, bovine serum albumin, etc are being used to prove their drug release behavior [91, 92, 94, 97].

## Implants

The use of hydrogels to replace damaged tissues is a positive approach because of their biocompatibility in contact with human tissues and could be an alternative for an implant. Breast reconstruction with an implant usually implies a sub muscular location to lower the incidence of capsular contracture, and that often necessitates a mastopexy or a reduction mammaplasty of the opposite breast [101, 102, 104, 108]. An ideal prosthesis would be so inert that it could be located subcutaneously without causing any visible contraction. Subcutaneous mastectomy and an immediate reconstruction with a subcutaneously located prosthesis give a good short term cosmetic result with symmetrical breasts. It can also be used in patients with drooping breasts. The development of a new prosthesis is therefore desirable. Silicon gel implants almost invariably cause capsular contraction when placed subcutaneously. Saline-filled implants with a textured rather than a smooth surface have a much lower incidence of such contraction. The Misti1 Gold II (Bioplasty, The Netherlands) breast implants were introduced in 1987. This implant has a textured surface and is prefilled with PVP-hydrogel and has been claimed by the manufacturer to contain a more viscous gel to provide excellent feeling and to be safely removed from the body in the event of traumatic rupture. It has a textured shell that interacts with host tissue to minimize capsular contracture [101]. Novagold breast implants were introduced to the UK in 1996 as a variable alternative to silicone breast implants to overcome the drawbacks of conventional silicone breast implants, like an increased risk of malignancy, autoimmune disease and connective tissue disorders associated with silicone implants. The Novagold used PVP, and guar gum hydrogels as filler, within a textured shell [104].

PVP based hydrogels are also being used as artificial cartilage replacement, because they contain high amount of water which gives good biocompatibility, high elasticity and mechanical strength, and good chemical stability and durability [103, 105, 107]. S.A. Maher and coworkers showed a study to explore the performance of nondegradable hydrogel implants in focal cartilage defects. Their hypothesis was that the structural integrity of a permanent plug and its surrounding tissue would be influenced by the compressive modulus of the material used, and that superior results would be obtained with the implantation of a more compliant material. To test this hypothesis, PVA-PVP hydrogel implants of two different moduli (0.6 MPa vs. 0.2 MPa) were manufactured and implanted into experimentally produced osteochondral defects in a rabbit model that was followed for up to 6 months. The implants were well tolerated by the animals, and no deaths directly related to the implant were recorded. A view of the implants has been shown in Figure 15 [103].

A B

Figure 15. A: Compliant implant at 6 weeks, demonstrating overgrowth of tissue, densification of bone under and around implant, and shrinkage of polymer. B: Stiffer implant at 6 weeks. New bone formation is evident surrounding the plug (Trichrome stain). Reprinted with permission from Ref. [103].Copyright 2007, John Wiley and Sons

## Other Applications of PVP-Based Blend Hydrogels

In spite of above mentioned applications PVP-based hydrogels have many other applications, like soft contact lens preparation using PVP and polyhydroxyethyl methacrylate (pHEMA) blend [109], as vitreous substitute [110], PVP coated polyurethane for urological use [111], PVP-acrylic acid (AAc) copolymer hydrogel for the removal of heavy metals from aqueous solution by chelate formation [138, 139], waste water treatment [140], electrically conductive hydrogel (polyaniline (PANI) nanoparticles dispersed in PVP hydrogel) [116], PVP hydrogel supported luminal chemiluminescence for the automatic determination of hydrogen peroxide [141], hydrocarbon degradation [142], superabsorbent nanogel [118], electrospun fiber and mat preparation and their application for remineralization of enamel tooth surfaces [115, 119], food packaging materials [143, 144], water retainers for agriculture [19], and so on.

# CONCLUSION

Due to the biocompatibility, transparency, gel forming ability, PVP has become one of the most desirable materials for hydrogel preparation and very demanding in biomedical applications. PVP can be easily blended with other polymers or polysaccharides to tailor the property of the hydrogel with respect to the application point of view. Crosslinking can be introduced in the PVP based hydrogels by means of chemical modifications, $\gamma$-radiation, electron beam radiation, UV-radiation, freezing thawing, moist heat treatment, etc. Radiation crosslinking is always preferable than chemically crosslinked products, to avoid side reactions and toxic byproducts. Moist heat treatment for the preparation of PVP based hydrogels is a new concept developed in our laboratory, which is simple and cost effective method for crosslinking, and involves a single step for the formation of crosslinking and sterilization. This method is more advantageous than radiation crosslinking, because no extra laboratory set up is required, comparatively cheaper, and there is no risk of hazard from radiation. With due time the applications of PVP based hydrogels have been increased intensively, the properties are modified, and broadened the field of applications despite of conventional biomedical applications.

## ACKNOWLEDGEMENT

This review article has been written with the support of Operational Programme Research and Development for Innovation co-funded by the European Regional Development Fund (ERDF) and national budget of Czech Republic within the framework of the Centre of Polymer Systems project (reg.number: CZ.1.05/2.1.00/03.0111). The authors are also thankful to Professor Petr Saha, Rector of Tomas Bata University in Zlin for providing excellent facilities to initiate research on hydrogel and encouragement to write review article on hydrogel.

## REFERENCES

[1] Zhao, L., Xu, L., Mitomo, H., Yoshii, F. *Carbohydr. Polym.* 2006, 64, 473-480.
[2] Gwon, H., Lim, Y., Nho, Y., Baik, S. *Radiat. Phys. Chem.* 2010, 79, 650-653.
[3] Mathur, A. M., Moorjani, S.K., Scranton, A.B. *Polym. Review.* 1996, 36, 405-430.
[4] Roy, N., Saha, N., Kitano, T., Saha, P. *J. Appl. Polym. Sci.* 2010, 117, 1703-1710.
[5] Roy, N., Saha, N., Kitano, T., Saha, P. *Soft Mater.* 2010, 8 (2), 130-148.
[6] Roy, N., Saha, N., Humpolicek, P., Saha, P. *Soft Mater.* 2010, 8 (4), 338-357.
[7] Roy, N., Saha, N., Kitano, T., Vitkova, E., Saha, P. *Prog. Colloid Polym. Sci.* 2011, 138, 127-130.
[8] Roy, N., Saha, N., Saha, P. *Int. J. Polym. Mater.* (in press).
[9] Saha, N., Saarai, A., Roy, N., Kitano, T., Saha, P. *J. Biomater. Nanobiotechnol.* 2011, 2, 85-90.
[10] Saha, P, Saha, N.; Roy, N. "Hydrogel Wound Covering" Patent no: 302405, Czech Republic, 2011.
[11] Wang, M., Xu, L., Hu, H., Zhai, M., Peng, J., et al. *Nucl. Instrum. Methods Phys. Res., Sect. B*, 2007, 265, 385-389.
[12] Risbud, M., Hardikar, A., Bhonde, R. *J. Biosci.* 2000, 25, 25-31.
[13] Risbud, M.V., Bhonde, M.R., Bhonde, R.R. *J. Biomed. Mater.* Res. 2001, 57, 300-305.
[14] Risbud, M.V., Bhat, S.V. *J. Mater. Sci. - Mater. Med.* 2001, 12, 75-79.
[15] Lorenz, D.H., et al. US Patent 5420197, 1995.
[16] Makuuchi, K. *Radiat. Phys. Chem.* 2010, 79, 267-271.

[17] Abashzadeh, S., Hajimiri, M.H., Atyabi, F., Amini, M., Dinarvand, R. *J. Appl. Polym. Sci.* 2011, 121, 2761-2771.
[18] Lu, S., Liu, M., Ni, B., Gao, C. *J. Polym. Sci. Part B: Polym. Phys*. 2010, 48, 1749-1756.
[19] Relleve, L., Yoshii, F., Rosa, A.D., Kume, T. *Die Angewandte Makromolekulare Chemie*. 1999, 273, 63-68.
[20] Maolin, Z., Hongfei, H., Yoshii, F., Makuuchi, K. *Radiat. Phys. Chem.* 2000, 57, 459-464.
[21] Abad, L., Relleve, L.S., Aranilla, C.T., Rosa, A.M.D. *Radiat. Phys. Chem.*2003, 68, 901-908.
[22] Prasad, K., Mehta, G., Meena, R., Siddhanta, A.K. *J. Appl. Polym. Sci.* 2006, 102, 3654-3663.
[23] Lugao, A.B., Machado, L.D.B., Miranda, L.F., Alvarez, M.R., Rosiak, J.M. *Radiat. Phys. Chem.*1998, 52, 319-322.
[24] Lopergolo, L.C., Lugao, A.B., Catalaini, L.H. *J. Appl. Polym. Sci.* 2002, 86, 662-666.
[25] Momesso, R.G.R.A.P., Moreno, C.S., Rogero, S.O., Rogero, J.R., Spencer, P.J., Lugao, A.B. *Radiat. Phys. Chem*. 2010, 79, 283-285.
[26] Isiklan, N., Inal, M., Yigitoglu, M. *J. Appl. Polym. Sci.* 2008, 110, 481-493.
[27] Baccaro, S., Pajewski, L.A., Scoccia, G., Volpe, R., Rosiak, J.M. *Nucl. Instrum. Methods Phys. Res., Sect. B*, 1995, 105, 100-102.
[28] Kaplan, H., Guner, A. *Adv. Polym. Technol*. 2000, 19, 210-217.
[29] Caykara, T., Oren, S., Kantoglu, O., Guven, O. *J. Appl. Polym. Sci.* 2000, 77, 1037-1043.
[30] Rosiak, J.M., Yoshii, F. *Nucl. Instrum. Methods Phys. Res., Sect. B*, 1999, 151, 56-64.
[31] Oren, S., Caykara, T., Kantoglu, O., Guven, O. *J. Appl. Polym. Sci.* 2000, 78, 2219-2226.
[32] Kaplan, H., Guner, A. *J. Appl. Polym. Sci.* 2000, 74, 994-1000.
[33] Muta, H., Ishida, K., Tamaki, E., Satoh, M. *Polymer*, 2002, 43, 103-110.
[34] Kaplan Can, H., Kavlak, S., Kirci, B., Guner, A. *Adv. Polym. Technol*. 2003, 22, 252-259.
[35] Ozyurek, C., Caykara, T., Kantoglu, O, Guven, O. Polym. Adv. Technol. 2002, 13, 87-93.
[36] Kaplan Can H., Kirci, B., Kavlak, S., Guner, A. *Radiat. Phys. Chem*. 2003, 68, 811-818.
[37] Kaplan Can, H. *Radiat. Phys. Chem*. 2005, 72, 703-710.
[38] Gottlieb, R., Schmidt, T., Arndt, K.F. *Nucl. Instrum. Methods Phys. Res., Sect. B*, 2005, 236, 371-376.
[39] Kaplan Can, H., Denizil, B.K., Kavlak, S., Guner, A. *Radiat. Phys. Chem*. 2005, 72, 483-488.
[40] Shariatpanahi, R.E., Orang, F., Emami, S.H., Naimi, T. J. *Biomater. Sci. Polymer Edn.* 2006, 17, 659-667.
[41] Benamer, S., Mahlous, M., Boukrif, A., Mansouri, B., Youcef, S.L. *Nucl. Instrum. Methods Phys. Res., Sect. B*, 2006, 248, 284-290.
[42] Ajji, Z. *Nucl. Instrum. Methods Phys. Res., Sect. B*, 2007, 265, 179-182.
[43] Dispenza, C., Fiandaca, G., Presti, C.L., Piazza, S., Spadaro, G. *Radiat. Phys. Chem.* 2007, 76, 1371-1375.

[44] Foroutan, H., Khodabakhsh, M., Rabbani, M. *Iran. J. Radiat. Res.* 2007, 5, 131-136.
[45] Wang, Y., Wang, H. *Radiat. Phys. Chem.* 2009, 78, 234-237.
[46] Nho, Y.C., Lim, Y.M., Gwon, H.J., Choi, E.K. *Korean J. Chem. Eng.* 2009, 26, 1675-1678.
[47] Zainuddin, Srrounina, E.V., Hill, D.J.T., Whittaker, A.K. *Polym. Int.* 2010, 59, 1520-1525.
[48] Ajji, Z., Mirjalili, G., Alkhatab, A., Dada, H. *Radiat. Phys. Chem.* 2008, 77, 200-202.
[49] Meinhold, D., schweiss, R., Zschoche, S., Janke, A, Baier, A., et al. *Langmuir*, 2004, 20, 396-401.
[50] Lopergolo, L.C., Lugao, A.B., Catalani, L.H. *Polymer*, 2003, 44, 6217-6222.
[51] Fechine, G.J.M., Barros, J.A.G., Catalani, L.H. *Polymer*, 2004, 45, 4705-4709.
[52] Chen, K.S., Ku, Y.A., Lin, H.R., Yan, T.R., Sheu, D.C., et al. *Mater. Chem. Phys.* 2005, 91, 484-489.
[53] Fechine, G.J.M., Barros, J.A.G., Alcantara, M.R., Catalani, L.H. *Polymer*, 2006, 47, 2629-2633.
[54] Bastos, E.L., Ciscato, L.F.M.L., Bartoloni, F.H., Catalani, L.H., Baader, W.J. *Luminescence*, 2007, 22, 113-125.
[55] D'Errico, G., Lellis, M.D., Mangiapia, G., Tedeschi, A., Ortona, O., et al. *Biomacromolecules*, 2008, 9, 231-240.
[56] Ricca, M., Fodera, V., Giacomazza, D., Leone, M., Spadaro, G., et al. *Colloid Polym. Sci.* 2010, 288, 969-980.
[57] Zhu, X., Lu, P., Chen, W., Dong, J. *Polymer*, 2010, 51, 3054-3063.
[58] Wang, J., Sun, F., Li, X. *J. Appl. Polym. Sci.* 2010, 117, 1851-1858.
[59] Devine, D.M., Higginbotham, C.L. *Polymer*, 2003, 44, 7851-7860.
[60] Barros, J.A.G., Fechine, G.J.M., Alcantara, M.R., Catalani, L.H. *Polymer*, 2006, 47, 8414-8419.
[61] Yu, H., Xu, X., Chen, X., Lu, T., Zhang, P., et al. *J. Appl. Polym. Sci.* 2007, 103, 125-133.
[62] Wu, M., Bao, B., Yoshii, F., Makuuchi, K. *J. Radioanal. Nucl. Chem.* 2001, 250, 391-395.
[63] Lugao, A.B., Rogero, S.O., Malmonge, S.M. *Radiat. Phys. Chem.* 2002, 63, 543-546.
[64] Ajji, Z., Othman, I., Rosiak, J.M. *Nucl. Instrum. Methods Phys. Res., Sect. B*, 2005, 229, 375-380.
[65] Langer, S., Botteck, N.M., Bosse, B., Reimer, K., Vogt, P.M., et al. Eur. *Surg. Res.* 2006, 38, 27-34.
[66] Yu, H., Xu, X., Chen, X., Hao, J., Jing, X. *J. Appl. Polym. Sci.* 2006, 101, 2453-2463.
[67] Vogt, P.M., Reimer, K., Hauser, J., Robbach, O., Steinau, H.U., et al. *Burns*, 2006, 32, 698-705.
[68] Patel, A., Mequanint, K., *Macromol. Biosci. 2007*, 7, 727-737.
[69] Sing, B., Pal, L. *Int. J. Biol. Macromol.* 2011, 48, 501-510.
[70] Reimer, K., Vogt, P.M., Broegmann, B., Hauser, J., Rossbach, O. *Dermatology*, 2000, 201, 235-241.
[71] Beukelman, C.J., Berg, A.J.J.V.D., Hoekstra, M.J., Uhl, R., Reimer, K., et al. *Burns*, 2008, 34, 845-855.
[72] Smith, T.J., Kennedy, J.E., Higginbotham, C.L. *J. Mater. Sci.* 2010, 45, 2884-2891.

[73] Razzak, M.T., Zainuddin, Erizal, Dewi, S.P., Lely, H., Taty, E., et al. *Radiat. Phys. Chem*. 1999, 55, 153-165.
[74] Nho, Y.C., Park, K.R. *J. Appl. Polym. Sci*. 2002, 85, 1787-1794.
[75] Razzak, M.T., Darwis, D., Zainuddin, Sukirno. *Radiat. Phys. Chem*. 2001, 62, 107-113.
[76] Thomas, J., Gomes, K., Lowman, A., Marcolongo, M. *J. Biomed. Mater. Res. Part B: Appl. Biomater*. 2004, 69B, 135-140.
[77] Thomas, J., Lowman, A., Marcolongo, M. *J. Biomed. Mater. Res.*2003, 67A, 1329-1337.
[78] Katta, J.K., Marcolongo, M., Lowman, A., Mansmann, K.A. *J. Biomed. Mater. Res.*2007, 83A, 471-479.
[79] Ma, R., Xiong, D., Miao, F., Zhang, J., Peng, Y. *J. Biomed. Mater. Res.* 2010, 93A, 1016-1019.
[80] Leone, G., Consumi, M., Greco, G., Bonechi, C., Lamponi, S., et al. *J. Biomed. Mater. Res. Part B: Appl. Biomater*. 2011, 97B, 278-288.
[81] Sahoo, S.K., De, T.K., Ghosh, P.K., Maitra, A. *J. Colloid Interface Sci*. 1998, 206, 361-368.
[82] Yaung, J.F., Kwei, T.K. *J. Appl. Polym. Sci*. 1998, 69, 921-930.
[83] Fang, J.A., Hsu, L.R., Huang, Y.B., Tsai, Y.H., *Int. J. Pharm*. 1999, 180, 137-149.
[84] Lordanskii, A.L., Feldstein, M.M., Markin, V.S., Hadgraft, J., Plate, N.A. *Eur. J. Pharm. Biopharm*. 2000, 49, 287-293.
[85] Risbud, M.V., Hardikar, A.A., Bhat, S.V., Bhonde, R.R. *J. Controlled Release*. 2000, 68, 23-30.
[86] Bajpai, A.K., Rajpoot, M., *J. Appl. Polym. Sci*. 2001, 81, 1238-1247.
[87] Bharali, D.J., Sahoo, S.K., Mozumdar, S., Maitra, A. *J. Colloid Interface Sci*. 2003, 258, 415-423.
[88] Chen, J., Yang, L., Liu, Y., Ding, G., Pei, Y., et al. *Macromol. Symp*. 2005, 225, 71-80.
[89] Adriane, K., Huang, J., Ding, G., Chen, J., Liu, Y. *J. Drug Targeting*. 2006, 14, 243-253.
[90] Sanli, O., Bicer, E., Isilkan, N. *J. Appl. Polym. Sci*. 2008, 107, 1973-1980.
[91] Mishra, R.K., Datt, M., Banthia, A.K. *AAPS Pharm. Sci. Tech*. 2008, 9, 395-403.
[92] Dergunov, S.G., Mun, G.A. *Radiat. Phys. Chem*. 2009, 78, 65-68.
[93] Alsarra, I.A., Hamed, A.Y., Mahrous, G.M., Maghraby, G.M.E., Al-Robayan, A.A., et al. *Drug Dev. Ind. Pharm*. 2009, 35, 352-362.
[94] Nielsen, A.L., Madsen, F., Larsen, K.L. *Drug Delivery*, 2009, 16, 92-101.
[95] Ma, L., Liu, M., Chen, J., Liu, H., Cui, D., et al. *Adv. Eng. Mater*. 2009, 11, B267-B274.
[96] Tarabukina, E., Zoolshoev, Z., Melenevskaya, E., Budtova, *T. Int. J. Pharm*. 2010, 384, 9-14.
[97] Tang, Q., Yu, J.R., Chen, L., Zhu, J., Hu, Z.M. *Curr. Appl Phys*. 2010, 10, 766-770.
[98] Bueno, V.B., Catalani, L.H., Daghastani, K.R.P., Cuccovia, I.M., Chaimovich, H. *Quim. Nova*, 2010, 10, 2083-2087.
[99] Alsarra, I.A., Hamed, A.Y., Alanazi, F.K., Neau, S.H. *Arch. Pharm. Res*. 2011, 34, 573-582.
[100] Alarifi, A.S. *J. Appl. Polym. Sci.* 2011, 120, 3484-3489.
[101] Benediktsson, K., Perbeck, L.G. *Scand. J. Reconstr. Hand Surg*. 2000, 34, 65-70.
[102] Bosch, G., Jacobo, O. *Aesth. Plast. Surg*. 2001, 25, 353-356.

[103] Maher, S.A., Doty, S.B., Torzilli, P.A., Thornton, S., Lowman, A.M., et al. *J. Biomed. Mater. Res.* 2007, 83A, 145-155.

[104] Hardwicke, J., Gaze, N.R., Laitung, J.K.G., *J. Plast. Recons. Aesth. Surg.* 2007, 60, 1313-1316.

[105] Zheng, Y., Huang, X., Wang, Y., Xu, H., Chen, X. *J. Appl. Polym. Sci.*2009, 113, 736-741.

[106] Su, J., Iomdina, E., Tarutta, E., Ward, B., Song, J., et al. *Exp. Eye Res.* 2009, 88, 445-457.

[107] Ma, R., Xiong, D., Miao, F., Zhang, J., Peng, Y. *Mater. Sci. Eng. C.* 2009, 29, 1979-1983.

[108] Garcia-Tutor, E., Alonso-Burgos, A., Marre, D., Adrian, C.A. Aesth. Plast. Surg. 2011, 35, 203-210.

[109] Yanez, F., Concheiro, A., Alvarez-Lorenzo, C., *Eur. J. Pharm. Biopharm.* 2008, 69, 1094-1103.

[110] Hong, Y., Chirila, T.V., Vijayasekaran, S., Dalton, P.D., Tahija, S.G., et al. *J. Biomed. Mater. Res.* 1996, 30, 441-448.

[111] Tunney, M.M., Gorman, S.P. *Biomaterials*, 2002, 23, 4601-4608.

[112] Jin, S., Liu, M., Zhang, F., Chen, S., Niu, A. *Polymer*, 2006, 1526-1532.

[113] Wang, W., Wang, A., *Carbohyd. Polym.* 2010, 80, 1028-1036.

[114] Jovasevic, J.S., Dimitrijevic, S.I., Filipovic, J.M, Tomic, S.L., Micis, M.M., et al. *Acta Physica Polonica A*, 2011, 120, 279-283.

[115] Fletcher, J., Walsh, D., Fowler, C.E., Mann, S. *Cryst. Eng. Comm.* 2011, 13, 3692-3697.

[116] Dispenza, C., Presti, C.L., Belfiore, C., Spadaro, G., Piazza, S. *Polymer*, 2006, 47, 961-971.

[117] Murthy, P.S.K., Mohan, Y.M., Varaprasad, K., Sreedhar, B., Raju, K.M. *J. Colloid Interface Sci.* 2008, 318, 217-224.

[118] Bueno, V.B., Cuccovia, I.M., Chaimovic, H., Catalani, L.H. *Colloid Polym. Sci.* 2009, 287, 705-713.

[119] Diaz, J.E., Barrero, A., Marquez, M., Fernandez-Nieves, A., Loscertales, I.G. *Macromol. Rapid Commun.* 2010, 31, 183-189.

[120] Viacell s.r.o., www.viacell.cz , accessed on December, 2011

[121] Nu-Gel Wound Dressing, Johnson and Johnson Consumer Products, http://www2.mooremedical.com/index.cfm?PG=CTL&CS=HOM&FN=ProductDetail &PID=2421 , accessed on December, 2011

[122] Neoheal, About the dressing, http://www.kikgel.com.pl/neoheal/ , accessed on December, 2011

[123] Hydrogel Burn And Injury Dressing, Health, technologies for better quality of life, http://www.dae.gov.in/publ/betrlife/health/hydrogel.pdf , accessed on December, 2011

[124] Park, K.R., Nho, Y.C. *Radiat. Phys. Chem.*2003, 67, 361-365.

[125] Hossen, M.K., Azim, M.A., Chowdhury, A.M.S., Dafader, N.C., Haque, M.E., et al. *Polym. Plast. Technol. Eng.* 2008, 47, 662-665.

[126] Joshi, A., Fussell, G., Thomas, J., Hsuan, A., Lowman, A., et al. *Biomaterials*, 2006, 27, 176-184.

[127] Lopes, C.M.A., Felisberti, M.I., *Biomaterials*, 2003, 24, 1279-1284.

[128] El-Hady, A.A., El-Rehim, H.A.A. *J. Biomed. Biotechnol.* 2004, 4, 219-226.

[129] Kadlubowski, S, Henke, A., Ulanski, P., Rosiak, J.M., Bromberg, L., et al. *Polymer*, 2007, 48, 4974-4981.

[130] Zhang, Y., Kato, S., Anazawa, T. *Smart Mater. Struct*. 2007, 16, 2175-2182.

[131] Zainuddin, Cooper-White, J.J., Hill, D.J.T. *J. Biomater. Sci. Polymer Edn.* 2002, 13, 1007-1020.

[132] Zainuddin, Hill, D.J.T., Strounina, E., Whittaker, A.K., Chirila, T.V. Soft Mater. 2004, 2, 195-212.

[133] Hill, D.J.T., Whittaker, A.K., Zainuddin. *Radiat. Phys. Chem.* 2011, 80, 213-218.

[134] Lakouraj, M.M., Tajbakhsh, M., Mokhtary, M. *Iran. Polym. J.* 2005, 14, 1022-1030.

[135] El-Mohdy, H.L.A., Ghanem,S. *J. Polym. Res*. 2009, 16, 1-10.

[136] Stammen, J.A., Williams, S., Ku, D.N., Guldberg, R.E. *Biomaterials*, 2001, 22, 799-806.

[137] Higa, O.Z., Rogero, S.O., Machado, L.D.B., Mathor, M.B., Lugao, A.B. *Radiat. Phys. Chem.* 1999, 55, 705-707.

[138] Ali, A.E.H., Shawky, H.A., El Rehim, H.A.A., Hegazy, E.A. *Eur. Polym. J.* 2003, 39, 2337-2344.

[139] Shawky, H.A., El- Sayed, M.H., Ali, A.E.H., Abdel Mottaleb, M.S. *J. Appl. Polym. Sci.* 2006, 100, 3966-3973.

[140] Hafez, H.S., Ali, A.E.H., Abdel-Mottaleb, M.S.A., *Int. J. Photoenergy*, 2005, 7, 181-185.

[141] Bastos, E.L., Ciscato, L.F.M.L., Bartoloni, F.H., Catalani, L.H., Romoff, P., et al. *Luminescence*, 2007, 22, 126-133.

[142] Djefal-Kerrar, A., Gais, S., Ouallouche, K., Khodja, A.N., Mahlous, M., et al. *Nucl. Instrum. Methods Phys. Res., Sect. B*, 2007, 265, 370-374.

[143] Roy, N., Saha, N., Sáha, P. Biodegradable Hydrogel Film for Food Packaging, Proceedings of *4th WSEAS International Conference on Recent Research in Geography, Geology, Energy, Environment and Biomedicine*, 2011, ISBN: 978-1-61804-022-0; page 329-334.

[144] Roy, N., Saha, N., Takeshi Kitano., Sáha, P. Biodegradation of PVP–CMC hydrogel film: A useful food packaging material. *Carbohydrate Polymers*, 2012, doi:10.1016/j.carbpol.2012.03.008

Hydrogels: Synthesis, Characterization and Applications ISBN: 978-1-61942-842-3
Editors: F. V. Câmara and L. J. Ferreira 

*Chapter 9*

# SOME NOVEL STRATEGIES TO FABRICATE HYDROGELS

***Jianquan Wang** and Dan Ren***
School of Materials Science and Engineering,
Beijing Institute of Technology, Beijing, China

## ABSTRACT

As one of hot research topics in polymer science, hydrogels attract increasing attentions. Following traditional preparation methods of hydrogels such as conventional chemical and physical crosslinks, some new strategies to fabricate hydrogels come forth recently. This review describes some novel hydrogels with excellent performances, which include slide-ring hydrogels, double-network hydrogels, nano-(micro-)composite hydrogels, ideal network hydrogels, self-oscillating hydrogels, non-covalently crosslinked hydrogels, and reversible covalently crosslinked hydrogels.

**Keywords:** hydrogels; fabrication; slide-ring; double-network; nano-(micro-) composite; ideal network; self-oscillating; non-covalent crosslinking; reversible covalent crosslinking.

## INTRODUCTION

Hydrogels, a class of three-dimensional hydrophilic macromolecular networks, are capable of absorbing and retaining large amounts of water, which impart them with some potential applications such as drug delivery systems, biosensors, contact lenses, artificial organs and so on. For a polymer hydrogel, several factors such as chemical composition and structure of macromolecules, crosslinking type, water, and sometimes other components inside cooperatively determine its comprehensive property. How much water absorption and whether or not stimuli response a certain hydrogel behaves are directly dependent on the

* Corresponding author, E-mail: jqwang@bit.edu.cn

chemical composition and structure of macromolecules inside. For example, the macromolecules including carboxyls can absorb more water than those including alcoholic hydroxyls; the polymer chain containing carboxyl and N-isopropylacrylamide (NIPAAm) endows pH- and temperature-sensitivity, respectively. As for the crosslinking type, it is well-known that covalent chemical and non-covalent physical approaches are usually adopted. Physical crosslinking concept involves ionic interaction, hydrophobic interaction, hydrogen bonding, microcrystal formation, helix formation and so on, so the physical gels are usually weaker than chemical ones except for some special cases. Generally, the polymer hydrogels are prepared by two methods: one is macromolecules crosslinking, and the other is polymerization of monomers in the presence of a crosslinker. The latter one is the most commonly used because of its convenient operability. In the last decade of the new century, some creative concepts and methods were applied to fabricate special hydrogels with extraordinary performances. In this article, some novel strategies to synthesize hydrogels are mainly reviewed. They include slide-ring gels, double-network gels, nano-(macro-)composite gels, ideal network gels, self-oscillating gels, non-covalently crosslinked gels, and reversible covalently crosslinked gels.

## SLIDE-RING HYDROGELS

Based on the polyrotaxane (PR) structure of poly(ethylene glycol) (PEG) and α-cyclodextrin (α-CD) supramolecular assembles, Ito and coworkers developed topological slide-ring hydrogels for the first time.[1] They performed inclusion reaction of PEG bisamine and α-CD, followed by end capping using 2,4-dinitrofluorobenzene to obtain PR, and then crosslinked PR with cyanuric chloride, deriving topological slide-ring gel (Figure 1 (a)). In this gel, some α-CD molecules on different PR chains were interlocked with each other, and thus figure-of-eight crosslinks formed. Benefit from the free mobility of α-CDs along PEG chains, the slide-ring gels showed excellent strechability and swellability up to 24 times in length and 24,000 times in weight, respectively.[2] Different from ordinary chemical gels with fixed crosslinking junctions, there exists the "pulley effect" of slide-ring crosslinks in this kind of gels. As shown in Figure 1(b), when the gel is subject to tension, the "pulley effect" disperses the tension in polymer chains automatically on tensile deformation.[2, 3] In addition, Small Angle Neutron Scatting (SANS), Small Angle X-Ray Scattering (SAXS) and Quasi-Elastic Light Scattering (QELS) were conducted to investigate the slide-ring gels in detail, further verifying the "pulley effect" within them.[4-7]

Besides pure PEG/α-CD PR based hydrogel, PEG/α-CD PR was extensively applied as a raw material to prepare other gels. α-CDs on PR chains had been grafted with vinyl, and thus the modified PR was utilized as a crosslinker for other vinyl monomers. Hydrophobic and hydrophilic PR crosslinkers were synthesized, respectively, used for fabricating temperature-sensitive poly(N-isopropylacrylamide) (PNIPAAm) hydrogels, i.e., $R_{wi}N$ gel and $R_{ws}N$ gel (Figure 2).[8, 9] In response to sudden temperature change from 20 °C to 40 °C, both types of PNIPAAm hydrogels underwent homothetic and unprecedentedly fast thermosensitivity, i.e. collapsed in ~10 min, in contrast with over 1 day for traditional N,N'-methylenebisacrylamide (BIS) crosslinked PNIPAAm hydrogels (TN gel). Consequently, PR crosslinked PNIPAAm gels also demonstrated very soft and flexible, and mechanically stable features. All the excellent properties were ascribed to the characteristic sliding effect of PNIPAAm-linked α-CDs along PEG chains, which gradually equalizes its experienced tension and gives the gel a very soft and mechanically stable nature.[8] Besides PR based hydrogels, Ito and coworkers also developed other PR-based materials, such as fiber, solvatochromic gel, polydimethylsiloxane based slide-ring gel, poly(butadiene) based PR, etc.[10-16]

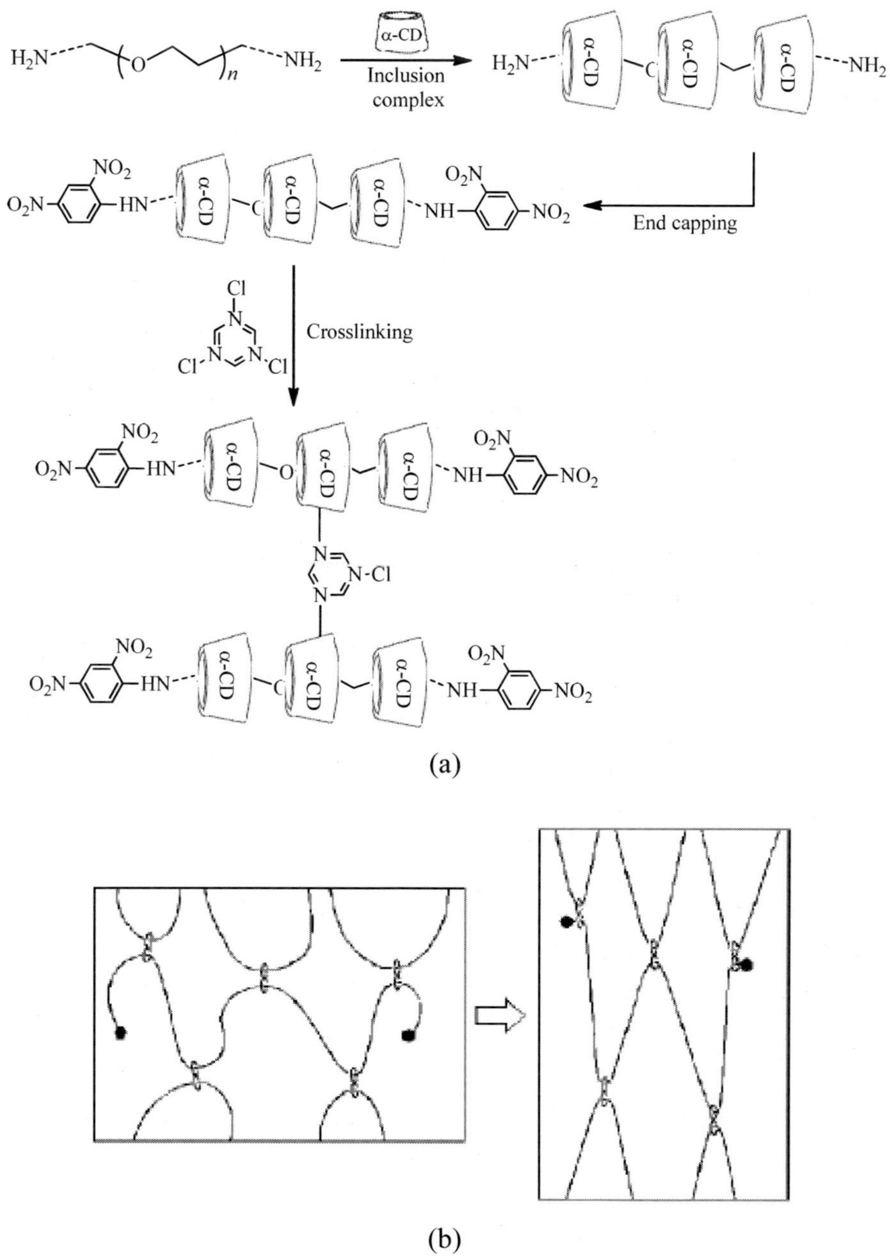


Figure 1. Schematic representation for topological slide-ring gel preparation (a), and model of slide-ring gels under tension (b)[1].

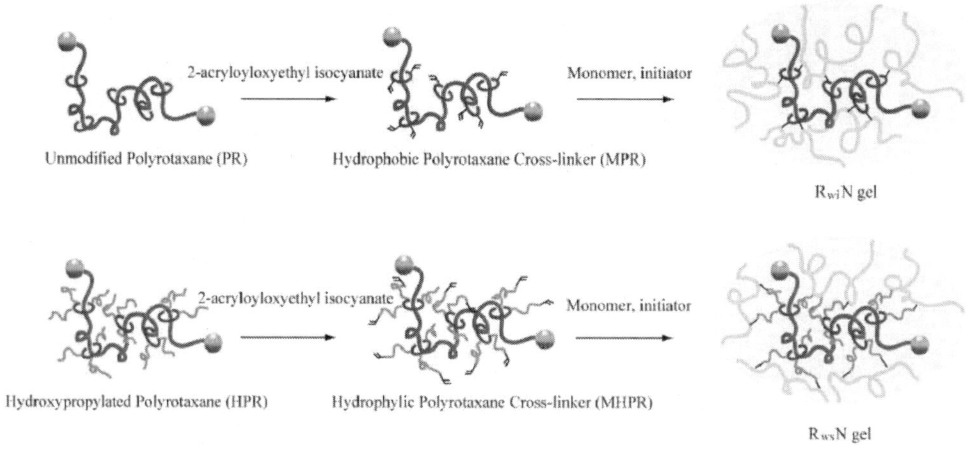


Figure 2. The preparation of hydrophobic and hydrophilic PR crosslinkers and their applications to fabricate hydrogels[9].

## DOUBLE-NETWORK HYDROGELS

In 2003, double-network (DN) hydrogels based on poly(2-acrylamido-2-methylpropansulfon acid)/polyacrylamide (PAMPS/PAAm) were firstly fabricated by Gong et al., which have similar structure as traditional interpenetrating polymer networks (IPN), but have superior mechanical properties to the latter.[17] For instance, PAMPS/PAAm DN hydrogels can sustain compressive strength of 17 MPa with 92% strain. For this type of DN hydrogel, the first network in DN gel is tightly crosslinked PAMPS, and the second loosely crosslinked PAAm network interpenetrates in the first one and fills the voids formed inside, as illustrated in Figure 3. The authors attributed the strength enhancement of DN gels to the stabilization of micro-cracks that form in the first network by the second lightly crosslinked one, deducing via dynamic light scattering (DLS) investigation.[18] Meanwhile, the entanglement between polymer chains also plays an important role in toughening of DN hydrogels, confirmed by SANS study.[19-24] Subsequently, Gong's group carried out numerous and detailed studies on DN gels, including natural bacterial cellulose (BC)/gelatin, poly(2,2'-disulfonyl-4,4'-benzidine terephthalamide) (PBDT)/PAAm, BC/PAAm DN gels,[25-28] and mechanical property studies.[29-43] Deep investigation to DN hydrogels disclosed that the true chemical structure wherein is the interconnection between both networks through covalent bonds. That is, unreacted vinyl groups from crossslinkers remained in first network participate in the formation of second network, thus both networks are interlocked actually, so the force applied to the DN gel is transferred from the first network to the second one through this interconnected structure.[44] While truly independent DN gels obtained by complete extinction of unreacted double bond before introduction of the second polymers didn't show such a high toughening, as revealed in Figure 4.[44] Extensively, inspired by common toughening methods of hard polymeric materials by soft fillers, Gong et al intentionally introduced some voids into the first PAMPS networks, and these voids were filled with soft PAAm components when the second network formed. Suitable size and number of voids imparted further toughening of improved DN gels.[45]

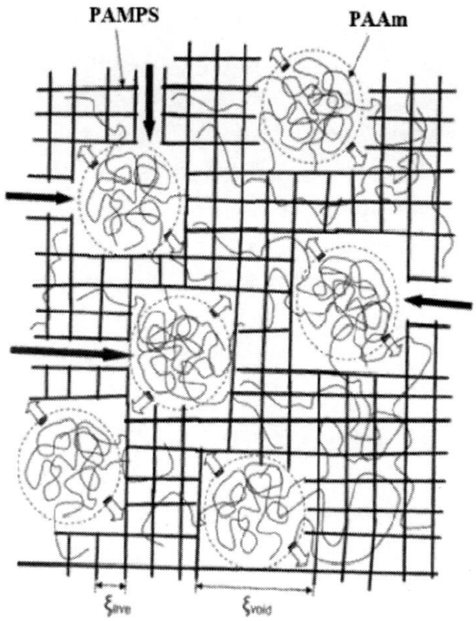


Figure 3. Structural model and mechanism to prevent crack development in PAMPS/PAAm DN gel[18].

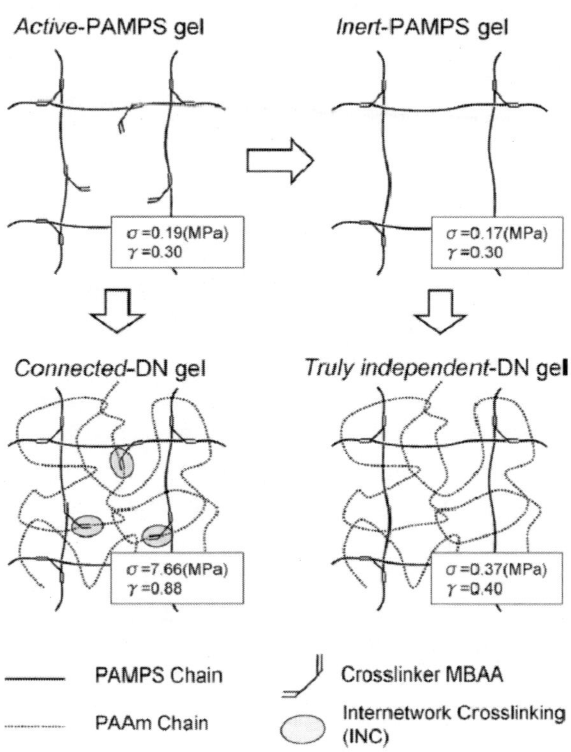


Figure 4. Chemical structures and the mechanical properties of active-PAMPS gels, inert-PAMPS gels, connected-DN gels, and truly independent-DN gels[44].

In addition to the abovementioned DN hydrogels developed by Gong's group, some other DN hydrogels were fabricated. For example, poly(ethylene glycol)/poly(acrylic acid) (PEG/PAA) hydrogel, tightly crosslinked neutral PEG network interpenetrated with loosely crosslinked anionic PAA network, showed strength improvement and mechanical property dependence on pH: initial modulus and fractural stress of PEG/PAA gel are 3-fold that of single PEG gel at pH7.4; higher swelling ratio at higher pH corresponds to higher initial modulus. This result was explained by that tightly crosslinked PEG chains were forced to highly extend due to swelling caused by PAA ionization at higher pH, and subsequently the modulus of PEG/PAA gels were increased.[46]

## NANO-(MICRO-)COMPOSITE HYDROGELS

Composition is an effective approach to increase material performance from two or more constituent materials with different physical or chemical properties. Here, composite hydrogels are classified into nanocomposition and microcomposition.

The most extensively researched nanocomposite (NC) hydrogels are exfoliated clay nanosheets filled hydrogels firstly developed by Haraguchi et al. NC hydrogels are easily prepared by dispersing exfoliated clays in a monomer solution followed by free radical polymerization. In NC gels, fully exfoliated clay nanoplatelets act as multifunctional crosslinker for the polymer despite that no additional organic crosslinker are used, as shown in Figure 5.[47] PNIPAAm/clay and poly(N, N-dimethyacrylamide)/clay (PDMAAm/clay) NC hydrogels were successively fabricated, and they revealed extraordinary mechanical and optical properties.[47-49] As for the mechanism of NC gel formation, the authors proposed that potassium persulfate (KPS) located near the clay surface through ionic interactions initiated the polymerization of monomers, and subsequently clay-polymer brushes were formed and concomitantly aggregated into clusters at about 7% monomer conversion, corresponding to the opaque appearance during early polymerization stage; as polymerization proceeded, the polymer chains became longer, and clay-polymer brush clusters disassembled again, so the system became transparent, and uniform nanoclay sheets dispersed hydrogel formed.[50] DLS and contrast variation SANS (CV-SANS) investigation further support above gelation mechanism, and suggest that polymer layer forms around clay sheets with thickness of 1 nm irrespective of clay content, and polymer layer fraction in NC gels is ca. 2.5 times that of clay and kept constant, which means that the microscopic structures of NC gels are almost the same regardless of clay content, contributing to the NC gels unique structure and properties as well as their significance in the mechanical properties, such as high deformability, high toughness, and high transparency.[51] Through analysis to deformation of NC gels by CV-SANS, it is inferred that under deformation, the clay platelets orient parallel to the stretching direction; polymer chains stretch along the stretching direction and peel off the clay platelets, resulting in the high extensibility and high modulus of the NC gels.[52] The effect of clay content on PNIPAAm/clay NC gels showed that increasing clay content favors the increment of modulus and strengths of gels without influence on their transparencies, while slower deswelling rates and smaller volume changes above lower critical solution temperature (LCST) were observed with higher clay content, and the LCST shifted to higher values with increasing clay content.[48, 53, 54] For NC gels with high clay content over

12mol% of NIPAAm, interesting phenomena occurred: NC gels behaved volume expansion but not contraction when transferred from 20 °C to 50 °C, and high transmittance remained unchanged.[54] This is because thermal motions of PNIPAAm chains are strongly restricted in the presence of so much clay, due to strong affinity between polymer and clay which has a pronounced effect on the phase separation in NC gels proved by CV-SANS and DLS research.[55] Furthermore, the effect of water content on NC gel properties was studied, and it was concluded that NC gels retained rubber tensile properties at high water content, whereas they changed to exhibit plastic-like deformations with decreasing water content.[56] Based on the abovementioned fundamental research on NC gels, their application to biomaterials and cosmetic were also explored.[57-60] Most recently, Haraguchi et al. reported that PNIPAAm/Clay NC hydrogels can self-heal without any additives for the same or different NC gel samples. The special self-reparation behavior is realized through autonomic reconstruction of crosslinks across a damaged interface, and can be achieved not only immediately after being cut but also after a long time.[61]

As expansions to NC gels detailedly studied by Haraguchi group, other NC gels based on (co)polymers of NIPAAm, AAm, AA, methyl acrylic acid (MA) and the sodium salt of AA and MA, as well as Pluronic F127 (PEO-PPO-PEO), were researched by others, all these NC gels demonstrated extraordinary performances owing to the composition of nano-scale clay sheets.[62-67] In early 2010, there published a NC hydrogel through non-covalent bonding of a telechelic dendritic macromolecule and clay in *Nature*. This special gel can be moulded into shape-persistent, free-standing objects owing to its exceptionally great mechanical strength, and rapidly and completely self-heals when damaged. Furthermore, it preserves biologically active proteins for catalysis.[68]

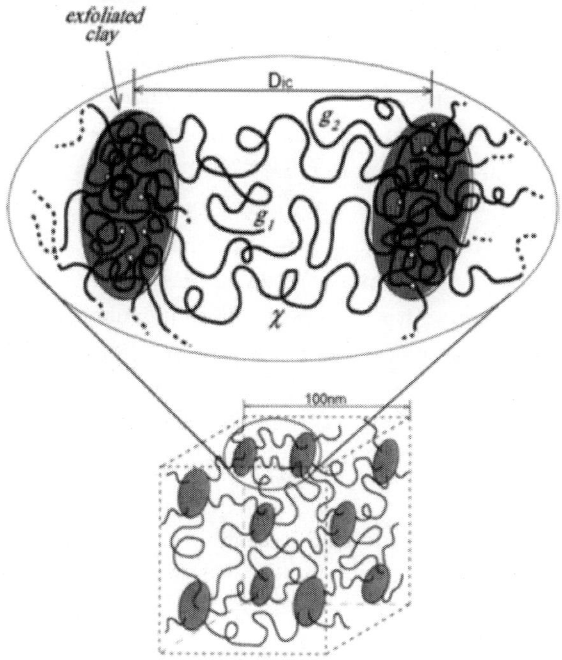


Figure 5. Network structure model for NC hydrogel[47].

As alternatives of nanoscale clay sheets, nanosphere and recently developed nanoscale graphene sheets were utilized to improve hydrogel properties. Wang et al. used ally modified starch-based nanospheres as polyfunctional crosslinker of AAm and NIPAAm, obtaining PAAm and PNIPAAm hydrogels, respectively. They both have high strength and the latter has rapid responsibility.[69-71] Besides, well-defined self-assembled block copolymer micelles (BCM) (diameter ca. 21 nm) containing vinyl groups served as crosslinker of AAm, giving rise to hybrid, elastomeric hydrogels whose mechanical properties can be readily tuned by varying the BCM concentration.[72] It was reported that silica nanoparticles can significantly enhance PDMAAm hydrogel, but have little effect on PAAm gel, because PDMAAm are strongly adsorbed onto silica particles while PAAm are not.[73-75] Hydroxyapatite nanoparticles can effectively increase the mechanical properties of poly(ethylene glycol) (PEG) hydrogel.[76] Ihara et al. fabricated a novel hybrid polymer hydrogels in a facile manner by simple mixing of a water-soluble copolymer having trimethoxysilyl side groups with silica nanoparticles used as multiple crosslinkers.[77] Alike to nanostructure of clay sheets, graphene oxide (GO) and graphene are expected to apply for preparing hydrogels. GO/PAAm hydrogels synthesized by *in situ* polymerization have uniform structure and significantly improved mechanical properties, along with response to external stimuli such as pH and electric fields.[78] PNIPAAm/graphene NC hydrogels was successfully fabricated through frontal polymerization. It was found that even small amounts of graphene increase swelling ratio of gels without shift of LCST value; G' modulus and complex viscosity were found to increase with increasing graphene concentration, while excessive amount of graphene showed a lubrication effect, lowering the rheological parameters and approaching the pseudoplastic behavior of the unfilled one.[79] Supramolecular hydrogel based on α-CD modified graphene (CD-G) and azo-PDMAAm-PNIPAAm showed rapid sol-gel transition at elevated temperature.[80]

Similar to the role of nanoparticles in hydrogels, microparticles can also produce some positive effect on gel performances. Wang et al. and Feng et al. adopted reactive microspheres as initiator and/or crosslinker, respectively, fabricating PAA and PAMPS hydrogels with high mechanical strength.[81, 82] Inspired by their PAMPS/PAAm DN hydrogels, Gong et al. designed and prepared quasi-monodispersed PAMPS microgels (with diameter ca. 5 μm) in a controlled manner, and then they dispersed the freeze-dried PAMPS microgels into AAm prepolymerization solution followed by UV-initiation, successfully fabricating microgel-reinforced gel films with excellent mechanical strength and toughness as comparable with the conventional bicontinuous DN hydrogels.[83]

## IDEAL NETWORK HYDROGELS

According to Hild, an ideal network can be defined as a collection of individual Gaussian elastic chains (linear chain elements) connecting $f$-functional crosslinks (junction points or branch points). For an ideal network, there are four basic requirements: (1) the lengths of all elastic chains of the network should be identical; (2) the network should be Gaussian; (3) the network should be homogeneous both micro- and macroscopically; (4) the functionality $f$ of the crosslinks should be known and constant throughout the entire network.[84] For an ideal network, micro- and macroscopically homogenous structure can endow materials with extraordinary performances, and such a well-defined structure make it easy to precisely study the relationship between property and structure. In hydrogel research field, many attempts were tried to make microscopically homogeneous ideal network. For example, the γ-ray irradiation produced more homogeneous PNIPAAm gel than chemical crosslinking.[85] "End linking" process of polymer precursors is also favorable to get more homogeneous networks.[86, 87] A crosslinker having a longer chain length produces more effective crosslinks during gelation and thus results in highly homogeneous hydrogels.[88] Malkoch et al. using tetraazide-functionalized PEG derivatives to crosslink PEG diacetylene through highly efficient Click reaction (Figure 6), generating more ideal PEG hydrogels with improved mechanical properties.[89] Interestingly, both tensile stress and elongation, along with swellability, of the hydrogels showed proportional to PEG molecular mass, consistent to Rubinstein theory, that is, the mechanical properties of ideal networks should improve with increasing molecular weight for gels prepared in a swollen state.

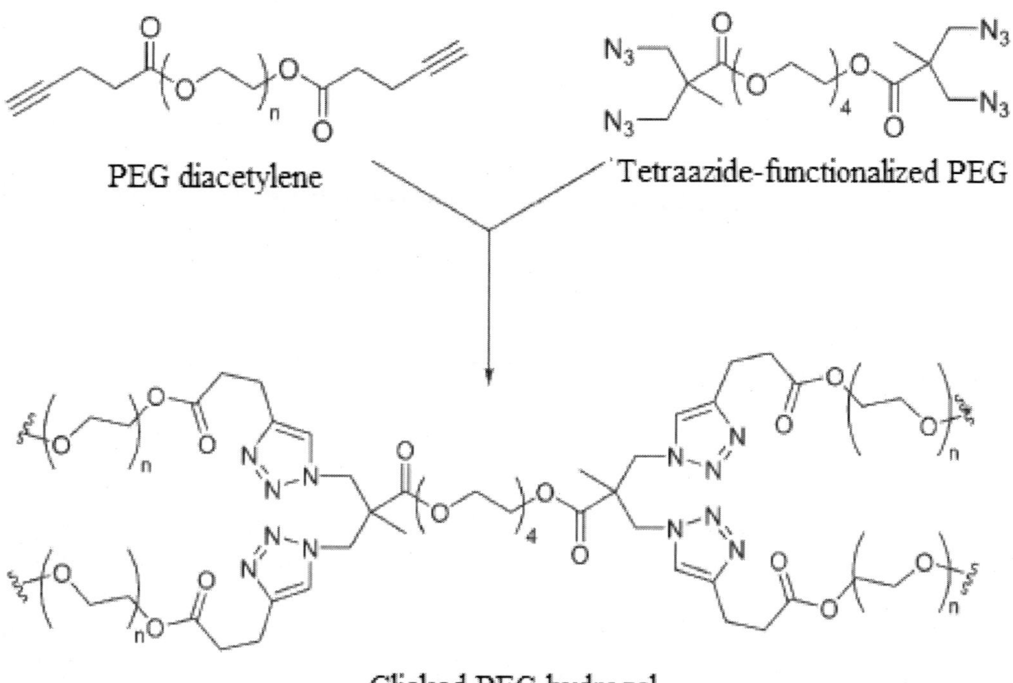


Figure 6. Synthesis scheme of PEG hydrogel by Click reaction[89].

Recently, Sakai et al. synthesized ideally homogeneous network hydrogel successfully. The hydrogel was fabricated by reaction of two structurally symmetrical tetrahedron-like PEG macromolecules of the same size, and so it was called tetra-PEG gel.[90] The synthesis scheme was shown in Figure 7. When the stoichiometric ratio of two macromonomers was 1, maximum compression strength of 2.5 MPa was achieved. The homogeneity of tetra-PEG gel was confirmed by characterization of DLS, SLS and SANS.[91, 92] Furthermore, the analysis to tetra-PEG gel using tree-like Miller-Macosko model and rubber elasticity theories,[93, 94] and the study on dried tetra-PEG gel as well,[95] also verified the extremely homogenous structure of this special gel.

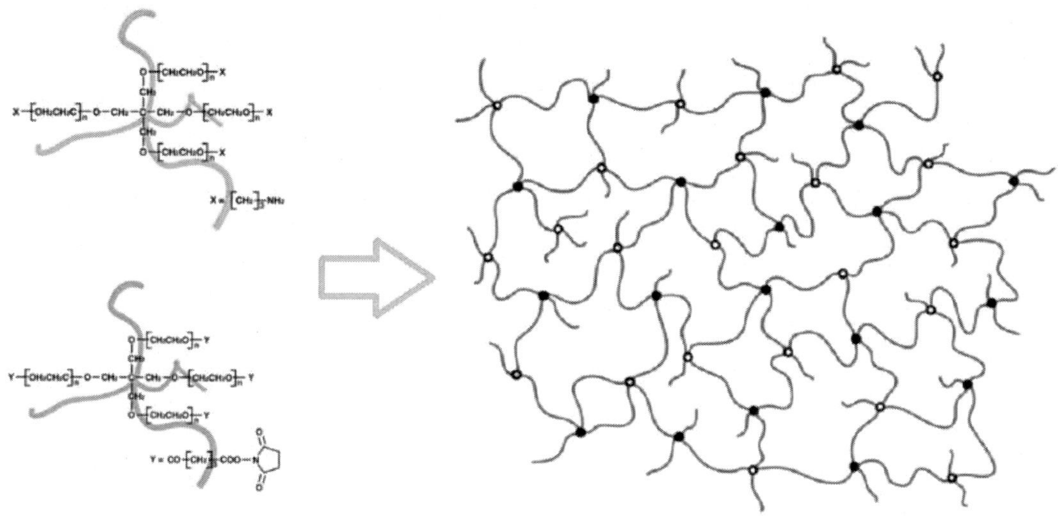

Figure 7. Formation scheme of Tetra-PEG gel[96].

## SELF-OSCILLATING HYDROGELS

On the basis of Belousov-Zhabotinsky (BZ) reaction exhibiting temporal and spatiotemporal oscillating phenomenon, Yoshida et al. created an autonomic swelling-deswelling self-oscillating hydrogel, successfully realizing spontaneous conversion from chemical energy to mechanical one without any on-off switching of external stimuli.[97] In this gel, $Ru(bpy)_3^{2+}$, a catalyst of BZ reaction, was chemically bonded to polymer chains through copolymerization of NIPAAm and vinyl functionalized Ru(II) tris-(2,2'-bipyridine) in the presence of N,N'- methylenebisacrylamide (BIS) crosslinker, as is shown in Figure 8. In swelling medium containing other constituents for BZ reaction (malonic acid or citric acid, bromate ion, and strong acid), Ru(II) undergoes spontaneous redox oscillation, inducing periodical swelling-deswelling alternation because Ru(III) is more hydrophilic than Ru(II). Using this characteristic self-oscillation, Yoshida et al. developed biomimetic micro-/nanoactuators such as ciliary-motion microactuator, mass transport surface and self-walking gel.[98-100]

In order to expand the application of self-oscillating gel to biomaterials, attempts were made to avoid using strong acid and/or oxidant in outside solution. By introducing correlative components of BZ reaction inside polymer chain, P(NIPAAm-co-Ru(bpy)$_3$-co-AMPS),[102, 103] P(NIPAAm-co-Ru(bpy)$_3$-co-MAPTAC),[104] and P(NIPAAm-co-Ru(bpy)$_3$-co-AMPS-co-MAPTAC) (structure is shown in Figure 9) [105] polymer gels were synthesized, and they also showed self-oscillating behavior under acid-free, oxidant-free and acid-/oxidant-free conditions, respectively, demonstrating potential utilization in biosystems. In addition, NIPAAm was replaced by other monomer, N,N'-ethylmethylacrylamide (EMAAm), generating self-oscillating gels with higher LCST (around body temperature), which revealed potential applications to biomaterials.[106]

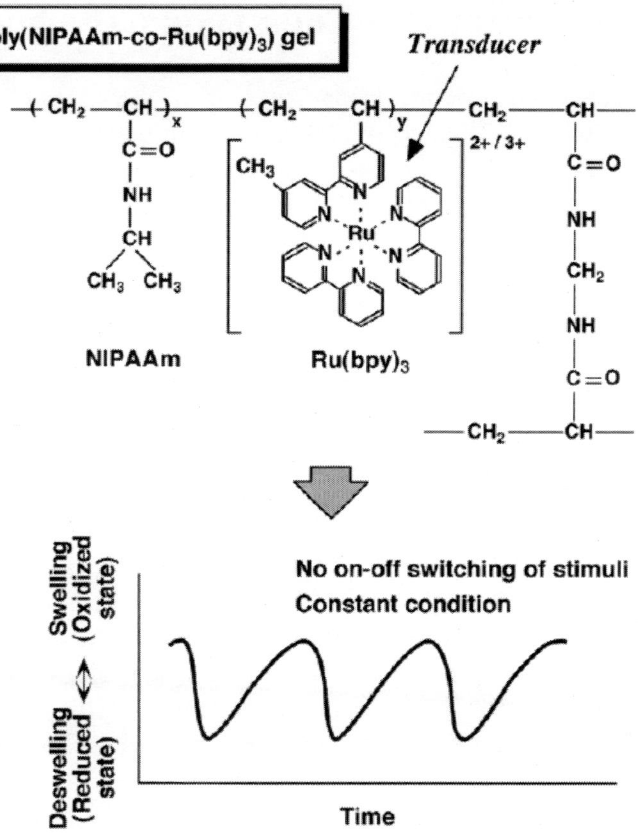


Figure 8. Structure of self-oscillating hydrogel and schematic periodical swelling-deswelling curve[101].

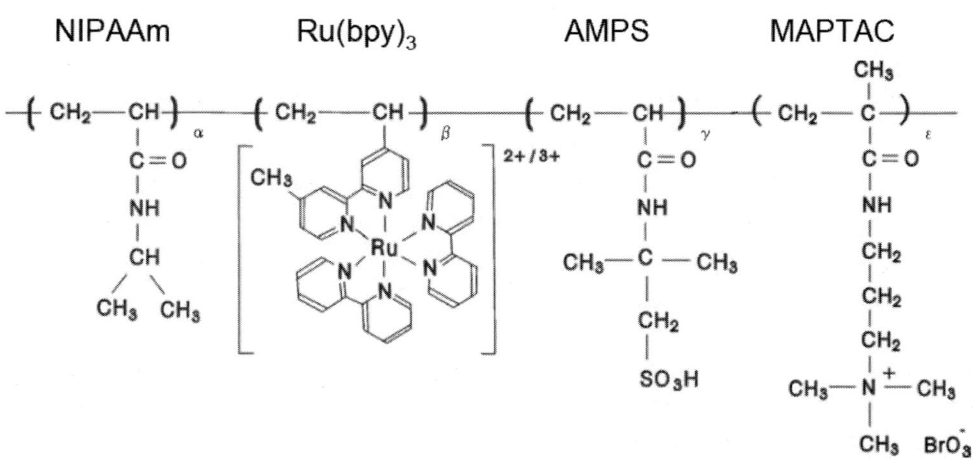


Figure 9. Structure of poly(NIPAAm-co-Ru(bpy)$_3$-co-AMPS-co-MAPTAC). MAPTAC means (methacrylamido)propyltrimethylammonium chloride, and in final hydrogel, chloride ion was exchanged by bromated ion[105].

## NON-COVALENTLY CROSSLINKED HYDROGELS

Some molecules with low molecular mass can self-assembled into macroscopic supramolecular hydrogels via the formation of 1D-fiber[107, 108] or 2D-sheet[109] nanostructures, owing to H-bonding, $\pi-\pi$ stacking and/or electrostatic interactions. And some macromolecules with block or graft structures behave sol-gel transition under special conditions, also achieving supramolecular hydrogels.[110-113] Usually these supramolecular hydrogels don't have enough mechanical strength because of the weakness of physical crosslinks, limiting their applications to some areas. Recently researchers managed to develop some non-covalently crosslinked hydrogels with comparable strength to covalently crosslinked one. It is known that hydrophobic associating water-soluble polymers have unique solution properties and have been used in coating, oil industry, etc.[114, 115] Now hydrophobic association gels (HA-gels) have been successfully fabricated through micellar polymerization of hydrophilic and hydrophobic monomer, and they revealed high mechanical strength. Liu et al. modified a octylphenol polyoxyethylene ether with acryloyl chloride (OP4-AC), followed by copolymerizing with AAm in the presence of sodium dodecyl sulfate (SDS), preparing mechanically improved and highly transparent HA-gels with self-reforming capacity.[116] In this HA-gel, hydrophobic association domains of OP4 side groups act as crosslinking junctions of gels (schemed in Figure 10), and OP4 content and SDS content (micelle amount) can directly influence mechanical and swelling properties of HA-gels. Further, they prepared AAm/AA/OP7-AC HA-gel with high strength and reversible phase transition nature.[117] Okay et al. adopted hydrophobic dodecyl methacrylate (C12), stearyl methacrylace (C18) and dococyl acrylate (C22) to copolymerize with AAm in SDS micellar solution, producing homogeneous tough PAAm-based HA-gels with maximum elongation ratios of 450%, 3600% and 1700%, respectively.[118, 119] The C12 hydrogels exhibit strain hardening behavior due to the organization of dodecyl blocks under the applied strain;[118] C18 hydrogels have about 100% self-healing efficiency to elongation at break, resulting from reconstruction of free nonassociated C18 blocks near the fracture surface of broken gel.[119] In some cases, hydrophobic association serves as a supplement and synergy to chemical crosslinks in hydrogel, effectively imparting more superior performances to original one, owing to stable junction zones of hydrophobic interaction. For example, about 300% elongation ratio at break was obtained by modification of chemically crosslinked PAAm network chains with 10 mol% N-octylacrylamide;[120] C12 side groups introduced to PAA backbones aggregated into hydrophobic cluster in PAA hydrogel, increasing fracture toughness without changing its elastic modulus;[121] strain of γ-ray irradiation crosslinked PNIPAAm hydrogel was improved by twice when introduced only 1wt% of non-ionic or cationic surfmer.[122, 123]

Ionic interaction is one of feasible crosslinking methods to fabricate hydrogels with appropriate strength, especially for polyelectrolytes. The typical ones include alginate-$Ca^{2+}$ gel and chitosan-tripolyphosphate gel.[124-127] Allcock et al. reported a redox responsive carboxyl-included polyphosphazene hydrogel crosslinked with redoxable metal ions such as $Cu^{2+}$ and $Fe^{3+}$, in which, the crosslinks dissociate when the cation crosslinker is reduced to a

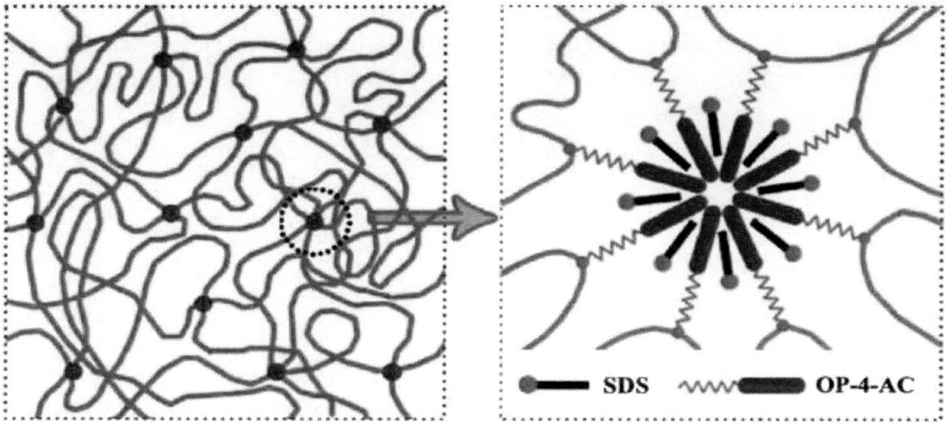


Figure 10. Schematic illstration of structural model with associating network in HA-gels[116].

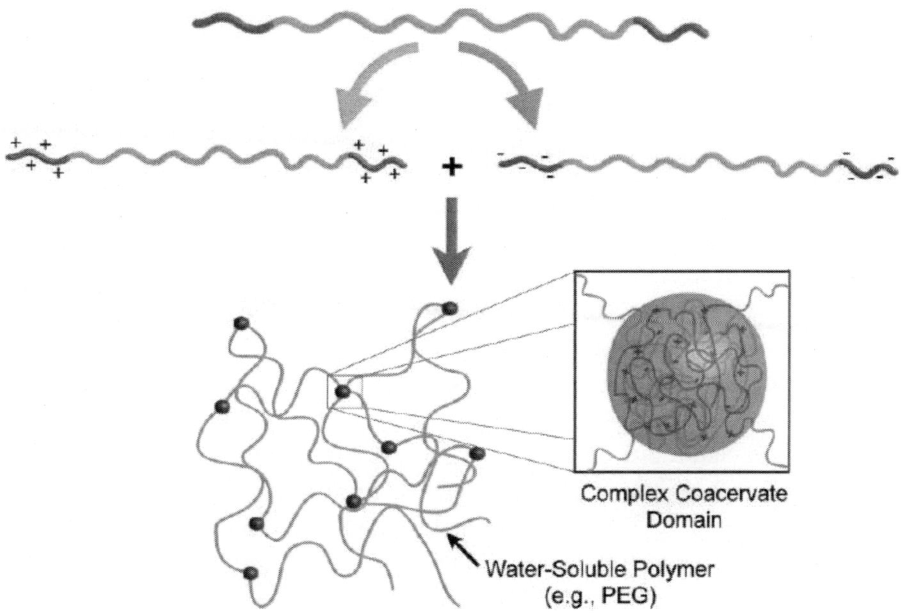


Figure 11. Schematic representation for coacervate-driven hydrogel formation[131].

lower oxidation state and reform following oxidation, which leads to a reversible and localized swelling-shrinking.[128] Poly(methyl methacrylate)-poly(methacrylic acid)-poly(methyl methacrylate) (PMMA-PMAA-PMMA) triblock copolymer hydrogels with high strength were fabricated through vapor phase solvent exchange and crosslinking by different divalent cations (Zn, Ca, Ni, Co, Cu), which revealed a dependence of mechanical behavior on ion absorption, pH, and cation identity.[129] The aforementioned clay-based non-covalent moldable high strength NC hydrogel in Part *Nano-(micro-)composite Hydrogels* was also formed via ionic intreraction of terminal guanidinium dendrimer and exfoliated clay sheets.[68] Moreover, ionic interaction between polyelectrolytes and opposite charges can drive hydrogel formation. When poly(potassium 3-sulfopropyl methacrylate)-poly(ethylene oxide)-poly(potassium 3-sulfopropyl methacrylate) ($PSPMA_{28}$-$PEO_{230}$-$PSPMA_{28}$) and poly(allyamine hydrochloride) ($PAH_{160}$) are mixed at higher concentrations, a highly viscous and transparent gel is formed spontaneously, indicating that an interconnected micellar network is indeed formed.[130] Hawker et al. modified ABA triblock poly(allyl glycidyl ether)-poly(ethylene glycol)-poly(allyl glycidyl ether) (PAGE-PEG-PAGE) to anion- and cation-functional ABA triblock copolymers, respectively, through thiol-ene click chemistry. In water solution, the obtained two oppositely charged ABA triblock copolymer polyelectrolytes coacervated into robust, yet highly tunable and dynamic and responsive hydrogels (Figure 11).[131]

## REVERSIBLE COVALENTLY CROSSLINKED HYDROGELS

Generally, polymer hydrogels are crosslinked either by reversible physical bond or by irreversible but stable chemical bond. Recently, the development of dynamic covalent chemistry opened a new way for producing novel chemically crosslinked hydrogels with both of reversibility and stability, which originally belong to physical and chemical crosslinking, respectively. And more importantly, this kind of hydrogel usually possesses self-healing characteristic under some conditions.

It is well-known that boronic ester can be reversibly formed and scissored via changing the pH value. And so far, the coordinate covalent bonding of phenylboronic acid (PBA) with polyol, salicylhydroxamic acid (SHA) and catechol has been used to fabricate novel pH-reversible covalently crosslinked hydrogels. Kiser et al. synthesized PAMPS-bound PBA and SHA, respectively, followed by crosslinking of the two polymers at an appropriate pH range, preparing pH-sensitive hydrogel (Figure 12). The result showed that hydrogel property and its pH-dependency were affected by polymer composition and solution ionic strength.[132, 133] Similarly, catechol terminated star-PEG was crosslinked by 1,3-benzenediboronic acid (Figure 13), yielding self-healing hydrogels exhibiting high stability at alkaline pH and low stability under acidic conditions, respectively.[134] Engbersen et al. synthesized two isomers of phenylboronic acid terminated PPO–PEO–PPO, and then used them as crosslinkers for PVA hydrogels (Figure 14). They found that the position of amino group with respect to boronic acid moiety have significant influence on the pH-dependency of boronic ester formation and gel properties.[135] In addition to pH-reversible boronic ester, acylhydrazone bond was also applied to synthesize dynamic covalent materials, owing to its reversibility with pH alteration.[136] Deng et al. applied acylhydrazone bond to construction of a pH-responsive hydrogel showing pH-stimulated reversible sol-gel transition (Figure 15); and more interestingly, the gel revealed autonomous self-healing property without any outside intervention.[137] In 2011, Takahara et al. prepared a dynamic covalent hydrogel using water-soluble polymers including alkoxyamine moiety (Figure 16).[138] The hydrogel was formed by heating polymer solution to 100 ℃ in a closed system, and it can be decomposed when adding hydrophilic alkoxyamine compound at 100 ℃. The gelation and decomposition arose from radical exchangement of alkoxyamine groups.

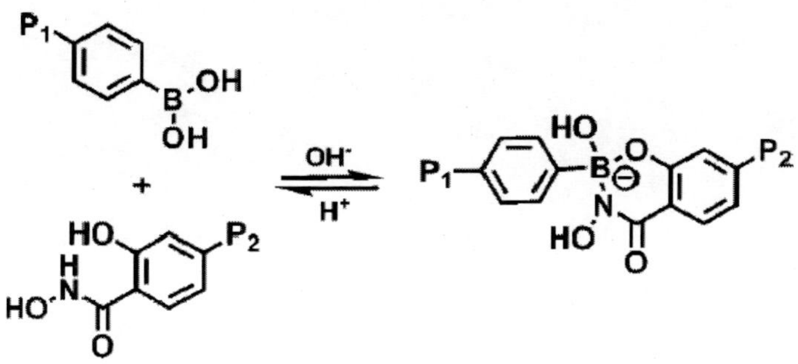

Figure 12. Scheme of reversible coordinate covalent phenyboronate[132, 133].

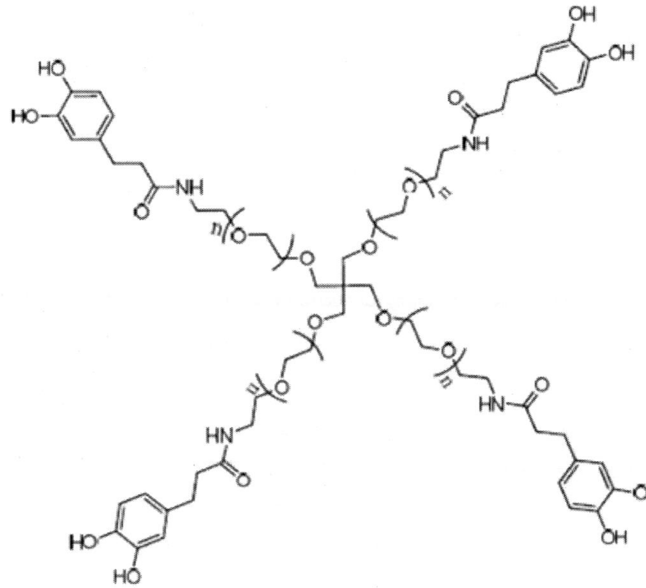

(a)

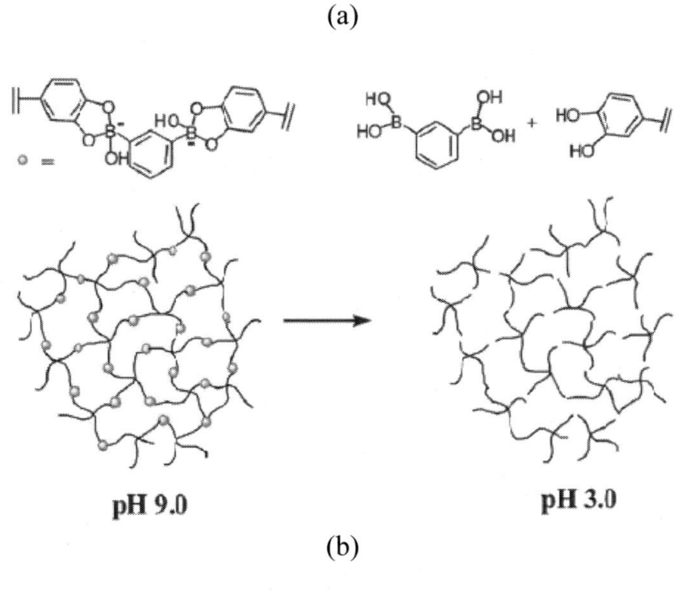


(b)

Figure 13. Structure of catechol terminated star-PEG (a), and schematic illustration of pH-responsive hydrogel (b)[134].

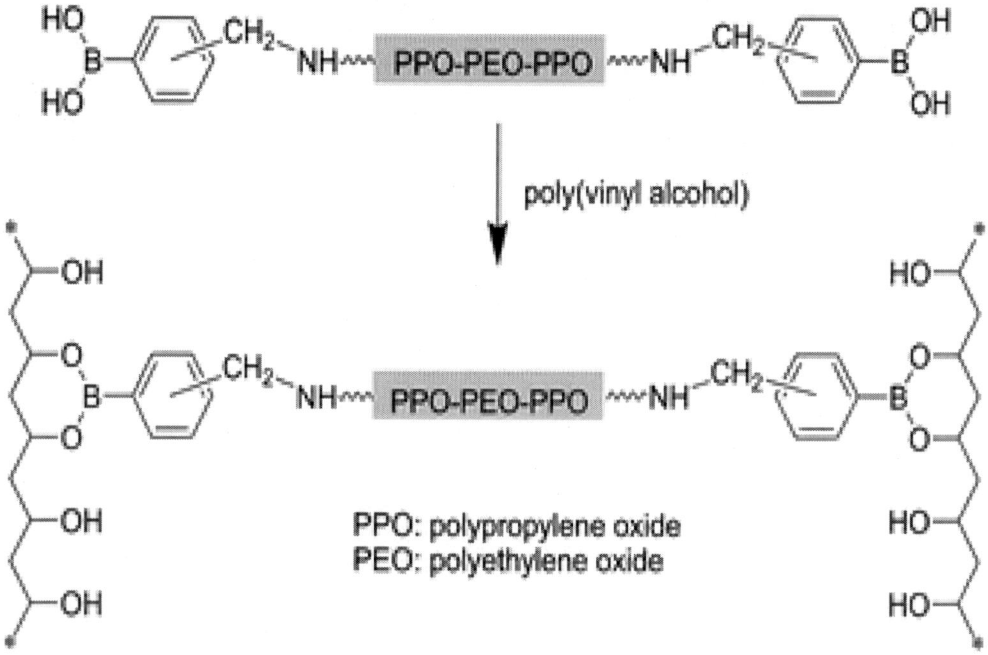


Figure 14. PVA hydrogels crosslinked by phenylboronic acid teminated PPO–PEO–PPO spacer[135].

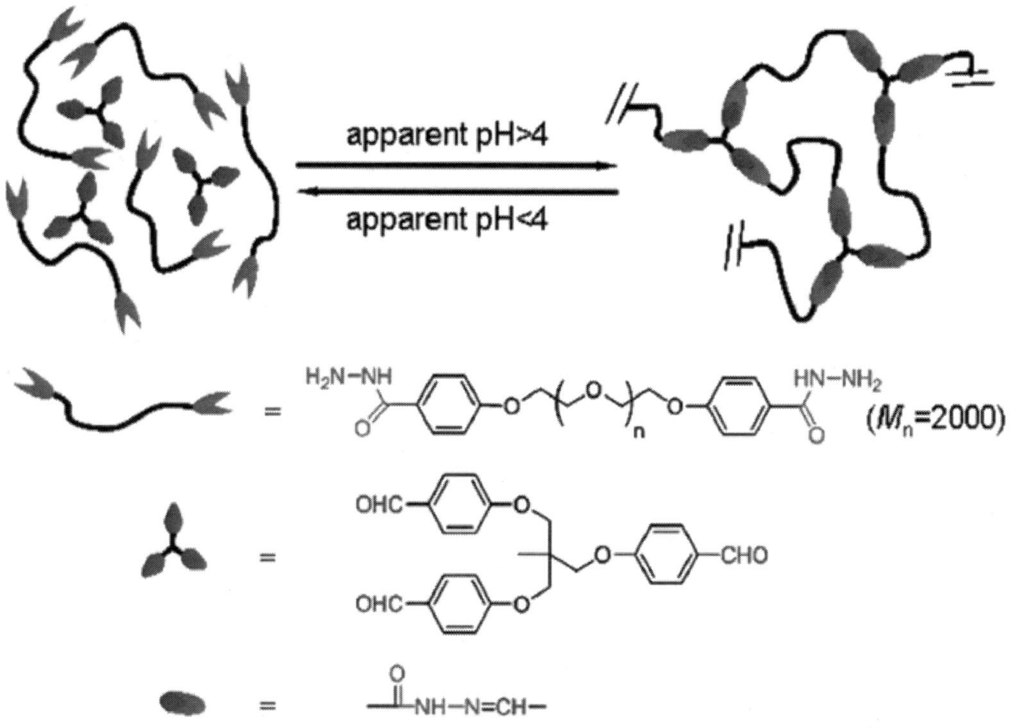


Figure 15. Construction of covalent cross-linked polymer gel based on reversible covalent acylhydrazone bond[137].

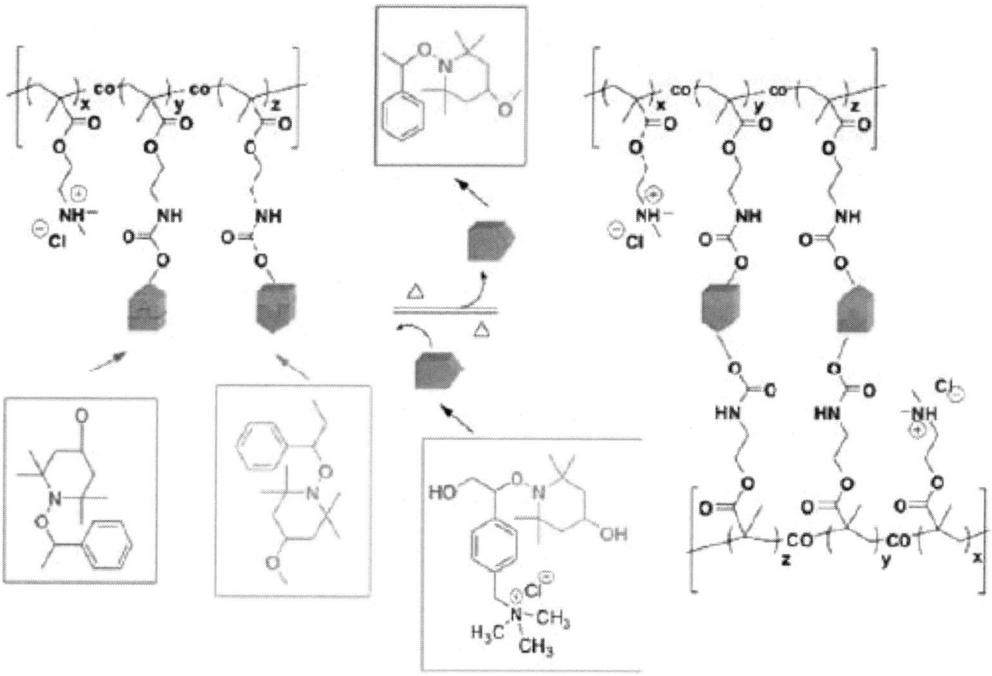

Figure 16. Crosslinking and decrosslinking reactions of dynamic covalent polymers via radical exchange reaction of alkoxyamine[138].

## Conclusions

The past decade has witnessed the exciting progress in prosperous hydrogels research field. The development of novel concepts and strategies gave rise to novel hydrogel materials with desirable functions and properties, allowing for exploiting wide applications in various fields. For instance, slide-ring hydrogels adopt removable crosslinks other than conventional fixed ones, providing extraordinary flexibility and strechability; DN hydrogels possess comparable mechanical performances to that of rubbers and some bio-tissues; clay-based NC hydrogels show high-strechable, robust, transparent and self-healing properties, along with their easy producing process; tetra-PEG ideal network hydrogels have extremely homogenous network structure with both of micro- and macroscopy, leading to improved mechanical property and providing an ideal model to precisely study the structure-property relationship; self-oscillating hydrogels open a new approach for smart materials with unnecessary of external on-off switching; non-covalently crosslinked hydrogels can achieve comparable and even superior properties to corresponding covalent ones in some cases; reversible covalently crosslinked hydrogels combine reversible and stable nature into one, and are usually imparted with self-healing ability. Although fundamental and applicational studies on the novel hydrogels described in this article are still limited in laboratory at present, it is believed that they must find their practical applications in the near future due to their so excellent performances. Meanwhile, there is full of great challenges and opportunities for researchers to use the existing strategies and/or create new ones to design and to develop novel hydrogels in order to satisfy multifarious practical demands.

## ACKNOWLEDGEMENT

This work was supported by National Natural Foundation of China (21004005) and Excellent Young Scholars Research Fund of Beijing Institute of Technology (2010CX04009).

## REFERENCES

[1] Okumura Y, Ito K. The polyrotaxane gel: A topological gel by figure-of-eight cross-links. *Advanced Materials*, 2001, 13(7): 485-487.

[2] Ito K. Novel cross-linking concept of polymer network: Synthesis, structure, and properties of slide-ring gels with freely movable junctions. *Polymer Journal*, 2007, 39(6): 489-499.

[3] Ito K. Slide-ring materials using topological supramolecular architecture. *Current Opinion in Solid State & Materials Science*, 2010, 14(2): 28-34.

[4] Karino T, Shibayama M, Okumura Y, et al. SANS studies on spatial inhomogeneities of slide-ring gels. *Macromolecules*, 2004, 37(16): p. 6177-6182.

[5] Karino T, Okumura Y, Zhao CM, et al. SANS studies on deformation mechanism of slide-ring gel. *Macromolecules*, 2005, 38(14): 6161-6167.

[6] Zhao CM, Domon Y, Okumura Y, et al. Sliding mode of cyclodextrin in polyrotaxane and slide-ring gel. *Journal of Physics-Condensed Matter*, 2005, 17(31): S2841-S2846.

[7] Shinohara Y, Kayashima K, Okumura Y, et al. Small-angle X-ray scattering study of the pulley effect of slide-ring gels. *Macromolecules*, 2006, 39(21): 7386-7391.

[8] Bin Imran A, Seki T, Kataoka T, et al. Fabrication of mechanically improved hydrogels using a movable cross-linker based on vinyl modified polyrotaxane. *Chemical Communications*, 2008(41): 5227-5229.

[9] Bin Imran A, Seki T, Ito K, et al. Poly(N-isopropylacrylamide) gel prepared using a hydrophilic polyrotaxane-based movable cross-linker. *Macromolecules*, 2010, 43(4): 1975-1980.

[10] Araki J, Kataoka T, Katsuyama N, et al. A preliminary study for fiber spinning of mixed solutions of polyrotaxane and cellulose in a dimethylacetamide/lithium chloride (DMAc/LiCl) solvent system. *Polymer*, 2006, 47(25): 8241-8246.

[11] Araki J, Kataoka T, Ito T. Preparation of a "sliding graft copolymer", an organic solvent-soluble polyrotaxane containing mobile side chains, and its application for a crosslinked elastomeric supramolecular film. *Soft Matter,* 2008, 4(2): 245-249.

[12] Murayama H, Bin Imran A, Nagano S, et al. Chromic slide-ring gel based on reflection from photonic bandgap. *Macromolecules*, 2008, 41(5): 1808-1814.

[13] Kato K, Inoue K, Masatoshi K, et al. Organic-inorganic hybrid slide-ring gels: Polyrotaxanes consisting of poly(dimethylsiloxane) and gamma-cyclodextrin and subsequent topological cross-linking. *Macromolecules*, 2009, 42(18): 7129-7136.

[14] Kato K, Komatsu H, Ito K. A versatile synthesis of diverse polyrotaxanes with a dual role of cyclodextrin as both the cyclic and capping components. *Macromolecules*, 2010, 43(21): 8799-8804.

[15] Inomata A, Sakai Y, Zhao C, et al. Crystallinity and cooperative motions of cyclic molecules in partially threaded solid-state polyrotaxanes. *Macromolecules*, 2010, 43(10): 4660-4666.
[16] Sakai Y, Ueda K, Katsuyama N, et al. Fabrication and structural analysis of polyrotaxane fibers and films. *Journal of Physics-Condensed Matter*, 2011, 23(28): 284108.
[17] Gong JP, Katsuyama Y, Kurokawa T, et al. Double-network hydrogels with extremely high mechanical strength. *Advanced Materials*, 2003, 15(14): 1155-1158.
[18] Yang HN, Kurokawa T, Katsuyama Y, et al. Structural characteristics of double network gels with extremely high mechanical strength. *Macromolecules*, 2004, 37(14): 5370-5374.
[19] Tsukeshiba H, Huang M, Yang HN, et al. Effect of polymer entanglement on the toughening of double network hydrogels. *Journal of Physical Chemistry B*, 2005, 109(34): 16304-16309.
[20] Huang M, Furukawa H, Tanaka Y, et al. Importance of entanglement between first and second components in high-strength double network gels. *Macromolecules*, 2007, 40(18): 6658-6664.
[21] Tominaga T, Tirumala VR, Eric KL, et al. The molecular origin of enhanced toughness in double-network hydrogels: A neutron scattering study. *Polymer*, 2007, 48(26): 7449-7454.
[22] Tominaga T, Tirumala VR, Lee S, et al. Thermodynamic interactions in double-network hydrogels. *Journal of Physical Chemistry B*, 2008, 112(13): 3903-3909.
[23] Tirumala VR, Tominaga T, Lee, et al. Molecular model for toughening in double-network hydrogels. *Journal of Physical Chemistry B*, 2008, 112(27): 8024-8031.
[24] Nakajima T, Furukawa H, Gong JP, et al. A Deformation Mechanism for Double-Network Hydrogels with Enhanced Toughness. *Polymer Networks: Synthesis, Properties, Theory and Applications*, 2010, 291-292: 122-126.
[25] Nakayama A, Kakugo A, Gong JP, et al. High mechanical strength double-network hydrogel with bacterial cellulose. *Advanced Functional Materials*, 2004, 14(11): 1124-1128.
[26] Yang W, Furukawa H, Gong JP. Highly extensible double-network gels with self-assembling anisotropic structure. *Advanced Materials*, 2008. 20(23): 4499-4503.
[27] Hagiwara Y, Putra A, Kakugo A, et al. Ligament-like tough double-network hydrogel based on bacterial cellulose. *Cellulose*, 2010, 17(1): 93-101.
[28] Wu ZL, Sawada D, Kurokawa T, et al. Strain-induced molecular reorientation and birefringence reversion of a robust, anisotropic double-network hydrogel. *Macromolecules*, 2011, 44(9): 3542-3547.
[29] Tanaka Y, Kuwabara R, Yang HN, et al. Determination of fracture energy of high strength double network hydrogels. *Journal of Physical Chemistry B*, 2005, 109(23): 11559-11562.
[30] Yang HN, Tanaka Y, Kawauchi Y, et al. Necking phenomenon of double-network gels. *Macromolecules*, 2006, 39(14): 4641-4645.
[31] Webber RE, Costantino C, Brown HR, et al. Large strain hysteresis and mullins effect of tough double-network hydrogels. *Macromolecules*, 2007. 40(8): 2919-2927.
[32] Tanaka, Y. A local damage model for anomalous high toughness of double-network gels. *Europhysics Letters*, 2007, 78(5): 56005.

[33] Tanaka Y, Kawauchi Y, Kurokawa T, et al. Localized yielding around crack tips of double-network gels. *Macromolecular Rapid Communications*, 2008, 29(18): 1514-1520.

[34] Yu QM, Tanaka Y, Furukawa H, et al. Direct observation of damage zone around crack tips in double-network gels. *Macromolecules*, 2009, 42(12): 3852-3855.

[35] Liang S, Wu ZL, Hu J, et al. Direct observation on the surface fracture of ultrathin film double-network hydrogels. *Macromolecules*, 2011, 44(8): 3016-3020.

[36] Furukawa H, Kuwabara, R., Tanaka, Y., et al., Tear velocity dependence of high-strength double network gels in comparison with fast and slow relaxation modes observed by scanning microscopic light scattering. *Macromolecules*, 2008, 41(19): 7173-7178.

[37] Azuma C, Yasuda K, Tanabe Y, et al. Biodegradation of high-toughness double network hydrogels as potential materials for artificial cartilage. *Journal of Biomedical Materials Research Part A*, 2007, 81A(2): 373-380.

[38] Tanabe Y, Yasuda K, Azuma C, et al. Biological responses of novel high-toughness double network hydrogels in muscle and the subcutaneous tissues. *Journal of Materials Science-Materials in Medicine*, 2008, 19(3): 1379-1387.

[39] Yasuda K, Kitamura N, Gong JP, et al. A novel double-network hydrogel induces spontaneous articular cartilage regeneration in vivo in a large osteochondral defect. *Macromolecular Bioscience*, 2009, 9(4): 307-316.

[40] Arakaki K, Kitamura N, Fujiki H, et al. Artificial cartilage made from a novel double-network hydrogel: In vivo effects on the normal cartilage and ex vivo evaluation of the friction property. *Journal of Biomedical Materials Research Part A*, 2010, 93A(3): 1160-1168.

[41] Arakaki K, Kitamura N, Kurokawa T, et al. Joint immobilization inhibits spontaneous hyaline cartilage regeneration induced by a novel double-network gel implantation. *Journal of Materials Science-Materials in Medicine*, 2011, 22(2): 417-425.

[42] Yokota M, Yasuda K, Kitamura N, et al. Spontaneous hyaline cartilage regeneration can be induced in an osteochondral defect created in the femoral condyle using a novel double-network hydrogel. *BMC Musculoskeletal Disorders*, 2011, DOI:10.1186/1471-2474-12-49.

[43] Arnold MP, Daniels AU, Ronken S, et al., Acrylamide polymer double-network hydrogels: candidate cartilage repair materials with cartilage-like dynamic stiffness and attractive surgery-related attachment mechanics. Cartilage, 2011, DOI: 10.1177/1947603511402320

[44] Nakajima T, Furukawa H, Tanaka Y, et al. True chemical structure of double network hydrogels. *Macromolecules*, 2009, 42(6): 2184-2189.

[45] Nakajima T, Furukawa H, Tanaka Y, et al. Effect of void structure on the toughness of double network hydrogels. *Journal of Polymer Science Part B-Polymer Physics*, 2011, 49(17): 1246-1254.

[46] Waters DJ, Engberg K, Parke-Houben R, et al. Structure and mechanism of strength enhancement in interpenetrating polymer network hydrogels. *Macromolecules*, 2011, 44(14): 5776-5787.

[47] Haraguchi K, Takehisa T. Nanocomposite hydrogels: A unique organic-inorganic network structure with extraordinary mechanical, optical, and swelling/de-swelling properties. *Advanced Materials*, 2002, 14(16): 1120-1124.

[48] Haraguchi K, Takehisa T, Fan S. Effects of clay content on the properties of nanocomposite hydrogels composed of poly(N-isopropylacrylamide) and clay. *Macromolecules*, 2002, 35(27): 10162-10171.
[49] Haraguchi K, Farnworth R, Ohbayashi A, et al. Compositional effects on mechanical properties of nanocomposite hydrogels composed of poly(N,N-dimethylacrylamide) and clay. *Macromolecules*, 2003, 36(15): 5732-5741.
[50] Haraguchi K, Li HJ, Matsuda K, et al. Mechanism of forming organic/inorganic network structures during in-situ free-radical polymerization in PNIPA-clay nanocomposite hydrogels. *Macromolecules*, 2005, 38(8): 3482-3490.
[51] Miyazaki S, Endo H, Karino T, et al., Gelation mechanism of poly(N-isopropylacrylamide)-clay nanocomposite gels. *Macromolecules*, 2007, 40(12): 4287-4295.
[52] Nishida T, Endo H, Osaka N, et al. Deformation mechanism of nanocomposite gels studied by contrast variation small-angle neutron scattering. *Physical review. E, Statistical, nonlinear, and soft matter physics,* 2009, 80(3 Pt 1): 030801.
[53] Haraguchi K, Li HJ. Mechanical properties and structure of polymer-clay nanocomposite gels with high clay content. *Macromolecules*, 2006, 39(5): 1898-1905.
[54] Haraguchi K. Stimuli-responsive nanocomposite gels. *Colloid and Polymer Science*, 2011, 289(5-6): 455-473.
[55] Osaka N, Endo H, Nishida T, et al. Microphase separation in nanocomposite gels. Physical Review E, 2009, 79(6): 060801(R)
[56] Haraguchi K, Li H-J. The effect of water content on the ultimate properties of rubbery nanocomposite gels. *Journal of Polymer Science Part B-Polymer Physics*, 2009, 47(23): 2328-2340.
[57] Zhou YM, Mizuno T, Takehisa T, et al. Nanocomposite hydrogels: a novel wound dressings. *2010 3rd International Conference on Biomedical Engineering and Informatics (Bmei 2010)*, 2010, 1-7: 432-434.
[58] Haraguchi K, Takehisa T. Characteristics of nanocomposite hydrogels and their applications to cosmetic materials. *FRAGRANCE Journal*, 2010, 38(7): 58-64.
[59] Haraguchi K. Research of new biomedical devices by the practical use of nanocomposite hydrogels. *WEB Journal*, 2009, 105: 43-46.
[60] Haraguchi K, Takehisa T, Ebato M. Control of cell cultivation and cell sheet detachment on the surface of polymer/clay nanocomposite hydrogels. *Biomacromolecules*, 2006, **7**(11): 3267-3275.
[61] Haraguchi K, Uyama K, Tanimoto H. Self-healing in nanocomposite hydrogels. *Macromolecular Rapid Communications*, 2011, 32(16): 1253-1258.
[62] Li B, Jiang Y, Liu Y, et al. Novel Poly(N-isopropylacrylamide)/clay/poly(acrylamide) IPN hydrogels with the response rate and drug release controlled by clay content. *Journal of Polymer Science Part B-Polymer Physics*, 2009, 47(1): 96-106.
[63] Zhang W, Liu Y, Zhu M, et al. Surprising conversion of nanocomposite hydrogels with high mechanical strength by posttreatment: From a low swelling ratio to an ultrahigh swelling ratio. *Journal of Polymer Science Part A-Polymer Chemistry*, 2006, 44(22): 6640-6645.
[64] Xiong L, Hu X, Liu X, et al. Network chain density and relaxation of in situ synthesized polyacrylamide/hectorite clay nanocomposite hydrogels with ultrahigh tensibility. *Polymer*, 2008, 49(23): 5064-5071.

[65] Xiong L, Zhu M, Hu X, et al. Ultrahigh deformability and transparence of hectonite clay nanocomposite hydrogels with nimble pH response. *Macromolecules*, 2009, 42(11): 3811-3817.

[66] Hu X, Xiong L, Wang T, et al. Synthesis and dual response of ionic nanocomposite hydrogels with ultrahigh tensibility and transparence. *Polymer*, 2009, 50(8): 1933-1938.

[67] Wu CJ, Gaharwar AK, Chan BK, et al. Mechanically tough pluronic F127/laponite nanocomposite hydrogels from covalently and physically cross-linked networks. *Macromolecules*, 2011, 44 (20): 8215–8224.

[68] Wang Q, Mynar JL, Yoshida M, et al. High-water-content mouldable hydrogels by mixing clay and a dendritic molecular binder. *Nature*, 2010. 463(7279): 339-343.

[69] Tan Y, Wang P, Xu K, et al. Designing starch-based nanospheres to make hydrogels with high mechanical strength. *Macromolecular Materials and Engineering*, 2009, 294(12): 855-859.

[70] Tan Y, Xu K, Wang P, et al. High mechanical strength and rapid response rate of poly(N-isopropylacrylamide) hydrogel crosslinked by starch-based nanospheres. *Soft Matter*, 2010, 6(7): 1467-1471.

[71] Morimoto N, Ohki T, Kurita K, et al. Thermo-responsive hydrogels with nanodomains: rapid shrinking of a nanogel-crosslinking hydrogel of poly(N-isopropyl acrylamide). *Macromolecular Rapid Communications*, 2008, 29(8): 672-676.

[72] Xiao L, Liu C, Zhu J, et al. Hybrid, elastomeric hydrogels crosslinked by multifunctional block copolymer micelles. *Soft Matter*, 2010, 6(21): 5293-5297.

[73] Lin WC, Marcellan A, Hourdet D, et al. Large strain and fracture properties of poly(dimethylacrylamide)/silica hybrid hydrogels. *Macromolecules*, 2010, 43(5): 2554-2563.

[74] Carlsson, L., et al., *Nano-hybrid self-crosslinked PDMA/silica hydrogels.* Soft Matter, 2010, 6(15): 3619-3631.

[75] Lin, W.-C., et al., *Effect of polymer-particle interaction on the fracture toughness of silica filled hydrogels.* Soft Matter, 2011. **7**(14): p. 6578-6582.

[76] Gaharwar AK, Dammu SA, Canter JM, et al. Highly extensible, tough, and elastomeric nanocomposite hydrogels from poly(ethylene glycol) and hydroxyapatite nanoparticles. *Biomacromolecules*, 2011, 12(5): 1641-1650.

[77] Takafuji M, Yamada S, Ihara H. Strategy for preparation of hybrid polymer hydrogels using silica nanoparticles as multifunctional crosslinking points. *Chemical Communications*, 2011, 47(3): 1024-1026.

[78] Zhang N, Li R, Zhang L, et al. Actuator materials based on graphene oxide/polyacrylamide composite hydrogels prepared by in situ polymerization. *Soft Matter*, 2011, 7(16): 7231-7239.

[79] Alzari V, Nuvoli D, Scognamillo S, et al. Graphene-containing thermoresponsive nanocomposite hydrogels of poly(N-isopropylacrylamide) prepared by frontal polymerization. *Journal of Materials Chemistry*, 2011, 21(24): 8727-8733.

[80] Liu JH, Chen GS, Jiang M. Supramolecular hybrid hydrogels from noncovalently functionalized graphene with block copolymers. *Macromolecules*, 2011, 44(19): 7682-7691.

[81] Huang T, Xu HG, Jiao KX, et al. A novel hydrogel with high mechanical strength: A macromolecular microsphere composite hydrogel. *Advanced Materials*, 2007, 19(12): 1622-1626.

[82] Qin X, Zhao F, Liu Y, et al. High mechanical strength hydrogels preparation using hydrophilic reactive microgels as crosslinking agents. *Colloid and Polymer Science*, 2009, 287(5): 621-625.

[83] Hu J, Hiwatashi K, Kurokawa T, et al. Microgel-reinforced hydrogel films with high mechanical strength and their visible meso-scale fracture structure. *Macromolecules*, 2011, 44(19): 7775–7781.

[84] Hild G. Model networks based on 'endlinking' processes: Synthesis, structure and properties. *Progress in Polymer Science*, 1998, 23(6): 1019-1149.

[85] Norisuye T, Masui N, Kida Y, et al. Small angle neutron scattering studies on structural inhomogeneities in polymer gels: irradiation cross-linked gels vs chemically cross-linked gels. *Polymer*, 2002, 43(19): 5289-5297.

[86] Gnanou Y, Hild G, Rempp P. Hydrophilic polyurethane networks based on poly(ethylene oxide)-synthesis, characterization, and properties-potential applications as biomaterials. *Macromolecules*, 1984, 17(4): 945-952.

[87] Gnanou Y, Hild G, Rempp P. Molecular-structure and elastic behavior of poly(ethylene oxide) networks swollen to equilibrium. *Macromolecules*, 1987, 20(7): 1662-1671.

[88] Abdurrahmanoglu S, Okay O. Homogeneous poly(acrylamide) hydrogels made by large size, flexible dimethacrylate cross-linkers. *Macromolecules*, 2008. 41(20): 7759-7761.

[89] Malkoch M, Vestberg R, Gupta N, et al. Synthesis of well-defined hydrogel networks using Click chemistry. *Chemical Communications*, 2006(26): 2774-2776.

[90] Sakai T, Matsunaga T, Yamamoto Y, et al. Design and fabrication of a high-strength hydrogel with ideally homogeneous network structure from tetrahedron-like macromonomers. *Macromolecules*, 2008, 41(14): 5379-5384.

[91] Matsunaga T, Sakai T, Akagi Y, et al. Structure characterization of tetra-PEG gel by small-angle neutron scattering. *Macromolecules*, 2009, 42(4): 1344-1351.

[92] Matsunaga T, Sakai T, Akagi Y, et al. SANS and SLS studies on tetra-arm PEG gels in as-prepared and swollen states. *Macromolecules*, 2009, 42(16): 6245-6252.

[93] Akagi, Y., Katashima, T., Katsumoto, Y., et al., Evaluation of topological defects in tetra-PEG gels. *Macromolecules*, 2010, 43(1): 488-493.

[94] Akagi, Y., et al., *Examination of the Theories of Rubber Elasticity Using an Ideal Polymer Network.* Macromolecules, 2011. **44**(14): 5817-5821.

[95] Nomoto Y, Matsunaga T, Sakai T, et al. Structure and physical properties of dried Tetra-PEG gel. *Polymer,* 2011, 52(18): 4123-4128.

[96] Sakai T, Matsunaga T, Akagi Y, et al. Highly elastic and deformable hydrogel formed from tetra-arm polymers. *Macromolecular Rapid Communications*, 2010, 31(22): 1954-1959.

[97] Yoshida R, Yamaguchi T, Ichijo H. Self-oscillating gel. *Journal of the American Chemical Society*, 1996, 118(21): 5134-5135.

[98] Tabata O, Hirasawa H, Aoki S, et al. Ciliary motion actuator using self-oscillating gel. in *14th IEEE International Conference on Micro Electro Mechanical Systems (MEMS 2001)*. 2001. Interlaken, Switzerland.

[99] Maeda S, Hara Y, Yoshida R, et al. Peristaltic motion of polymer gels. *Angewandte Chemie-International Edition*, 2008, 47(35): 6690-6693.

[100] Maeda S, Hara Y, Sakai T, et al., Self-walking gel. *Advanced Materials*, 2007, 19(21): 3480-3484.

[101] Yoshida R, Sakai T, Hara Y, et al. Self-oscillating gel as novel biomimetic materials. *Journal of Controlled Release*, 2009, 140(3): 186-193.

[102] Hara Y, Yoshida R. Self-oscillation of polymer chains induced by the Belousov-Zhabotinsky reaction under acid-free conditions. *Journal of Physical Chemistry B*, 2005, 109(19): 9451-9454.

[103] Hara Y, Yoshida R. Control of oscillating behavior for the self-oscillating polymer with pH-control site. *Langmuir*, 2005, 21(21): 9773-9776.

[104] Hara Y, Sakai T, Maeda S, et al. Self-oscillating soluble-insoluble changes of a polymer chain including an oxidizing agent induced by the Belousov-Zhabotinsky reaction. *Journal of Physical Chemistry B*, 2005, 109(49): 23316-23319.

[105] Hara Y, Yoshida R. Self-oscillating polymer fueled by organic acid. *Journal of Physical Chemistry B*, 2008, 112(29): 8427-8429.

[106] Hidaka M, Yoshida R. Self-oscillating gel composed of thermosensitive polymer exhibiting higher LCST. *Journal of Controlled Release*, 2011, 150(2): 171-176.

[107] Saha A, Roy B, Garai A, et al. Two-component thermoreversible hydrogels of melamine and gallic acid. *Langmuir*, 2009, 25(15): 8457-8461.

[108] Roy B, Bairi P, Saha A, et al. Variation of physical and mechanical properties in the bicomponent hydrogels of melamine with positional isomers of hydroxybenzoic acid. *Soft Matter*, 2011, 7(18): 8067-8076.

[109] Du R, Xu Y, Luo Y, et al. Synthesis of conducting polymer hydrogels with 2D building blocks and their potential-dependent gel-sol transitions. *Chemical Communications*, 2011, 47(22): 6287-6289.

[110] Hu J, Ge Z, Zhou Y, et al. Unique thermo-induced sequential gel-sol-gel transition of responsive multiblock copolymer-based hydrogels. *Macromolecules*, 2010, 43(12): 5184-5187.

[111] O'Lenick TG, Jiang X, Zhao B. Thermosensitive aqueous gels with tunable sol gel transition temperatures from thermo- and pH-responsive hydrophilic ABA triblock copolymer. *Langmuir*, 2010, 26(11): 8787-8796.

[112] Yu J, Fan H, Huang J, et al. Fabrication and evaluation of reduction-sensitive supramolecular hydrogel based on cyclodextrin/polymer inclusion for injectable drug-carrier application. *Soft Matter*, 2011, 7(16): 7386-7394.

[113] Park CW, Kim SC. Sol-gel transition behavior of amphiphilic comb-like poly (PEG-b-PLGA)acrylate block copolymers. *Journal of Polymer Science Part A-Polymer Chemistry*, 2010, 48(6): 1287-1297.

[114] Castelletto V, Hamley IW, Xue W, et al. Rheological and structural characterization of hydrophobically modified polyacrylamide solutions in the semidilute regime. *Macromolecules*, 2004, 37(4): 1492-1501.

[115] Shashkina YA, Zaroslov YD, Smirnov VA, et al. *Hydrophobic aggregation in aqueous solutions of hydrophobically modified polyacrylamide in the vicinity of overlap concentration. Polymer*, 2003, 44(8): 2289-2293.

[116] Jiang G, Liu C, Liu XL, et al. Network structure and compositional effects on tensile mechanical properties of hydrophobic association hydrogels with high mechanical strength. *Polymer*, 2010, 51(6): 1507-1515.

[117] Yang M, Liu C, Li ZY, et al. Temperature-Responsive Properties of Poly(acrylic acid-co-acrylamide) hydrophobic association hydrogels with high mechanical strength. *Macromolecules*, 2010, 43(24): 10645-10651.

[118] Abdurrahmanoglu S, Cilingir M, Okay O. Dodecyl methacrylate as a crosslinker in the preparation of tough polyacrylamide hydrogels. *Polymer*, 2011, 52(3): 694-699.

[119] Tuncaboylu DC, Murat S, Wilhelm O, et al. Tough and Self-Healing Hydrogels Formed via Hydrophobic Interactions. *Macromolecules*, 2011, 44(12): 4997-5005.

[120] Abdurrahmanoglu S, Can V, Okay O. Design of high-toughness polyacrylamide hydrogels by hydrophobic modification. *Polymer*, 2009, 50(23): 5449-5455.

[121] Miquelard-Garnier G, Hourdet D, Creton C. Large strain behaviour of nanostructured polyelectrolyte hydrogels. *Polymer*, 2009, 50(2): 481-490.

[122] Friedrich T, Tieke B, Stadler FJ, et al. Thermoresponsive Copolymer Hydrogels on the Basis of N-Isopropylacrylamide and a Non-Ionic Surfactant Monomer: Swelling Behavior, Transparency and Rheological Properties. *Macromolecules*, 2010, 43(23): 9964-9971.

[123] Friedrich T, Tieke B, Stadler FJ, et al. Improvement of elasticity and strength of poly(N-isopropylacrylamide) hydrogels upon copolymerization with cationic surfmers. *Soft Matter*, 2011,**7**(14): 6590-6597.

[124] Kong HJ, Wong E, Mooney DJ. Independent control of rigidity and toughness of polymeric hydrogels. *Macromolecules*, 2003. 36(12): 4582-4588.

[125] Lee K Y, Mooney DJ. Alginate: Properties and biomedical applications. *Progress in Polymer Science*, 2012, 37(1): 106-126

[126] Sun P, Li P, Li YM, et al. A pH-sensitive chitosan-tripolyphosphate hydrogel beads for controlled glipizide delivery. *Journal of Biomedical Materials Research Part B-Applied Biomaterials*, 2011, 97B(1): 175-183.

[127] Vodna L, Bubenikova S, Bakos D. Chitosan based hydrogel microspheres as drug carriers. *Macromolecular Bioscience*, 2007, 7(5): 629-634.

[128] Fei ST, Phelps MVB, Wang Y, et al. A redox responsive polymeric gel based on ionic crosslinking. *Soft Matter,* 2006, 2(5): 397-401.

[129] Henderson KJ, Zhou TC, Otim KJ, et al. Ionically cross-linked triblock copolymer hydrogels with high strength. *Macromolecules*, 2010, 43(14): 6193-6201.

[130] Lemmers M, Joris S, Voets IK, et al. Multiresponsive reversible gels based on charge-driven assembly. *Angewandte Chemie-International Edition,* 2010, 49(4): 708-711.

[131] Hunt JN, Feldman KE, Lynd NA, et al. Tunable, high modulus hydrogels driven by ionic coacervation. *Advanced Materials,* 2011, 23(20): 2327-2331.

[132] Roberts MC, Hanson MC, Massey AP, et al. Dynamically restructuring hydrogel networks formed with reversible covalent crosslinks, *Advanced Materials.* 2007, 19(18): 2503-2507.

[133] Roberts MC, Mahalingam A, Hanson MC, et al. Chemorheology of Phenylboronate-salicylhydroxamate cross-linked hydrogel networks with a sulfonated polymer backbone. *Macromolecules*, 2008, 41(22): 8832-8840.

[134] He L, Fullenkamp DE, Rivera JG, et al. pH responsive self-healing hydrogels formed by boronate-catechol complexation, *Chemical Communications*. 2011, 47(26): 7497-7499.

[135] Piest M, Zhang XL, Trinidad J, et al. pH-responsive, dynamically restructuring hydrogels formed by reversible crosslinking of PVA with phenylboronic acid functionalised PPO–PEO–PPO spacers (Jeffamines). *Soft Matter*, 2011, DOI: 10.1039/c1sm06230a

[136] Skene WG, Lehn JMP. Dynamers: polyacylhydrazone reversible covalent polymers, component exchange, and constitutional diversity. *Proceedings of the National Academy of Sciences of the United States of America,* 2004, 101(22): 8270-8275.
[137] Deng G, Tang CM, Li FY, et al. Covalent Cross-Linked Polymer Gels with Reversible Sol-Gel Transition and Self-Healing Properties. *Macromolecules*, 2010, 43(3): 1191-1194.
[138] Su J, Amamoto Y, Nishihara M, et al. Reversible cross-linking of hydrophilic dynamic covalent polymers with radically exchangeable alkoxyamines in aqueous media. *Polymer Chemistry*, 2011, 2(9): 2021-2026.

Hydrogels: Synthesis, Characterization and Applications
Editors: F. V. Câmara and L. J. Ferreira
ISBN: 978-1-61942-842-3


*Chapter 10*

# SYNTHESIS AND CHARACTERIZATION OF NOVEL pH SENSITIVE PAAM-PMAA-CMC HYDROGELS AND THEIR APPLICATIONS IN THE CONTROLLED RELEASE OF FERTILIZER

***Fauze A. Aouada[1,4*], Adriel Bortolin[2], Márcia R. de Moura[3], Elson Longo[1] and Luiz H. C. Mattoso[1]***

[1]Chemistry Institute, São Paulo State University, Araraquara campus, Araraquara, SP, Brazil
[2] Chemistry Department, Federal University of São Carlos, São Carlos, SP, Brazil
[3] Physics Institute of São Carlos, University of São Paulo, São Carlos, SP, Brazil
[4] National Nanotechnology Laboratory for Agriculture – LNNA - Embrapa Instrumentation - CNPDIA, São Carlos, SP, Brazil

## ABSTRACT

Recently, the applications of hydrogels as a carrier vehicle for the controlled release of fertilizers in the agricultural field have been extensively proposed due to the unique properties of these materials, such as their biodegradability, low-toxicity, high water uptake and reversible properties. The major benefit of these systems when applied in the agricultural setting is the gradual release of a loading, which maximizes that loading's concentration in the soil over a long time period. As a direct consequence, the number and frequency of the fertilizer applications to the soil are reduced. This diminishes the cost of agricultural production as well as the risks of human contamination and environmental pollution. This chapter is intended to describe the preparation of novel hydrogels composed of poly(acrylamide) (PAAm) / poly(methacrylic acid) (PMAA) / carboxymethyl cellulose (CMC) synthesized by a free-radical polymerization reaction catalyzed by temperature for applications in the controlled release of fertilizers. Fourier transform infrared (FTIR) spectroscopy confirmed the formation of the PAAm-PMAA-

* E-mail address: faouada@yahoo.com.br

CMC hydrogels. The hydrophilic properties, determined by the degree of swelling, were dependent on the pH of the external swelling medium, indicating that this hydrogel is pH sensitive. The pKa value of the hydrogel estimated from the sigmoid-curve was approximately 4.0. It was also possible to control the amount and rate of water absorbed by controlling the monomer and polysaccharide compositions. The amount of water absorbed is directly related to the network rigidity. The Freundlich model performed better than the Langmuir model in describing the adsorption isotherm of the PAAm-PMAA-CMC/$(NH_4)_2SO_4$ fertilizer system. The $K_F$ and 1/n Freundlich constants were 3.242 L $g^{-1}$ and 0.4143, respectively. Thermodynamic investigations proved that the adsorption of the fertilizer in the hydrogels is spontaneous. An investigation of the cumulative desorption confirmed that these hydrogels have potential applications in the agricultural field.

# 1. INTRODUCTION

Hydrogels can be defined as materials formed by hydrophilic polymers that are capable of retaining a large amount of water [1-6] due to their high physicochemical stability [7]. Fan et al. [8] defined hydrogels as materials formed by a three-dimensional (3D) network of polymer chains, where some parts are solvated by water molecules but other parts are chemically or physically linked with each other.

Recently, hydrogels have been extensively investigated as carrier vehicles for controlled release applications in the agricultural setting due to their unique properties, such as their biodegradability [9], low-toxicity [10], high water uptake [11] and their sorption and desorption capabilities for agrochemicals [12].

Amongst the various polysaccharides, cellulose derivatives have frequently been applied in the preparation of hydrogels [13-20]. Their presence in synthetic hydrogels may increase the hydrophilic, adsorption and desorption properties of the lattices without considerably diminishing their mechanical properties [21].

Carboxymethyl cellulose (CMC) is a cellulose derivative with polar carboxyl and hydroxyl groups bonded to the cellulose backbone, providing water solubility, chemical reactivity and high hydrophilicity. These characteristics give CMC attractive and promising properties in its various applications, including absorbent-superabsorbent fields [22], drug controlled release systems [23] and agrochemical controlled release systems. For instance, Li et al. [24] investigated the water sorption behavior and the release of the herbicide metolachlor from CMC/bentonite composite hydrogels. The authors observed a considerable gain in the release time of 50% of the herbicide ($t_{50}$) from the composite ($t_{50}$ = 158 h) in comparison with the pure CMC gel ($t_{50}$ = 61 h).

The preparation of the first synthetic hydrogel was reported by Wichterle and Lim [25], and since this discovery, research into hydrogels has expanded exponentially. One of the most important research areas is in the applications of the hydrogels as carrier vehicles for the controlled release of drugs [26], pesticides [27] and fertilizers [28-31].

In this chapter, we will focus on a brief description of fertilizer controlled release from hydrogels. Tyliszczak et al. [28] investigated the release of nitrogen-containing fertilizers entrapped in a hydrogel structure prepared from acrylic acid and cross-linked by N,N'-methylene-bisacrylamide. According to these authors, the release of compounds from hydrogels involves several processes: penetration of the release medium into the polymeric network, dissolution of the dispersed fertilizers and release of the compounds from the hydrogel under swelling conditions. In this study, the authors observed that the ammonium ion release was at its highest during the first 24 h. The release then reached a plateau because after this time the ionic concentration increased in the swelling medium, reducing the rate of release. Other authors have also studied the use of hydrogels as carrier vehicles for the controlled release of fertilizers. Jamnongkan et al. [29] investigated a possible controlled release system based on hydrogels formed from poly(vinyl alcohol) – (PVA), PVA/chitosan and chitosan using glutaraldehyde as a crosslinker. The authors studied the release behavior of potassium (with potassium nitrate as the potassium source) in deionized water and soil. They observed that the hydrogels formed by the PVA chains exhibited the highest cumulative release of all of the prepared hydrogels because the potassium release behavior was related to the degree of swelling of the hydrogels. The potassium release in soil was more efficient with the chitosan hydrogel. The authors explained this behavior by invoking the facile degradation of the chitosan by the microorganisms and ions in the soil. The study reported by Ma et al. [30] discussed the desorption processes of $NH_4^+$ and $PO_4^{3-}$ from a new controlled release agent produced through graft copolymerization with wheat straw, acrylic acid, acrylic amide and dimethyl diallyl ammonium chloride. The main results of this study indicated that the amounts of $NH_4^+$ and $PO_4^{3-}$ released were 2.73 mg $NH_4^+$/g hydrogel (after 300 min) and 0.17 mg $PO_4^{3-}$/g hydrogel (after 240 min), and both release processes varied differently with the pH of the medium. The $NH_4^+$ and $K^+$ desorption processes were quantified by spectrometry (using a flow injection analysis system) and by flame photometry from polyacrylamide hydrogels that have also been studied by our group [31]. In conclusion, this study demonstrated that the hydrogel synthesized with 21.7% acrylamide had a higher nutrient release potential, especially for $NH_4^+$.

The main goals of this chapter are: 1) to present the potential applications of hydrogels in controlled release technology and 2) to present the synthesis and characterization of novel hydrogels composed of poly(acrylamide) (PAAm) / poly(methacrylic acid) (PMAA) / carboxymethyl cellulose (CMC) synthesized by a free-radical polymerization reaction catalyzed by temperature intended for use in the controlled release of fertilizers.

## 2. The Controlled Release of Fertilizers from Hydrogels

Controlled release was initially proposed for drug delivery systems, aiming to reduce the costs and minimize the collateral effects that excessive doses may provoke in the target organism. In addition, controlled release drug delivery systems allow therapeutic concentrations of a specific drug for extended times inside in the organism.

Hydrogels emerged as a promising carrier vehicle for applications in the controlled release of drug substances because of their biocompatibility and non-toxicity. More recently, because of their biodegradability and other important properties, these hydrogels are being tested in the controlled release of agrochemicals. Besides the aforementioned abilities, the application of the hydrogels improves the physicochemical properties of the soil, acts as a soil conditioner and aids in erosion control. Other potential uses in agriculture are in aiding the germination of seeds and plant growth due improvement of soil moisture retention [32].

Fertilizers are one of the most important products of the agrochemical industry [33], and together with water, they are the primordial factors that often limit agricultural production [34]. However, the amount of the applied fertilizers actually absorbed by the crops is very low. It is estimated that approximately 40-70% of the nitrogen, 80-90% of the phosphorus, and 50-70% of the potassium in the standard application of fertilizers is lost to the environment and cannot be absorbed by plants [35]. The rapid dissolution of the water-soluble inorganic salts present in most fertilizers almost immediately leads to local nutrient concentration levels that are excessively high for effective absorption by crops [36].

Thus, the main problem with conventional agrochemical applications is the use of greater amounts of agrochemicals over a long period than is actually needed, possibly leading to crop damage and environmental contamination [37]. Researchers have attempted to overcome these major disadvantages through the use of controlled release technology. The control of fertilizer release maintains the fertilizer concentration at effective levels in the soil solution and releases the fertilizer when the plant most needs it [28] by protecting the supply of the agent and allowing automatic release over time.

Some advantages of controlled release technology over conventional fertilizer applications are listed in the work described by Guo et al. [38]: i) a decreased rate of removal of the fertilizer from the soil by rain or irrigation water; ii) a sustained supply of nutrition for a prolonged period; iii) an increased efficiency of the fertilizer; iv) a lower frequency of application relative to the normal crop requirement; v) reduced toxicity and vi) minimized potential negative effects associated with overdosages.

## 3. Preparation and Application of Hydrogels in the Controlled Release of Fertilizers

### 3.1. PAAm-MC Hydrogels [39]

In previously reported work, our group investigated poly(acrylamide) and methylcellulose (PAAm-MC) hydrogels as potential delivery vehicles for the extended controlled release of ammonium sulfate $(NH_4)_2SO_4$ and potassium phosphate ($KH_2PO_4$) fertilizers. We observed that the amount of $(NH_4)_2SO_4$ and $KH_2PO_4$ loaded and the time of release for both fertilizers were extended when the MC was incorporated in the PAAm chains. In this study, the concentration effects of AAm and MC on the sorption capacity of equilibrium hydrogels were investigated, and the results are shown in Table 1.

**Table 1. Values of the maximum load of $(NH_4)_2SO_4$ and $KH_2PO_4$ fertilizers for different hydrogels after 48 hours of immersion time [39]**

| (AAm/MC)hydrogel* | mg $(NH_4)_2SO_4$ /g hydrogel | mg $KH_2PO_4$/g hydrogel |
|---|---|---|
| (6.0 – 1.0) | 58.0 ± 6.7 | 124.0 ± 15.9 |
| (9.0 – 0) | 13.2 ± 1.1 | 31.5 ± 0.8 |
| (9.0 – 0.5) | 20.2 ± 2.2 | 43.9 ± 4.2 |
| (9.0 – 1.0) | 36.8 ± 6.0 | 76.4 ± 7.7 |
| (12.0 – 1.0) | 15.2 ± 2.2 | 26.3 ± 1.3 |

* [AAm] and [MC] in wt. %.

The sorption capacity of these hydrogels was greater for $KH_2PO_4$ than for $(NH_4)_2SO_4$, indicating that the interaction of the former with the polymer matrix is more effective. The overall behavior exhibited an increase in the loading rates of the fertilizer with increasing concentrations of MC due to the increased number of hydroxyl groups on the MC chains. However, the increase in the AAm concentration led to a decrease in sorption due to reduced water absorption by the hydrogel (from the greater rigidity of the polymer chains).

The release of the chemicals entrapped in a hydrogel occurs only after water penetrates the network to swell the polymer and dissolve the chemicals. This is followed by diffusion of the chemical along the aqueous pathways to the surface of the device. The release of the chemicals is closely related to the swelling characteristics of the hydrogels, which in turn is a function of the chemical architecture of the hydrogels [40]. Thus, when a hydrogel loaded with agrochemical species is placed in contact with water, by because of chemical potential differences (chemical spontaneity) [41] (Eq. 1) and the osmotic pressure defined by Donnan equilibrium theory [42] (Eq. 2), the water molecules tend to diffuse into the polymer matrix through interactions with the hydroxyl groups, resulting in a chemical competition between the agrochemicals/water and the matrix. As a result, the agrochemical molecules diffuse from the matrix (i.e., the aqueous medium), achieving controlled-extended desorption, as shown in Figure 1.

$$\mu_i = \left(\frac{\partial G}{\partial n_i}\right)_{T,P,n_j} \qquad (1)$$

where G is the Gibb's free energy, $n_i$ is the amount of component *i*, T is the absolute temperature, V is the volume and P is the pressure. The subscripts indicate that the temperature, pressure and the amounts of all other components are held constant.

$$\pi_{ion} = RT\sum_i (C_i^g - C_i^s) \qquad (2)$$

where R is the universal gas constant, $C_i$ is the mobile ion concentration of species *i*, and superscripts ‘g’ and ‘s’ represent the gel and solution phase, respectively.

Figures 2*a* and 2*b* show the dependence of the released quantities of $KH_2PO_4$ as a function of the concentration of methylcellulose and the concentration of acrylamide, respectively. Figures 3*a* and 3*b* show the same kinetic studies performed for $(NH_4)_2SO_4$ for the same hydrogels indicated in Figure 2. The hydrogels exhibited distinct desorption kinetics, reflecting a significant variation in the total amount of fertilizer desorbed. Thus, by controlling the synthesis of the hydrogel, it is possible to control the release process of the fertilizer, releasing more or less depending on the desired application. This makes the PAAm-MC hydrogels extremely interesting as carrier vehicles for controlled-extended desorption systems.

## 3.2. PAAm-PMAA-CMC Hydrogels

The aim of this study was to develop a fast, simple and reproducible preparation for hydrogels based on polyacrylamide (PAAm), polymethacrylic acid (PMAA) and carboxymethyl cellulose (CMC) using a one-step synthesis and biodegradable, non-toxic, and low-cost materials. The other objective was to study in detail the spectroscopic and hydrophilic properties of a novel pH sensitive hydrogel with applications in the controlled release of fertilizers as a function of the hydrogel constituents.

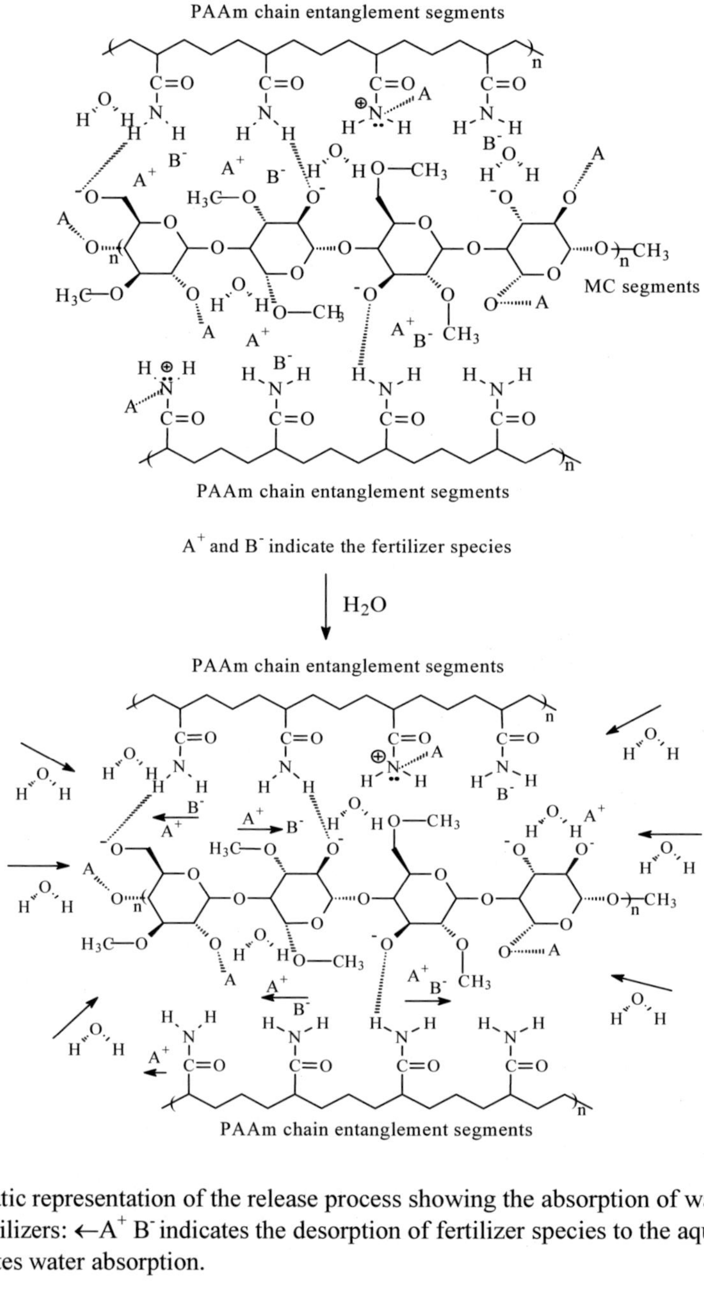


Figure 1. Schematic representation of the release process showing the absorption of water and the desorption of fertilizers: ←$A^+$ $B^-$ indicates the desorption of fertilizer species to the aqueous phase, and ← H-O-H indicates water absorption.

### *3.2.1. Experimental Section*

#### 3.2.1.1. Preparation of the PAAm-PMAA-CMC Hydrogels

Novel hydrogels composed of PAAm / PMAA / CMC were synthesized through a free radical polymerization reaction. Briefly, the AAm monomer (5.0 or 1.0 wt. %), MAA monomer (5.0 or 1.0 wt. %), CMC polysaccharide (0, 0.25, 0.5, 0.75 or 1.0 wt. %) and MBAAm were placed in a bottle and homogenized by stirring for 30 min at 25 °C. After homogenization, the sodium persulfate (SP) initiator was added and the resulting solution was quickly heated to 65 °C. The gelling process was observed after approximately 150 min. At the end of reaction, the hydrogel was immersed in deionized water to remove the unreacted chemicals (i.e., the monomers and initiator). The water was changed every 10-12 h for 7-10 days. The notation (X-Y-Z) was used to identify the hydrogels, where X, Y and Z correspond to the concentrations of AAm, MAA and CMC in wt. %.

#### 3.2.1.2. Swelling Degree (SD)

The swelling capability of the hydrogels was investigated by measuring their water uptake at 25.0 °C. Solutions of different pH and deionized water were used as the external media. SD values were obtained by taking the mass ratio of the swollen hydrogel to that of the dried hydrogel. Before these measurements, the hydrogels were removed from the swelling media and their surface was dried with soft paper to eliminate excess water.

#### 3.2.1.3. Fourier Transform Infrared (FTIR) Spectroscopy

Spectra of the PAAm-PMAA-CMC hydrogels in range of 400 to 4000 $cm^{-1}$ were recorded on a Fourier transform infrared (FTIR) instrument from Perkin Elmer (model Spectrum Paragon 1000). The samples were dried under vacuum for 24 h or until reaching a constant weight and were mixed with KBr (1 wt. %) to form pellets. Before the measurements, the pellets were conditioned at 100 °C for 24 h.

#### 3.2.1.4. Adsorption and Desorption Studies

Experimental methods for the adsorption and desorption processes of fertilizers from hydrogels were recently published by our group (Bortolin et al. [39]). Essentially, the amounts of adsorbed or released fertilizer were quantified by conductimetric measurements (Horiba conductivity meter, model ES-12) applying a typical standardization curve ($R^2$ > 0.99).

#### 3.2.1.5. Mathematical Modeling of the Adsorption

Adsorption isotherms were applied to analyze the equilibrium stage after an adsorption process, i.e., how a specific molecule interacts with a specific adsorbent. Therefore, the analysis of the isotherm data is of great importance for optimizing the adsorption system [43].

The Langmuir model and the Freundlich model are the most common isotherm models employed to investigate the adsorption of fertilizer onto hydrogels. These models are described by Eqs. 3 and 4, respectively:

$$\frac{C_{eq}}{q_{eq}} = \frac{1}{K_L} + \frac{\alpha_L}{K_L} C_{eq} \tag{3}$$

where $C_{eq}$ is the solution fertilizer concentration at equilibrium (mg $L^{-1}$), $q_{eq}$ is the maximum fertilizer adsorption on the hydrogel (mg $g^{-1}$) and $K_L$ (L $g^{-1}$) and $\propto_L$ (L $mg^{-1}$) are the Langmuir isotherm constants.

$$\ln q_{eq} = \ln K_F + \frac{1}{n} \ln C_{eq} \quad (4)$$

where $K_F$ (L $g^{-1}$) and n (mg $L^{-1}$) are constants incorporating all of the factors affecting the adsorption process (i.e., the adsorption capacity and the intensity of adsorption).

Isotherm studies were conducted with a constant hydrogel weight (approximately 30 - 40 mg) and varying the initial concentration ($C_0$) of fertilizer from 58 – 1893 mg $L^{-1}$.

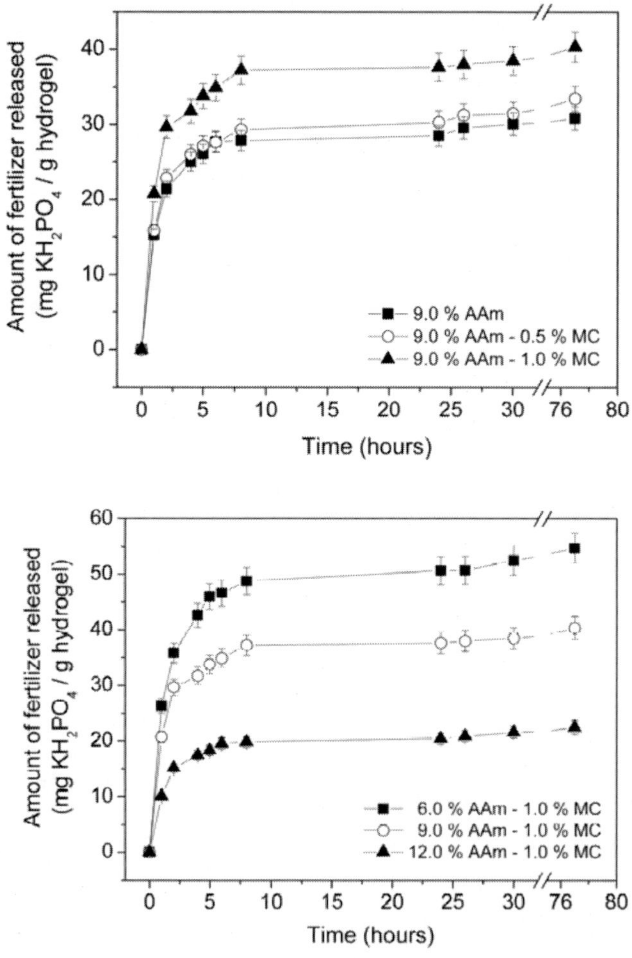


Figure 2. Profiles of potassium phosphate ($KH_2PO_4$) release from hydrogels synthesized from PAAm and MC as a function of time in different conditions: a) [AAm] = 9.0 wt. % with 0, 0.5 and 1.0 wt. % MC, and b) [MC] = 1.0 wt. % with 6.0, 9.0 and 12.0 wt. % AAm [39].

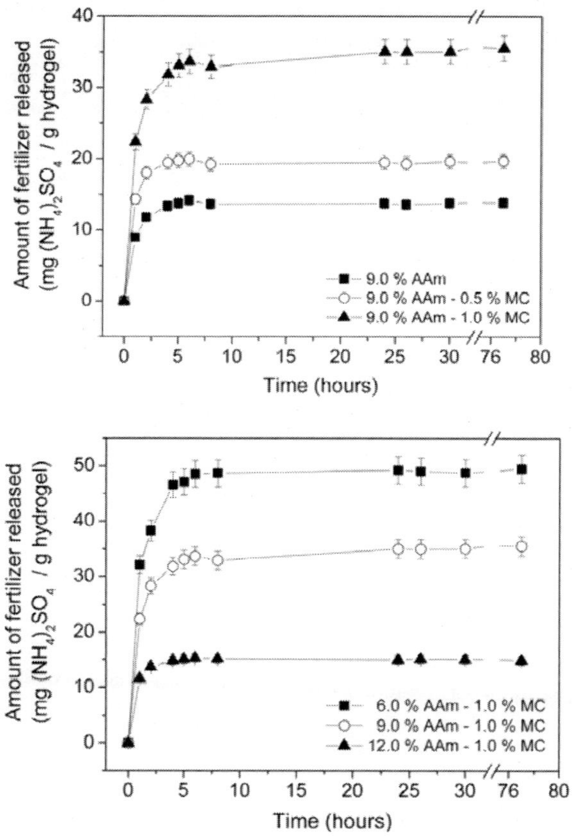


Figure 3. Profiles of ammonium sulfate $(NH_4)_2SO_4$ release from hydrogels synthesized from PAAm and MC as a function of time in different conditions: a) [AAm] = 9.0 wt. % with 0, 0.5 and 1.0 wt. % MC, and b) [MC] = 1.0 wt. % with 6.0, 9.0 and 12.0 wt. % AAm [39].

### *3.2.2. Results and Discussion*

#### 3.2.2.1. Fourier Transform Infrared (FTIR) Spectroscopy

The characterization of the PAAm-PMAA-CMC hydrogel was performed by FTIR spectroscopy, as shown in Figure 4.

Narrow bands at 1710 and 1264 $cm^{-1}$ corresponding to the C=O moiety confirmed the presence of PMAA [44] in the hydrogel. The band at 1183 $cm^{-1}$ was attributed to the C–O stretching of the ether group or the ether linkage [1,4-β-D-glucoside] of cellulose, and the bands at 1603 and 3410 $cm^{-1}$ were attributed to C=O and OH stretching frequencies from the CMC polysaccharide [45]. The main characteristic bands of the PAAm unit [2] are as follows: 3425-3193 $cm^{-1}$ (N-H stretch) and 800-664 $cm^{-1}$ (N-H out-of-plane bend) corresponding to the amide groups; 1668 and 1466 $cm^{-1}$ corresponding to the axial deformation of the carbonyl and C-N bonds and the $CH_2$ asymmetric and symmetric stretching vibrations appearing from 2925-2860 $cm^{-1}$. In addition, the strong bands observed at 1120-1030 $cm^{-1}$ (symmetric) and 466 $cm^{-1}$ (asymmetric) were assigned to C-C stretching modes.

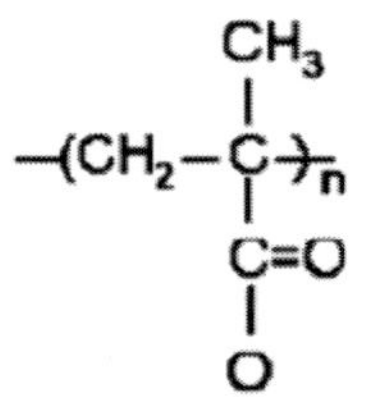


PMAA

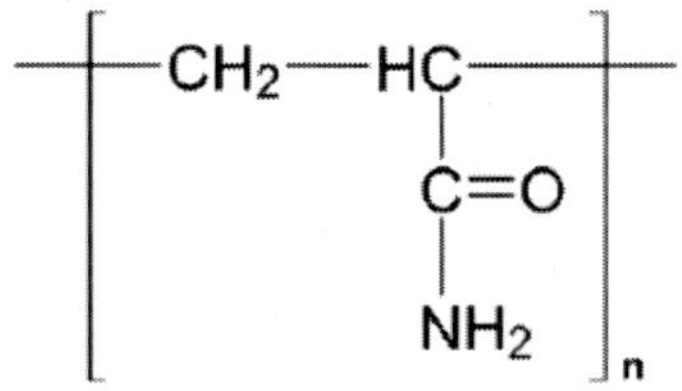


PAAm

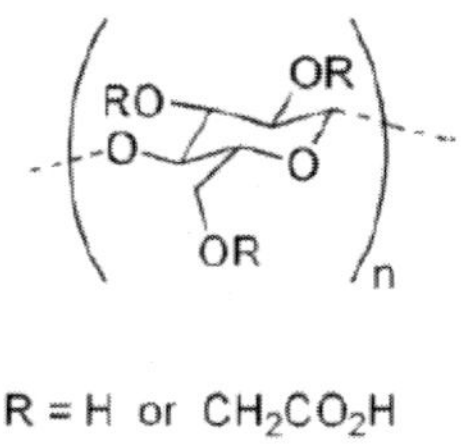


CMC

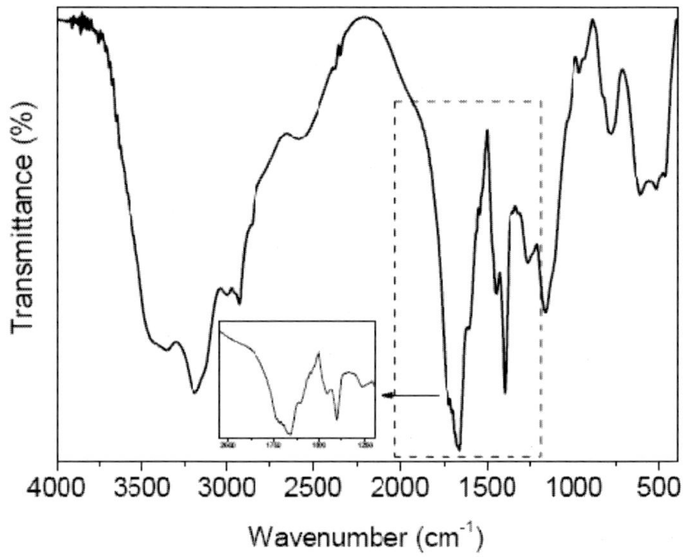


Figure 4. FTIR spectra of the PAAm-PMAA-CMC hydrogel swollen in a buffer solution at pH 7.0 at 25.0 °C. Inset: expanded view of the region between 1200 and 2050 $cm^{-1}$.

### 3.2.2.2. Hydrophilic Properties

#### *3.2.2.2.1. Determination of the pKa Values of the PAAm-PMAA-CMC Hydrogels*

The variation of the hydrophilic properties of the hydrogels with changes in pH was evaluated in order to determine the hydrogel pKa, as shown in Figure 5. The pKa value of the hydrogel was estimated from the sigmoid-curve and was approximately 4.0. This value is between the pKa values of individual components, i.e., the pKa of CMC ≈ 3.2 [46] and the pKa of PMAA ≈ 4.9 [47]. When the pH of the external medium is lower than the hydrogel pKa, the SD was low. However, when the pH was increased to 6.5 (pH > hydrogel pKa), the SD increased sharply to 9 g $H_2O$ / g hydrogel. These behaviors are interpreted as follows: when the pH < hydrogel pKa, $–COO^-$ groups are converted to –COOH groups, and hydrogen bonding is formed among the –OH, –COOH (present in PMAA and CMC), and $–CONH_2$ (present in PAAm and MBAAm) groups; when the pH > hydrogel pKa: the COOH groups dissociate to $COO^-$, and the electrostatic repulsion between the CMC molecular chains contributes to the expansion of the polymeric network, as shown in Scheme 1.

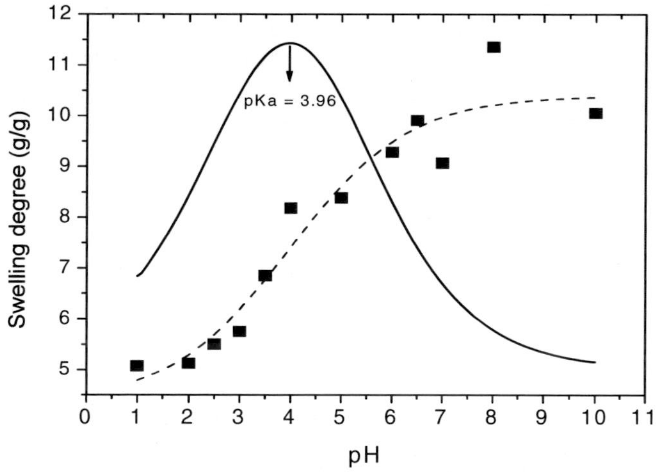


Figure 5. Swelling degree of the PAAm-PMAA-CMC hydrogels as a function of pH: [ionic strength] = 0.10 mol $L^{-1}$, T = 25.0 °C, [AAm] = 5.0 wt. %, [MAA] = 5.0 wt. % and [CMC] = 1.0 wt. %.

The production of pH-sensitive PAAm-PMAA-CMC hydrogels is of key importance for several applications, such as the controlled release of drugs, nutraceuticals, pesticides and fertilizers. Additionally, according to Liang et al. [48], the release of solute from fertilizer controlled systems is controlled by the degradation rate, which in turn is affected by various factors, such as the pH and the microorganisms in the soil.

### *3.2.2.2.2. Influence of the MAA and AAm Concentrations*

Figure 6 shows the swelling degree as a function of time for the hydrogels prepared with different MAA and AAm concentrations. The hydrophilic properties were greatly influenced by both the AAm and MAA monomer concentrations. Despite the hydrophilic characteristics of MAA, the increase of its concentration from 5.0 to 10.0 wt. % induced a pronounced decrease in the water uptake of the resulting hydrogel. This behavior is possibly related to increased network rigidity, which diminishes potential flexibility, and consequently, the volume of water that the hydrogel may absorb also decreases. A similar trend in the hydrophilic properties of PAAm hydrogel-derivatives was observed by Aouada et al. [2].

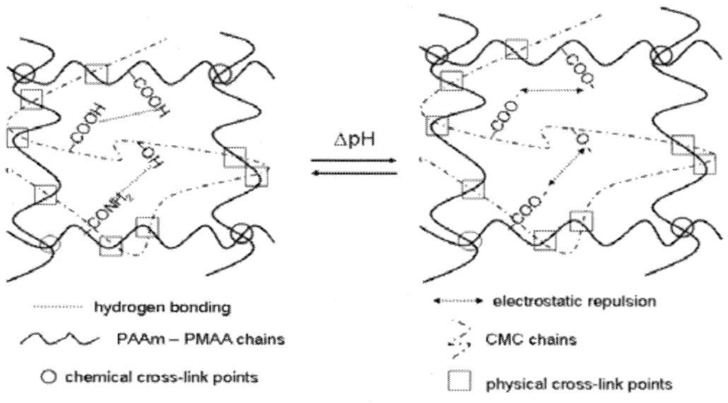


Scheme 1. Schematic illustration of the swelling behavior of the PAAm-PMAA-CMC hydrogel at different pHs.

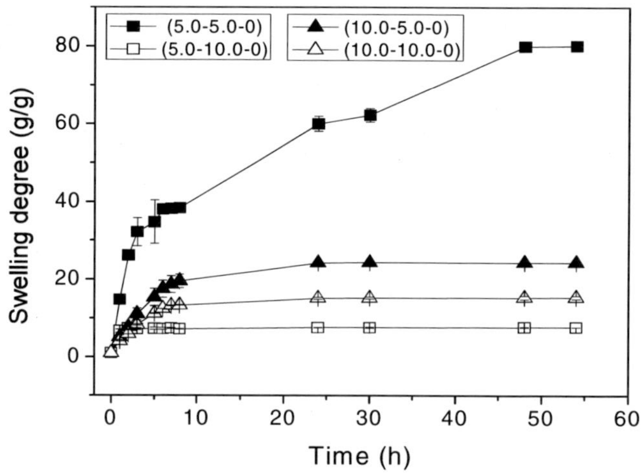


Figure 6. Dependence of the swelling degree as a function of time for hydrogels prepared with different AAm and MAA concentrations.

On the other hand, it was also observed that the water uptake was facilitated by the increased network rigidity induced by increasing the MAA content. For instance, the (5.0-10.0-0) hydrogel with a MAA:AAm ratio of approximately 2:1 reached its equilibrium water uptake after 1 h of immersion. This tendency may be explained by the fact that the MAA monomer is more hydrophilic than the AAm monomer.

In summary, increasing the MAA concentration simultaneously induced a decrease in the SD and an increase in the rate of water uptake. These properties are very relevant to industrial applications, where materials need to reach their equilibrium water uptake quickly with intermediate SD values.

The influence of the AAm concentration on the swelling degree for hydrogels was very similar to the MAA dependence. Here, the decrease in SD was also attributed to an increase in the hydrogel compaction. For example, increasing the AAm concentration from 5.0 to 10.0 wt. %, i.e., from a (5.0-5.0-0) to a (10.0-5.0-0) hydrogel, decreased the SD values from 80.1 ± 0.4 to 24.3 ± 0.1 g $g^{-1}$. Unexpectedly, the SD of the (5.0-10.0-0) hydrogels was lower than that of the (10.0-10.0-0) hydrogels. If this hydrogel series followed the expected behavior, the hydrogels prepared with higher AAm concentrations would exhibit low swelling degrees, as discussed earlier. In these cases, it is likely that the combination of the high AAm and MAA contents induced a decrease in the number of chemical entanglements between the monomers (AAm, MAA and MBAAm). Theoretically, the expansion of the less entangled chains is facilitated, and the theoretical water volume that may remain inside is higher when compared with more entangled chains.

#### *3.2.2.2.3. The Influence of CMC Concentration*

The effect of the CMC content on the swelling degree of the hydrogel containing 5.0 wt. % AAm and 5.0 wt. % MAA was studied by varying the CMC concentration in the range of 0-1.0 wt. % in the feed solution. The results are shown in Figure 7.

When the CMC concentration increased up to 0.5 wt. %, a pronounced increase in the SD values was observed due to an increase in the hydrogel hydrophilicity (and thus an increase in the water absorption capacity). In this case, the increase in the hydrogel hydrophilicity was attributed to the incorporation of the hydroxyl and carboxyl groups from the CMC chains. Similar results have also been observed in other hydrogel systems containing CMC hydrophilic polysaccharides [23,49,50]. On the other hand, for CMC concentrations above 0.5 wt. %, a pronounced decrease in the SD was observed. Molecularly, the number of possible intra- and inter-chain interactions between the CMC chains and/or interaction points between the CMC chains and the PMAA chains increases as the amount of CMC in the hydrogels increases. Bajpai & Giri [37] attributed this drop in the SD to the reduced diffusion of solvent molecules and the relaxation of the macromolecular chains because the increase in the network chains is facilitated by the high concentration of CMC.

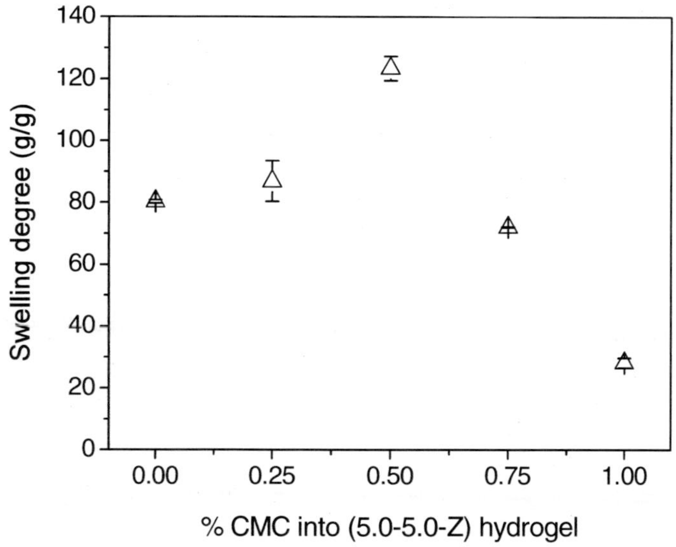


Figure 7. Influence of the CMC content on the swelling degree for the (5.0-5.0-Z) series hydrogels.

### 3.2.2.3. Equilibrium Adsorption Isotherms

Adsorption isotherm models help to understand the interaction between the adsorbate and the adsorbent [51] in the adsorption process. Here, we investigated the equilibrium sorption of ammonium sulfate ($(NH_4)_2SO_4$) fertilizer by applying the Langmuir and Freundlich isotherm models to the observed data, as shown in Figure 8.

The constants for the equilibrium isotherm models are shown in Table 2. The Langmuir model did not show a good fit to the experimental data, and the linear regression coefficient ($R^2 = 0.8976$) had a low value. On the other hand, the experimental data could be well fitted by a Freundlich isotherm ($R^2 = 0.9213$). This indicates that the surface of the PAAm-PMAA-CMC had a heterogeneous profile because the application of this model assumes that the adsorption process occurs on a heterogeneous surface. The $K_F$ parameter is generally defined as an adsorption or distribution coefficient. It represents the quantity of fertilizer adsorbed by the hydrogel. The 1/n parameter is also a measure of the adsorption intensity. The 1/n value of 0.4143 obtained from the Freundlich isotherm application showed that fertilizer adsorption is favorable.

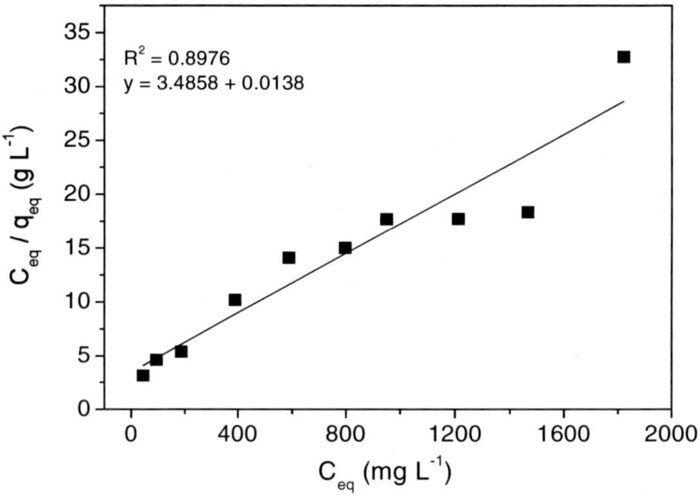


a

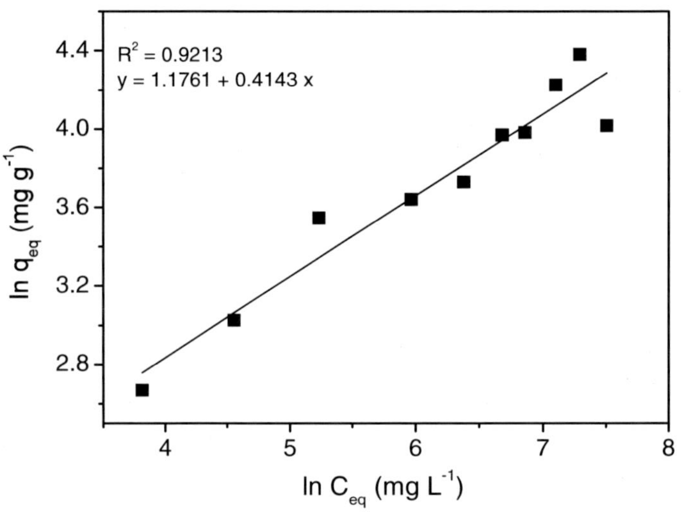


b

Figure 8. Fitted lines for the data modeled using the (a) Langmuir and (b) Freundlich isotherms for the adsorption of ammonium sulfate $(NH_4)_2SO_4$ for the (5.0-5.0-0.5) hydrogel.

**Table 2. Constants for the equilibrium isotherm models**

| Langmuir isotherm | $K_L$ (L $g^{-1}$) | $\propto_L$ (L $mg^{-1}$) | $R^2$ | $Q_0{}^a = K_L / \propto_L$ (mg $g^{-1}$) |
|---|---|---|---|---|
| | 0.2869 | 0.00396 | 0.8976 | 72.45 |
| Freundlich isotherm | $K_F$ (L $g^{-1}$) | 1/n | $R^2$ | |
| | 3.242 | 0.4143 | 0.9213 | |

[a] theoretical monolayer saturation capacity.

Milosavljevic et al. [52] determined the spontaneity of the adsorption of $Cd^{2+}$ ions onto hydrogels composed of chitosan - itaconic acid - methacrylic acid through thermodynamic studies by applying Eq. 5, a standard thermodynamic equation modified by the Langmuir model:

$$\Delta G = -RT \ln K_L \tag{5}$$

where R is the universal gas constant (8.314 J $mol^{-1}$ $K^{-1}$), T is the temperature (298.15 K) and $K_L$ is the Langmuir constant or Langmuir parameter.

The application of Eq. 5 in our study of the $(NH_4)_2SO_4$ sorption showed that ΔG has a positive value (+ 3.1 kJ $mol^{-1}$). Thus, the Langmuir model would indicate that the adsorption process is not spontaneous. The low $R^2$ for the fitting also corroborates the conclusion that this model describes the system poorly. When the $K_L$ constant is changed to $K_F$, Eq. 5 incorporates the Freundlich model. The negative value of ΔG (-2.9 kJ $mol^{-1}$) found using the Freundlich constant proves that this adsorption process is thermodynamically spontaneous. This spontaneity corroborates the results found for the Freundlich constant because an adsorption process is favorable for a 1/n value less than 1.

### 3.2.2.4. Desorption Studies

The potential applications of these hydrogels as carriers for controlled fertilizer release were also investigated. The cumulative desorption (CD) was determined by applying Eq. 6:

$$\mathrm{CD}(\%) = \frac{\mathrm{AR_t}}{\mathrm{AL_\infty}} \mathrm{x100} \tag{6}$$

where $AR_t$ is the amount of fertilizer released by the hydrogel at time t, and $AL_\infty$ is the total amount of fertilizer loaded by the hydrogel: $AL_\infty$ = 21.8 ± 1.8 mg of $(NH_4)_2SO_4$ / g of (5.0-10.0-1.0) hydrogel.

The CD and the amount of $(NH_4)_2SO_4$ released are shown in Figure 9 as a function of time. They exhibited three different behavioral regimes. Firstly, in the initial stage until t = 5 h, the $(NH_4)_2SO_4$ release was relatively fast. Possibly, $(NH_4)_2SO_4$ fertilizer on or close to the surface quickly diffused from the hydrogel. In this stage, the initial hydration of the hydrogel occurs, and the adsorption of the water molecules aids in the desorption of the fertilizer molecules (see SD section above). Secondly, between t = 5 - 24 h, the CD (%) increased, but the rate (shown in Figure 6 as a dotted line that indicates the tangent or angular coefficient) decreased. It is likely that the fertilizer molecules that are deeper in relation to the surface or the molecules that are more strongly linked with the polymeric matrix diffused from the hydrogel during this period. Thirdly, the CD (%) changes in a very controlled and sustained manner, with the concentration gradually increasing after 24 h and remaining constant up to the end of experiment. The final CD and amount of fertilizer released were 60.6 % and 13.2 mg $g^{-1}$, respectively.

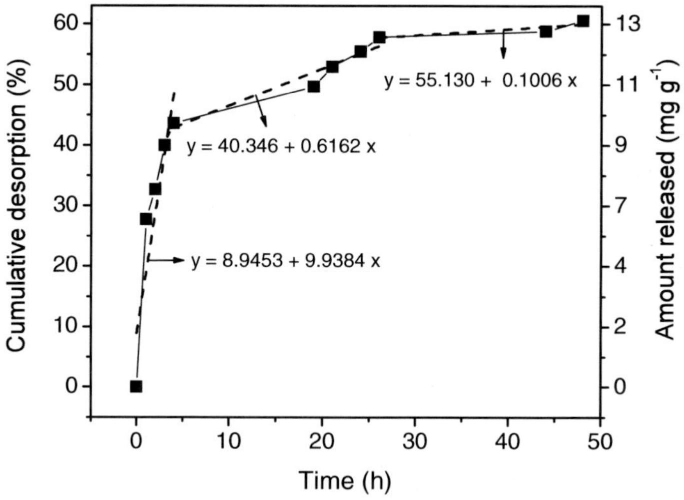


Figure 9. Profiles of ammonium sulfate $(NH_4)_2SO_4$ released from the (5.0-10.0-1.0) hydrogels as a function of time.

## Conclusion

Over the past decades, hydrogel polymers have attracted a great deal of attention as potential materials for agricultural applications because of their biodegradability, non-toxicity, high water uptake, optimal swelling/de-swelling cycles and pH and ionic-strength sensitivities. The production of pH-sensitive biodegradable hydrogels is of key importance for use in several applications, such as use as a carrier vehicle for the controlled release of pesticides and fertilizers. The purpose of hydrogel-fertilizer controlled release systems is to protect the supply of the fertilizer and to allow the automatic release of the fertilizer to the target at a controlled rate.

Therefore, this chapter aimed to contribute to the improvement and development of controlled release technology intended to reduce the risks of human contamination and environment pollution inherent to conventional agrochemical applications. In this sense, novel series of hydrogels were successfully prepared by a free radical polymerization reaction, and the spectroscopic and hydrophilic properties were investigated in relation to the monomer and polysaccharide compositions. The hydrophilic properties, determined by the swelling degree, were dependent on the pH of the external swelling medium, indicating that these hydrogels are pH sensitive. The pKa value of the PAAm-PMAA-CMC hydrogel was estimated to be approximately 4.0. The $(NH_4)_2SO_4$ adsorption data can be well fitted by a Freundlich isotherm, indicating that the surface of the hydrogel is heterogeneous. The Freundlich constants $K_F$ and $1/n$ were 3.242 L $g^{-1}$ and 0.4143, respectively. Thermodynamic investigations proved that the adsorption process of the fertilizer in the hydrogels is spontaneous. The potential applications of these hydrogels as vehicles for controlled release were confirmed by analyzing the cumulative desorption curves for the $(NH_4)_2SO_4$ fertilizer.

## ACKNOWLEDGMENTS

The authors are grateful to FAPESP/CMDMC, FINEP/MCT, CNPq/INCTMN, Embrapa (Rede AgroNano) and CAPES for their financial support and fellowships.

## REFERENCES

[1] Aouada, F. A., Moura, M. R. de, Orts, W. J. & Mattoso, L. H. C. (2011). Preparation and characterization of novel micro- and nanocomposite hydrogels containing cellulosic fibrils. *Journal of Agricultural and Food Chemistry, 59,* 9433-9442.

[2] Aouada, F. A., Moura, M. R. de, Silva, W. T. L. da, Muniz, E. C. & Mattoso, L. H. C. (2011). Preparation and characterization of hydrophilic, spectroscopic, and kinetic properties of hydrogels based on polyacrylamide and methylcellulose polysaccharide. *Journal of Applied Polymer Science, 120,* 3004-3013.

[3] Aouada, F. A., Moura, M. R. de, Orts, W. J. & Mattoso, L. H. C. (2010). Polyacrylamide and methylcellulose hydrogel as delivery vehicle for the controlled release of paraquat pesticide. *Journal of Materials Science, 45,* 4977-4985.

[4] Moura, M. R. de, Aouada, F. A., Favaro, S. L., Radovanovic, E., Rubira, A. F. & Muniz, E. C. (2009). Release of BSA from porous matrices constituted of alginate-$Ca^{2+}$ and PNIPAAm-interpenetrated networks. *Materials Science & Engineering C - Materials for Biological Applications, 29,* 2319-2325.

[5] Wang, W., Wang, J., Kang, Y. & Wang, A. (2011). Synthesis, swelling and responsive properties of a new composite hydrogel based on hydroxyethyl cellulose and medicinal stone. *Composites Part B: Engineering, 42,* 809-818.

[6] Butun, S. & Sahiner, N. (2011). A versatile hydrogel template for metal nano particle preparation and their use in catalysis. *Polymer, 52,* 4834-4840.

[7] Pang, S. C., Chin, S. F., Tay, S. H. & Tchong, F. M. (2011). Starch–maleate–polyvinyl alcohol hydrogels with controllable swelling behaviors. *Carbohydrate Polymers, 84,* 424-429.

[8] Fan, T., Li, M., Wu, X., Li, M. & Wu, Y. (2011). Preparation of thermoresponsive and pH-sensitivity polymer magnetic hydrogel nanospheres as anticancer drug carriers. *Colloids and Surfaces B: Biointerfaces, 88,* 593-600.

[9] Pourjavadi, A., Ayyari, M. & Amini-Fazl, M. S. (2008). Taguchi optimized synthesis of collagen-g-poly(acrylic acid)/kaolin composite superabsorbent hydrogel. *European Polymer Journal, 44,* 1209-1216.

[10] Hawkins, A. M., Milbrandt, T. A., Puleo, D. A. & Hilt, J. Z. (2011). Synthesis and analysis of degradation, mechanical and toxicity properties of poly(β-amino ester) degradable hydrogels. *Acta Biomaterialia, 7,* 1956-1964.

[11] Gallego Ferrer, G., Monleón Pradas, M., Gómez Ribelles, J. L., Romero Colomer, F., Castilla-Cortázar, I. & Vidaurre, A. (2010). Influence of the nature of the porous confining network on the sorption, diffusion and mechanical properties of hydrogel IPNs. *European Polymer Journal, 46,* 774-782.

[12] Tao, S., Liu, J., Jin, K., Qiu, X., Zhang, Y., Ren, X. & Hu, S. (2011). Preparation and characterization of triple polymer-coated controlled-release urea with water-retention property and enhanced durability. *Journal of Applied Polymer Science, 120,* 2103-2111.

[13] Bao, Y., Ma, J. & Li, N. (2011). Synthesis and swelling behaviors of sodium carboxymethyl cellulose-g-poly(AA-co-AM-co-AMPS)/MMT superabsorbent hydrogel. *Carbohydrate Polymers, 84*, 76-82.

[14] Peng, Z. Y. & Chen, F. G. (2010). Synthesis and properties of temperature-sensitive hydrogel based on hydroxyethyl cellulose. *International Journal of Polymeric Materials, 59*, 450-461.

[15] Hagiwara, Y., Putra, A., Kakugo, A., Furukawa, H. & Gong, J. P. (2010). Ligament-like tough double-network hydrogel based on bacterial cellulose. *Cellulose*, *17,* 93-101.

[16] Fan, L. -H., Pan, X. -R., Zhou, Y., Chen, L. -Y., Xie, W. -G., Long, Z. -H. & Zheng, H. (2011). Preparation and characterization of crosslinked carboxymethyl chitosan–oxidized sodium alginate hydrogels. *Journal of Applied Polymer Science*, *122,* 2331-2337.

[17] Eyholzer, C., Borges de Couraça, A., Duc, F., Bourban, P. E., Tingaut, P., Zimmermann, T., Manson, J. A. E. & Oksman, K. (2011). Biocomposite hydrogels with carboxymethylated, nanofibrillated cellulose powder for replacement of the nucleus pulposus. *Biomaterials, 12,* 1419-1427.

[18] Van Vlierberghe, S., Dubruel, P. & Schacht, E. (2011). Biopolymer-based hydrogels as scaffolds for tissue engineering applications: A review. *Biomaterials, 12,* 1387-1408.

[19] Serrero, A., Trombotto, S., Cassagnau, P., Bayon, Y., Gravagna, P., Montanari, S. & David, L. (2010). Polysaccharide gels based on chitosan and modified starch: Structural characterization and linear viscoelastic behavior. *Biomacromolecules, 11*, 1534-1543.

[20] Zhang, J. & Wang, A. (2010). Adsorption of Pb(II) from aqueous solution by chitosan-g-poly(acrylic acid)/attapulgite/sodium humate composite hydrogels. *Journal of Chemical and Engineering Data, 55,* 2379-2384.

[21] Aouada, F. A., Chiou, B. S., Orts, W. J. & Mattoso, L. H. C. (2009). Physicochemical and morphological properties of poly(acrylamide) and methylcellulose hydrogels: Effects of monomer, crosslinker and polysaccharide compositions. *Polymer Engineering and Science, 49,* 2467-2474.

[22] Wang, W. & Wang A. (2010). Nanocomposite of carboxymethyl cellulose and attapulgite as a novel pH-sensitive superabsorbent: Synthesis, characterization and properties. *Carbohydrate Polymers, 82,* 83-91.

[23] El-Din, H. M. N., Abd Alla, S. G. & El-Naggar, A. W. M. (2010). Swelling and drug release properties of acrylamide/carboxymethyl cellulose networks formed by gamma irradiation. *Radiation Physics and Chemistry, 79,* 725-730.

[24] Li, J., Lu, J. & Li, Y. (2009). Carboxylmethylcellulose/bentonite composite gels: Water sorption behavior and controlled release of herbicide. *Journal of Applied Polymer Science, 112,* 261-268.

[25] Wichterle, O. & Lim, D. (1960). Hydrophilic gels for biological use. *Nature, 185,* 117-118.

[26] Perale, G., Rossi, F., Santoro, M., Marchetti, P., Mele, A., Castiglione, F., Raffa, E. & Masi, M. (2011). Drug release from hydrogel: A new understanding of transport phenomena. *Journal of Biomedical Nanotechnology, 7,* 476-481.

[27] Abd El-Mohdy, H. L., Hegazy, E. A., El-Nesr, E. M. & El-Wahab, M. A. (2011). Control release of some pesticides from starch/(ethylene glycol-co-methacrylic acid) copolymers prepared by γ-irradiation. *Journal of Applied Polymer Science, 122,* 1500-1509.

[28] Tyliszczak, B., Polaczek, J. & Pielichowski, K. (2009). PAA-based hybrid organic-inorganic fertilizers with controlled release. *Polish Journal of Environmental Studies, 18,* 475-479.

[29] Jamnongkan, T. & Kaewpirom, S. (2010). Potassium release kinetics and water retention of controlled-release fertilizers based on chitosan hydrogels. *Journal of Polymers and the Environment, 18,* 418-423.

[30] Ma, Z., Li, Q., Yue, Q., Gao, B., Li, W., Xu, X. & Zhong, Q. (2011). Adsorption removal of ammonium and phosphate from water by fertilizer controlled release agent prepared from wheat straw. *Chemical Engineering Journal, 171,* 1209-1217.

[31] Aouada, F. A., Moura, M. R. de, Menezes, E. D., Nogueira, A. R. D., Mattoso, L. H. C. (2008). Hydrogel synthesis and kinetics of ammonium and potassium release. *Revista Brasileira de Ciência do Solo, 32,* 1643-1649.

[32] Tamjmoto, T., Sato, H. & Matsushita, S. Eur Pat 0686340 (1995).

[33] Han, X., Chen, S. & Hu, X. (2009). Controlled-release fertilizer encapsulated by starch/polyvinyl alcohol coating. *Desalination,* 240, 21-26.

[34] Wu, L. & Liu, M. (2008). Preparation and characterization of cellulose acetate-coated compound fertilizer with controlled-release and water-retention. *Polymer for Advanced Technologies, 19,* 785-792.

[35] Akelah, A. (1996). Novel utilizations of conventional agrochemicals by controlled release formulations. *Materials Science & Engineering C-Biomimetic Materials Sensors and Systems, 4,* 83-98.

[36] Calabria, L., Vieceli, N., Bianchi, O., Oliveira, R. V. B., Filho, I. N. & Schmidt, V. (2011). Soy protein isolate/poly(lactic acid) injection-molded biodegradable blends for slow release of fertilizers. *Industrial Crops and Products, 36,* 41-46.

[37] Bajpai, A. K. & Giri, A. (2003). Water sorption behaviour of highly swelling (carboxy methylcellulose-g-polyacrylamide) hydrogels and release of potassium nitrate as agrochemical. *Carbohydrate Polymers, 53,* 271-279.

[38] Guo, M., Liu, M., Liang, R. & Niu, A. (2006). Granular urea-formaldehyde slow-release fertilizer with superabsorbent and moisture preservation. *Journal of Applied Polymer Science, 99,* 3230-3235.

[39] Bortolin, A., Aouada, F. A., Moura, M. R. de, Ribeiro, C., Longo, E. & Mattoso, L. H. C. (2012). Application of polysaccharide hydrogels in adsorption and controlled-extended release of fertilizers processes. *Journal of Applied Polymer Science, 123,* 2291-2298.

[40] Singh, B., Sharma, D. K. & Gupta, A. (2008). In vitro release dynamics of thiram fungicide from starch and poly(methacrylic acid)-based hydrogels. *Journal of Hazardous Materials, 154,* 278-286.

[41] Shang, L., Zhang, S., Du, H. & Venkatraman, S. S. (2008). A novel approach for the control of drug release rate through hydrogel membrane: II. Thermodynamic modeling of the partition control scheme. *Journal of Membrane Science, 321,* 331-336.

[42] Liang, R., Yuan, H., Xi, G. & Zhou, Q. (2009). Synthesis of wheat straw-g-poly(acrylic acid) superabsorbent composites and release of urea from it. *Carbohydrate Polymers, 7,* 181-187.

[43] Liu, Y., Zheng, Y. & Wang, A. (2010). Enhanced adsorption of methylene blue from aqueous solution by chitosan-g-poly (acrylic acid)/vermiculite hydrogel composites. *Journal of Environmental Sciences – China,* 22, 486-493.

[44] Moura, M. R. de, Aouada, F. A. & Mattoso, L. H. C. (2008). Preparation of chitosan nanoparticles using methacrylic acid. *Journal of Colloid and Interface Science, 321,* 477-483.

[45] Hebeish, A., Higazy, A., El-Shafei, A. & Sharaf, S. (2010). Synthesis of carboxymethyl cellulose (CMC) and starch-based hybrids and their applications in flocculation and sizing. *Carbohydrate Polymers, 79,* 60-69.

[46] Hoogendam, C. W., de Keizer, A., Cohen Stuart, M. A. & Bijsterbosch, B. H. (1998). Persistence length of carboxymethyl cellulose as evaluated from size exclusion chromatography and potentiometric titrations. *Macromolecules, 31,* 6297-6309.

[47] Pohlmeier, A. & Haber-Pohlmeier, S. (2004). Ionization of short polymethacrylic acid: titration, DLS, and model calculations. *Journal of Colloid and Interface Science, 273,* 369-380.

[48] Liang, R., Liu, M. & Wu, L. (2007). Controlled release NPK compound fertilizer with the function of water retention. *Reactive and Functional Polymers, 67,* 769-779.

[49] Roy, N., Saha, N., Kitano, T. & Saha, P. (2010). Novel Hydrogels of PVP–CMC and their swelling effect on viscoelastic properties. *Journal of Applied Polymer Science, 117,* 1703-1710.

[50] Lü, S., Liu, M., Ni, B. & Gao, C. (2010). A novel pH- and thermo-sensitive PVP/CMC semi-IPN hydrogel: Swelling, phase behavior, and drug release study. *Journal of Polymer Science Part B - Polymer Physics, 48,* 1749-1756.

[51] Chatterjee, S. & Woo, S. H. (2009). The removal of nitrate from aqueous solutions by chitosan hydrogel beads. *Journal of Hazardous Materials, 164,* 1012-1018.

[52] Milosavljević, N. B., Ristić, M. D., Perić-Grujić, A. A., Filipović, J. M., Strbac, S. B., Rakočević, Z. L. & Kalagasidis Krušić, M. T. (2010). Hydrogel based on chitosan, itaconic acid and methacrylic acid as adsorbent of $Cd^{2+}$ ions from aqueous solution. *Chemical Engineering Journal, 165,* 554-562.

Hydrogels: Synthesis, Characterization and Applications ISBN: 978-1-61942-842-3
Editors: F. V. Câmara and L. J. Ferreira 

*Chapter 11*

# THERMORESPONSIVE CHITOSAN AND POLOXAMER 407 HYDROGELS FOR DRUG AND VACCINE DELIVERY

***T. Kojarunchitt and S. Hook***
School of Pharmacy, University of Otago, Dunedin, New Zealand

## ABSTRACT

Thermoresponsive hydrogels have been widely investigated in the past few years for various biomedical purposes including controlled delivery of drugs and vaccines. This is due to their unique characteristics of being able to exist as a solution at low temperature and form a solid gel at physiological temperature. Among thermoresponsive hydrogels, chitosan and poloxamer 407 are the two polymers that have attracted most attention for drug and vaccine delivery. Chitosan is a polysaccharide derived from crustacean shells and poloxamer 407 consists of the non-ionic triblock co-polymers, poly(ethylene oxide) (PEO) and poly(propylene oxide) (PPO). The following review focuses on the synthesis and characterisation of hydrogels based on these polymers or modified polymer/co-polymer systems. Data on the use of these systems and future prospects of hydrogels for drug and vaccine delivery will be presented.

## 1. INTRODUCTION

Hydrogels are a special class of gel, consisting of polymeric networks with a three-dimensional configuration that are able to swell in water or biological fluids [1]. Natural polymers are commonly utilized in preparation of hydrogels as they offer the characteristics of being biodegradable, biocompatible and of having similar physical properties to living tissue [2], thus these polymers are widely investigated for drug delivery applications [2]. The release of proteins and peptides can be controlled by the use of such polymeric systems which respond to changes in environmental conditions such as temperature and pH.

Thermosensitive polymers that respond to temperature changes are useful in drug and vaccine delivery applications as the change in temperature from room to physiological conditions is the only stimulus required for initiation of gelation. Thermosensitive hydrogels undergo a phenomenon commonly referred to as phase transition at critical temperature which can be either at the lower critical solution temperature (LCST) or the upper critical solution temperature (UCST). For the purpose of drug delivery the systems that have received most attention are the LCST systems in which the polymers are soluble below the LCST but become increasingly insoluble and hydrophobic when above the LCST, leading to the formation of gel.

Advantages of thermosensitive hydrogels as potential delivery systems include:

- Ease of administration to the body without the requirement for surgery, as the liquid solutions can be easily introduced through injection
- High loading of active into hydrogels
- Ease of preparation of formulation and ease of loading of the active by a simple mixing process
- Additional properties of the sol-gel itself, for example mucoadhesion, haemostasis, antimicrobial activity, inflammatory activity
- Sustained release of active from hydrogels may reduce the requirement for multiple dosing leading to better patient compliance

Thermoresponsive hydrogels can be formed from natural or synthetic polymers. Two of the most commonly used polymers used to formuate thermoresponsive sol-gels are chitosan and poloxamer. These polymers have been investigated in a variety of biomedical applications either alone or in combination with copolymers to formulate thermally sensitive hydrogels with the desired properties for the specific conditions.

## 2. SYNTHESIS AND CHARACTERISATION OF THERMOGELS

Chitosan is produced predominantly through the alkaline deacetylation of chitin, a naturally occurring abundant polysaccharide found in the cell wall of many invertebrates. It is a biocompatible and non-toxic [3] copolymer of D-glucosamine and N-acetyl-D-glucosamine in a β-(1–4) linkage (Figure 1).

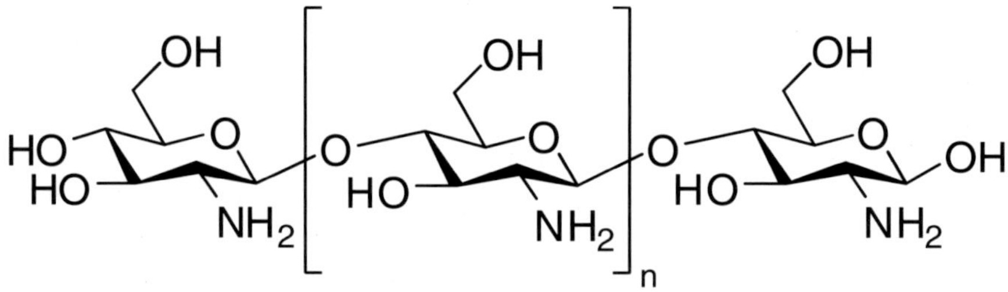


Figure 1. Chemical structure of chitosan.

For commercial purposes it is commonly prepared from the exoskeleton waste of processed crabs, shrimps and prawns [3]. Contaminating minerals and protein are removed through sequential treatment with 1 - 2 M NaOH (high temperature, 70 °C) and 1.25 M HCl (room temperature). This processing is harsh and will result in some hydrolysis of the polymeric chitin chains. Pigments are removed by solvent extraction or using oxidising agents. Once the contaminants are removed alkaline deacetylation is carried out [3]. The conditions for deacetylation vary depending on the required end product, for example the required degree of acetylation (DA) and molecular weight (MW) of the final product. Both of these characteristics and the levels of contaminants will impact on the physical and biological properties of the final product. Some research groups will reacetylate highly deacetylated (> 90%) chitosan preparations in order to generate a product with the required properties [4].

Poloxamers are synthetic polymers with thermoresponsive behaviour. They are tri-block copolymers consists of poly(ethylene oxide) (PEO) and poly(propylene oxide) (PPO) (Figure 2).

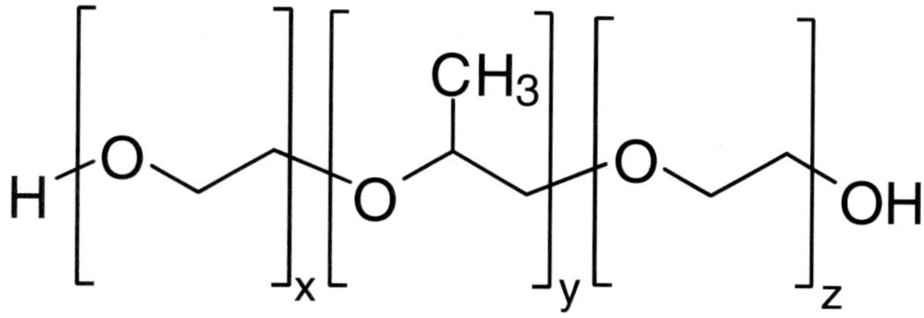


Figure 2. Chemical structure of poloxamer (PEO-PPO-PEO).

They are synthesized by sequentially adding PPO and then PEO to propylene glycol by oxylalkylation in the presence of an alkaline catalyst [5]. The number of blocks and the ratio of PPO to PEO can vary and many different poloxamers are commercially available (also known as Pluronics® or Lutrol®). The different poloxamers have different physical and surface properties.

Important characteristics of thermogels that can be examined include the rheology of both the solution and the gel, the kinetics and temperature of the sol-gel transition, swelling, release of the active, gel degradation and toxicity (of both the parent compound and any degradation products). As chitosan has documented biological activities (for example haemostasis, mucoadhesion, immunogenicity, anti-microbial activities) these should also be examined.

Numerous methods are available to determine these characteristics ranging from very simple tests such as the test-tube inversion method for determining gelation [6] to nuclear magnetic resonance (NMR) for examining the molecular structure of formulations [5].

## 3. Thermoresponsive Chitosan-based Delivery Systems

Chitosan preparations vary in the DA, MW and purity. Chitosan is only soluble in weakly acidic environments, up to a pH of approximately 6.2 (lower than the pKa), as protonation of the amine groups reduces interchain interactions and increases solubility [7]. Solubility generally increases with increasing DA. Solubility is also affected by MW with a lower MW being associated with increased solubility. In addition to affecting solubility, DA and MW have also been shown to impact on other characteristics such as biocompatibility, degradation, mucoadhesion and immunogenicity [8]. As chitosan is a natural product impurities (proteins, inorganic material, lipids, pigments) can be present, these may also impact on physical and biological properties.

**Table 1. Examples of thermoresponsive chitosan-GP hydrogels as sustained release delivery systems**

| Incorporated actives | Outcomes | Reference |
|---|---|---|
| Chondrocytes and bone protein | Showed a successful thermoresponsive injectable system for delivery of living cells and therapeutic proteins for tissue engineering applications | Chenite *et al.*[9] |
| Calcein, chlorpheniramine maleate, dextran and albumin | Sustained release of macromolecules from several hours to a few days. | Ruel-Gariepy *et al.*[10] |
| Paclitaxel | *In vitro* studies showed sustained release of paclitaxel over 1 month and one shot of paclitaxel loaded chitosan gel is effective as 4 i.v. injection of Taxol in inhibiting the growth of cancer cells | Ruel-Gariepy *et al.*[11] |
| Doxorubicin | Effectively inhibited tumor growth and when given in combination with vacinia virus-based vaccine, the anti-tumor effects were observed up to 60 days after tumor challenge | Han *et al.*[12] |
| 5-(and-6)carboxyfluorescein (CF) loaded in liposomes | Liposome loaded in chitosan gel can sustained the release of CF, a low MW hydrophilic compounds | Ruel-Gariepy *et al.*[13] |
| Ovalbumin (OVA) | Sustained release of antigen (OVA), with <10% released after 10 days. Induction of both humoral and cell-mediated immunity | Gordon *et al.* [14] |
| Silica nanoparticles (SNP) loaded with antigen, ovalbumin | *In vitro*, sustained antigen release over a 14-day period and demonstrated ability to stimulate cellular and humoral immune responses *in vivo* | Gordon *et al.*[15] |
| Ellagic acid loaded in PLGA nanoparticles | Sustained release of ellagic acid from PLGA loaded in chitosan gel with 20% released after 360 hours | Sharma *et al.*[16] |
| Cytarabine-loaded liposome | Sustained the release of cytarabine up to 60 hours in comparison to liposomes suspension | Nulik *et al.*[17] |

## 3.1. Chitosan – Glycerol Phosphate Thermogels

Chitosan is extremely versatile in terms of formulation. It can be formulated as gels, micro- and nanoparticles, films and scaffolds. All of these have utility as far as drug and vaccine delivery are concerned. This review will focus on chitosan thermogels. Chitosan is soluble in weakly acidic media and at neutral or alkaline pH a hydrated gel-like precipitate is formed. Chenite *et al.* found that the addition of a glycerol-phosphate (GP) disodium salt while increasing the pH of the solution to around physiological levels, allows the chitosan to stay in solution. Moreover the chitosan-GP solution exhibits reversible temperature dependent gelation at physiological temperatures [9, 18]. The thermal basis of the sol-gel transition is due to the glycerol groups removing water at higher temperatures, allowing interactions (ionic, hydrophobic) to occur between chitosan molecules.

Chitosan-GP thermogels have been widely investigated *in vitro* and *in vivo* to deliver both large and small molecules (Table 1). Shamji *et al.* investigated the loading and release of a variety of different compounds and found that the rate of release of the different molecules was variable, suggesting that simple diffusion was not solely responsible for release and that other factors such as interactions between the active and chitosan may influence release kinetics [19]. They additionally found that the amount of active loaded had an impact on gelation and release. Interestingly, loading the active into a particulate delivery system such as liposomes before incorporation into the gel, as well as increasing gel strength, prolonged release [13]. Release was dependent upon the size and composition of the liposome but not upon the charge. These studies highlight the importance of determining gelation and release experimentally for each new application.

In addition to its thermogelling and sustained release characteristics, chitosan has a number of other interesting features. Some of these activities include being mucoadhesive, facilitating cellular uptake [20], being haemostatic, being able to stimulate wound revascularization and repair, being immunogenic (in some studies) and having antimicrobial activity [8]. In a rabbit cartilage repair model a commercial chitosan-GP formulation, BST-CarGel®, was used as a scaffold to stabilise blood clots and was shown to improve the repair tissue quality and integration [21]. A multicenter trial using the BST-CarGel® formulation to treat articular cartilage lesions in the knee has recently been completed (clinical trials identifier NCT00314236). The inflammatory and immunostimulatory properties of chitosan-based delivery systems have been extensively investigated and debated. Mucosal administration of chitosan (oral or nasal) activates dendritic cells and T cells and stimulates antibody production [20]. Chitosan hydrogels have also been reported to stimulate the infiltration of neutrophils [22] and macrophages [23]. Chenite *et al.* found that chitosans with a low DA formed gels that stimulated considerable inflammatory responses, whereas more highly deacetylated chitosans form gels with a longer residence time and caused no appreciable inflammation [9]. A slower rate of degradation is thought to lead to mild or negligible inflammatory reactions, due to a less rapid production of breakdown products [24]. Porporatto *et al.* reported a relationship between the MW of chitosan and immune reactivity, with low MW formulations being more reactogenic and suggested that C-type lectins such as the mannose receptor could be involved in mediating the immune response to chitosan [25].

For the purpose of drug delivery, inflammatory immune responses are not wanted and should be avoided. However if the formulation is being used for vaccine delivery then controlled inflammation, including the influx of polymorphonuclear and mononuclear cells, is highly advantageous. In addition the sustained release of vaccine antigens from a gel will likely result in a prolonged antigen presentation and extend the interaction of antigen with the immune cells [26] potentially leading to a more effective immune responses [27] that may reduce the requirement for multiple immunisations [28].

Gordon *et al*. studied the ability of chitosan-GP sol-gels loaded either directly with a model protein antigen (ovalbumin) or with nanoparticles (liposomes, cubosomes, silica nanoparticles) to stimulate immune responses in mice after subcutaneous injection. They reported that chitosan sol-gels were able to induce humoral and cell-mediated immunity when delivered subcutaneously [14, 15, 26]. Interestingly, while chitosan sol-gels exhibited inherent immunogenicity, chitosan nanoparticles made using the same chitosan were not immunogenic, suggesting the importance of a sustained release formulation for immune activation. They also found that inclusion of the vaccine antigen in a particle before loading into the sol-gel had an impact both on antigen release and the resulting immune response.

## 3.2. Modified Chitosan Thermogels

Chitosan hydrogels, while having many useful properties also have some limitations. These include the aqueous solubility, the viscosity of the sol and the degradation kinetics of the gel. This has led researchers to modify chitosan and to include copolymers in the formulations in order to modify these characteristics.

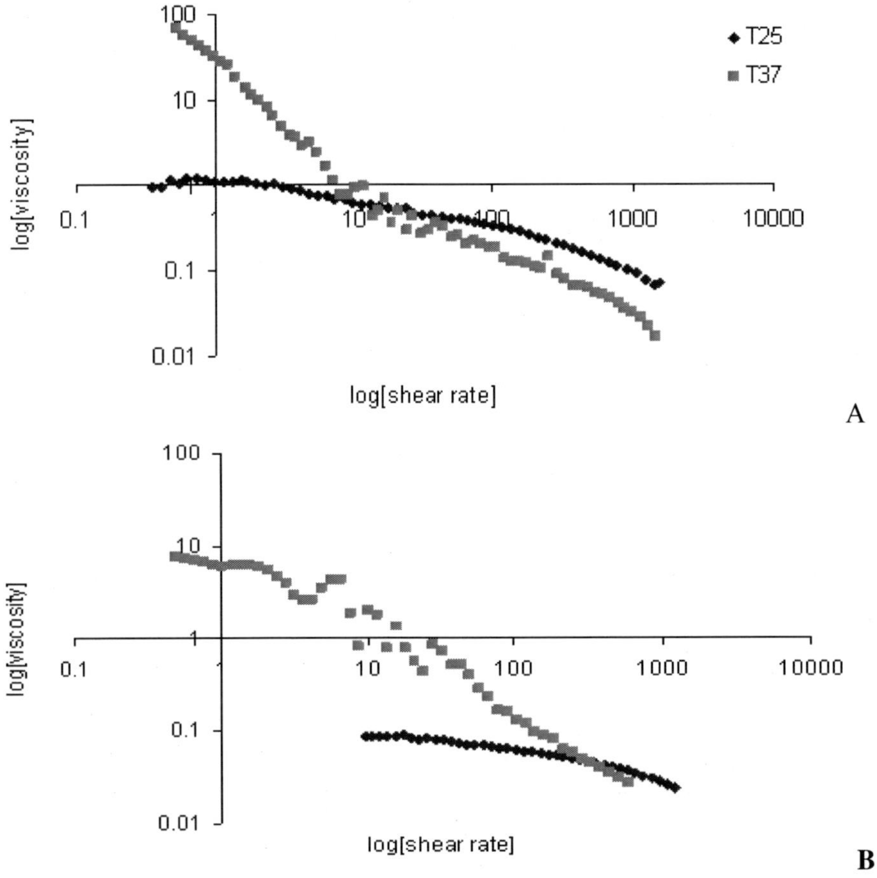


Figure 3. Effect of viscosity as a function of shear rate at 25 °C (■) and 37 °C (◆) for 1% chitosan: GP formulations modified with 0.5% MC (A) compared to 1% chitosan: GP (B) (T. Kojarunchitt, unpublished data).

The optimal thermogel formulation would have Newtonian flow characteristics when in the sol state. This would facilitate manipulations such as drug or vaccine loading and delivery of the formulation, for example by injection. The viscosity of the sol is concentration dependent, therefore decreasing chitosan concentration reduces viscosity. Generally the concentration of chitosan in hydrogels varies between 2 - 3%. This can be decreased to around 1%, however this impacts on the speed of the sol-gel transition and on the strength of the resulting gel (unpublished data). Inclusion of co-polymers such as methyl cellulose and dextran (0.2 - 0.5%) into the formulation can improve the speed of the sol-gel transition and the strength of the resulting gel (Figure 3).

A number of researchers have investigated ways to improve the poor aqueous solubility of chitosan at physiological pH. These include the chemical modification of chitosan itself or grafting on additional functional groups. Ji *et al.* investigated water soluble quaternized chitosan as a thermogel and found that temperature dependent gelation could be achieved at around physiological temperatures using formulations containing a 5:1 ratio of chitosan to quaternized chitosan, however the gelation was not reversible [29]. Another approach to improving solubility at physiological pH has been to utilise chitosan with a low DA and a random distribution of acetyl groups. Viscosan™ (50% DA) when formulated as a thermogel (Viscogel™) vaccine stimulated both cellular and humoral immune responses to *H. influenzae* type b after subcutaneous injection [30]. The gel induced a leukocyte infiltration 1-2 days post injection and was completely cleared by 3 weeks. The authors argued that the robust neutrophil infiltrate and speedy clearance of the gel was due not only to the low DA but also to the increased solubility.

Another approach to improving chitosan solubility is to graft on additional functional groups such as poly(ethylene glycol) (PEG) [31]. Bhattarai *et al.* reported that grafting 40% PEG onto chitosan resulted in a thermogelling formulation with aqueous solubility at neutral pH. Other modified chitosan-based formulations which show thermoresponsive property include poly(N-isopropylacrylamide) (PNIPAAM) grafted chitosan [32], poly(ethylene glycol)-poly(alanine) grafted chitosan [33] and maleic anhydride and 2-(dimethylamino)ethyl methacrylate grafted chitosan [34].

## 4. Poloxamer-Based Delivery Systems

Poloxamers are commercially available (Pluronic® and Lutrol®) in various composition ratios, molecular weights and forms. The most widely investigated poloxamer is poloxamer 407 (P407) also commercially known as Pluronic F127®. It has a molecular weight of approximately 12 600 (9 840 – 14 600) and the lengths of PEO and PPO blocks are 95 - 105 and 54 - 60 monomeric units respectively [35]. Poloxamers have amphiphilic properties that are highly dependent on the length of the PPO and PEO chains. Given that the PPO is less hydrophilic than the PEO, the polymer easily self-assembles forming micelles. This process is both concentration and temperature dependent and is due to the dehydration of hydrophobic PPO blocks, which represents the first step in the gelling process [36]. The micelles are spherical with a dehydrated PPO core and an outer layer of hydrated PEO chains. Micellization is followed by gelation, which is attributed to the packing of the micelles, if the solution is sufficiently concentrated [36-40].

P407 is listed by the FDA as an inactive ingredient for various pharmaceutical and cosmetic applications [41]. It is non-toxic [42], water soluble (especially in cold water), cleared by renal and biliary excretion [43], is highly versatile with thermoresponsive characteristics and has the ability to entrap both hydrophilic and hydrophobic drugs in micelles to provide sustained release [44]. Aqueous solutions of P407 at a concentration of 20 - 30% (w/w) can be easily prepared with thermoreponsive characteristics where they are a solution at a low temperature (4 - 5 °C) but gel upon an increase to room temperature [39].

P407 sol-gels have been investigated extensively for the delivery of peptides, proteins, hydrophilic and hydrophobic drugs (reviewed in detail by Dumortier *et al.* [45] and Escobar-Chávez *et al.* [46]).

P407 has additionally been investigated as a potential system for transdermal delivery as it has been reported that P407 can influence the permeation of drug through the skin [47]. Moreina *et al.* evaluated P407 gels for transdermal delivery of lumiracoxib, a non-steroidal anti-inflammatory drug. *In vitro* skin permeation and retention studies using porcine ear skin found that a 25% P407 formulation, also containing the penetration enhancer oleic acid, promoted transdermal delivery of lumiracoxib. Furthermore, *in vivo* skin irritation tests in rabbits showed no visual evidence of any changes on the skin [48]. Autunes *et al.* developed a sustained release P407 formulation containing niosomes loaded with diclofenac. They hypothesized that the particle bilayers and gel networks would create barriers slowing release, resulting in a low permeation of drug though the skin which could be beneficial for some drugs [47].

## 4.1. Modified Poloxamer-Based Thermogels

Although P407 has the beneficial properties of being a low viscosity solution at room temperature and transforming rapidly into a gel at physiological temperatures, its use in biomedical applications is limited at least in part by its low mechanical strength, high permeability and limited stability in aqueous environments [44, 49], caused by the dilution of polymer by body fluids to a concentration below the LCST. To overcome these limitations, various methods (physical and chemical) of modifying poloxamer sol-gels have been investigated (Table 2).

Hydrophilic polymers such as carrageenan [50], dextran [51] and methyl cellulose (MC) [52] physically interact with the PEO blocks through entanglements, facilitating scaffolding of the micelle network. Inclusion of these co-polymers can influence viscosity and the mucoadhesive and gelling behaviour of P407-based systems. Salem developed a novel formulation in which a progesterone nanosuspension was dispersed in a P407-MC sol-gel. Sustained release of progesterone for up to 36 hours was observed in rats [53]. Similarly Liu *et al.* demonstrated that P407 modified with carrageenan could retard the release of acyclovir over 10 hours *in vitro* and prolong the drug *in vivo* residence time more than 6 hours. Recently Lee *et al.* developed tyramine-conjugated P407. Upon *in vitro* exposure to tyrosinase, enzymatic oxidation of the terminal tyramine groups occurs resulting in P407 cross-linking. Tyrosinase treated gels exhibit increased strength and sustained release of FITC-dextran over 13 days, demonstrating the potential of this system for the sustained release of therapeutic agents [54].

**Table 2. Examples of various physical and chemical modifications of P407**

| Modification | Influence on drug release or gel stability +/- gel strength and mucoadhesive force | Ref. |
|---|---|---|
| Carbopol | ↑ gel strength and mucoadhesive force | Jones *et al.* [55]; Majithiya *et al.* [56] |
| Hydroxypropyl cellulose | ↑ gel strength, prolong drugs release (ibuprofen released over 8 hours, ceftiofur released up to ~6 hours),no influenced on lidocaine released | Paavola *et al.* [51]; Zhang *et al. [57]* |
| Methyl cellulose | ↑ mucoadhesive force | Das *et al.*[58] |
| Polyvinyl pyrrolidone | Prolong ceftiofur up to ~6 hours | Zhang *et al.* [57] |
| Carboxy methyl cellulose | Prolong ibuprofen release over 8 hours | Paavola *et al.* [51] |
| Carrageenan | Prolong acyclovir release over 10 hours | Liu *et al.* [50] |
| Conjugation of 3,4-dihydroxyphenyl-L-alanine to P407 | ↑ mucoadhesive force | Huang *et al.*[59] |
| poly(lactic acid)-F127-poly(lactic acid) pentablock | ↑ *in vitro* stability (over 65 days) | Cohn *et al.*[60] |
| Acyrlate/thiol modified P407 | ↑ in vitro stability and prolong release up of methylene blue up to 10 days | Niu *et al.* [61] |
| Tyramine-conjugated P407 | ↑ in vitro stability, mucoadhesive force and prolong FITC-dextran release for 13 days | Lee *et al.* [54] |

One of the most widely investigated modified P407 system is P407 in combination with Poloxamer 188 (P188), an analogue of P407 which has low toxicity and the ability to increase the temperature at which P407 gelation occurs [62]. This formulation is widely used for occular drug delivery. Thermoresponsive sol-gels are ideal delivery systems for ocular delivery due to their simple preparation and sol-gel properties in which an eye drop can be easily administered into the eye in a liquid form which will then gel to provide sustained or controlled release. Drug release from such *in situ* sol-gels is prolonged due to longer precorneal contact times compared with conventional eye drops which are subjected to a rapid precorneal elimination by the tears. Quian *et al.* formulated P188 modified P407 sol-gels containing methazolamide for the treatment of glaucoma. *In vitro* release demonstrated a diffusion-controlled release of methazolamide over a period of 10 hours, with a higher AUC compared to methazolamide formulated in stimulated tear fluid. However, no increase in the mean residence time was observed *in situ* with the sol-gel formulation as compared to drug in stimulated tear fluid [63]. P407 modified with P188 and loaded with a nanoemulsion containing dorzolamide was also investigated for the treatment of glaucoma. The release studies showed prolonged drug release for up to 8 hours for the sol-gel formulation as compared to 3 hours from drug solution potentially facilitating a reduction in the number of applications per day leading to a better patient compliance [64]. Cao *et al.* added carbopol to

P188 modified P407 sol-gels containing azithromycin to develop a sustained release formulation for the treatment of bacterial conjunctivitis. The addition of carbopol enhanced the mucoadhesive force prolonging the pre-corneal residence time. *In vitro* studies demonstrated sustained release of azithromycin over 24 hours and *in vivo* studies demonstrated that *in situ* gelling prolonged precorneal drug resident time and improved ocular drug bioavailability [65].

P188 modified P407 sol-gels have also been investigated for parenteral drug delivery. A P188 modified P407 sol-gel was developed for the delivery of piroxican. The plasma concentration of piroxican measured after intramascular injection of the modified P407 gel in rabbits was significantly higher than after injection of piroxicam in solution with the drug concentration remaining elevated for 4 days (1.5- 2.2 μg/mL) before falling to less than 20 μg/mL at the end of 7 days [66]. The same group also studied another injectable system in which doxorubicin was loaded in a P188 modified P407 hydrogel. Doxorubicin formulated in the thermogel had 5-fold better bioavailability than doxorubicin in solution [67].

Another area of modification has been to increase the mucoadhesive properties of P407 based sol-gels. Baloglu *et al.* reported that a P188 modified P407 sol-gel exhibited an appropriate gelation temperature and mucoadhesion suitable for vaginal delivery. *In vivo* studies showed that a single-application of an econazole nitrate P407:P188 sol-gel was effective in the treatment of vaginal candidiasis in rats [68]. Koffi *et al.* added the bioadhesive polymer HPMC to P407 to increase mucoadhesion, however the addition of even small amounts of HPMC resulted in an immediate phase separation. The inclusion of propanediol 1,2 in the formulation prevented phase separation and resulted in a sol-gel with increased mucoadhesion, increased elasticity and a decreased transition temperature (possibly as a result of HPMC forming bridges between poloxamer micelles). Further studies on drug release, toxicity and pharmacokinetic parameters are required to determine the potential of this formulation for rectal drug delivery [69]. The mucoadhesive and tissue scaffolding capacity of P407 thermogels has also been modified through grafting onto chitosan. Chondrocytes encapsulated in P407:chitosan sol-gels remained viable for 28 days in culture and the formation of cartilage tissue by cells over this period was superior as compared to the control (alginate hydrogel). Thus, this study demonstrated the possible benefit of grafting P407 onto chitosan for cartilage regeneration [70].

Poloxamer based thermogels have been investigated for vaccine delivery. Reverse poloxamer modified P407 thermogel vaccines were investigated by Kojarunchitt *et al.* [71]. Reverse-poloxamer (Pluronic 25R4) was included in the formulation in order to enhance the stability of P407 in aqueous environments. Although P407 in combination with Pluronic 25R4 showed enhance stability *in vitro*, this system still appeared to have low stability *in vivo*.

P407:chitosan combination sol-gel vaccines have also been investigated, perhaps with the idea of utilising the best aspects of each system. In one such study P407:chitosan sol-gels containing tetanus toxoid were evaluated as novel mucosal vaccine delivery systems where P407 provided a sustained release vehicle while chitosan was utilised as a penetration enhancing agent. An enhancement in antibody responses against tetanus toxoid was demonstrated [72] and it was suggested that P407 and chitosan may have an additive or synergistic effect on the immune system. Enhanced antibody responses were also observed upon parenteral delivery of P407-chitosan vaccines [73].

## 5. FUTURE PROSPECTS

Thermoresponsive hydrogels offer an elegant opportunity for both drug and vaccine delivery. Chitosan and poloxamer are very different polymers and form gels with different characteristics. Chitosan gels are more stable *in vivo*, exhibit more sustained release of the active and are in themselves biologically active, which may or may not be beneficial. However there are issues with the immunogenicity of chitosan sol-gels which may limit use for drug delivery but advance their use for vaccine delivery. On the other hand, P407 gels have weak mechanical strength leading to rapid erosion. Many studies have attempted to improve P407 mechanical strength, with chemical modifications showing more success than physical modifications. Finally thermoresponsive formulations will face problems in countries where the ambient temperature is over 30 °C and vaccine preparations can not be kept cool. Research is progressing with other environmentally responsive systems (pH, light) which may be more appropriate in this situation.

## REFERENCES

[1] Peppas, N. and Mikos, A., *Preparation methods and structure of hydrogels*, in *Hydrogels in Medicine and Pharmacy*, N. Peppas, Editor. 1986, CRC Press: Florida. p. 1-27.

[2] Hamidi, M., Azadi, A., and Rafiei, P., *Hydrogel nanoparticles in drug delivery.* Advanced Drug Delivery Reviews, 2008. 60(15): p. 1638-49.

[3] Barbosa, M.A., Pego, A.P., and Amaral, I.F., *Chitosan*, in *Materials of Biological Origin*. 2011, Elsevier. p. 222-237.

[4] Doan, T., Melvold, R., and Waltenbaugh, C., *Concise medical immunology*. 1st ed. 2005, Baltimore, MD: Lippincott Williams & Wilkins.

[5] Bonacucina, G., Cespi, M., Mencarelli, G., Giorgioni, G., and Palmieri, G.F., *Thermosensitive Self-Assembling Block Copolymers as Drug Delivery Systems.* Polymers, 2011. 3(2): p. 779-811.

[6] Chung, Y.-M., Simmons, K.L., Gutowska, A., and Jeong, B., *Sol-Gel Transition Temperature of PLGA-g-PEG Aqueous Solutions.* Biomacromolecules, 2002. 3(3): p. 511-516.

[7] Sorlier, P., Viton, C., and Domard, A., *Relation between Solution Properties and Degree of Acetylation of Chitosan:‚Äâ Role of Aging.* Biomacromolecules, 2002. 3(6): p. 1336-1342.

[8] Aranaz, I., Mengíbar, M., Harris, R., Paños, I., Miralles, B., Acosta, N., Galed, G., and Heras, Á., *Functional Characterization of Chitin and Chitosan.* Current Chemical Biology, 2009. 3: p. 203-230.

[9] Chenite, A., Chaput, C., Wang, D., Combes, C., Buschmann, M.D., Hoemann, C.D., Leroux, J.C., Atkinson, B.L., Binette, F., and Selmani, A., *Novel injectable neutral solutions of chitosan form biodegradable gels in situ.* Biomaterials, 2000. 21(21): p. 2155-2161.

[10] Ruel-Gariépy, E., Chenite, A., Chaput, C., Guirguis, S., and Leroux, J.C., *Characterization of thermosensitive chitosan gels for the sustained delivery of drugs.* International Journal of Pharmaceutics, 2000. 203(1-2): p. 89-98.

[11] Ruel-Gariépy, E., Shive, M., Bichara, A., Berrada, M., Le Garrec, D., Chenite, A., and Leroux, J.-C., *A thermosensitive chitosan-based hydrogel for the local delivery of paclitaxel.* European Journal of Pharmaceutics and Biopharmaceutics, 2004. 57(1): p. 53-63.

[12] Han, H.D., Song, C.K., Park, Y.S., Noh, K.H., Kim, J.H., Hwang, T., Kim, T.W., and Shin, B.C., *A chitosan hydrogel-based cancer drug delivery system exhibits synergistic antitumor effects by combining with a vaccinia viral vaccine.* International Journal of Pharmaceutics, 2008. 350(1-2): p. 27-34.

[13] Ruel-Gariépy, E., Leclair, G., Hildgen, P., Gupta, A., and Leroux, J.C., *Thermosensitive chitosan-based hydrogel containing liposomes for the delivery of hydrophilic molecules.* Journal of Controlled Release, 2002. 82(2-3): p. 373-383.

[14] Gordon, S., Saupe, A., McBurney, W., Rades, T., and Hook, S., *Comparison of chitosan nanoparticles and chitosan hydrogels for vaccine delivery.* Journal of Pharmacy and Pharmacology, 2008. 60: p. 1591-1600.

[15] Gordon, S., Teichmann, E., Young, K., Finnie, K., Rades, T., and Hook, S., *In vitro and in vivo investigation of thermosensitive chitosan hydrogels containing silica nanoparticles for vaccine delivery.* European Journal of Pharmaceutical Sciences, 2010. 41(2): p. 360-368.

[16] Sharma, G., Italia, J.L., Sonaje, K., Tikoo, K., and Ravi Kumar, M.N.V., *Biodegradable in situ gelling system for subcutaneous administration of ellagic acid and ellagic acid loaded nanoparticles: Evaluation of their antioxidant potential against cyclosporine induced nephrotoxicity in rats.* Journal of Controlled Release, 2007. 118(1): p. 27-37.

[17] Mulik, R., Kulkarni, V., and Murthy, R.S.R., *Chitosan-Based Thermosensitive Hydrogel Containing Liposomes for Sustained Delivery of Cytarabine.* Drug Development and Industrial Pharmacy, 2009. 35(1): p. 49-56.

[18] Chenite, A., Buschmann, M., Wang, D., Chaput, C., and Kandani, N., *Rheological characterisation of thermogelling chitosan/glycerol-phosphate solutions.* Carbohydrate Polymers, 2001. 46(1): p. 39-47.

[19] Shamji, M.F., Hwang, P., Bullock, R.W., Adams, S.B., Nettles, D.L., and Setton, L.A., *Release and activity of anti-TNFα therapeutics from injectable chitosan preparations for local drug delivery.* Journal of Biomedical Materials Research Part B: Applied Biomaterials, 2009. 90B(1): p. 319-326.

[20] Porporatto, C., Bianco, I.D., and Correa, S.G., *Local and systemic activity of the polysaccharide chitosan at lymphoid tissues after oral administration.* Journal of Leukocyte Biology, 2005. 78(1): p. 62-69.

[21] Hoemann, C.D., Sun, J., McKee, M.D., Chevrier, A., Rossomacha, E., Rivard, G.E., Hurtig, M., and Buschmann, M.D., *Chitosan-ìglycerol phosphate/blood implants elicit hyaline cartilage repair integrated with porous subchondral bone in microdrilled rabbit defects.* Osteoarthritis and Cartilage, 2007. 15(1): p. 78-89.

[22] Vandevord, P.J., Matthew, H.W.T., Desilva, S.P., Mayton, L., Wu, B., and Wooley, P.H., *Evaluation of the biocompatibility of a chitosan scaffold in mice.* Journal of Biomedical Materials Research, 2002. 59(3): p. 585-590.

[23] Peluso, G., Petillo, O., Ranieri, M., Santin, M., Ambrosic, L., Calabró, D., Avallone, B., and Balsamo, G., *Chitosan-mediated stimulation of macrophage function.* Biomaterials, 1994. 15(15): p. 1215-1220.

[24] Tomihata, K. and Ikada, Y., *In vitro and in vivo degradation of films of chitin and its deacetylated derivatives.* Biomaterials, 1997. 18(7): p. 567-575.

[25] Porporatto, C., Bianco, I.D., Riera, C.M., and Correa, S.G., *Chitosan induces different l-arginine metabolic pathways in resting and inflammatory macrophages.* Biochemical and Biophysical Research Communications, 2003. 304(2): p. 266-272.

[26] Gordon, S., Young, K., Wilson, R., Rizwan, S., Kemp, R., Rades, T., and Hook, S., *Chitosan hydrogels containing liposomes and cubosomes as particulate sustained release vaccine delivery systems* Journal of Liposome Research, 2011. in press.

[27] McBurney, W.T., Lendemans, D.G., Myschik, J., Hennessy, T., Rades, T., and Hook, S., *In vivo activity of cationic immune stimulating complexes (PLUSCOMs).* Vaccine, 2008. 26(35): p. 4549-56.

[28] Shari, L., *Immunological aspects of controlled antigen delivery.* Advanced Drug Delivery Reviews, 2002. 54(6): p. 863-870.

[29] Amidi, M., Mastrobattista, E., Jiskoot, W., and Hennink, W.E., *Chitosan-based delivery systems for protein therapeutics and antigens.* Advanced Drug Delivery Reviews, 2010. 62(1): p. 59-82.

[30] Neimert-Andersson, T., Hallgren, A.-C., Andersson, M., Langeback, J., Zettergren, L., Nilsen-Nygaard, J., Draget, K.I., van Hage, M., Lindberg, A., Gafvelin, G., and Gronlund, H., *Improved immune responses in mice using the novel chitosan adjuvant ViscoGel, with a Haemophilus influenzae type b glycoconjugate vaccine.* Vaccine, 2011. 29(48): p. 8965-8973.

[31] Bhattarai, N., Ramay, H.R., Gunn, J., Matsen, F.A., and Zhang, M., *PEG-grafted chitosan as an injectable thermosensitive hydrogel for sustained protein release.* Journal of Controlled Release, 2005. 103(3): p. 609-624.

[32] Fang, J.Y., Chen, J.P., Leu, Y.L., and Hu, J.W., *Temperature-sensitive hydrogels composed of chitosan and hyaluronic acid as injectable carriers for drug delivery.* European Journal of Pharmaceutics and Biopharmaceutics, 2008. 68(3): p. 626-636.

[33] Cho, H.-J., Balakrishnan, P., Park, E.-K., Song, K.-W., Hong, S.-S., Jang, T.-Y., Kim, K.-S., Chung, S.-J., Shim, C.-K., and Kim, D.-D., *Poloxamer/cyclodextrin/chitosan-based thermoreversible gel for intranasal delivery of fexofenadine hydrochloride.* Journal of Pharmaceutical Sciences. 100(2): p. 681-691.

[34] Guo, B., Yuan, J., and Gao, Q., *Preparation and characterization of temperature and pH-sensitive chitosan material and its controlled release on coenzyme A.* Colloids and Surfaces B: Biointerfaces, 2007. 58(2): p. 151-156.

[35] Kabanov, A.V., Batrakova, E.V., and Alakhov, V.Y., *Pluronic® block copolymers as novel polymer therapeutics for drug and gene delivery.* Journal of Controlled Release, 2002. 82(2-3): p. 189-212.

[36] Cabana, A., Aït-Kadi, A., and Juhász, J., *Study of the Gelation Process of Polyethylene Oxidea-Polypropylene Oxideb-Polyethylene OxideaCopolymer (Poloxamer 407) Aqueous Solutions.* Journal of Colloid and Interface Science, 1997. 190(2): p. 307-312.

[37] Habas, J.-P., Pavie, E., Lapp, A., and Peyrelasse, J., *Understanding the complex rheological behavior of PEO--PPO--PEO copolymers in aqueous solution.* Journal of Rheology, 2004. 48(1): p. 1-21.

[38] Hvidt, S., Joergensen, E.B., Brown, W., and Schillen, K., *Micellization and Gelation of Aqueous Solutions of a Triblock Copolymer Studied by Rheological Techniques and Scanning Calorimetry.* The Journal of Physical Chemistry, 1994. 98(47): p. 12320-12328.

[39] Lenaerts, V., Triqueneaux, C., Quartern, M., Rieg-Falson, F., and Couvreur, P., *Temperature-dependent rheological behavior of Pluronic F-127 aqueous solutions.* International Journal of Pharmaceutics, 1987. 39(1-2): p. 121-127.

[40] Miller, S.C. and Drabik, B.R., *Rheological properties of poloxamer vehicles.* International Journal of Pharmaceutics, 1984. 18(3): p. 269-276.

[41] Rowe, R., P. Sheskey, and S. Owen, *Handbook of pharmaceutical excipients.* . 5th ed. 2005: Pharmaceutical, London, UK and American Pharmaceutical Association, Washington, USA.

[42] Lee, S.H., Lee, J.E., Baek, W.Y., and Lim, J.O., *Regional delivery of vancomycin using pluronic F-127 to inhibit methicillin resistant Staphylococcus aureus (MRSA) growth in chronic otitis media in vitro and in vivo.* Journal of Controlled Release, 2004. 96(1): p. 1-7.

[43] Wang, Z.Y. and Stern, I.J., *Disposition in rats of a polyoxypropylene-polyoxyethylene copolymer used in plasma fractionation.* Drug Metabolism and Disposition, 1975. 3(6): p. 536-542.

[44] Ruel-Gariépy, E. and Leroux, J.-C., *In situ-forming hydrogels—review of temperature-sensitive systems.* European Journal of Pharmaceutics and Biopharmaceutics, 2004. 58(2): p. 409-426.

[45] Dumortier, G., Grossiord, J., Agnely, F., and Chaumeil, J., *A Review of Poloxamer 407 Pharmaceutical and Pharmacological Characteristics.* Pharmaceutical Research, 2006. 23(12): p. 2709-2728.

[46] Escobar-Chavez, J.J., Lopez-Cervantes, M., Naik, A., Kalia, Y.N., Quintanar-Guerrero, D., and Ganem-Quintanar, A., *Applications of thermo-reversible pluronic F-127 gels in pharmaceutical formulations.* Journal of Pharmacy and Pharmaceutical Sciences, 2006. 9(3): p. 339-58.

[47] Antunes, F.E., Gentile, L., Oliviero Rossi, C., Tavano, L., and Ranieri, G.A., *Gels of Pluronic F127 and nonionic surfactants from rheological characterization to controlled drug permeation.* Colloids and Surfaces B: Biointerfaces, 2011. 87(1): p. 42-48.

[48] Moreira, T.S.A., *A Novel Transdermal Delivery System for the Anti-Inflammatory Lumiracoxib: Influence of Oleic Acid on In Vitro Percutaneous Absorption and In Vivo Potential Cutaneous Irritation.* AAPS PharmSciTech, 2010. 11(2): p. 621-629.

[49] Klouda, L. and Mikos, A.G., *Thermoresponsive hydrogels in biomedical applications.* European Journal of Pharmaceutics and Biopharmaceutics, 2008. 68(1): p. 34-45.

[50] Liu, Y., Zhu, Y.-y., Wei, G., and Lu, W.-y., *Effect of carrageenan on poloxamer-based in situ gel for vaginal use: Improved in vitro and in vivo sustained-release properties.* European Journal of Pharmaceutical Sciences, 2009. 37(3-4): p. 306-312.

[51] Paavola, A., Yliruusi, J., and Rosenberg, P., *Controlled release and dura mater permeability of lidocaine and ibuprofen from injectable poloxamer-based gels.* Journal of Controlled Release, 1998. 52(1-2): p. 169-178.

[52] El-Kamel, A.H., *In vitro and in vivo evaluation of Pluronic F127-based ocular delivery system for timolol maleate.* International Journal of Pharmaceutics, 2002. 241(1): p. 47-55.

[53] Salem, H.F., *Sustained-release progesterone nanosuspension following intramuscular injection in ovariectomized rats.* International Journal of Nanomedicine, 2010. 5: p. 943-954.

[54] Lee, S.H., Lee, Y., Lee, S.-W., Ji, H.-Y., Lee, J.-H., Lee, D.S., and Park, T.G., *Enzyme-mediated cross-linking of Pluronic copolymer micelles for injectable and in situ forming hydrogels.* Acta Biomaterialia, 2011. 7(4): p. 1468-1476.

[55] Jones, D.S., Bruschi, M.L., de Freitas, O., Gremião, M.P.D., Lara, E.H.G., and Andrews, G.P., *Rheological, mechanical and mucoadhesive properties of thermoresponsive, bioadhesive binary mixtures composed of poloxamer 407 and carbopol 974P designed as platforms for implantable drug delivery systems for use in the oral cavity.* International Journal of Pharmaceutics, 2009. 372(1-2): p. 49-58.

[56] Majithiya, R., Ghosh, P., Umrethia, M., and Murthy, R., *Thermoreversible-mucoadhesive Gel for nasal delivery of sumatriptan.* AAPS PharmSciTech, 2006. 7(3): p. E80-E86.

[57] Zhang, L., Parsons, D.L., Navarre, C., and Kompella, U.B., *Development and in-vitro evaluation of sustained release Poloxamer 407 (P407) gel formulations of ceftiofur.* Journal of Controlled Release, 2002. 85(1-3): p. 73-81.

[58] Das, N., Madan, P., and Lin, S., *Development and in vitro evaluation of insulin-loaded buccal Pluronic F-127 gels.* Pharmaceutical Development and Technology, 2010. 15(2): p. 192-208.

[59] Huang, K., Lee, B.P., Ingram, D.R., and Messersmith, P.B., *Synthesis and Characterization of Self-Assembling Block Copolymers Containing Bioadhesive End Groups.* Biomacromolecules, 2002. 3(2): p. 397-406.

[60] Cohn, D., Lando, G., Sosnik, A., Garty, S., and Levi, A., *PEO–PPO–PEO-based poly(ether ester urethane)s as degradable reverse thermo-responsive multiblock copolymers.* Biomaterials, 2006. 27(9): p. 1718-1727.

[61] Niu, G., Du, F., Song, L., Zhang, H., Yang, J., Cao, H., Zheng, Y., Yang, Z., Wang, G., Yang, H., and Zhu, S., *Synthesis and characterization of reactive poloxamer 407s for biomedical applications.* Journal of Controlled Release, 2009. 138(1): p. 49-56.

[62] Wei, G., Xu, H., Ding, P.T., Li, S.M., and Zheng, J.M., *Thermosetting gels with modulated gelation temperature for ophthalmic use: the rheological and gamma scintigraphic studies.* Journal of Controlled Release, 2002. 83(1): p. 65-74.

[63] Qian, Y., Wang, F., Li, R., Zhang, Q., and Xu, Q., *Preparation and evaluation of in situ gelling ophthalmic drug delivery system for methazolamide.* Drug Development and Industrial Pharmacy, 2010. 36(11): p. 1340-1347.

[64] Ammar, H.O., Salama, H.A., Ghorab, M., and Mahmoud, A.A., *Development of dorzolamide hydrochloride in situ gel nanoemulsion for ocular delivery.* Drug Development and Industrial Pharmacy, 2010. 36(11): p. 1330-1339.

[65] Cao, F., Zhang, X., and Ping, Q., *New method for ophthalmic delivery of azithromycin by poloxamer/carbopol-based in situ gelling system.* Drug Delivery, 2010. 17(7): p. 500-507.

[66] Xuan, J.-J., Balakrishnan, P., Oh, D.H., Yeo, W.H., Park, S.M., Yong, C.S., and Choi, H.-G., *Rheological characterization and in vivo evaluation of thermosensitive*

*poloxamer-based hydrogel for intramuscular injection of piroxicam.* International Journal of Pharmaceutics, 2010. 395(1-2): p. 317-323.

[67] Xuan, J.-J., Yan, Y.-D., Oh, D.H., Choi, Y.K., Yong, C.S., and Choi, H.-G., *Development of thermo-sensitive injectable hydrogel with sustained release of doxorubicin: Rheological characterization and in vivo evaluation in rats.* Drug Delivery, 2011. 18(5): p. 305-311.

[68] Baloglu, E., Karavana, S.Y., Senyigit, Z.A., Hilmioglu-Polat, S., Metin, D.Y., Zekioglu, O., Guneri, T., and Jones, D.S., *In-situ gel formulations of econazole nitrate: preparation and in-vitro and in-vivo evaluation.* Journal of Pharmacy and Pharmacology, 2011. 63(10): p. 1274-1282.

[69] Koffi, A.A., Agnely, F., Ponchel, G., and Grossiord, J.L., *Modulation of the rheological and mucoadhesive properties of thermosensitive poloxamer-based hydrogels intended for the rectal administration of quinine.* European Journal of Pharmaceutical Sciences, 2006. 27(4): p. 328-335.

[70] Park, K.M., Lee, S.Y., Joung, Y.K., Na, J.S., Lee, M.C., and Park, K.D., *Thermosensitive chitosan–Pluronic hydrogel as an injectable cell delivery carrier for cartilage regeneration.* Acta Biomaterialia, 2009. 5(6): p. 1956-1965.

[71] Kojarunchitt, T., Hook, S., Rizwan, S., Rades, T., and Baldursdottir, S., *Development and characterisation of modified poloxamer 407 thermoresponsive depot systems containing cubosomes.* International Journal of Pharmaceutics, 2011. 408(1-2): p. 20-6.

[72] Julie Westerink, M.A., Louise Smithson, S., Srivastava, N., Blonder, J., Coeshott, C., and Rosenthal, G.J., *ProJuvant™ (Pluronic F127®/chitosan) enhances the immune response to intranasally administered tetanus toxoid.* Vaccine, 2001. 20(5-6): p. 711-723.

[73] Coeshott, C.M., Smithson, S.L., Verderber, E., Samaniego, A., Blonder, J.M., Rosenthal, G.J., and Westerink, M.A.J., *Pluronic® F127-based systemic vaccine delivery systems.* Vaccine, 2004. 22(19): p. 2396-2405.

Hydrogels: Synthesis, Characterization and Applications ISBN: 978-1-61942-842-3
Editors: F. V. Câmara and L. J. Ferreira 

*Chapter 12*

# ADVANCED BIOMATERIALS: HYDROGELS THAT ALLOW DIVISION AND GROWTH OF ENCAPSULATED CELLS

***Mercedes Perullini,* Cecilia Spedalieri, Matías Jobbágy and Sara A. Bilmes***
INQUIMAE-DQIAQF, Universidad de Buenos Aires,
Facultad de Ciencias Exactas y Naturales,
Ciudad Universitaria Pabellon II,
Buenos Aires, Argentina

## ABSTRACT

Sol-gel encapsulation of living cells within inorganic matrices is an area of great scientific and technological interest. However, only recently a two-step procedure allowing population growth inside the hydrogel matrix has been proposed. In conventional one-pot encapsulations, space limitation suffered by confined cells makes their division completely impossible, and only intrinsic resistant microorganisms (mainly bacteria and yeasts) were encapsulated using extremely biocompatible synthesis procedures in order to obtain an acceptable viability in the resulting biomaterial.

The two-step procedure based on a pre-encapsulation of the biological guest within Ca(II)-alginate, has proved to be a versatile method for advanced biomaterials development. For instance, in applications were it is necessary to keep a large number of metabolically active cells to render them efficient, or where the encapsulation of actively growing cells may be required. Moreover, it opened up many different possibilities to biomaterials, allowing the encapsulation of higher organisms cells (more sensitive than microorganism) as carrot phloematic tissue and tobacco BY2 Line or even of multicellular organisms, such as filamentous fungi.

On the other hand, while one-pot encapsulations uses almost exclusively silica hydrogel hosts, the protection conferred to the cells by the pre-encapsulation step during the sol-gel synthesis procedure allows to extend this approach to other sol–gel materials, as has recently been addressed for Al(III) and Zr(IV) oxohydroxide hydrogels. This work

* corresponding author: mercedesp@qi.fcen.uba.ar

constitutes the first observation of *Escherichia coli* growth within such metal oxide hydrogels, giving clear evidence that sol–gel based cell encapsulation can now be envisioned within a wide variety of metal oxide hosts through the optimization of the pre-encapsulation environment.

## INTRODUCTION

Several natural protective strategies developed by microorganisms are based on the construction of a shell by means of a biomineralization process; among them, diatomea develop their protective shell with silica.[1] Inspired by such a complex process, the encapsulation of living cells into silica materials derived from the sol-gel process has resulted in novel materials with exciting functionalities, combining the advanced functions of the host matrix and the biological functions of the cellular guest.[2-5] The possible applications of these Materials with Biological Activity (MBAs) have an enormous potential, given the complexity of a living cell and the wide biodiversity of microorganisms with very different metabolisms, i.e. different functionalities. However, despite the many successes in this field,[6,7] until almost recently the majority of the published works were limited to the encapsulation of bacteria or yeasts in silica matrices. Traditional one-pot encapsulation does not allow population growth, since space limitation suffered by confined cells makes their division completely impossible.[8-10] In this scenario, many interesting alternatives have been explored to extend the viability of cells submitted to entrapment, which include the addition of complexing agents to reduce silanol aggressiveness,[11] the removal of alcohol before silica gelation,[12] and the use of biological mechanisms such as quorum sensing molecules to promote cell cycle arrest, conferring cell adaptation to confinement.[9] Although considerable attention has been focused on the minimization of natural cell damage during and after sol-gel synthesis, low long-term viability and activity of encapsulated cells posed a significant obstacle for the production of practical application devices. Moreover, the synthesis of the host matrix was limited to extremely biocompatible procedures restricting the quality of material properties obtained and, nevertheless, the whole encapsulation procedure could only be withstood by intrinsically resistant microorganisms, dramatically limiting the number of possible biological guests.

Pioneer work in whole cell immobilization aimed to exploit the bioactivity of an overexpressed enzyme of interest,[13] rendering the physiological status of entrapped cells completely irrelevant since enzymatic activity could be retained or even improved by many folds even in nonviable cells.[14] But while macromolecules retain their activity within a given range of pH, ionic strength, or water activity, presenting an on/off response type, living organisms show more sophisticated response patterns. External environmental factors sensed by cells are integrated by signal transduction pathways and converge to regulate cell phenotype, modulating the biosynthesis of metabolic products.[15] The encapsulation process required a change of focus to face this challenge, in order to take advantage of the complex biological activities of living cells.

For applications in which the entire biosynthetic machinery is compromised, such as biosensors, the physiological status of cells becomes of great importance and typical responses of living cells in nature were not to be reproduced unless in near-natural environments. On the other hand, the high biocompatibility claimed in sol-gel synthesis procedures was so only in terms of high initial viability and degree of survival of certain microorganisms.[16-18] Far from an on/off response type, living cells may remain viable but still develop a high stress level, turning unproductive. Another constraint arises from the fact that for many useful applications, such as bioreactors, it is necessary to keep a large number of metabolically active cells. In addition, certain products of biotechnological interest are synthesized only by actively growing cells.

Moreover, the development of MBAs based on multicellular organisms was hindered in traditional one-pot encapsulations, since confinement inhibits cell growth and cell division, stimulating apoptosis (cellular death).[8-10] Therefore, creating space inside these matrices was the stepping stone toward the development of new MBA's useful applications based on the encapsulation of sensitive eukaryotic cells, growing cultures and/or multicellular organisms. [19]

## Two-Step Encapsulation Procedure

In order to overcome the aforementioned limitations, a new approach to allow cellular division and growth inside a silica matrix was the application of a two-step encapsulation process. It constituted a method for an entire culture encapsulation (instead of a single cell), giving rise to the possibility of culturing cells in near-natural environmental conditions, even enabling cell-to-cell contact. It is worth mentioning that many other strategies that allow cellular growth that use macroporous scaffolds have been proposed.[20] However, these approaches fail to prevent the exit of biological material of interest as well as the entrance of contaminating agents, which is one of the benefits pursued in certain applications. That is to say, the biological guest is immobilized but not encapsulated inside the matrix.

Based on a procedure proposed by Boninsegna *et al.*[21] in which groups of animal or vegetal cells are encapsulated inside a siliceous layer and the inorganic synthesis occurs in the periphery only affecting closely-in-contact cells, a two-step encapsulation procedure was designed.[19] In the former case, the encapsulation in thin siliceous layers did not confer the mechanical stability required for bioreactor applications. The upgrade proposed allowed the encapsulation of sensitive eukaryotic cells inside a biocompatible silica matrix that allowed free diffusion of nutrients and metabolic products, provided mechanical support but permitted at the same time cell proliferation inside the material. Such devices also succeeded in preventing cells escape from the matrix, as well as the entrance of contaminating agents to their bosom. [16]

The strategy to obtain cells lodged in macrocavities inside the silica matrix is based in a previous immobilization of the cells of interest within a natural Ca(II)-alginate polymer. These alginate cell-containing beads are subsequently trapped in a sol-gel monolith obtained by silicate sol-gel condensation, avoiding any harmful contact of cells with silica precursors. Once silica consolidation is completed, alginate is liquefied by means of a diffusing Ca(II)-chelating agent through the silica matrix, leaving the cells confined into macrocavities containing a liquid medium inside a solid silica monolith. To provide necessary nutrients to the immobilized cells, Ca(II)-chelating agent solution is further replaced by the specific culture medium according to encapsulated cell strain requirements. This synthesis strategy is schematically shown in Figure 1.

The utilization of Ca(II)-alginate complex for cell encapsulation is a common step used in microbiological procedures.[22] This pre-encapsulation procedure is performed by dropwise addition of a sodium alginate cells suspension in a $CaCl_2$ solution. As stated before, thin silica coating of such soft biocompatible polymer has been previously described.[23,24] However, the procedure explained hereafter enhances the mechanical strength since the cavity is fully surrounded by a continuous monolithic phase, providing effective resistance properties.

This procedure was first employed for yeast (*Sacharomyces cerevisiae*) and bacteria (*Escherichia coli* and *Bacillus subtilis*), demonstrating cell division and growth inside a "mineral host", overcoming the limits imposed by initial cell survival after the encapsulation procedure. MAB two-scale porosity is depicted in Figure 2. The diffusion time of nutrients within the nanoporous silica monolith proved to be fast enough to allow growing rates almost indistinguishable from a traditional culture (control). The inorganic matrix provided an effective mechanical barrier for the isolation of contiguous growing cultures.

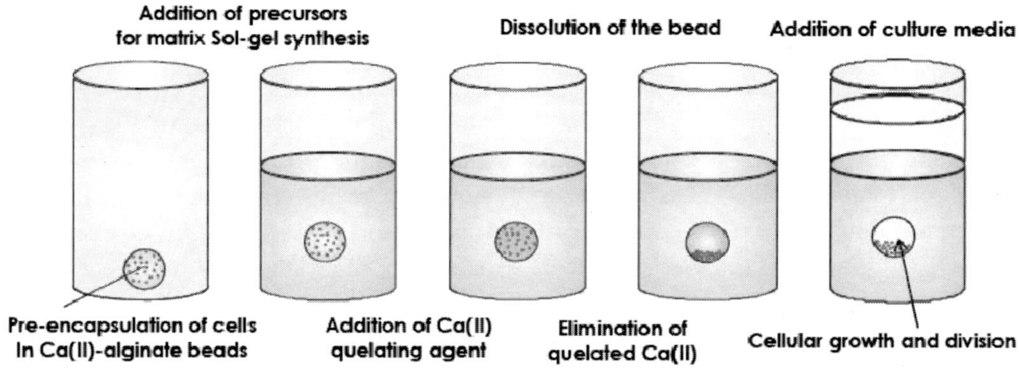


Figure 1. Schematic representation of the two-step synthesis strategy.

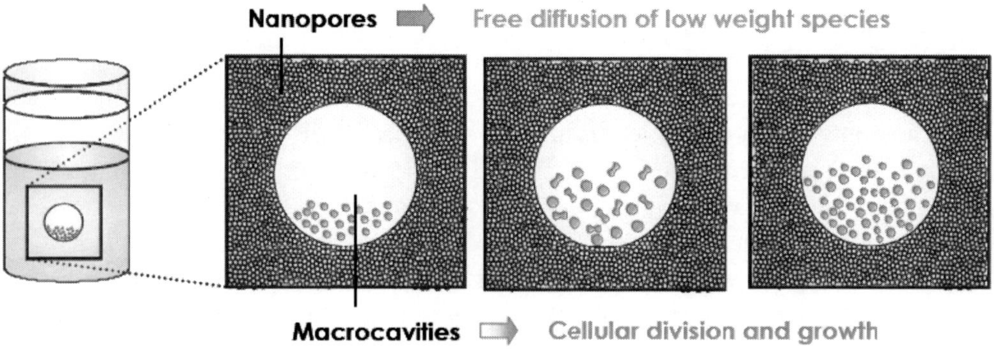


Figure 2. Schematic representation of the two-scale porosity allowing free diffusion of oxygen, nutrients and other low molecular weight molecules through nanopores and cellular division and growth inside macrocavities.

## REDUCING CELLULAR STRESS – APPLICATION TO BIOSENSORS

Another important highlight of cellular-silica hybrid systems is the design of biosensing devices. These small devices can detect the presence of virus or toxins in the environment.[25,26] The most relevant advantage of encapsulation resides in allowing the stabilization of biological materials and making them more resistant to environmental changes. On the other hand, manipulation of the biological included systems turns out to be simpler and generates scarce risk of biological contamination. The main constraint is that a high level of cellular stress can mask the signal to be sensed by the immobilized cell strain. In standard procedures (i.e., one pot encapsulation) not only are cells exposed to an unnatural environment during the inorganic host synthesis, but also by being tightly encapsulated they are kept in permanent contact with the gel's surface silanol groups throughout the immobilization process.

Thus, if the cell is already stressed for the conditions of confinement inside the host material, it cannot give a signal in the presence of the toxin or the signal is not reliable.[24,27]

In this direction, a fast and noninvasive method to assess the cellular stress that silica entrapment exerts over living cells during and after the encapsulation process [28,29] was recently developed.[30] Measurement of the cellular stress status of genetically modified *Sacharomyces cerevisiae*, a true biological probe, based on Confocal Microscopy imaging, allowed us to perform a quantitative analysis of cellular stress in a short time basis (compared to conventional long-term viability tests), opening the gate for a more sophisticated approach to optimize the synthesis conditions minimizing the stress caused to MBAs biological guests. The cellular stress measured as a function of the main preparation variables such as the nature of the silica source, protecting functional groups or total solid concentration, were compared with the main synthetic strategies developed to date (one pot or two step encapsulation). The relation of cellular stress with initial viability was also discussed.

In contrast to straight observation of cell damage by means of *ex situ* or destructive methods (cryo-SEM textural observation),[17] Confocal Microscopy based methods are noninvasive, allowing the immediate quantification of stress of both isolated cells or colonies as function of time or external stimuli.

Due to the inherent fast response and low sampling volumes required, this technique arises as the gate for a combinatorial approach to the formulation of optimized cell entrapping biomaterials.

In principle, the obtained results of the influence of the principal synthesis parameters over the cellular stress status of *S. cerevisiae* submitted to a wide range of silica encapsulation procedures can be extended to other cell types since yeasts have been used as a model for the study of stress response in eukaryotic cells [31]

In summary, the two step strategy offers an optimum response in diminishing cellular stress level. The use of adequate concentrations of sodium alginate in the starting cell suspension grants a minimal stress before and after the Ca(II)-alginate bead formation. However, certain conditions must be carefully taken into account. Using high sodium alginate concentrations in the bead formation generates ionic stress. On the other hand, the lax structures that result from the crosslinking of low concentration alginate cannot confer an effective protection to immobilized cells during the next step of silica synthesis. Either way, the two step method opens the gate for the preparation of novel biomaterials with improved mechanical and optical properties, granting a minimum cellular stress which is essential for MBAs biosensor applications.

## Encapsulation of Sensitive Eukariotic Cells – Application to Bioreactors

MBAs applications in bioreactors offer an alternative to batch culture. One of the clear advantages of this technology is the easier separation between biosynthesis products and the producing cells. Further benefits arise from the use of superior eukaryotic cells encapsulating devices for bioreactor applications, since eukaryotic cells are generally weak against hydrodynamic stress generated by agitation or aeration in conventional bioreactor designs. [32]

Encapsulation limits the risk of contamination of the cells. This represents an important advantage for culturing low metabolism cells that require weeks or even months to reach the productive cellular density for the biosynthesis of value-added products. Plant cells have a significantly slower growth than most microbial organisms, with doubling times ranging from 20 to 100 h. [33]

The existence of a universal genetic code allows the production of heterologous recombinant proteins from genetically transformed microorganisms, like bacteria and yeasts. Nevertheless, the biological activity of many proteins of eukaryotic origin can be affected due to absence or different specificity of posttranslational modifications in the mentioned systems. This loss of activity is a limiting factor for biotechnological applications. On the other hand, many secondary metabolites produced by eukaryotic cells, mainly by plant cells, have a great interest in the agricultural and pharmaceutical industries. These metabolic products are synthesized only by actively growing cells and their synthesis is triggered in high density cultures, requiring cell-to-cell contact. [34,35]

For these reasons, the biosynthetic machinery of superior eukaryotic cells offers better possibilities of biotechnological application than that of prokaryotic ones. Nevertheless, slow metabolism of plant cells limits the design of large-scale systems, where contamination by microorganisms implies enormous losses, both in delay of production and in costs of decontamination. [36]

The immobilization systems employed so far with plant cells, which include gelation in natural [37] or synthetic [38] organic polymers, cannot solve many of the raised limitations. On the other hand, sol–gel chemistry of silica has emerged as an outstanding strategy to obtain biomaterials in which living cells are truly protected. However, even mild thermal and chemical conditions employed in sol–gel synthesis may result harmful to cells of higher organisms, [39] greatly affecting cell viability and resulting completely inappropriate for bioreactor applications. On the contrary, by extending the two-step encapsulation procedure described above to plant cell lines, [40] assessment of cell viability conducted at the beginning of the assay revealed that (97±1)% of carrot cells and (95±2)% of tobacco BY2 cells remained metabolically active after the encapsulation procedure. Furthermore, a doubling time of 250 h was measured for the encapsulated cells throughout a 6-month period.

Though being slow, cell division and *calli* (vegetal non-differentiated cell aggregates) formation within the inorganic material was established. The MBA had good mechanical stability and proved to be an effective barrier against biological contamination, and in turn could be employed for long-term plant cell entrapment applications. The high initial viability obtained suggests that this technique can be extended to higher eukaryotic cells.

## ENCAPSULATION OF MULTICELLULAR ORGANISMS – APPLICATION TO BIOREMEDIATION DEVICES

As part of their biological cycle, many organisms degrade compounds that can be harmful to themselves as well as to other species, helping to diminish environmental negative impact of pollution. Bioremediation is a process of controlled biodegradation promoted by man and has been proposed as a good alternative for environmental remediation as well as for pre-treatment of effluents. [41-46] Nevertheless, there are several disadvantages related to the use of foreign organisms in direct contact with the natural ambience. [47] For instance, the introduction of exotic species implies an ecological inherent risk, promoting imbalances with regard to the growth of indigenous species. In contrast, the chosen organism can be inhibited by the native ones. Additionally, an inherent limitation of bioremediation arises from the tolerance limits in which the active microorganism can degrade the pollutant. Above some critical level of pollutant concentration there is a drastic decline in viability, limiting the biodegradation process. These limitations usually hinder the possibility of straight *in situ* treatment. [48]

Efforts have been made in order to improve bioremediation processes by immobilization of active organisms in different supports such as spheres of Ca(II)-alginate, [49] synthetic biodegradable materials [50] and steel sponges.[51] Nevertheless, to achieve a significant advance it is necessary to assure the isolation of the organism, which is achieved by means of the encapsulation in a physical and chemically stable matrix.

As the two step encapsulation procedure allows cellular growth and division, it provides optimal conditions for the encapsulation of living multicellular organisms. This opened up the possibility of designing an assembly for bioremediation of polluted water based on the immobilization of alginate beads loaded with filamentous fungus *Stereum hirsutum* inside nanoporous silica hydrogels. [52] The resulting MAB exhibits good physical, chemical and biological stability, and also was effective in the removal and degradation of a textile dye, even with high pollutant concentration. The efficiency and performance of each remediating MBA is a consequence of the synergic effect between the metabolic functions of the biological guest and the transport properties of the silica matrix. The optimal structure of the hydrogel for an efficient dye–enzyme encounter resulted from the fine adjustment of synthesis conditions in order to achieve a suitable porosity, regulation of the adsorption and transport of the dye, as well as the retention of dye degradation enzymes inside the hydrogel.

Within this context, the challenge is the design and synthesis of a tough non-biodegradable matrix with tuned porosity in order to avoid accumulation of contaminants above the toxicity level accepted by the working biological host. This was accomplished through the encapsulation of *Stereum hirsutum*, a white rot basidiomycete distributed worldwide, which is known to degrade lignin in wood by action of extracellular oxidases. [53-55] This model organism has been successfully employed for the degradation of several organics due to the broad substrate specificity of its ligninolytic enzymes such as lignin peroxidases, manganese peroxidases and laccases. [56-60] To the best of our knowledge, this was the first proposal of bioremediation of organic pollutants with microorganisms encapsulated in mineral cages.

These results open the possibility of bioremediation without dissemination of exotic organisms to the environment, and can be extended to a vast variety of strains [37,61] due to the inherent high biocompatibility of the present procedure. Although natural or waste waters are complex multicomponent systems, it is possible to select a specific microorganism for degradation of a certain pollutant and control the delivery of noxious compounds below toxicity levels. In addition, physical isolation of microorganisms also makes the bioremediation of complex systems possible by simultaneous use of different species, each being sensitive to particular pollutants.

## IMPROVING THE PROPERTIES OF THE HOST MATRIX

The encapsulation of living cells inside inorganic matrices implies an intrinsic limitation in synthesis conditions, as the overall process must evolve within biocompatible chemical constraints imposed by the encapsulated cell strain.[39] On the other hand, depending on the application for which the biomaterial is developed, some macroscopic properties of such inorganic hydrogels are mandatory. For instance, materials meant for long-term biotechnological applications have to maintain their mechanical properties for several months. Absence of cracking, shrinkage or syneresis above the nanoscale is a prerequisite to guarantee the effectiveness of silica gel as a barrier to prevent both the entrance of contaminating microorganisms and the exit of cells of interest to the external medium. In the design of certain applications such as biosensors based on optical detection systems or devices based on encapsulation of photosynthetic cells, optical properties of the host matrix are of a crucial importance.[62] Porosity, in turn, plays an important role in the input and output of nutrients, biodegradable species and metabolic products. Moreover, as stated in the preceding section, the advanced functionality of the MBA combines the biological functions of the guest with specific functions of the host matrix. The latter, far from being an inert support, can also play a key role in recognition, discrimination, selective permeation and adjustment in response to changes in the environment.

The hydrogel properties depend on its microstructure, which can be controlled by sol–gel synthesis variables.[63] In particular, transport and adsorption can be tuned by changing the surface charge of the inorganic hydrogel, which is achieved by adjusting the synthesis pH or modifying the surface with additives. On the other hand, the achievement of materials with good macroscopic properties often produces death or stress due to the release of toxic by-products originated during the synthesis.[30] Thus, development of sol-gel based processes for the encapsulation of living cells is a challenging task where the host material must be optimized both in terms of intrinsic properties and impact on cell viability. Optimization of the sol–gel process in a cellular context is possible through a combination of chemical, physical and biological approaches.[16,27,64-68]

By means of the indirect procedure, and due to the protection provided by the Ca(II)-alginate pre-encapsulation, silica synthesis can be attempted in more cytotoxic conditions, allowing the use of a wider range of synthesis pH, ionic strength and precursor's concentrations than in one-pot encapsulation procedures. This approach should contribute to the design of MBAs with exquisitely regulated properties in order to adapt to specific requirements of particular applications. Furthermore, as the biological system is spatially separated from the inorganic sol-gel synthesis, the incorporation of toxic synthesis additives is also possible. In a recent work, the incorporation of $CeO_2$ nanoparticles as UV absorptive entities within the inorganic matrix, acting as an inner cutoff filter was proposed. These (photo)-chemically stable absorbing entities can result toxic in direct contact with cells, but being retained within the silica matrix they offer safe protection to encapsulated cells both in terms of harmful UV light filtering and reactive free radicals scavengeing, while ensuring their viability or even growth for periods of several months.[69]

The present findings open the gate to the formulation of more robust hydrogels probing that it is possible to further widen the structural and chemical diversity of inorganic matrices that are compatible with the encapsulation of living cells.

## BEYOND SILICA

As the variety of entrapped strains and targeted applications is expanding, novel host materials are necessary, which may require the use of other metal oxides.[16,70] Iron or titanium oxides are envisaged as good candidates for encapsulation, especially if positive charged surfaces are desired in a wide pH range. [71,72] The possibility to extend this approach to other sol–gel materials has recently been addressed, with promising yet moderate success.[73]

Inorganic matrices made from boehmite and ferrihydrite colloids allowed a long-term preservation of *Escherichia coli* bacteria but suffer from an important loss of viability in the first hour post-encapsulation. [71,72] Titanium oxide and zirconia matrices have been successfully applied to yeast cells encapsulation. [74-76] Finally, zirconium oxyhydroxide colloids were demonstrated to be unsuitable for bacteria encapsulation due to rapid cell death. [73] Indeed, several parameters can influence cell viability, such as gelation time, [77] ionic strength [30] or cell/matrix interactions.[16] However, in the case of non-silica hosts, there is an additional detrimental effect arising from the cytotoxicity of the initial inorganic precursor solution. [73]

Recently, the two-step encapsulation process relying on cell encapsulation in alginate beads followed by inorganic gelation from colloidal metal oxides was successfully applied to the immobilization of *Escherichia coli* in the presence of boehmite and zirconium oxyhydroxide particles.[78]

The pre-encapsulation in Ca(II)-alginate avoids the detrimental effect of metal oxide colloidal solutions on *E. coli* viability during encapsulation. This is possible through an optimization of the bead structure and composition, both having an impact on its diffusion properties, as nanotoxicity arises not only by the direct contact between cells and the nanoparticle itself but also with its solubility products. In the case of Al(III)-based gels, the alginate bead obtained at low biopolymer content provides an efficient barrier against the encapsulation stress.

In contrast, an increase in the alginate concentration together with the phosphate induced mineralization of the biopolymer bead is found necessary to maintain the viability of entrapped bacteria in Zr(IV)-based gels. Diffusion studies using model molecular and colloidal species put in evidence that positively charged Zr(IV)-oligomers and $ZrO_2$ nanoparticles may be involved in the cytotoxicity of the precursor solution. This optimization of the encapsulation process allows, for the first time to our knowledge, the growth of *E. coli* within the non-silica hybrid matrix, opening the route to the design of novel functional cell-based bioreactors.

These results give a clear evidence that sol–gel based cell encapsulation can now be envisioned within a wide variety of metal oxide hosts through the optimization of the pre-encapsulation environment. Therefore, it becomes possible to extend the range of inorganic environments for the design of cell-based bioreactors based on this technology. In particular, it is now important to study alternative host materials that are currently not obtained in cytocompatible conditions (i.e. far from neutral pH, high ionic strength, organic additives) to establish whether this two-step process can, after further optimization, provide an efficient protection allowing long-term preservation of cellular activity.

## Conclusion

Encapsulation of cells and microorganisms in silica matrices has reached a status that puts up the biological machinery at work for producing value-added chemicals, or for designing high performance devices for remediation or sensing.[79,80] The synthesis of silica by the sol-gel route can be made fulfilling the green chemistry principles,[81] thus encapsulation in these matrices has the additional advantage of an environmentally friendly procedure with a non-biodegradable cage. In particular, the two step procedure provides a steady state population of an ensemble of living cells, allowing cell-to-cell communication. These high cellular density MBAs can be envisaged for the design of modular bioreactors for the development of medicinal products from secondary metabolites synthesised by transgenic organisms.

Additionally, the two-step procedure creates a robust MBA for applications in open environments where foreign microorganisms are not predated by those present in the eco-system, and prevents any possible alteration produced by the introduction of foreign species. By this way, bioremediating devices can be thought for in-situ cleaning up of pollutants in rivers, lakes and seas, as for example oil spills or mining toxic wastes by choosing the appropriate microorganisms.

Among the possible applications envisaged for MBAs, the encapsulation of photosynthetic organisms opens the possibility to exploit the action of thylakoids, chloroplasts, and whole plant cells for the conversion of $CO_2$ into value added biomass, [82] as well as for the production of hydrogen [83] or gold nanoparticles. [84]

Encapsulation in silica is an intrinsic biomimetic approach. Like diatoms, MBAs can be conceived as modules with wide nanobiotechnological applications in sensors, solar cells, batteries, and electroluminescent devices.[85] The protection provided by Ca(II)-alginate pre-encapsulation step broadens the biological diversity of possible cellular guests, as schematized in Figure 3.

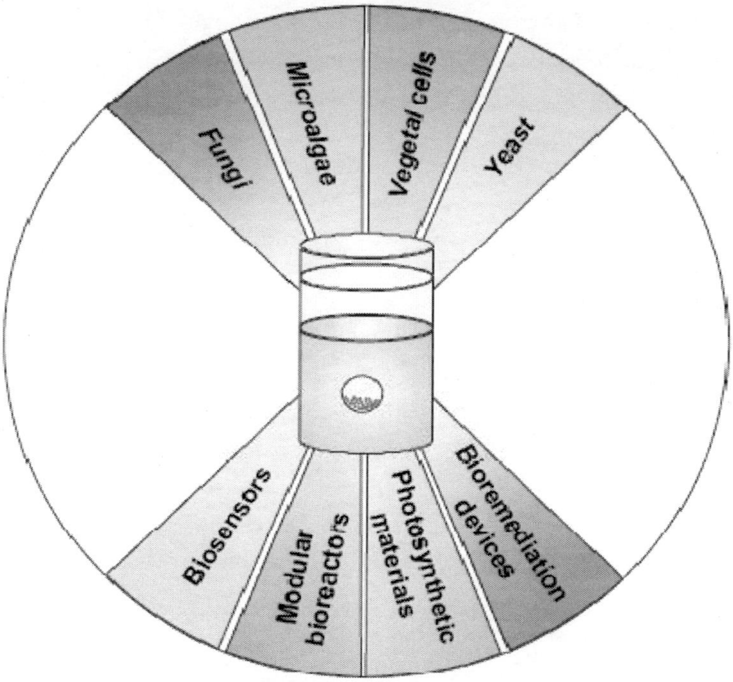


Figure 3. Examples of application of MBAs synthesized by the two-step procedure as biotechnological tools. [30, 40, 52, 69].

Nowadays, the knowledge in sol-gel processes provides a valuable tool for choosing synthesis precursors in order to obtain a high viability and a tailored mineral structure for optimal transport of nutrients and metabolites. Within this framework we can now conceive a scope of biotechnological developments that are mostly limited by our imagination.

## REFERENCES


[1] Nassif, N.; Livage, J. *Chemical Society Reviews* 2011, *40*, 849.

[2] Pierre, A. C. *Biocatalysis and Biotransformation* 2004, *22*, 145.

[3] Gill, I.; Ballesteros, A. *Trends in Biotechnology* 2000, *18*, 282.

[4] Livage, J.; Coradin, T.; Roux, C. *Journal of Physics Condensed Matter* 2001, *13*, R673.

[5] Avnir, D.; Lev, O.; Livage, J. *Journal of Materials Chemistry* 2006, *16*, 1013.

[6] Böttcher, H.; Soltmann, U.; Mertig, M.; Pompe, W. *Journal of Materials Chemistry* 2004, *14*, 2176.

[7] Davis, M. E. *Nature* 2002, *417*, 813.

[8] Premkumar, J. R.; Sagi, E.; Rozen, R.; Belkin, S.; Modestov, A. D.; Lev, O. *Chemistry of Materials* 2002, *14*, 2676.

[9] Coiffier, A.; Coradin, T.; Roux, C.; Bouvet, O. M. M.; Livage, J. *Journal of Materials Chemistry* 2001, *11*, 2039.

[10] Nassif, N.; Roux, C.; Coradin, T.; Bouvet, O. M. M.; Livage, J. *Journal of Materials Chemistry* 2004, *14*, 2264.

[11] Nassif, N.; Roux, C.; Coradin, T.; Rager, M. N.; Bouvet, O. M. M.; Livage, J. *Journal of Materials Chemistry* 2003, *13*, 203.

[12] Ferrer, M. L.; Yuste, L.; Rojo, F.; Del Monte, F. *Chemistry of Materials* 2003, *15*, 3614.

[13] Carturan, G.; Campostrini, R.; Diré, S.; Scardi, V.; De Alteriis, E. *Journal of Molecular Catalysis* 1989, *57*.

[14] Pressi, G.; Toso, R. D.; Monte, R. D.; Carturan, G. *Journal of Sol-Gel Science and Technology* 2003, *26*, 1189.

[15] Klipp, E.; Nordlander, B.; Kruger, R.; Gennemark, P.; Hohmann, S. *Nature Biotechnology* 2005, *23*, 975.

[16] Nassif, N.; Bouvet, O.; Rager, M. N.; Roux, C.; Coradin, T.; Livage, J. *Nature Materials* 2002, *1*, 42.

[17] Ferrer, M. L.; García-Carvajal, Z. Y.; Yuste, L.; Rojo, F.; Del Monte, F. *Chemistry of Materials* 2006, *18*, 1458.

[18] Baca, H. K.; Ashley, C.; Carnes, E.; Lopez, D.; Hemming, J.; Dunphy, D.; Singh, S.; Chen, Z.; Liu, N.; Fan, H.; López, G. P.; Brozik, S. M.; Werner-Washburne, M.; Brinker, C. J. *Science* 2006, *313*, 337.

[19] Perullini, M.; Jobbagy, M.; Soler-Illia, G.; Bilmes, S. A. *Chemistry of Materials* 2005, *17*, 3806.

[20] Kataoka, K.; Nagao, Y.; Nukui, T.; Akiyama, I.; Tsuru, K.; Hayakawa, S.; Osaka, A.; Huh, N. H. *Biomaterials* 2005, *26*, 2509.

[21] Boninsegna, S.; Bosetti, P.; Carturan, G.; Dellagiacoma, G.; Dal Monte, R.; Rossi, M. *Journal of Biotechnology* 2003, *100*, 277.

[22] Bienaime, C.; Barbotin, J. N.; Nava-Saucedo, J. E. *Journal of Biomedical Materials Research Part A* 2003, *67A*, 376.

[23] Boninsegna, S.; Dal Toso, R.; Dal Monte, R. *Journal of Sol-Gel Science and Technology* 2003, *26*, 1151.

[24] Coradin, T.; Nassif, N.; Livage, J. *Applied Microbiology and Biotechnology* 2003, *61*, 429.

[25] Premkumar, J. R.; Rosen, R.; Belkin, S.; Lev, O. *Analytica Chimica Acta* 2002, *462*, 11.

[26] Power, M.; Hosticka, B.; Black, E.; Daitch, C.; Norris, P. *Journal of Non-Crystalline Solids* 2001, *285*, 303.

[27] Depagne, C.; Roux, C.; Coradin, T. *Analytical Bioanalytical Chemistry* 2011, *400*, 965.

[28] Avnir, D.; Braun, S.; Lev, O.; Ottolenghi, M. *Chemistry of Materials* 1994, *6*, 1605.

[29] Premkumar, J. R.; Lev, O.; Rosen, R.; Belkin, S. *Advanced Materials* 2001, *13*, 1773.

[30] Perullini, M.; Jobbagy, M.; Moretti, M. B.; Garcia, S. C.; Bilmes, S. A. *Chemistry of Materials* 2008, *20*, 3015.

[31] Fernandes, P. M. B. *Brazilian Journal of Medical and Biological Research* 2005, *38*, 1239.

[32] Moon, K. H.; Honda, H.; Kobayashi, T. *Journal of Bioscience and Bioengineering* 1999, *87*, 661.

[33] Kieran, P. M.; MacLoughlin, P. F.; Malone, D. M. *Journal of Biotechnology* 1997, *59*, 39.

[34] Shuler, M. L.; Kargi, F. *Bioprocess engineering*; Prentice Hall PTR: New York, 2002.

[35] Su, W. W.; He, B. J.; Liang, H.; Sun, S. *Journal of Biotechnology* 1996, *50*, 225.

[36] Bourgaud, F.; Gravot, A.; Milesi, S.; Gontier, E. *Plant Science* 2001, *161*, 839.

[37] Onishi, N.; Sakamoto, Y.; Hirosawa, T. *Plant Cell Tissue and Organ Culture* 1994, *39*, 137.
[38] Vaněk, T.; Valterová, I.; Vănková, R.; Vaisar, T. *Biotechnology Letters* 1999, *21*, 625.
[39] Kuncova, G.; Podrazky, O.; Ripp, S.; Trögl, J.; Sayler, G. S.; Demnerova, K.; Vankova, R. *Journal of Sol-Gel Science and Technology* 2004, *31*, 335.
[40] Perullini, M.; Rivero, M. M.; Jobbagy, M.; Mentaberry, A.; Blimes, S. A. *Journal of Biotechnology* 2007, *127*, 542.
[41] Bai, R. S.; Abraham, T. E. *Bioresource Technology* 2003, *87*, 17.
[42] Bettmann, H.; Rehm, H. J. *Applied Microbiology and Biotechnology* 1984, *20*, 285.
[43] Field, J. A.; Dejong, E.; Costa, G. F.; Debont, J. A. M. *Applied and Environmental Microbiology* 1992, *58*, 2219.
[44] Cabana, H.; Jones, J. P.; Agathos, S. N. *Engineering in Life Sciences* 2007, *7*, 429.
[45] McMullan, G.; Meehan, C.; Conneely, A.; Kirby, N.; Robinson, T.; Nigam, P.; Banat, I. M.; Marchant, R.; Smyth, W. E. *Applied Microbiology and Biotechnology* 2001, *56*, 81.
[46] Reddy, C. A. *Current Opinion in Biotechnology* 1995, *6*, 320.
[47] Boopathy, R. *Bioresource Technology* 2000, *74*, 63.
[48] Cabana, H.; Alexandre, C.; Agathos, S. N.; Jones, P. *Bioresource Technology* 2009, *100*, 3447.
[49] Arica, M. Y.; Kaçar, Y.; Genç, Ö. *Bioresource Technology* 2001, *80*, 121.
[50] Bennett, J.; Childress, A.; Wunch, K.; Connick Jr., W. US 6204049, 2001; Vol. 6204049.
[51] Rodríguez Couto, S.; Sanromán, M. A.; Hofer, D.; Gübitz, G. M. *Bioresource Technology* 2004, *95*, 67.
[52] Perullini, M.; Jobbagy, M.; Mouso, N.; Forchiassin, F.; Bilmes, S. A. *Journal of Materials Chemistry* 2010, *20*, 6479.
[53] Mouso, N.; Diorio, L.; Forchiassin, F. *Acción de Stereum hirsutum (Wild) Pers. en la degradación de colorantes* 2007, *24*, 294.
[54] Barr, D. P.; Aust, S. D. *Environmental Science and Technology* 1994, *28*, 78A.
[55] Schreiner, K. M.; Filley, T. R.; Blanchette, R. A.; Bowen, B. B.; Bolskar, R. D.; Hockaday, W. C.; Masiello, C. A.; Raebiger, J. W. *Environmental Science and Technology* 2009, *43*, 3162.
[56] Levin, L.; Viale, A.; Forchiassin, A. *International Biodeterioration and Biodegradation* 2003, *52*, 1.
[57] Levin, L.; Castro, M. A. *Iawa Journal* 1998, *19*, 169.
[58] Pointing, S. B. *Applied Microbiology and Biotechnology* 2001, *57*, 20.
[59] Wesenberg, D.; Kyriakides, I.; Agathos, S. N. *Biotechnology Advances* 2003, *22*, 161.
[60] D'Annibale, A.; Rosetto, F.; Leonardi, V.; Federici, F.; Petruccioli, M. *Applied and Environmental Microbiology* 2006, *72*, 28.
[61] Green, D. W.; Leveque, I.; Walsh, D.; Howard, D.; Yang, X.; Partridge, K.; Mann, S.; Oreffo, R. O. C. *Advanced Functional Materials* 2005, *15*, 917.
[62] Hanh, N. N.; Durrieu, C.; Canh, T. M. *Ecotoxicology and Environmental Safety* 2009, *72*, 316.
[63] Perullini, M.; Jobbágy, M.; Bilmes, S. A.; Torriani, I. L.; Candal, R. J. *Journal of Sol-Gel Science and Technology* 2011, *59*, 174.
[64] Perullini, M.; Amoura, M.; Roux, C.; Coradin, T.; Livage, J.; Japas, M. L.; Jobbagy, M.; Bilmes, S. A. *Journal of Materials Chemistry* 2011, *21*, 4546.

[65] Kempf, B.; Bremer, E. *Archives of Microbiology* 1998, *170*, 319.
[66] Brisson, D.; Vohl, M. C.; St-Pierre, J.; Hudson, T. J.; Gaudet, D. *Bioessays* 2001, *23*, 534.
[67] Cayley, S.; Record, M. T. *Biochemistry* 2003, *42*, 12596.
[68] Capp, M. W.; Pegram, L. M.; Saecker, R. M.; Kratz, M.; Riccardi, D.; Wendorff, T.; Cannon, J. G.; Record, M. T. *Biochemistry* 2009, *48*, 10372.
[69] Sicard, C.; Perullini, M.; Spedalieri, C.; Coradin, T.; Brayner, R.; Livage, J.; Jobbagy, M.; Bilmes, S. A. *Chemistry of Materials* 2011, *23*, 1374.
[70] Toncheva-Panova, T.; Ivanova, J.; Sholeva, M.; Chernev, G.; Samuneva, B. *Comptes Rendus De L Academie Bulgare Des Sciences* 2008, *61*, 211.
[71] Amoura, M.; Brayner, R.; Perullini, M.; Sicard, C.; Roux, C.; Livage, J.; Coradin, T. *Journal of Materials Chemistry* 2009, *19*, 1241.
[72] Amoura, M.; Nassif, N.; Roux, C.; Livage, J.; Coradin, T. *Chemical Communications* 2007, 4015.
[73] Amoura, M.; Roux, C.; Masse, S.; Steunou, N.; Coradin, T. *Comptes Rendus de l'Academie des Sciences - Series IIc: Chemistry* 2010, *13*, 52.
[74] Kessler, V. G.; Seisenbaeva, G. A.; Unell, M.; Håkansson, S. *Angewandte Chemie - International Edition* 2008, *47*, 8506.
[75] Sannino, F.; Pirozzi, D.; Aronne, A.; Fanelli, E.; Spaccini, R.; Yousuf, A.; Pernice, P. *Environmental Science and Technology* 2010, *44*, 9476.
[76] Pirozzi, D.; Fanelli, E.; Aronne, A.; Pernice, P.; Mingione, A. *Journal of Molecular Catalysis B-Enzymatic* 2009, *59*, 116.
[77] Alvarez, G. S.; Desimone, M. F.; Diaz, L. E. *Applied Microbiology and Biotechnology* 2007, *73*, 1059.
[78] Perullini, M.; Amoura, M.; Jobbagy, M.; Roux, C.; Livage, J.; Coradin, T.; Bilmes, S. A. *Journal of Materials Chemistry* 2011, *21*, 8026.
[79] Lèonard, A.; Dandoy, P.; Danloy, E.; Leroux, G.; Meunier, C. F.; Rooke, J. C.; Su, B. L. *Chemical Society Reviews* 2011, *40*, 860.
[80] Brayner, R.; Coute, A.; Livage, J.; Perrette, C.; Sicard, C. *Analytical and Bioanalytical Chemistry* 2011, *401*, 581.
[81] Baccile, N.; Babonneau, F.; Bejoy, T.; Coradin, T. *Journal of Materials Chemistry* 2009, *19*, 8537.
[82] Rooke, J. C.; Meunier, C.; Lèonard, A.; Su, B. L. *Pure and Applied Chemistry* 2008, *80*, 2345.
[83] Dickson, D. J.; Page, C. J.; Ely, R. L. *International Journal of Hydrogen Energy* 2009, *34*, 204.
[84] Sicard, C.; Brayner, R.; Margueritat, J.; Hémadi, M.; Couté, A.; Yéprémian, C.; Djediat, C.; Aubard, J.; Fiévet, F.; Livage, J.; Coradin, T. *Journal of Materials Chemistry* 2010, *20*, 9342.
[85] Jeffryes, C.; Campbell, J.; Li, H.; Jiao, J.; Rorrer, G. *Energy and Environmental Science* 2011, *4*, 3930.

Hydrogels: Synthesis, Characterization and Applications
Editors: F. V. Câmara and L. J. Ferreira
ISBN: 978-1-61942-842-3


*Chapter 13*

# TRIBOLOGICAL PROPERTIES OF POLYVINYL ALCOHOL HYDROGELS AS ARTIFICIAL ARTICULAR CARTILAGE

***V. M. Sardinha[1], W. D. Belangero[2,4], V. P. Bavaresco[3,4] and J. R. Gomes[1,5]***

[1]Centro de Tecnologias Mecânicas e de Materiais (CT2M), Guimarães, Portugal

[2]Universidade Estadual de Campinas (UNICAMP), Departamento de Ortopedia e Traumatologia, Faculdade de Medicina, Campinas – SP, Brasil

[3]Universidade Estadual de Campinas (UNICAMP), Colégio Técnico de Campinas, Departamento de Plásticos, Campinas – SP, Brasil

[4]Instituto Nacional de Biofabricação - INCT-BIOFABRIS, Campinas – SP, Brasil

[5]Universidade do Minho, Departamento de Engenharia Mecânica, Guimarães, Portugal

## ABSTRACT

An exhaustive scientific investigation has been performed on polymeric hydrogels as potential materials to be used in cartilage tissue repair or replacement. Polyvinyl alcohol (PVA) possesses excellent properties as biocompatibility, low coefficient of friction, low interfacial tension and high permeability to fluids, emerging as an eligible material to be studied for biomedical applications.

The main goal of this study was the tribological characterization of PVA hydrogels in lubricated sliding with distilled water and phosphate buffered saline solution (PBS). The influence of hydrogels mechanical properties and operating conditions were considered. Tribological tests were performed on a pin-on-plate tribometer with a linear reciprocating geometry using the PVA/stainless steel 316L and PVA/bovine articular cartilage pairs. Experiments were performed at 37 °C with a constant reciprocating sliding frequency of 1 Hz and applied load of 1 MPa, 2 MPa and 3.5 MPa. For each experiment the friction coefficient was determined and the dominating wear mechanisms analysed by scanning electron microscopy and EDS. PVA showed an excellent performance as possible material for the repair of articular cartilage due to the high level of surface preservation and the very low friction coefficient presented when PBS was used as fluid lubricant. The performed tribological study revealed the high potentiality of PVA hydrogels to be used as artificial articular cartilage.

## 1. INTRODUCTION

The articular cartilage has a very limited self-repair ability due to the low vascularity of the tissue and low mitotic activity of chondrocytes, which turns out to be an extremely attractive target of study. The rupture of cartilage manifests itself through severe pain, reduction or loss of articular mobility and high medical costs. Every year around 3 millions of articular cartilage surgeries are performed owing to sportive injury, accidents or other traumatic problems on articular joints [1-3]. Unfortunately, the invasiveness of surgery processes and the complex properties of articular cartilage make difficult to perform success surgical techniques failing in the total repair of the tissue as well as on its function [1].

Recently an intensive study over polymeric hydrogels as synthetic polymers to be used as artificial articular cartilage has being followed according to the excellent properties of these materials. Biocompatiblity, good compliance and high swelling capacity are crucial properties to reproduce the human tissue and to ensure an excellent performance [4-8].

The viscoelastic behaviour of crosslinked and non degradable polyvinyl alcohol (PVA) make this polymer an optimal option as bioartificial material. PVA has outstanding properties as biocompatibility, chemical stability and good elasticity, hydrophilicity, low cost and many others [3,5,9,10]. PVA hydrogel is then of great matter on biomedical applications and has been studied for many authors.

According to Ratner et al., Bray and Merrill produced a synthetic articular cartilage based on PVA to repair a sinovial joint [4]. Covert et al. [11] studied the mechanical properties of a PVA cryogel as a potential biomaterial for regeneration of articular cartilage, considering the influence of parameters like temperature, lubricant, hardness of the material, load, sliding speed and surface roughness. Bavaresco et al. [12] showed the contribution of the swelling capacity of PVA hydrogels on the reduction of friction coefficient and the dependence of this behaviour on sliding speed. The reinforcement of PVA hydrogels with hydroxyapatite nanoparticles was performed by Gang et al. [6]. The author studied the tribological behaviour of the hydrogel against stainless steel using a pin-on-disc geometry tribometer, with distilled water as the lubricant fluid. Many other studies were developed around PVA [9,13,14] including different methods of crosslinking as a way to reduce the friction coefficient of the hydrogel.

Considering the many existing classes of polymeric hydrogels, the electron beam radiation crosslinked PVA hydrogel has showed good physico-chemical and mechanical properties on many performed studies [4,12,15]. These properties can be controlled, in part, through the crosslinking method and swelling capacity. The PVA hydrogels compliance is the main responsible for shock absorbance, also depending on crosslinking degree [15-17]. The radiation crosslinking method is well known for being a simple method which ensure the sterilization of the membranes and does not require a crosslinking agent, possibly toxic, expanding the variety of applications of the polymer on biological and biomedical areas [18-22].

The primary goal of this research was to investigate the tribological behaviour of electron beam radiation crosslinked membranes of PVA hydrogels as potential artificial articular cartilage when sliding against 316L stainless steel and bovine articular cartilage plates. The specific objectives of this research included the evaluation of friction coefficient evolution using an oscillating pin-on-plate device, determination of the effect of normal load and lubrication fluid on friction and wear of PVA hydrogels, and analysis of predominant wear mechanisms.

## 2. EXPERIMENTAL

In this study PVA hydrogel membranes were produced and many techniques were used in order to characterise the mechanical and tribological properties of PVA hydrogels, including water swelling, indentation creep test, reciprocating sliding experiments, scanning electron microscopy (SEM) and energy dispersive X-ray spectroscopy (EDS).

### 2.1. Production of PVA Hydrogel Membranes

PVA hydrogel membranes were processed from PVA aqueous solutions (Aldrich Mw 89000-98000 g/mol, 99% hydrolized) with 10% (w/w) concentration of polymer in solution. Thc polymer was added to distilled water at 80 °C and the solution was poured into Petri plates with a 30 cm diameter until total cool down. The membranes were then submitted to a chemical treatment of acetalization by immersion in a solution of formaldehyde, sulphuric acid and anhydrous sodium (60:50:300 g). After this, the produced PVA membranes were crosslinked by ionizing radiation of electrons beam in oxygen free atmosphere using a dose of 25 kGy in a Dynamitron electron accelerator (Radiation Dynamics, Inc). The aim of the process was to ensure the formation of chemical bonds between polymer molecules closing the polymer chains and improving the membranes mechanical properties. PVA hydrogel membranes, having approximately 1 mm thickness and 8 mm diameter were obtained and used in the present study.

### 2.2. Equilibrium Water Content

The equilibrium water content (*EWC*) of the used PVA hydrogel membranes was determined by immersing dried test samples previously weighed, in distilled water at room temperature until reaching a constant weight measure, meaning the equilibrium water content achievement. Samples were then removed from the swelling medium and the weight of the swollen membranes was measured allowing the calculation of the *EWC* through Eq. (1):

$$EWC(\%) = \frac{W_{hh}-W_{dh}}{W_{hh}} \times 100 \tag{1}$$

Where $W_{hh}$ is the weight of hydrated hydrogel and $W_{dh}$ is the weight of dried hydrogel. EWC was determined by the average of five measurements.

## 2.3. Indentation Creep Test

To perform the indentation creep test an MTS – Teststar II equipment was used. The PVA hydrogel samples were immersed in distilled water and indented using a spherical tip with a diameter of 3.2 mm. The load of 4.91 N (0.5 kgf) was constant and normally applied on the sample for 180 s. After this period of time the load was immediately removed. The indentation height ($h$) was measured during the test. The indentation creep modulus was calculated based on Kempson's equation [16,23] (Eq. (2)):

$$E = \frac{9\times10^4 p}{16\sqrt{r}}\left[\frac{1-\exp\,(-0.42\times e/a}{h}\right]^{3/2} \qquad (2)$$

where $E$ is the creep modulos (kgf/cm$^2$); $p$ the applied load (kgf); $r$ the indenter radius (cm); $e$ the sample thickness (cm), $h$ the indenter height (cm) and $a = \sqrt{(2rh - h^2)}$.

## 2.4. Morphological Characterization

The produced PVA hydrogel membranes were morphologically characterized by Scanning Electron Microscopy (SEM - FEI Nova 200). The analysis was performed on samples gold sputter-coated and the magnifications ranged between 250, 500 and 2000 times to a magnification equal or superior to 4000 times for relevance details.

## 2.5. Preparation of Pins and Plates

Pin samples for friction and wear tests were prepared by using a stainless steel 316L cylindrical substrate with 8 mm diameter, cut with 35 mm length for each pin. Pins were then partially coated with PVA hydrogel pellicles with 4 mm diameter, previously washed in distilled water, using a small drop of powerful glue (Loctite® 401). Coated pins were immersed in distilled water in order to keep the level of hydration of the hydrogel until tribological tests.

Plates of stainless steel 316L with 6 mm thickness were machined and the surface destined to tribological interaction was polished finishing with 1 µm diamond paste. The surface roughness of the plates was measured showing an average value of 0,028 µm $R_a$. Before tribological tests plates were washed with water and detergent and then with isopropanol. After, the excess solvent was evaporated in hot air flow.

The articular cartilage plates were harvested aseptically from adult bovine knee joints being tested in a maximum period of 24 h after slaughter. The samples were washed in distilled water for approximately 5 minutes and kept at 4 °C until use.

## 2.6. Tribological Tests

The tribological tests were performed in a reciprocating pin-on-plate tribometer, Plint TE67/R (*Phoenix Tribology Ltd,* UK), at physiological temperature, around 37 °C ± 1 °C, in order to reproduce the human body conditions. Temperature was measured and controlled using a thermocouple located in the lubricant bath. Two different lubricants were used, namely distilled water and phosphate buffered saline (PBS) (Sigma-Aldrich). The sliding occurred for two distinct contact pairs: hydrogel/bovine articular cartilage and hydrogel/316L stainless steel. An aluminum container was used to maintain the samples submerged in the lubricant medium. To perform the tribological tests, the frequency of the reciprocating motion of the plate and the stroke length were fixed at 1 Hz and 8 mm, respectively. Different normal applied loads were used in order to obtain contact pressure values ranging from 1 MPa to 3.5 MPa. Figure 1 shows the schematic representation of mounting and positioning of the different elements involved in the tribological arrangement.

Tribological tests had duration of 2 hours and the friction coefficient was continuously monitored during sliding. After testing, the worn surfaces were analysed by SEM in order to identify the dominant wear mechanisms. Reference tribological tests with bovine articular cartilage against itself were also done following exactly the conditions described above.

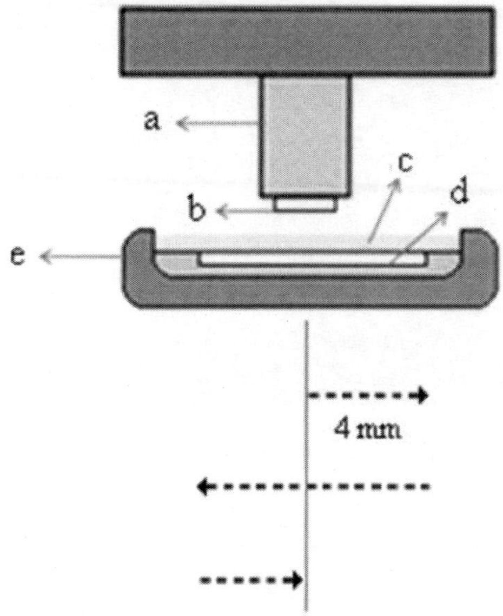


Figure 1. Schematic representation of mounting and positioning of the different elements for tribological tests and illustration of reciprocating movement with a stroke length of 8 mm. (a) 316L stainless steel pin, (b) PVA hydrogel membrane, (c) lubricant fluid, (d) counterface and (e) aluminum container.

# 3. Results and Discussion

An ideal material to replace articular cartilage should be biocompatible and present physico-mechanical properties similar to the tissue [14], which are responsible for its compressive resistance and for the ability to expulse fluids to act as lubricants on contact surfaces. One great advantage of using PVA as synthetic polymer has to do with the possibility to control the synthesis and crosslinking processes of the material, and then, the hardness and swelling capacity of these membranes. Articular cartilage mechanical properties are mainly determined by its water content, since it is the major component of the tissue [24,25]. According to this, the indentation creep modulus and the equilibrium water content of PVA samples were calculated, achieving a modulus of 1.61±0.4 MPa and *EWC* of 32% ±0.5%. The relatively low modulus ensures the necessary compliance for perfect adaptation of hydrogel membranes to counterface. On the other hand, the *EWC* is related to the absorption of an enough volume of water which is responsible for lubrication and protection of the opposing surface.

A morphological characterization of the unworn surface of PVA samples was performed by SEM with the purpose to evaluate the hydrogel surface and to have a basis of comparison for tested membranes. Figure 2 shows an unworn PVA membrane with a smooth and monotonous surface, revealing a vitreous appearance which allows the characterization of PVA as an amorphous material.

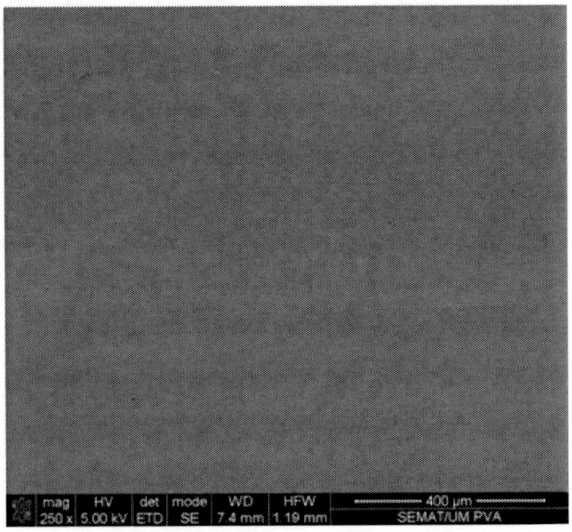


Figure 2. SEM micrograph of the unworn PVA hydrogel showing an homogeneous and amorphous appearance.

The use of compliant materials to replace natural articular cartilage seems to be beneficial on friction and wear reduction. Polymeric hydrogels are characterized for its high water content which ensures a very low friction coefficient due to water release from polymeric net when hydrogels are compressed against an opposing surface, leading to the formation of a protective film. For *in vitro* testing, when tribological sliding begins, a start-up friction tends to occur characterized by relatively high friction values, particularly if the hydrogel contacts a metallic counterface. This phenomenon is attributed to the rupture of asperities joints between the mating surfaces. After this, the friction coefficient decreases suddenly and a steady-state regime is established giving rise to very low friction values.

This standard friction behaviour occurred for the two tribological pairs under study: hydrogel/316L stainless steel and hydrogel/bovine articular cartilage. Table 1 presents the average values of start-up friction for PVA/316L stainless steel and PVA/bovine articular cartilage pairs, when sliding on distilled water ($dH_2O$) and PBS. Increasing contact pressure from 1 MPa to 2 MPa leads to a notorious decrease on start-up friction coefficient, which slightly increases when maximum contact pressure of 3.5 MPa is applied, independently of used lubricant fluid. The difference between distilled water and PBS on start-up friction coefficient evaluation is quite evident considering that PBS has higher viscosity than $dH_2O$, contributing to its own retention for longer periods of time on hydrogel matrix, retarding its expulsion as a result of load appliance. This helps reducing the direct contact between surfaces and consequently outcome in a smaller start-up friction coefficient. The difference observed between the contacts involving stainless steel and bovine articular cartilage is remarkable, which can be explained by the different physico-mechanical properties of the plate surface that interacts with the hydrogel membrane. In fact, articular cartilage is a viscoelastic tissue allowing a better adaptation to the PVA hydrogel counterface, while stainless steel is a rigid material, with no compliance, which increases the interaction between the mating surfaces resulting in higher values of start-up friction when compared to the PVA/bovine articular cartilage pair.

The steady-state friction coefficient ($\mu$) for different values of contact pressure (p=1 MPa; p=2 MPa and p=3.5 MPa) is shown in Figure 3 (a) for the pair PVA/316L stainless steel and in Figure 3 (b) for the pair PVA/bovine articular cartilage. From Figure 3 (a) it can be observed that for hydrogel/316L stainless steel pair in presence of distilled water the friction coefficient tended to stay high ($\mu \approx 0.24$ to 0.32). For 2 MPa contact pressure some plastic deformation was noticed at the contact surface of tested hydrogel membranes which started to elongate by the reciprocating action of the friction force. When the contact pressure increased for 3.5 MPa a very early rupture of the hydrogel membrane occurred at the region under tribological interaction, leading to the end of the tribological test and thus preventing the measurement of the steady-state friction coefficient. On the other hand, for the reciprocating sliding of hydrogel membrane against stainless steel in the presence of PBS as lubricant fluid, better friction results were achieved, particularly for 1 MPa and 2 MPa ($\mu \approx 0.01$) (Figure 3 (a)). In addition, a tendency for an increase on the steady-state friction coefficient for maximum contact pressure (3.5 MPa) was evidenced, which is attributed to a mechanical degradation of the hydrogel membrane, considering the superior hardness of 316L stainless steel counterface and the damage that can be imposed to the compliant hydrogel surface under high contact pressure.

**Table 1. Average values of start-up friction for tribological tests performed with PVA hydrogel against 316L stainless steel and bovine articular cartilage, sliding in distilled water and PBS for contact pressure values of 1 MPa, 2 MPa and 3.5 MPa**

| | PVA | | | |
|---|---|---|---|---|
| | 316L stainless steel | | Bovine articular cartilage | |
| Lubricant fluid / Contact pressure | $dH_2O$ | PBS | $dH_2O$ | PBS |
| 1 MPa | 0.34 ± 0,13 | 0.24 ± 0,09 | 0.09 ± 0,03 | 0.16 ± 0,07 |
| 2 MPa | 0.25 ± 0,09 | 0.14 ± 0,02 | 0.08 ± 0,03 | 0.07 ± 0,04 |
| 3.5 MPa | 0.26 ± 0,05 | 0.15 ± 0,02 | 0.10 ± 0,05 | 0.08 ± 0,04 |

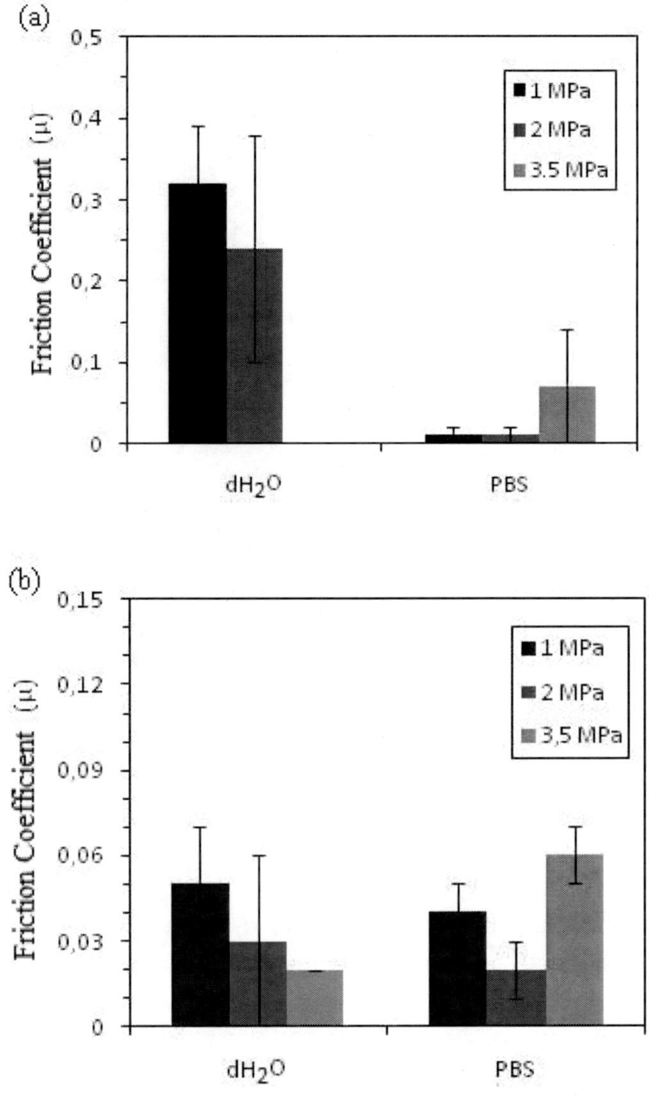


Figure 3. Steady-state friction coefficient versus lubricant fluid (distilled water ($dH_2O$) and PBS) according to different applied contact pressure values of 1 MPa, 2 MPa and 3.5 MPa for PVA hydrogel tested against (a) 316L stainless steel and (b) bovine articular cartilage.

Bovine articular cartilage was tribologicaly tested against itself in order to have an experimental value of friction coefficient to be used as basis of comparison. A contact pressure of 1 MPa was applied and PBS was used as the lubricant fluid. Very low friction coefficient values ranging between 0.001 and 0.01 were obtained, tending to the results reported in the literature for tests conducted with self-mated articular cartilage, where friction coefficient values as low as 0.003 were found [26,27].

From Figure 3 (b) it can be observed that when PVA hydrogel membranes were tested against bovine articular cartilage in distilled water or PBS the friction values are relatively close to the result previously mentioned for self-mated articular cartilage in PBS. In addition, there is a trend for the decreasing of steady-state friction coefficient with the increasing contact pressure.

The only exception occurred when the hydrogel/articular cartilage pair was tested in the presence of PBS under a contact pressure of 3.5 MPa, giving rise to a slightly higher friction value (μ≈0.06). This friction response can be justified by premature mechanical degradation of bovine articular cartilage plate since the subchondral bone, and then the resistance and support of the tissue, was lacking. Finally, comparing PBS and distilled water, smaller average friction coefficient values were achieved in the presence of PBS for contact pressures of 1 MPa (μ≈0.04) and 2 MPa (μ≈0.02). These results show the advantage of using PBS instead of distilled water as lubricant fluid on tribological tests to better simulate the natural tribosystem existing in natural articular joints.

Taking into consideration the very low steady-state friction coefficient values presented by both sliding pairs under study, it is expected that PVA hydrogel membranes also present good wear response.

Through SEM analysis, morphological wear features of contact surfaces were characterized and described. The type and nature of lubricant fluid in use directly influences the establishment of the lubrication regime so, the use of distilled water and PBS resulted in distinct damage effects at the hydrogel contact surfaces. The SEM micrograph presented in Figure 4 (a) is representative of the worn surface of PVA after sliding against 316L stainless steel in the presence of distilled water under 1 MPa contact pressure. The surface is relatively preserved, showing some abrasion marks with a slightly grooved surface. The magnification image (Figure 4 (b)) reveals the occurrence of an intensive plastic deformation with adhesive and abrasive wear promoted by the hard mating surface, which justifies the high friction coefficient obtained for this test condition (μ≈0.32). Similar wear features were obtained for 2 MPa contact pressure, but with the intensification of the degree of plastic deformation. For the higher contact pressure (3.5 MPa), the hydrogel suffered large scale deformation and rupture, the reason why no friction values were presented in Figure 3 (a) for this test condition and no morphological analysis of the hydrogel contact surface was performed.

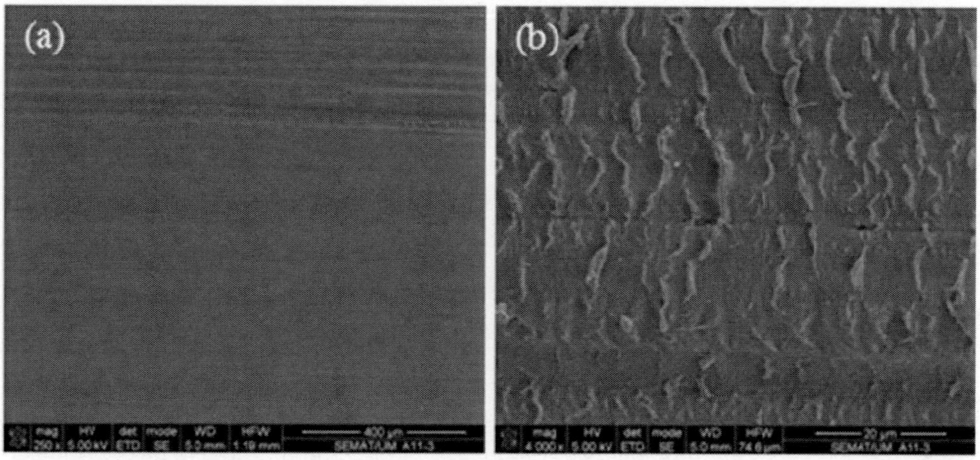


Figure 4. Worn surface of the PVA hydrogel tested against 316L stainless steel with distilled water as fluid lubricant under contact pressure of 1 MPa: (a) general appearance and (b) magnification view showing plastic deformation and abrasion grooves.

SEM micrographs of worn surfaces of PVA hydrogel membranes after reciprocating sliding against 316L stainless steel in the presence of PBS as lubricant medium for contact pressure values of 1 MPa, 2 MPa and 3.5 MPa are presented in Figure 5. From Figure 5 (a) and (b) it can be observed that the hydrogel contact surface remains well preserved, justifying the relatively low friction coefficient value obtained for 1 MPa contact pressure ($\mu \approx 0.01$). The magnified image presented in Figure 5 (b) shows signs of shallow abrasion grooves aligned parallel to the sliding direction. These grooves resulted from the tribological interaction with the harder opposing stainless steel surface. For the intermediate contact pressure of 2 MPa, a smooth surface alternates with grooved areas aligned parallel to the sliding direction revealing that the abrasion mechanism intensifies with the increasing of contact pressure (Figure 5 (c)). In fact, the increase of contact pressure induced surface damage leading to some removal and accumulation of polymeric wear particles on hydrogel surface. During sliding, free wear particles are dragged by reciprocating movement conducting to the formation of tribological rolls perpendicularly aligned to sliding direction (Figure 5 (c)). Tribological rolls present a high level of compactness and behave like micro-bearings, thus enabling the low friction values obtained in this test condition ($\mu \approx 0.01$). A high magnification of one of those tribological rolls is presented in Figure 5 (d), where an intensive plastic deformation and accumulation of wear particles can be observed around the roll. For the maximum contact pressure of 3.5 MPa, the worn surface of the hydrogel membrane (Figure 5 (e)) is characterized by similar features as described for sliding under 2 MPa contact pressure. Abrasion grooves aligned parallel to the sliding direction can be observed (Figure 5 (e) and (f)) as well as tribological rolls perpendicularly oriented to sliding direction. However, for this test condition, the presence of tribological rolls was scarcer than for the intermediate contact pressure and its aspect ratio (length/diameter) tends to be smaller (Figure 5 (c) and (e)), corroborating the increase of the steady-state friction coefficient for sliding under 3.5 MPa ($\mu \approx 0.07$) (Figure 3 (a)).

For PVA/bovine articular cartilage sliding pair only the morphology of worn surfaces corresponding to contact pressures of 1 MPa and 3.5 MPa are presented as similar features were observed independently of the contact pressure. From Figure 6 (a) it can be observed that when PVA was tested against bovine articular cartilage in distilled water with a contact pressure of 1 MPa, the hydrogel surface is well preserved, corroborating the low steady-state friction coefficient value around 0.05 (Figure 3 (b)) obtained for this test condition. With the increase of contact pressure from 1 MPa to 3.5 MPa the hydrogel surface remains preserved (Figure 6 (b)) and a decrease on friction coefficient was evidenced ($\mu \approx 0.02$). Lighter spots can be observed in Figure 6 (b) corresponding to thin adhesive tribolayers with origin in wear particles from PVA hydrogel. Those tribolayers with smooth appearance can act as protective films which help reducing the friction coefficient and preserving the smooth appearance of the whole contact surface. Thus, no significant changes were observed in terms of surface damage of the hydrogel, with the increasing of contact pressure. In fact, for both contact pressures, the hydrogel membranes exhibited a smooth and homogeneous worn surface without any signs of abrasion.

When the hydrogel was tested against bovine articular cartilage in PBS, the PVA membrane also presented a high level of surface preservation regardless of contact pressure. Figure 7 (a) and (b) illustrates SEM micrographs of PVA worn surfaces after testing against bovine articular cartilage in the presence of PBS for contact pressures of 1 MPa and 3.5 MPa, respectively. In both cases the high level of preservation of the hydrogel contact surface is notorious and comparable to the morphological features that characterize the unworn hydrogel surface presented in Figure 2. These observations corroborate the low steady-state friction value of 0.04 obtained for 1 MPa contact pressure (Figure 3 (b)) as well as the slightly higher friction value of 0.06 measured for sliding under 3.5 MPa. Thus, based on the obtained results, it can be suggested that PVA hydrogel/articular cartilage pair has high potential to be considered in future biomedical applications where the hydrogel is used as artificial articular cartilage.

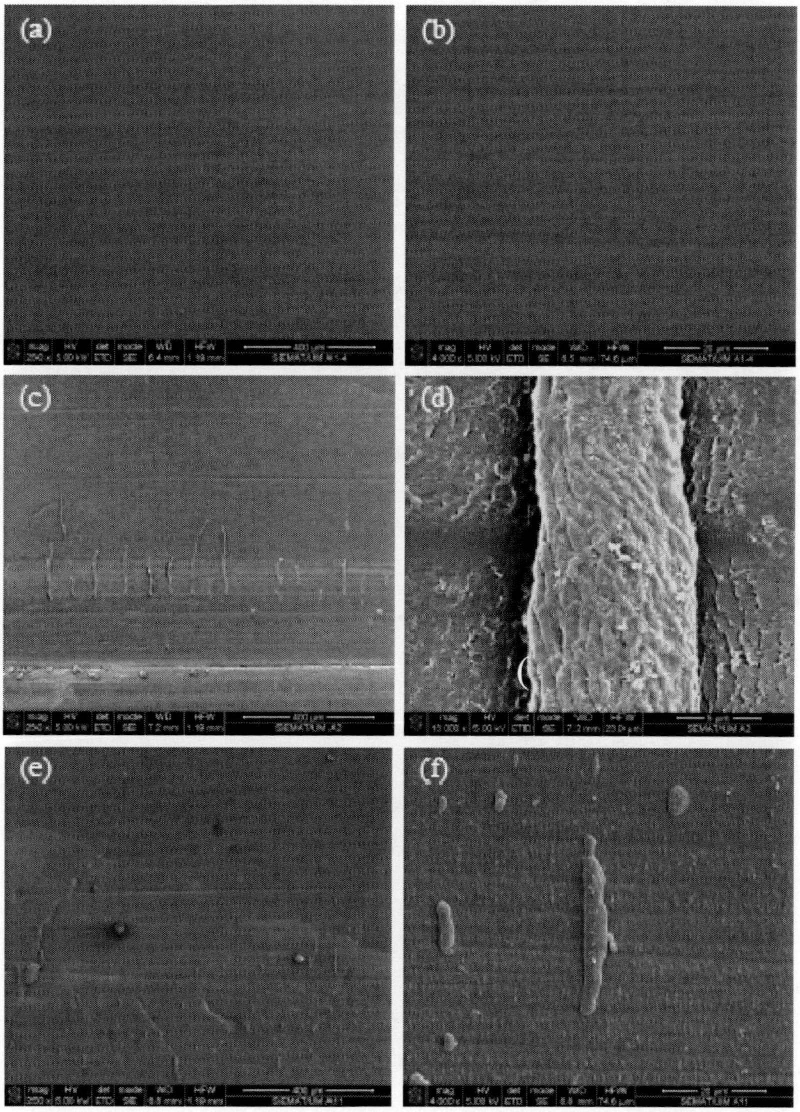


Figure 5. Worn surface of the PVA hydrogel tested against 316L stainless steel with PBS as lubricant fluid and contact pressure of 1 MPa (a) and (b), 2 MPa (c) and (d), 3.5 MPa (e) and (f). General appearance of tested surfaces (a), (c) and (e) and magnification of particular details (b), (d) and (f).

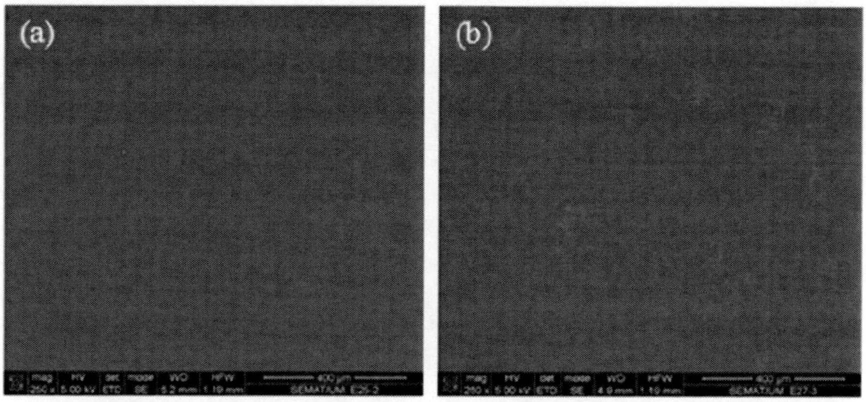


Figure 6. Worn surface of the PVA hydrogel tested against bovine articular cartilage with distilled water as lubricant fluid and contact pressure of 1 MPa (a) and 3.5 MPa (b).

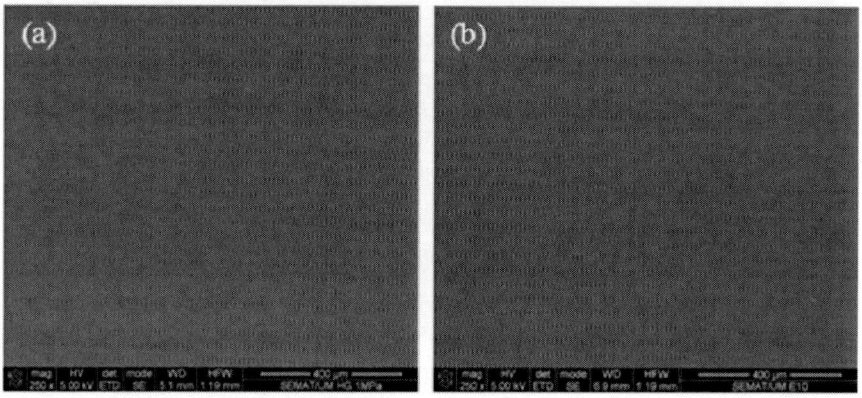


Figure 7. Worn surface of the PVA hydrogel tested against bovine articular cartilage with PBS as lubricant fluid and contact pressure of 1 MPa (a) and 3.5 MPa (b).

## CONCLUSION

In this work, the tribological behaviour of PVA hydrogel was investigated for the reciprocating sliding against 316L stainless steel and bovine articular cartilage under contact pressures from 1 MPa to 3.5 MPa in the presence of distilled water and PBS as lubricating media. In comparison to distilled water, PBS revealed to be a better lubricant since lower values of friction coefficient were achieved due to its higher viscosity that assists hydrodynamic lubrication conditions.

For both tribological pairs submitted to contact pressures of 1 MPa and 2 MPa, a decrease on friction coefficient with the increase of contact pressure was observed independently of the lubricant fluid in use. In the case of PVA hydrogel/316L stainless steel pair under the high contact pressure of 3.5 MPa in PBS the friction coefficient increased due to plastic deformation of the hydrogel surface caused by the opposing stainless steel rigid surface and the dominant wear mechanism of the hydrogel was abrasion. On the other hand, for PVA hydrogel/bovine articular cartilage pair in PBS, the increase of friction coefficient for the maximum contact pressure was justified by mechanical degradation evidenced at the articular cartilage surface, while a high level of surface preservation was observed for the opposing PVA hydrogel.

The pair PVA hydrogel/articular cartilage showed exceptional tribological properties with very low friction coefficient values ($\mu \approx 0.02$ to 0.06) combined with the preservation of the hydrogel contact surface.

Crosslinked PVA hydrogel can be considered as a potential biomaterial to be used as artificial articular cartilage for the repair of articular joints, being suggested future investigation. This should include the use of articular cartilage samples with subchondral bone in order to ensure the natural properties and resistance of the tissue and the evaluation of indentation creep modulus and equilibrium water content (*EWC*) after each tribological test to characterise possible changes on hydrogels mechanical behaviour.

## References


[1] Athanasiou, K.A., Darling, E.M., Hu, J.C. *Articular Cartilage Tissue Engineering*, Morgan and Claypool Publishers, 2010.

[2] Mansour, J.M. Biomechanics of Cartilage. (C.A. Oatis, Ed.) *Arthritis and Rheumatism.*Lippincott Williams and Wilkins, 2008, Vol. 58, 7, pp. 2065-2074.

[3] Seal, B.L., Otero, T.C., Panitch, A. Review: Polymeric biomaterials for tissue and organ regeneration. *Materials Science and Engineering R.* 2001, Vol. 34, pp. 147-230.

[4] Ratner, B.D., Hoffman, A.S. Synthetic hydrogels for biomedical applications. *Hydrogels for medical and related applications,* ACS *Symposium Series*, Washington, DC, 1976, Vol. 31, pp. 1-36.

[5] Jiang, S., Liu, S., Feng, W. PVA hydrogel properties for biomedical application. *Journal of the Mechanical Behaviourof Biomedical Materials,* 2011, Vol. 4, 1228-1233.

[6] Gang, W.U. Wenguang, Z., Chengtao, W. Tribological properties of polyvinyl alcohol hydrogel reinforced with nanometer hydroxyapatite. *Journal of Wuhan University of Technology-Mater. Sci.*2008, Vol. 23, 1.

[7] Rosiak, J.M., Ulanski, P., Pajewski, L.A., Yoshi, F., Makuuchi, K. Radiation formation of hydrogels for biomedical purposes. Some remarks and comments. *Radiation Physics and Chemistry.*1995, Vol. 46, 2, 161-168.

[8] Holloway, J.L., Spiller, K.L., Lowman, A.M., Palmese, G.R. Analysis of the in vitro swelling behavior of poly(vinyl alcohol) hydrogels in osmotic pressure solution for soft tissue replacement. *Acta Biomaterialia.* 2011, Vol. 7, 2477-2482.

[9] Pan, Y.S., Xiong, D.S., Ma, R.Y. A study on the friction properties of poly(vinyl alcohol) hydrogel as articular cartilage against titanium alloy. *Wear,* 2007, Vol. 262, 1021-1025.

[10] Pan, Y., Xiong, D. Friction properties of nano-hydroxyapatite reinforced poly(vinyl alcohol) gel composites as an articular cartilage. *Wear,* 2009, Vol. 266, 699-703.

[11] Covert, R.J., Ott, R.D., Ku, D.N. Friction characteristics of a potential articular cartilage biomaterial. *Wear,* 2003, Vol. 255, 1064-1068.

[12] Bavaresco, V.P.Tribological study of polymeric hydrogels for use as artificial articular cartilage. *PhD Thesis,* State University of Campinas, SP, Brazil, 2004.

[13] Zheng, Y., Huang, X., Wang, Y., Xi, T., Chen, X., Xu, H. The surface lubricative properties of PVA/PVP hydrogels treated with radiation used as artificial cartilage. *Applied Surface Science,* 2008, Vol. 255, 568-570.

[14] Bonakdar, S., Emami, S.H., Shokrgozar, M.A., Farhadi, A., Ahmadi, S.A., Amanzadeh, A. Preparation and characterization of polyvinyl alcohol hydrogels crosslinked by biodegradable polyurethane for tissue engineering of cartilage. *Materials Science and Engineering C.* 2010, Vol. 30, 636-643.

[15] Krumova.M., López, D., Benavente, R., Mijangos, C., Pereña, J.M. Effect of crosslinking on the mechanical and thermal properties of poly(vinyl alcohol). *Polymer,* 2000, Vol. 41, 9265-9272.

[16] Bavaresco, V.P., Zavaglia, C.A., Reis, M.C., Gomes, J.R. Study on the tribological properties of pHEMA hydrogels for use in artificial articular cartilage. *Wear,* 2008, Vol. 265, 269-277.

[17] Gong, J.P., Osada, Y. Surface friction of polymer gels. *Progress in Polymer Science.* 2002, Vol. 27, 3-38.

[18] Bolto, B., Tran, T., Hoang, M., Xie, Z. Crosslinked poly(vinyl alcohol) membranes. *Progress in Polymer Science.* 2009, Vol. 34, 969-981.

[19] Hennink, W.E., Nostrum, C.F. Novel crosslinking methods to design hydrogels. *Advanced Drug Delivery Reviews,* 2002, Vol. 54, 12-36.

[20] Ajji, Z. Preparation of poly(vinyl alcohol) hydrogels containing citric or succinic acid using gamma radiation. *Radiation Physics and Chemistry.* 2005, Vol. 74, 36-41.

[21] Wang, B., Mukataka, S., Kokufuta, E., Kodama, M. The influence of polymer concentration on the radiation-chemical yield of intermolecular crosslinking of poly(vinyl alcohol) by γ-rays in deoxygenated aqueous solution. *Radiation Physics and Chemistry.* 2000, Vol. 59, 91-95.

[22] Wang, B., Kodama, M., Mukataka, S., Kokufuta, E. On the intermolecular crosslinking of PVA chains in an aqueous solution by γ-ray irradiation. *Polymer Gels and Networks.*1998, Vol. 6, 71-81.

[23] Kempson, G.E., Freeman, M.A.R., Swanson, S.A.V. The determination of a creep modulus for articular cartilage from indentation tests on the human femoral head. *Journal of mechanics.* 1971, Vol. 4, 239-250.

[24] Gong, J.P. Friction and lubrication of hydrogels – its richness and complexity. *Soft Matter.* 2006, Vol. 2, 544-552.

[25] Katta, J., Jin, Z., Ingham, E., Fisher, J. Biotribology of articular cartilage – A review of the recent advances. *Medical Engineering and Physics.* 2008, Vol. 30, 1349-1363.

[26] Graindorge, S.L., Stachowiak, G.W. Changes occurring in the surface morphology of articular cartilage during wear. *Wear*, 2000, Vol. 241, 143-150.

[27] Mow, V.C., Ateshian, G.A. Lubrication and wear of diarthrodial joint. *Basic Orthopedic Biomechanics.* 2nd edition. Philadelphia : Lippincott-Raven Publishers, 1997, 275-315.

Hydrogels: Synthesis, Characterization and Applications
Editors: F. V. Câmara and L. J. Ferreira
ISBN: 978-1-61942-842-3


*Chapter 14*

# HYDROGELS WITH 3D GRADIENT OF MECHANICAL PROPERTIES

***G. Orsi*[1,2] , *V. Carta*[1] *and G. Vozzi*[1,2]**
[1] Interdepartmental Research Center "E. Piaggio", Faculty of Engineering, University of Pisa, Pisa, Italy
[2] Department of Chemical Engineering, Industrial Chemistry and Materials Science, Pisa, Italy

## ABSTRACT

This chapter presents the development of a hydrogel matrix with 3D gradient of mechanical properties. It can have several applications to the study of phenomena related to cell mechanotaxis (cells tend to migrate from soft regions to more rigid ones when they are seeded on a substrate with a gradient of stiffness), or differentiation on different stiff substrates. We performed mechanical characterization of alginate hydrogels at different concentrations, cross-linked with several concentrations of calcium chloride. The samples, produced by spraying the calcium chloride solution on a layer of alginate solution, were subjected to compression tests to obtain a "map" of elastic modulus as function of alginate and calcium chloride concentration. Then we realized alginate matrices characterized by a gradient of mechanical properties. To do that we used a device purposely developed, able to impose a three-dimensional gradient of calcium chloride concentration to alginate solution, allowing a differential crosslinking in the matrix. The concentration profile was predicted before with computational fluid dynamics software. Compression tests were made on these matrices, and the the formation of an elastic modulus gradient was observed. However hydrogel skin formation effects limit the whole efficiency of the device, that worked properly only in the middle of the matrix. The important aspect of this research is that we can construct hydrogel matrix with a gradient of mechanical properties, in which cells can be embedded in order to analyze cell behavior as function of substrate mechanical properties.

## INTRODUCTION

What are the mechanisms that lead human body tissues form? How these mechanisms do act? Cell response to substrate stiffness has been suggested playing an important role in cell processes, ranging from developmental morphogenesis [1] to pathogenesis of disease states, and may have deep implications in Tissue Engineering research. Great attention has been dedicated to engineer biocompatible extra-cellular matrices, named scaffolds, realized with controlled 3D architectures and using biomaterials mimicking the mechanical properties of natural tissues. Therefore, several reports [2-10] have shown that biomaterials can be used in the form of scaffolds as active components in cell cultures. The material should allow adhesion, cell growth, and formation of new tissue in a 3D architecture. The scaffolds can be made from most widely used biodegradable synthetic polymers including: poly-acrylic-acid (PAA) and poly-vinyl-alcohol) (PVA). Another class of materials used in this area is represented by hydrogels. The use of hydrogels in Tissue Engineering is popular due to their viscoelastic characteristics, biocompatibility, and their inner structure that allow gases and nutrients transfer. It has been shown that they are quite promising, providing a homogeneous, structure-less soft 3D environment, which is ideal to induce cell differentiation. [11]. Among the most commonly used hydrogels there are natural polymers such as agarose, alginate, chitosan, collagen, fibrin and hyaluronic acid. Hydrogels are particularly useful for engineering of soft tissues, and have been used to replicate cartilage, cornea, skin and vascular tissue. One of the principal features of hydrogels is the ability to replicate tissues mechanical and viscoelastic properties. [12], [13]. The mechanical strength of hydrogel can improve the mechanical stimulation of seeded cells, improving tissue remodeling within the construct [14], allowing the development of a tissue equivalent. Thus, it is important to know perfectly the mechanical properties of hydrogel in order to optimize cell stimulation. 3D environments can also provide more controlled spatial information to cells, providing a stable framework for cell adhesion and spatial spreading. In this view, for example, different studies were performed on alginate hydrogels subjected to several concentrations of calcium, in order to analyze its influence on response of primary bovine chondrocytes seeded in these alginate scaffolds [15]. These researches showed how the repair of cartilage tissue is affected by calcium level; in particular as calcium concentration increase the scaffold porosity decrease, reducing the average inflow of nutrients. This result induces an extracellular matrix (ECM) production and reduces cell proliferation. Another important aspect of mechano-structural environment is chemical nature of substrates, which controls and modulates cell adhesion and structural support in natural tissues [16]. In human body the range of measured elastic modulus varies from low to high values. For example in heart Tissue Engineering, the most studied mechano-structural feature of substrates or scaffolds is the elastic modulus (or stiffness), and its effect on cell differentiation. Engler at al. [17] have pioneered studies on matrix elasticity and its effect on mesenchymal stem cell (MSC), observing that cell differentiation was independent by inductive biochemical factors, but strictly dependent by substrate elastic modulus. In the past few years a wide number of biomaterials have been employed in an attempt to generate tissue-like scaffolds, recreating the complexity of biological environment. These biomaterials (synthetic or naturally derived) have been mechanically characterized to reproduce elasticity and deformability of ECM. Therefore relationships between cell response and scaffold can be modulated to guide and control cell fate. Many cellular observations in modern biology have been carried out on glass slides often functionalized with a thin layer of ECM. Unfortunately, this layer does not reproduces typical mechanical environment perceived by cells [18], and can lead to undesired cell differentiation, or a loss of cell function. The stiffness of substrate on which cells can adhere varies greatly depending by tissue. In our bodies we have deformable tissues like the brain (E = 0.1 kPa) and hard tissues such as bone (>30 kPa). Small changes of the elastic modulus, due to time and spatial variations in the matrix during development [19], can stimulate cells during their differentiation and maturation process. From previous works it has been observed that fibroblasts, epithelial cells and smooth muscle cells decreases the cell adhesion as decrease substrate stiffness, and cells tend to migrate from soft regions to more rigid area when they are seeded on a substrate with a gradient of stiffness, this phenomenon is called durotaxis (or mechanotaxis) [20], [21]. The use of hydrogels, combined with the concept of gradient, allowed us to develop a material with a 3D gradient of stiffness. This chapter deals the realization of a matrix with 3D gradient of mechanical properties, which will allow studying cell behavior under a gradient of mechanical and biochemical stimuli. Alginate was chosen as a highly biocompatible material, cross-linked with an aqueous solution of calcium chloride at various concentrations, and its mechanical properties were analyzed with compression test. Different concentrations of alginate and calcium chloride were analyzed. In this way it was possible to obtain a "map" that relates hydrogel elastic modulus with alginate and calcium chloride concentrations. We also realized alginate 3D matrices, characterized by a gradient of mechanical properties. It was necessary use a device able to impose a three-dimensional gradient of calcium chloride concentration to alginate solution, allowing a differential crosslinking. Compression tests were made on obtained samples, and it was observed the formation of an elastic modulus gradient. The correspondence between theoretical and experimental results allowed us to evaluate the stiffness corresponding to a specific concentration of alginate and calcium chloride. In this way it will be possible to realize a matrix with mechanical properties gradient just changing working parameters of the 3D concentration gradient maker. Knowing the stiffness of substrate at any point in 3D matrix, it will be possible to analyze cell behavior of cells embedded in 3D matrix.

## MATERIALS AND METHODS

In order to obtain a hydrogel with adequate stiffness, it was decided to use alginate crosslinked with calcium chloride. For the preparation of hydrogel matrix were followed two different protocols: spray calcium chloride solution on a layer of alginate solution, in order to obtain non mechanical-graded samples; use a device able to create a 3D calcium chloride gradient on pre cross-linked alginate, in order to obtain samples with a gradient of mechanical properties.

### Non Mechanical-Graded Samples

The concentrations of alginate chosen were 1%, 2%, 3%, 4%, and 5% (w/v) in water while the aqueous solution of calcium was prepared at 0.1 M, 0.25 M, 0.5 M, 0.75 M and 1

M. Because reaction-diffusion process between alginate and calcium is almost instantaneous, in the order of few seconds, there is large volume shrinkage and formation of spheroidal domains, if the two solution and mixed directly. The samples obtained in this way are not suitable for mechanical tests. For this reason it was used an airbrush to spray calcium chloride solution on the upper layer of alginate. Matrices with smooth surfaces were obtained, which it was possible to obtain cylindrical samples (Figure 1) with a diameter of 9.3 mm and thickness of 3 mm. Cross-linking has been set at one hour, after which the samples were washed in deionizer water to remove calcium excess, and stored in a fridge to allow the maintenance of physical and chemical characteristics.

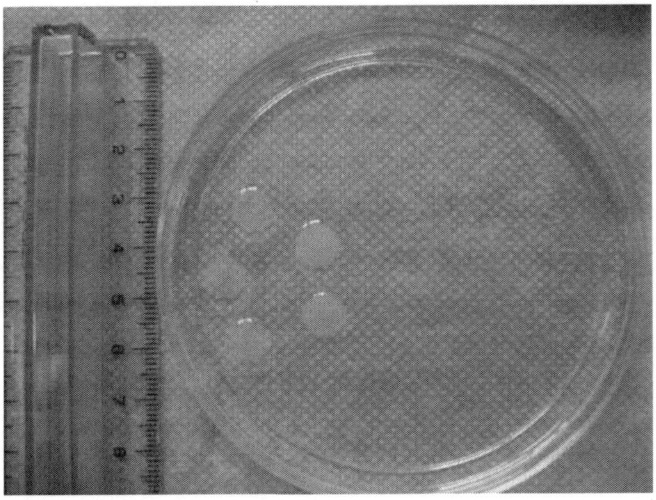

Figure 1. Cylindrical samples of alginate, obtained spraying calcium chloride solution on a upper surface of a volume of alginate solution.

## Mechanical-Graded Samples

Samples with graded mechanical properties were prepared by using a 3D gradient maker. This is a device which working principle is based on an extension of common microfluidic devices able to apply a concentration gradient in a 2D environment [22-26]. 3D concentration gradient maker device is composed of a series of microchannels connected to a chamber in which the concentration gradient is formed. Laminar streams of liquid with different concentrations continuously flow side-by-side and as the stream flow down the device, with diffusive mixing occurring perpendicular to the direction of flow. By this way a wide variety of concentration profiles can be created in a single device. However previous gradient makers were able to realize a desired concentration gradient only in a 2D chamber. This device is able to realize a concentration gradient in a three-dimensional environment (Figure 2), thanks to two series of serpentine channels perpendicular one to each other. The series of serpentines are mirrored to obtain a symmetric gradient in the main chamber. The device working principles are better detailed in another paper [27].

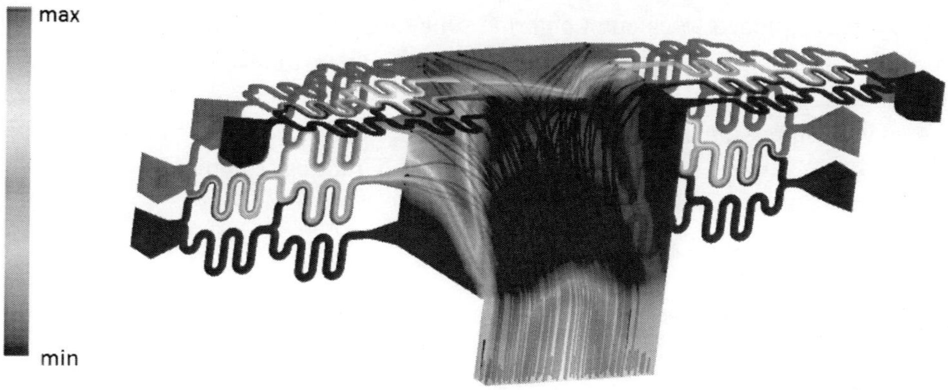


Figure 2. Simulation of gradient maker: concentration profile in serpentines and chamber. Scale color goes from maximum concentration to minimum concentration. In transparency are shown the streamlines.

Figure 3. Concentration profile imposed by the bioreactor. Red arrows represent the inlets with maximum concentration of calcium chloride solution, while blue arrows represent inlets with minimum concentration. White arrows indicate where the cross-linked 3D hydrogel matrix was sliced before mechanical tests, and the number above identifies the slice.

The bioreactor-pump system was installed to ensure a maximum concentration of calcium at the inner areas of 3D hydrogel matrix (red arrows, Figure 3) and a minimum concentration in the outer (blue arrows, Figure 3). Before perfusion in 3D gradient maker, alginate gel was prepared by mixing in volume ratio 2:1 alginate solution and calcium chloride at 0.075 or 0.1M in a cubic box. This pre-crosslinking was necessary to give right consistency to hydrogel, not allowing it to flow out from 3D gradient maker chamber. However 3D matrix was still permeable to calcium chloride solution. The pre-crosslinking has been set for 15 minutes, after which precrosslinked 3D hydrogel was removed from box, and placed in the gradient maker chamber. To visualize the diffusion of crosslink solution two different dyes were added to calcium chloride solutions: red for calcium chloride 0.1 M and blue for calcium chloride 1 M. The experiment in gradient maker was performed for two different times (1h and 2h), in order to evaluate the optimal time of crosslinking. Samples were tested in compression and to do this were cut longitudinally to obtain four slices for each sample (Figure 4). From each slice we retailed five cylindrical samples with diameter of 5 mm, and thickness of 6 mm were cut (Figure 4).

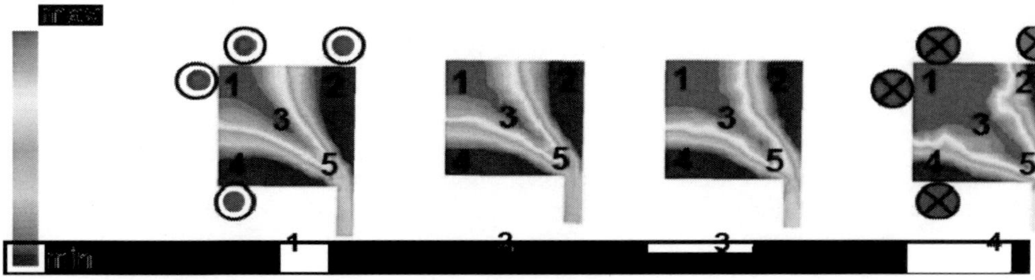

Figure 4. Normalized calcium chloride concentration profile through the four slices. Blue represents the maximum concentration of calcium chloride, while red represents minimum. Numbers indicates the zones where compression tests were performed. Crossed and concentric circles represent the inlets of our system. Concentric circles represent a flow "exiting" from the sheet, while crossed circles a flow "entering" in the sheet.

## Mechanical Tests

Hydrogels stiffness was quantified with uniaxial compression tests, performed with a double-column dynamometer (Z005, Zwick Roell, Germany). Compression tests were performed at 1 mm/min, with sample maximum deformation of 20%. Young modulus was estimated with a linear regression of first linear trait of stress-strain curve.

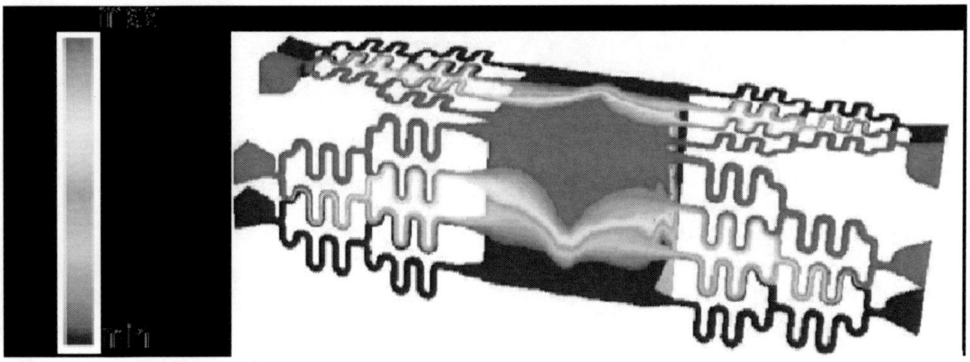

Figure 5. Front view of the concentration profile within sample. The color scale represents the normalized concentration of calcium chloride.

# RESULTS

## Characterization of Non Mechanical-Graded Samples

We analyzed the trends of elastic modulus as a function of calcium concentration (Figure 6).

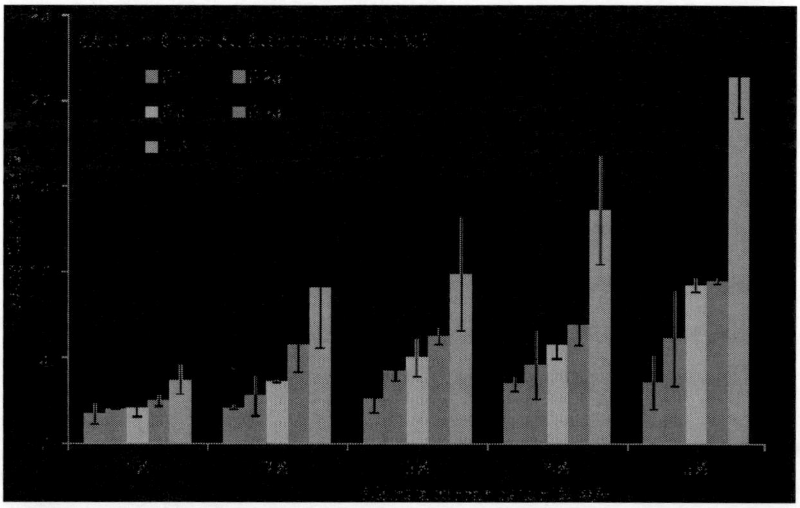

Figure 6. Young's modulus for 1%, 2%,3%, 4%, 5% (w/v) in deionizer water as a function of calcium concentration.

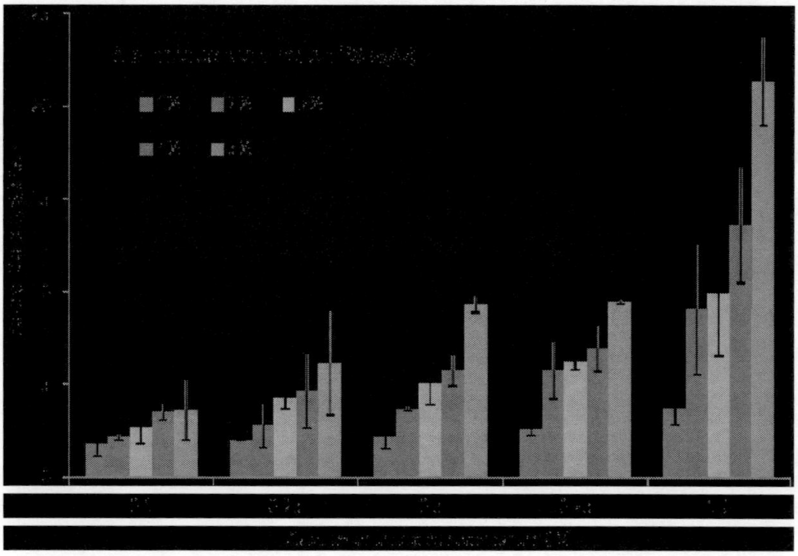

Figure 7. Young modulus for 0.1, 0.25,0.5,0.75, 1 M calcium chloride solution, as function of alginate concentration.

These data were studied considering the modeling of viscoelastic bodies [28-30]. There was a general trend of increasing stiffness as calcium chloride increase (Figure 6). The data that do not follow this trend can be explained by the formation of a dense skin during the early phase of crosslinking. This limits the diffusion of $Ca^{2+}$ ions inside alginate matrix, so the crosslinking can be not uniform in the matrix.

Even increasing the concentration of alginate, the elastic modulus of the gel tends to increase (Figure 7). From the graphs it should be noted, however, that at low values of calcium, such as 0.25 M and for high concentrations of alginate, there is a strong decrease in material stiffness. This is probably due to a low concentration of calcium in the center of the sample due to the effect of skin on calcium chloride diffusion, dependent from alginate/calcium chloride ratio. From these results it was constructed a map of the elastic modulus by varying the concentration of alginate and calcium chloride (Figure 8).

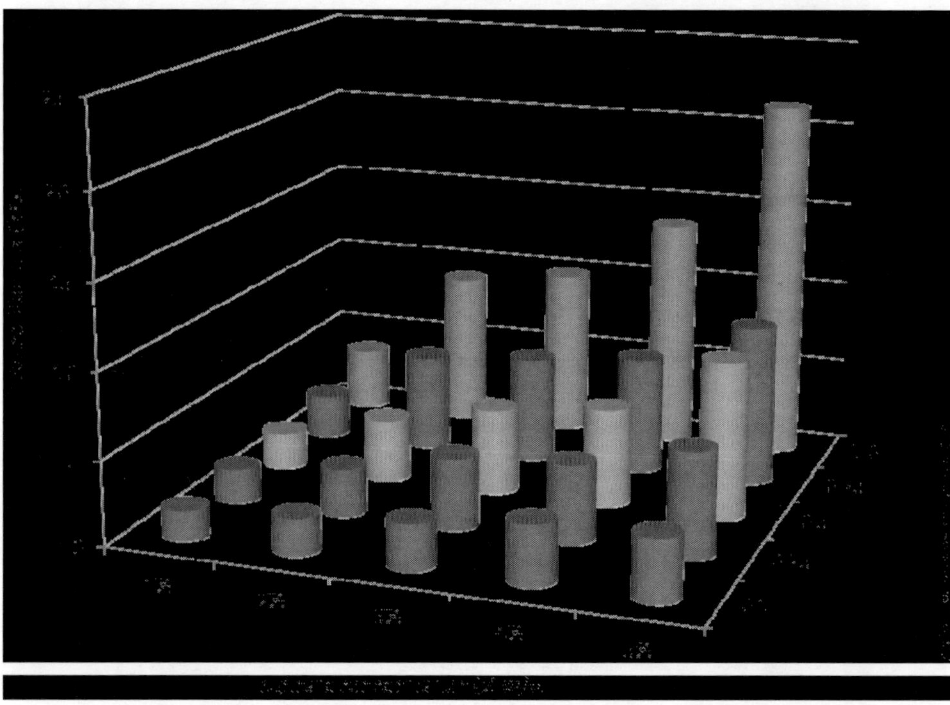

Figure 8. Map of stiffness derived from airbrush treated samples. While moving in positive sense on alginate and calcium chloride axis we can find an increase on stiffness, as expected.

## Characterization of 3D Mechanical-Graded Hydrogel Matrix

From mechanical analysis was possible to obtain the trend of elastic modulus in the five selected areas, for samples of alginate 1% and 5% (w/v) pre cross-linked with calcium 0.1 M and diffusion time of one hour. It was observed that the external slices show a higher elastic modulus than the inner ones. The same phenomena was observed using two different concentration of calcium chloride solution (0.075 and 0.1M) to precrosslink the hydrogel matrix, and either keeping the perfusion for two different times (1h and 2h). We used a mathematical inverse function on the "map" of elastic modulus obtained in the previous paragraph to calculate the effective concentrations of calcium chloride that crosslinked the matrix in sampling sites. Therefore, we compared experimental values derived from 3D mechanical-graded matrix and the values obtained from stiffness map in order to assess data convergence. Figure 9 is related to Alginate 1% precrosslinked with calcium chloride 0.1 M, and kept into perfusion for 1h. It is possible to notice that in outer zones of the slices calcium chloride concentration is lower than that expected from the model.

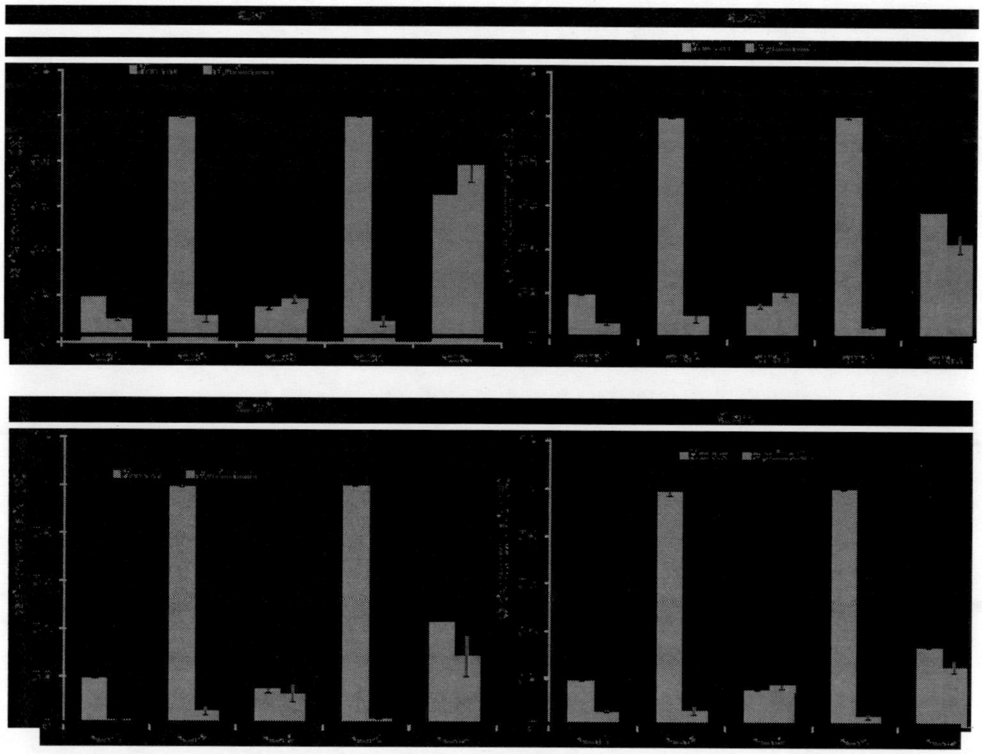

Figure 9. Comparison of experimental behavior versus predicted behavior. Alginate 1% (w/v), pre cross-linked with calcium chloride 0.1 M. Perfusion time was set to 1 h. Calcium chloride concentration is lower than that expected from the model in outer zones (1, 2, 4). This effect is due to formation of a dense skin in the outer part of the sample that not allowed to calcium to diffuse properly.

This result can be explained by the formation of the skin during cross-linking, as observed from the results obtained in previous paragraphs. This skin acts as a barrier to diffusion, twisting the local shape of 3D concentration gradient. In the central area of the slices, instead, it is observed that the experimental data almost agree with theoretical data. The same analysis was done varying the alginate concentration (1%-5%), precrosslinker concentration (0.075 M, 0.1 M), and perfusion time (1h – 2h). The trend was confirmed even for alginate 5%, and it showed to be independent of precrosslinker concentration and perfusion time. Comparing CFD model and experimental data it was possible to note that they are similar in the center of 3D hydrogel matrix (Figure 12). The area was found by confronting the concentration data coming from experimental data and simulated data. We interpolated experimental data in a 3D grid, obtaining a continue function in the 3D space, and then we calculated the relative percentage error on each point as

$e_r = 100(1 - d_m/d_e)$, where $d_m$ is the data coming from the model prediction, and $d_e$ is the experimental data. We set the threshold for properly working zone as $e_r < 10\%$.

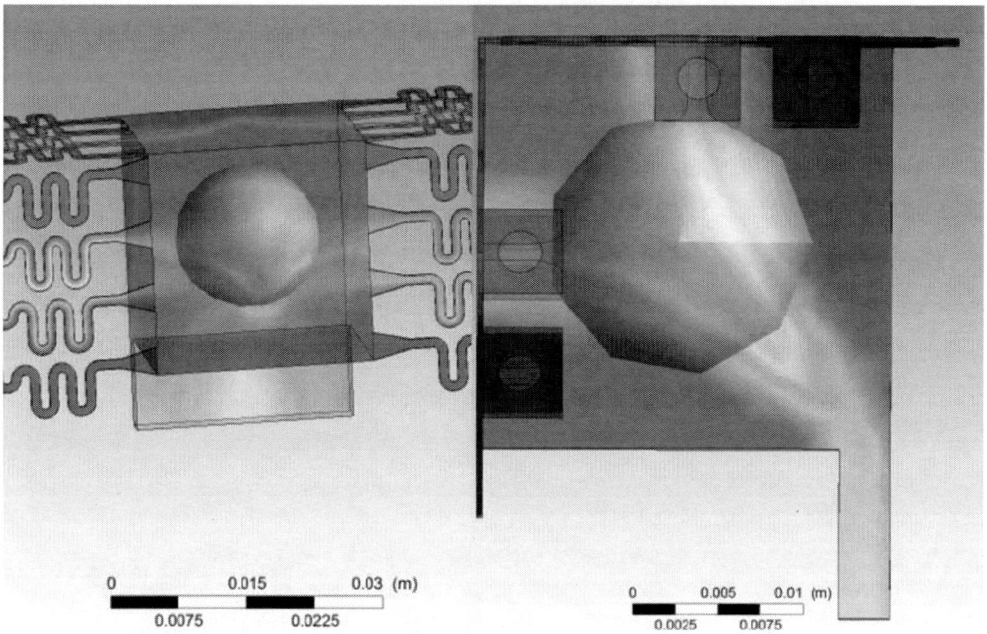


Figure 10. Front view and side view of the effective working area of the device. The area was found by interpolating the concentration data coming from experimental data in a 3D grid, and then confronting these data with those simulated.

## CONCLUSION

We presented a first attempt to create a 3D hydrogel matrix with a desired 3D gradient of mechanical properties. The 3D gradient maker generator was capable to produce a desired gradient of mechanical properties, even if in a selected area of itself. This result could be applied to the study of behavior of cells seeded or embedded in substrates with graded mechanical properties, to evaluate cell migration and differentiation in a 3D stiffness varying material.

## REFERENCES

[1] W. Driever and C. Nüsslein-Volhard, "A gradient of bicoid protein in Drosophila embryos," *Cell*, vol. 54, no. 1, pp. 95-104, Jul. 1988.

[2] C. K. Kuo and P. X. Ma, "Ionically crosslinked alginate hydrogels as scaffolds for tissue engineering: part 1. Structure, gelation rate and mechanical properties.," *Biomaterials*, vol. 22, no. 6, pp. 511-21, Mar. 2001.

[3] N. O. Dhoot, C. A. Tobias, I. Fischer, and M. A. Wheatley, "Peptide-modified alginate surfaces as a growth permissive substrate for neurite outgrowth.," *Journal of biomedical materials research. Part A*, vol. 71, no. 2, pp. 191-200, Nov. 2004.

[4] J. L. Drury, R. G. Dennis, and D. J. Mooney, "The tensile properties of alginate hydrogels.," *Biomaterials*, vol. 25, no. 16, pp. 3187-99, Jul. 2004.

[5] N. E. Fedorovich, J. Alblas, J. R. de Wijn, W. E. Hennink, A. J. Verbout, and W. J. A. Dhert, "Hydrogels as extracellular matrices for skeletal tissue engineering: state-of-the-art and novel application in organ printing.," *Tissue engineering*, vol. 13, no. 8, pp. 1905-25, Aug. 2007.

[6] L. Klouda and A. G. Mikos, "Thermoresponsive hydrogels in biomedical applications.," *European journal of pharmaceutics and biopharmaceutics : official journal of Arbeitsgemeinschaft für Pharmazeutische Verfahrenstechnik e.V*, vol. 68, no. 1, pp. 34-45, Jan. 2008.

[7] P. X. Ma, "Scaffolds for tissue fabrication," *Materials Today*, vol. 7, no. 5, pp. 30-40, May 2004.

[8] L. N. Novikova, A. Mosahebi, M. Wiberg, G. Terenghi, J.-O. Kellerth, and L. N. Novikov, "Alginate hydrogel and matrigel as potential cell carriers for neurotransplantation.," *Journal of biomedical materials research. Part A*, vol. 77, no. 2, pp. 242-52, May 2006.

[9] P. Tayalia, E. Mazur, and D. J. Mooney, "Biomaterials Controlled architectural and chemotactic studies of 3D cell migration," *Biomaterials*, vol. 80, pp. 1-8, 2011.

[10] S. R. Van Tomme, G. Storm, and W. E. Hennink, "In situ gelling hydrogels for pharmaceutical and biomedical applications.," *International journal of pharmaceutics*, vol. 355, no. 1-2, pp. 1-18, May 2008.

[11] M. Ahearne, Y. Yang, A. J. El Haj, K. Y. Then, and K.-K. Liu, "Characterizing the viscoelastic properties of thin hydrogel-based constructs for tissue engineering applications.," *Journal of the Royal Society, Interface / the Royal Society*, vol. 2, no. 5, pp. 455-63, Dec. 2005.

[12] E. J. Orwin, M. L. Borene, and A. Hubel, "Biomechanical and optical characteristics of a corneal stromal equivalent.," *Journal of biomechanical engineering*, vol. 125, no. 4, pp. 439-44, Aug. 2003.

[13] H. A. Awad, M. Q. Wickham, H. A. Leddy, J. M. Gimble, and F. Guilak, "Chondrogenic differentiation of adipose-derived adult stem cells in agarose, alginate, and gelatin scaffolds.," *Biomaterials*, vol. 25, no. 16, pp. 3211-22, Jul. 2004.

[14] T. Wakatsuki and E. L. Elson, "Reciprocal interactions between cells and extracellular matrix during remodeling of tissue constructs," *Biophysical Chemistry*, vol. 100, no. 1-3, pp. 593-605, Jan. 2003.

[15] L. Q. Wan, J. Jiang, D. E. Arnold, X. E. Guo, H. H. Lu, and V. C. Mow, "Calcium Concentration Effects on the Mechanical and Biochemical Properties of Chondrocyte-Alginate Constructs.," *Cellular and molecular bioengineering*, vol. 1, no. 1, pp. 93-102, Mar. 2008.

[16] M. Ahearne, Y. Yang, and K.-k Liu, "Mechanical Characterisation of Hydrogels for Tissue Engineering Applications," *Tissue Engineering*.

[17] A. J. Engler, S. Sen, H. L. Sweeney, and D. E. Discher, "Matrix elasticity directs stem cell lineage specification.," *Cell*, vol. 126, no. 4, pp. 677-89, Aug. 2006.

[18] A. Engler, L. Richert, J. Wong, C. Picart, and D. Discher, "Surface probe measurements of the elasticity of sectioned tissue, thin gels and polyelectrolyte multilayer films: Correlations between substrate stiffness and cell adhesion," *Surface Science*, vol. 570, no. 1-2, pp. 142-154, 2004.

[19] T. Rozario, B. Dzamba, G. F. Weber, L. A. Davidson, and D. W. DeSimone, "The physical state of fibronectin matrix differentially regulates morphogenetic movements in vivo.," *Developmental biology*, vol. 327, no. 2, pp. 386-98, Mar. 2009.

[20] C. M. Lo, H. B. Wang, M. Dembo, and Y. L. Wang, "Cell movement is guided by the rigidity of the substrate.," *Biophysical journal*, vol. 79, no. 1, pp. 144-52, Jul. 2000.

[21] N. Zaari, P. Rajagopalan, S. K. Kim, a J. Engler, and J. Y. Wong, "Photopolymerization in Microfluidic Gradient Generators: Microscale Control of Substrate Compliance to Manipulate Cell Response," *Advanced Materials*, vol. 16, no. 23-24, pp. 2133-2137, Dec. 2004.

[22] S. K. W. Dertinger, D. T. Chiu, N. L. Jeon, and G. M. Whitesides, "Generation of Gradients Having Complex Shapes Using Microfluidic Networks," *Society*, vol. 73, no. 6, pp. 1240-1246, 2001.

[23] N. L. Jeon, S. K. W. Dertinger, D. T. Chiu, I. S. Choi, A. D. Stroock, and G. M. Whitesides, "Generation of Solution and Surface Gradients Using Microfluidic Systems," *Processing*, no. 15, pp. 8311-8316, 2000.

[24] X. Jiang, Q. Xu, S. K. W. Dertinger, A. D. Stroock, T.-may Fu, and G. M. Whitesides, "A General Method for Patterning Gradients of Biomolecules on Surfaces Using Microfluidic Networks," *Society*, vol. 77, no. 8, pp. 2338-2347, 2005.

[25] K. Campbell and A. Groisman, "Generation of complex concentration profiles in microchannels in a logarithmically small number of steps," *Lab on a Chip*, pp. 264-272, 2007.

[26] D. Irimia, D. A. Geba, and M. Toner, "Universal Microfluidic Gradient Generator," *Society*, vol. 78, no. 10, pp. 3472-3477, 2006.

[27] G. Papini and G. Vozzi, "Design and realization of a 3D gradient concentration maker.", To be published

[28] "Polyelectrolyte intelligent gels_v6 - 2°+ fig I." .

[29] "Polyelectrolyte intelligent gels_v6 - 2°+ fig II." .

[30] "Polyelectrolyte intelligent gels_v6 - 2°+ fig -III." .

# INDEX

## A

## B

## C

## D

## E

## F

## G

## H



## I

## J



## K

## L



## M

## N



## O



## P

## Q

## R



## S

## T

## U



## V